T0342691

JACARANDA
SCIENCE QUEST 10
AUSTRALIAN CURRICULUM | THIRD EDITION

GRAEME LOFTS | MERRIN J. EVERGREEN

jacaranda
A Wiley Brand

Third edition published 2018 by
John Wiley & Sons Australia, Ltd
42 McDougall Street, Milton, Qld 4064

First edition published 2012
Second edition published 2015

Typeset in 11/14 pt Times LT Std

ISBN: 978-0-7303-4685-2

Front cover image: Yimmyphotography / Shutterstock

Cartography by Spatial Vision, Melbourne and MAPgraphics Pty Ltd, Brisbane

Illustrated by various artists, diacriTech and the Wiley Art Studio.

Typeset in India by diacriTech

All activities have been written with the safety of both teacher and student in mind. Some, however, involve physical activity or the use of equipment or tools. **All due care should be taken when performing such activities**. Neither the publisher nor the authors can accept responsibility for any injury that may be sustained when completing activities described in this textbook.

NATIONAL LIBRARY OF AUSTRALIA

A catalogue record for this book is available from the National Library of Australia

Printed in Singapore
M WEP196765 180423

CONTENTS

7 Global systems 256

8 Forces, energy and motion 303

9 Science quests 339

10 Psychology

11 Forensics

OVERVIEW

Jacaranda Science Quest 10 Australian Curriculum Third Edition has been completely revised to help teachers and students navigate the Australian Curriculum Science syllabus. The suite of resources in the *Science Quest* series is designed to enrich the learning experience and improve learning outcomes for all students.

Science Quest is designed to cater for students of all abilities: no student is left behind and none is held back. *Science Quest* is written with the specific purpose of helping students deeply understand science concepts. The content is organised around a number of features, in both print and online through Jacaranda's *learnON* platform, to allow for seamless sequencing through material to scaffold every student's learning.

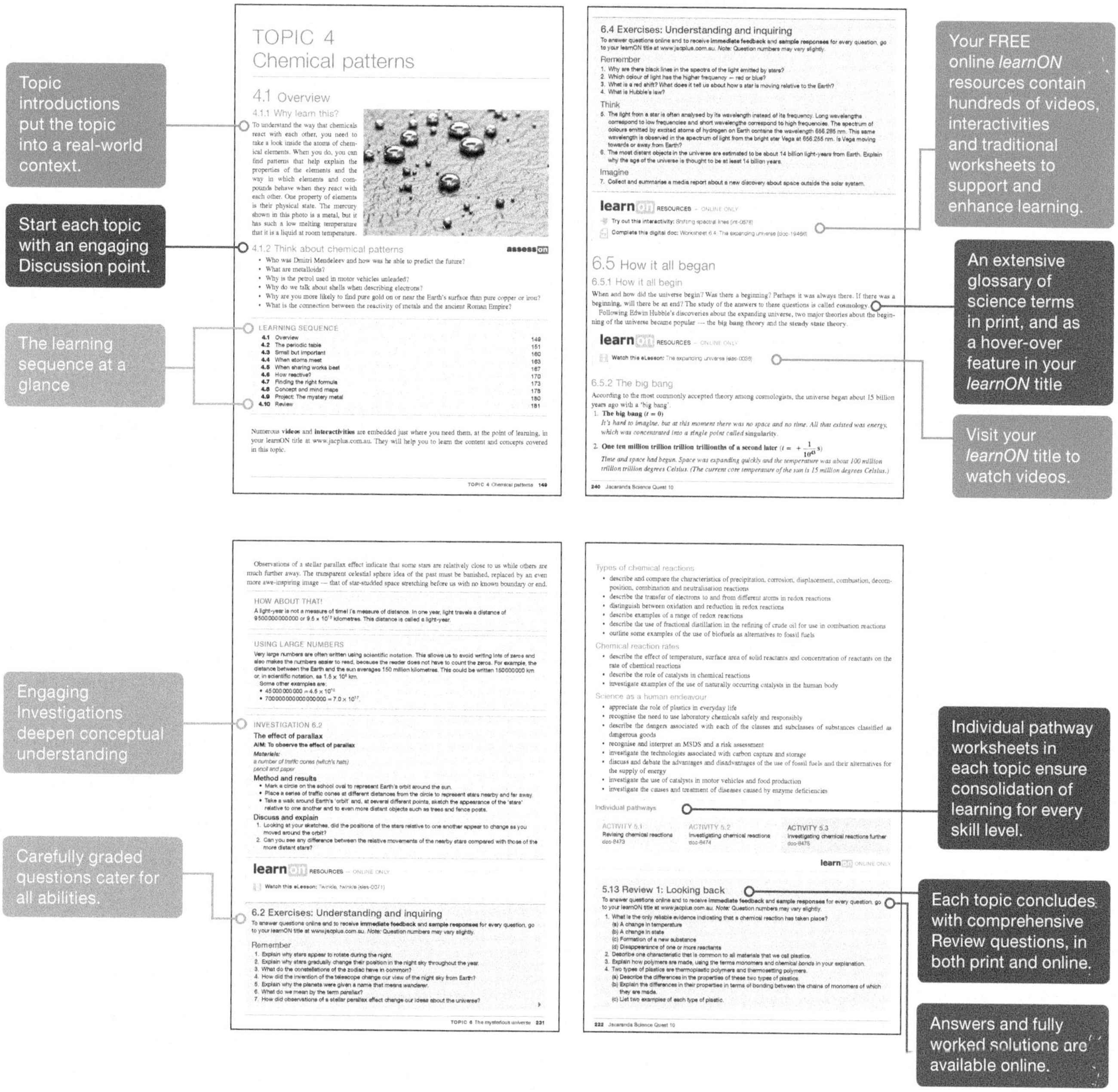

Topic introductions put the topic into a real-world context.
Start each topic with an engaging Discussion point.
The learning sequence at a glance
Your FREE online learnON resources contain hundreds of videos, interactivities and traditional worksheets to support and enhance learning.
An extensive glossary of science terms in print, and as a hover-over feature in your learnON title
Visit your learnON title to watch videos.
Engaging Investigations deepen conceptual understanding
Carefully graded questions cater for all abilities.
Individual pathway worksheets in each topic ensure consolidation of learning for every skill level.
Each topic concludes with comprehensive Review questions, in both print and online.
Answers and fully worked solutions are available online.

learnON

LearnON is Jacaranda's immersive and flexible digital learning platform that transforms trusted Jacaranda content to make learning more visible, personalised and social. Hundreds of engaging videos and interactivities are embedded just where you need them — at the point of learning. At Jacaranda, our 'learning made visible' framework ensures immediate feedback for students and teachers, with customisation and collaboration to drive engagement with learning.

Science Quest contains a free activation code for *learnON* (please see instructions on the inside front cover), so students and teachers can take advantage of the benefits of both print and digital, and see how *learnON* enhances their digital learning and teaching journey.

learnON includes:

- Students and teachers connected in a class group
- Hundreds of videos and interactivities to bring concepts to life
- Fully worked solutions to every question
- Immediate feedback for students
- Immediate insight into student progress and performance for teachers
- Dashboards to track progress
- Collaboration in real time through class discussions
- Comprehensive summaries for each topic
- Dynamic interactivities help students engage with and work through challenging concepts.
- Formative and summative assessments
- And much more …

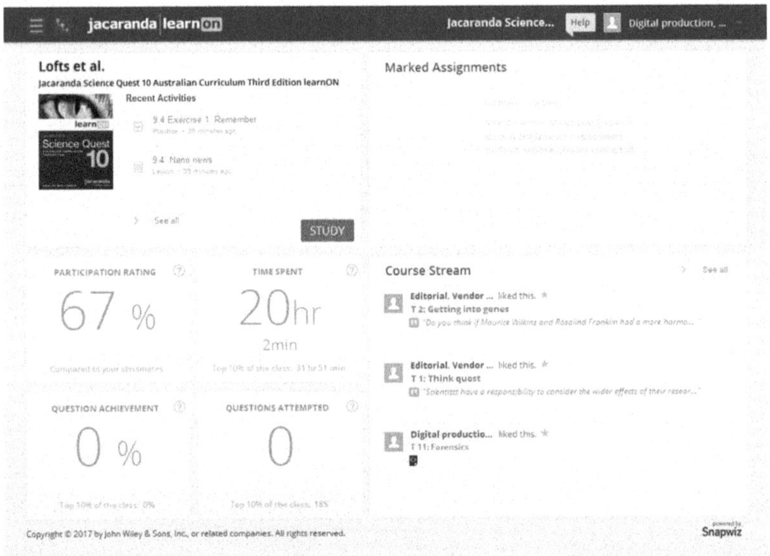

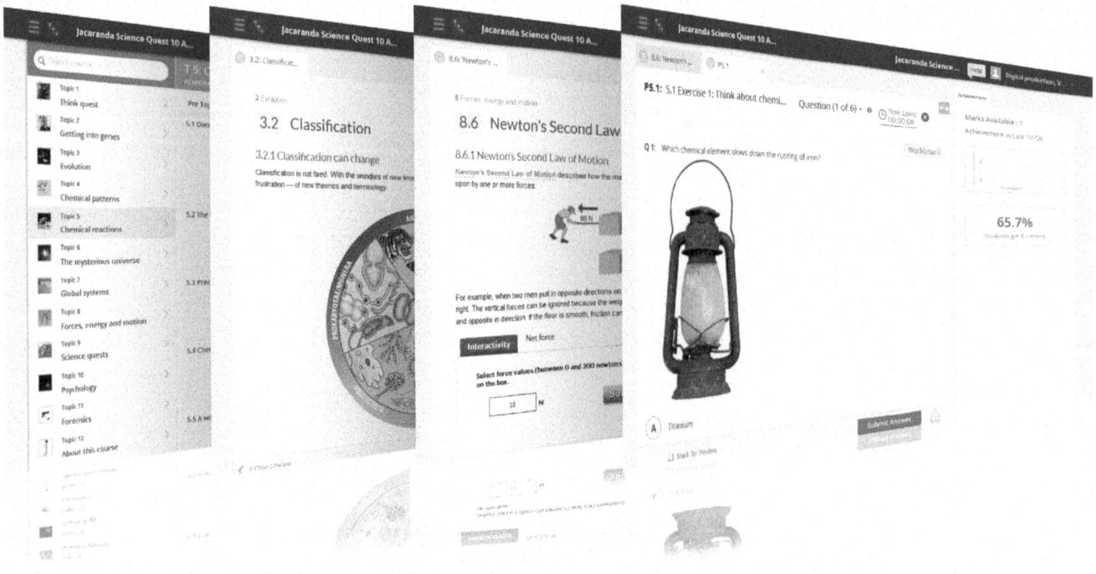

PREFACE

To the science student

Science is both a body of knowledge and a way of learning. It helps you to understand the world around you: why the sun rises and sets every day, why it rains, how you see and hear, why you need a skeleton and how to treat water to make it safe to drink. You can't escape the benefits of science. Whenever you turn on a light, eat food, watch television or flush the toilet, you are using the products of scientific knowledge and scientific inquiry.

Global warming, overpopulation, food and resource shortages, pollution and the consequences of the use of nuclear weapons are examples of issues that currently challenge our world. Possible solutions to some of these challenges may be found by applying our scientific knowledge to develop new technologies and creative ways of rethinking the problems. It's not just scientists who solve these problems; people with an understanding of science, like you, can influence the future. It can be as simple as using a recycling bin or saving energy or water in your home.

Scientific inquiry is a method of learning. It can involve, for example, investigating whether life is possible on other planets, discovering how to make plants grow faster, finding out how to swim faster and even finding a cure for cancer. You are living in a period in which knowledge is growing faster than ever before and technology is changing at an incredible rate.

Learning how to learn is becoming just as important as learning itself. *Science Quest* has been designed to help you learn how to learn, enable you to 'put on the shoes of a scientist' and take you on a quest for scientific knowledge and understanding.

To the science teacher

This edition of the *Science Quest* series has been developed in response to the Australian curriculum for Science. The Australian curriculum focuses on seven **General capabilities** (literacy, numeracy, ICT competence, critical and creative thinking, ethical behaviour, personal and social competence, and intercultural understanding). The history and culture of Aboriginal and Torres Strait Islanders, Australia's engagement with Asia, and sustainability have been embedded with the general capabilities where relevant and appropriate.

Science Quest interweaves **Science understanding** with **Science as a human endeavour** and **Science inquiry skills** under the umbrella of six **Overarching ideas** that 'represent key aspects of a scientific view of the world and bridge knowledge and understanding across the disciplines of science'.

The Australian Science curriculum provides the basis for the development of a Science curriculum in schools throughout Australia. However, it does not specify what you do in your classroom and how to engage individual classes and students.

We have attempted to make the *Science Quest* series a valuable asset for teachers, and interesting and relevant to the students who are using it. *Science Quest* comes complete with online support for students, including answers to questions, interactivities to help students investigate concepts, and video eLessons featuring real scientists and real-world science.

Exclusively for teachers, the online *Science Quest* teacher resources provides teaching advice and suggested additional resources, testmaker questions with assessment rubrics, and worksheets and answers.

Graeme Lofts and Merrin J. Evergreen

ACKNOWLEDGEMENTS

The authors and publisher would like to thank the following copyright holders, organisations and individuals for their assistance and for permission to reproduce copyright material in this book.

Images

• 123RF: **4**/Ozgur Coskun; **37**/Dan Kosmayer; **107** (bottom c)/Rahmat Nugroho; **109** (f)/Visarute Angkatavanich; **118**/ikarasi • AAP Newswire: **9**/EPA / TONI ALBIR • Al Rowland, Dr: **8** (top right) • Alamy Australia Pty Ltd: **98** (top)/The Print Collector; **112**, **120**/© Lourens Smak; **118**/© Sabena Jane Blackbird; **158**/John Wellings; **217** (bottom)/Juniors Bildarchiv GmbH; **234**/Universal Images Group North America LLC; **270**/jeremy sutton-hibbert • Alamy Stock Photo: **138** (middle)/Andrey Nekrasov; **343** (left), **350**/Moviestore collection Ltd; **345**/ScreenProd / Photononstop; **345** (bottom)/AF archive • ANU E Press: **249**, **249** (top)/Stuart Hay • Auscape International Pty Ltd: **105**/Copyright Biosphoto • Australian Curriculum: **1** • Australias Virtual Herbarium: **117** • Byeong Chun Lee: **355** (top left), **355** (top) • Centre for Australian National Biodiversity Research: **117** • Corbis Australia: **123**/Richard T. NowitzRichard T. Nowitz • Creative Commons: **13** (bottom a), **87**, **94**, **95** (top), **226**, **235**, **246**, **249** (bottom), **249** (middle), **357** • Dan Harley: **133** • Department of Industry, Innovation and Science: **140** (bottom left); **140** (bottom right)/Prime Ministers Prizes for Science, administered by the Australia Government Department of Industry, Innovation and Science • Doris Taylor: **353** (bottom)/Provided courtesy of Texas Heart Institute • Emmanuel Buschiazzo: **73** (a) • Fabiano Ximenes: **278** (middle) • Gabor Forgacs: **354** (top left)/University of Missouri; **354** (top left) • Getty Images: **319**/3DSculptor • Getty Images Australia: **6** (bottom)/ANDREW SYRED; **14** (middle b), **15** (top left), **16**, **49**, **121** (bottom right), **131**, **147**, **215** (bottom)/Science Photo Library; **98** (bottom)/Wolf Suschitzky / Time & Life Pictures; **129** (top)/De Agostini; **223**; **227** (left)/Science Photo Library/ NASA/STSCI; **243** (top)/Bettman; **257**/Science Photo Library / SPL Creative; **293** (bottom), **294**/Roger Ressmeyer/Corbis/VCG; **349** (bottom)/Stocktrek Images; **351**/Coneyl Jay; **353** (top)/ISM / Phototake Science; **354**(a bottom)/Dr Torsten Wittmann • Guang Shi: **139** • John Wiley & Sons Australia: **155**/Photo by Coo-ee Picture Library • Keith McLean: **354**(b bottom) • Mike Kuiper: **74**/Vlsci • Museum Victoria: **135** (bottom) • NASA: **184**, **185** (b), **227**, **248** (top), **284**, **286**, **358**; **244**/WMAP; **283**/http://ozonewatch.gsfc.nasa.gov/ • National Library of Medicine: **13** (bottom b), **14** (middle a), **15** (top right) • Newspix: **310**/Brett Costello • Out of Copyright: **135** (top) • Patrick Krug : **138** (bottom a) • Perth Zoo: **107** (middle d) • Photodisc: **37** (top right) • Public Domain: **218** (a), **248** • Ray Baughman: **350** • Sangbae Kim: **347** • Science Photo Library: **67**/Dr. Brad Mogen, VISUALS UNLIMITED • Shoji Takeuchi: **354** (top right) • Shutterstock: **5**/Wavebreak Media Ltd; **36** (a)/Elena Kharichkina; **36** (b)/Martin Higgs Photography; **36** (c), **290** (middle)/bikeriderlondon; **36** (d)/Ocskay Bence; **37** (top left)/Champion studio; **37** (middle left a)/Dan Kosmayer; **37** (middle left b)/Carl Stewart; **51** (bottom)/wonderisland; **51** (middle)/© Sebastian Kaulitzki; **59**/Sofiaworld; **73** (middle b)/Broadbelt; **73** (middle c)/John A. Anderson; **77**/Viorel Sima; **82**/aquatic creature; **88**/Erik Lam; **90**/sunsetman; **100**/Wilm Ihlenfeld; **101** (a)/Rich Carey; **101** (b)/Mares Lucian; **101** (c)/FiledIMAGE; **107** (middle a)/Patsy A. Jacks; **107** (middle b)/Parkol; **107** (middle c)/Mr. SUTTIPON YAKHAM; **107** (bottom a)/Kristian Bell; **107** (bottom b)/hallam creations; **107** (bottom d)/Filipe Frazao; **109** (a)/Kristina Vackova; **109** (top a)/Howard Sandler; **109** (top b)/Arto Hakola; **109** (b)/frantisekhojdysz; **109** (e)/Jason Mintzer; **109** (g)/Jody Ann; **111**/Naypong; **111**/© Timothy Craig Lubcke; **112**/Ambient Ideas; **112**/panda3800; **121** (middle left)/kerstiny; **121** (bottom left)/Baciu; **122** (top a)/Zadiraka Evgenii; **122** (top b)/mark higgins; **122** (middle)/Claude Huot; **125** (bottom a)/Joe Belanger; **125** (bottom b)/Vangert; **130**/Aaron Amat; **132**/Elenarts; **142** (bottom)/Julien Tromeur; **142** (top)/MilousSK; **149**/Ventin; **152**/magnetix; **157**/Olivier Le Queinec; **160**/Martin Novak; **166** (a)/Vasilyev; **166** (b)/molekuul.be; **166** (c)/Michael Coddington; **170**/Steffen Foerster; **180** (bottom)/jarous; **180** (top)/Iakov Filimonov; **185** (a)/NorGalNorGal; **185** (c)/efecreata mediagroup; **189**/Lindsey Moore; **197**/Paula Cobleigh; **199**/Albert RussAlbert Russ; **204**/Maxal Tamor; **207** (bottom)/Matthijs Wetterauw; **207** (top)/Ververidis Vasilis; **211**/Adam J; **212**/BGSmith; **214** (top a)/Shcherbakov Ilya; **214** (top b)/Evikka; **214** (top c)/design56; **214** (bottom)/Chay Talanon; **217** (middle)/bluesnote; **218** (b)/Richard A McMillin; **223**/Dirk Ercken; **228**, **228**/tobkatrina; **228**/dalmingo; **230**/peresanz; **233**, **252** (bottom)/Charles Lillo; **235** (bottom)/Reinhold Wittich; **247**/Manamana; **256**/Mikadun; **260**/David Hyde; **261**/David Sprott; **262**/Thomas Barrat; **264**/pixfly; **269**/Kzenon; **272**/Anette Holmberg; **277**/Andrew Zarivny; **278** (bottom)/Susan Flashman; **279**/Againstar; **296**/RAYphotographerRAYphotographer; **299**/EastVillage Images; **303**/Racheal Grazias; **304** (a)/Colette3; **304** (b)/homydesign; **304** (bottom)/BestPhotoStudio; **304** (c)/Iakov Kalinin; **304** (d)/elina; **304** (e)/Stefan Schurr; **304** (f)/gorillaimages; **305** (top)/IM_photo; **308** (bottom)/chartphoto; **309**/Zeljiko Radojko; **312**/Action Sports Photography; **312**/Francisco Turnes; **316**/© David Hilcher; **321** (bottom)/Jamie Roach; **321** (top)/FCG; **323** (b)/ESB Professional; **323** (c)/Dmitri Melnik; **325**/wellphoto; **326**/© sonya etchison, 2010.; **329**/Lorelyn Medina; **330**, **330**/conrado; **334**/© Racheal Grazias; **338**/thieury; **339**, **342**/Ociacia; **339**/r.classen; **343** (right)/Family Business; **344**/ikayaki; **346**/Fred

Kamphues; **346** (bottom)/Lightspring; **346** (bottom)/© stocksnapp; **348**, **349** (top)/Andrea Danti; **359**/SergeyDV; **360**/Jan Kaliciak; **362**/marekuliasz; **323**(a)/S.Buchnev; **355**(top left)/FikMik • Space Science and Engineering Center: **285**/Provided courtesy of University of Wisconsin-Madison Space Science and Engineering Center • Specialist Diagnostic Services Pty Ltd trading as Genomic Diagnostics: **58** • University Corporation for Atmospheric Research: **276** • University of Illinois: **75** (bottom)/Photo used with permission of the College of Liberal Arts and Sciences, University of Illinois • Walter & Eliza Hall Institute: **61** (bottom) • Wikipedia: **39**, **96** (middle left), **109** (h), **138** (bottom b), **293**/Creative Commons; **57**, **97** (bottom), **129** (bottom left), **129** (bottom right), **134**, **269**/Public Domain; **94** (top), **96** (middle right), **96** (top), **97** (top), **109** (c), **290** (top)

Text

• © Australian Curriculum, Assessment and Reporting Authority (**ACARA**) 2010 to present, unless otherwise indicated. This material was downloaded from the Australian Curriculum website (www.australiancurriculum.edu.au) (**Website**) (accessed October, 2017) and was not modified. The material is licensed under CC BY 4.0 (https://creativecommons.org/licenses/by/4.0). Version updates are tracked on the 'Curriculum version history' page (www.australiancurriculum.edu.au/Home/CurriculumHistory) of the Australian Curriculum website.

Every effort has been made to trace the ownership of copyright material. Information that will enable the publisher to rectify any error or omission in subsequent reprints will be welcome. In such cases, please contact the Permissions Section of John Wiley & Sons Australia, Ltd.

TOPIC 1
Think quest

1.1 Overview

1.1.1 Why learn this?

Are you ethical? Does it matter? What influences your opinions, values and beliefs? How do your attitudes affect when, how and why you learn? How and why do you think the way that you do? Is it ever worth changing your mind? Why doesn't everyone think the same way as you do? Who are you and who are you yet to become?

Einstein's theories were used to develop nuclear weapons — something he ethically opposed. Should scientists be responsible for how their inventions are used?

1.1.2 Think about these **assessON**

- What are the ABCs of attitude?
- Are you obeying the proximate rules of others?
- What are four key ways of knowing?
- What have Socrates, Karl Popper and Thomas Kuhn got to do with thinking about knowledge?
- What scientific event occurred the year that Einstein was born?
- Is all news about radioactivity bad?
- Can unethical behaviour ever be justified?
- Who owns genetic material?

LEARNING SEQUENCE

Numerous **videos** and **interactivities** are embedded just where you need them, at the point of learning, in your learnON title at www.jacplus.com.au. They will help you to learn the content and concepts covered in this topic.

1.1.3 What makes you, you?

Who are you? What do you need? Why do you react in the ways that you do? Possible answers to questions about the essence of who you are may be related to:

- the chemical instructions in the DNA that you inherited from your parents
- your experiences and the environment in which you live
- a combination of both of these.

Are you a product of only your genes or your environment, or do they both contribute to make you who you are? Scientists have been involved in this 'nature versus nurture' debate for many years. Which do you think is the key contributing factor to why you are you?

1.1.4 Bombarded by the media

We are in an age of information. In fact, you are continually being bombarded by it! How can you begin to make sense of all the information you receive? How can you better evaluate it? How can you incorporate this new information into what you already know to develop a better understanding of the world in which you live?

To effectively evaluate articles in the media you need to be able to determine what the facts are, and consider the type of journalism, the quality of writing and the article's ability to effectively present its message.

WHAT MAKES GOOD NEWS?

Read the article headlines and opening paragraphs below, and then answer the following questions.

1. For each article, consider the following.
   (a) What do you think the article is about?
   (b) What type of article do you think it is? Is it:
      (i) sensational
      (ii) informative
      (iii) entertaining
      (iv) thought provoking?
   (c) Use the internet to find further content from each article and find out more about the story by using search parameters such as the article headline and newspaper source.
   (d) Analyse the language and style of writing used in the article. What kind of audience do you think this article was written for?
   (e) Do you think you need to be a scientist to understand what the author is writing about?
   (f) Did the article headline grab your attention and make you want to read more? If not, how could it be improved?
   (g) Research one of the events or issues mentioned and write your own article about it. Collate the class articles into a journal or newspaper.
2. The first of these articles was written more than ten years ago.
   (a) What types of environmental and scientific problems did people face at the time?
   (b) Are they similar or different to those we face today?
   (c) Use the internet to find out more about the following issues mentioned in the articles:
      (i) carbon tax
      (ii) China syndrome
      (iii) nuclear power
      (iv) millennium bug.
   (d) How do you think people's opinions of the above issues have changed in the past ten years? Justify your answer.

That white-hot ball-bearing in the sky

Our supposedly middle-aged sun has been behaving like an adolescent of late, hurling huge clouds of particles at us after its face broke out in spots.

Sydney Morning Herald

'Bang' when a nuclear reactor fails

There is no such thing as fail-safe nuclear power, science commentator Karl Kruszelnicki said yesterday.

'Nuclear reactors are not fail-safe. They won't fail in a safe way. They can go bang as Chernobyl did,' Dr Kruszelnicki said.

Herald Sun

Nuclear crisis is no longer fiction

The nightmare scenario for Japan's crippled nuclear power plants is the so-called China syndrome.

The Hollywood movie *The China Syndrome* portrayed a near-meltdown of nuclear fuel rods in a US reactor.

Herald Sun

Millennium bug melee misses the true degree of our challenge

Tim Flannery's 1000-year carbon concession is a straw man that will no doubt burn brightly throughout the highly contested carbon tax debate.

The Australian

TOPIC 2
Getting into genes

2.1 Overview

2.1.1 Why learn this?

Can you roll your tongue? Did you know that your genes determine whether you can? Do you fit into your genes or do they fit into you? The characteristics of living things are determined by both the genetic information that they contain and the environment in which they live. New technologies have harnessed genetic machinery in order to change or create new organisms. What are the implications of manipulating the raw material of life?

2.1.2 Think about genes

assessON

- Do you have a Darwin's point?
- Why don't hairs grow on stomach linings?
- What have monks, peas, mathematics and genes got in common?
- What do Xs and Ys have to do with sex?
- Designer babies — should we or shouldn't we?
- Is the junk in your DNA actually a treasure?

LEARNING SEQUENCE

Numerous **videos** and **interactivities** are embedded just where you need them, at the point of learning, in your learnON title at www.jacplus.com.au. They will help you to learn the content and concepts covered in this topic.

Genes

Think, share and discuss

In your team, look at the pictures of the individuals in the families shown at right and share your observations.

1. Record any patterns that you notice.
2. If the following couples had another child, suggest what their eye colour may be and give a reason for your suggestion.
   (a) Ken and Margaret Davis
   (b) Kevin and Gwenda Swift
   (c) Geoff and Linda Davis
3. Suggest a reason why Geoff (brown eyes) and Linda (blue eyes) had brown-eyed and blue-eyed children, whereas Ken (brown eyes) and Margaret (blue eyes) had only children with brown eyes.
4. Martin Swift's fiancée, Justine, has blue eyes, but both of her parents have brown eyes. If Justine and Martin have children together, what colour eyes do you think may be possible? Discuss reasons for your response.
5. Bring to school a collection of photographs from as many members of your family as you can.
   (a) For at least one member of your team, carefully observe each photograph of the family members, looking for similarities.
   (b) Construct a table with the family features that you have observed and indicate which family members have these features.
   (c) Can you see any patterns or make any interesting suggestions on these observations? If so, discuss and record these.
   (d) Discuss which characteristics may be passed from parent to child and which may not.
   (e) Make a summary of your discussion to share with other teams. Add any other interesting points from these discussions to your team summary.

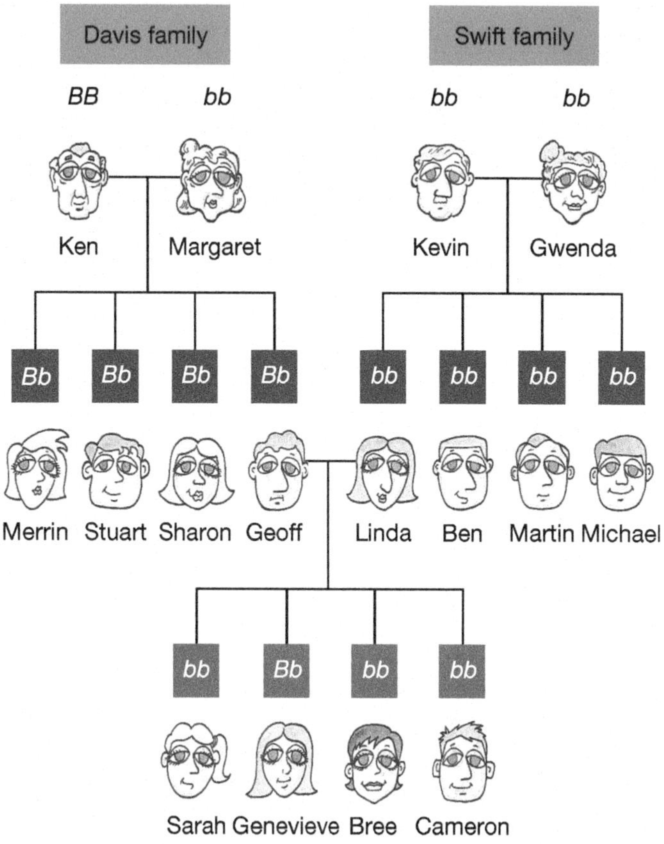

Think and discuss

6. Read and think about each of the following statements, then state whether you agree, disagree or don't know. Discuss your decisions with your team.
   (a) Because June and Frank have five sons, the chance of their next child being a daughter is increased.
   (b) People who have committed very violent crimes should be sterilised because their children will also be violent.
   (c) Parents should be allowed access to technologies that enable them to select the gender and specific characteristics of their children.
   (d) Technologies that alter the gametes (sperm and ova) should be illegal.

2.2 Nuclear matters

2.2.1 DNA: past, present and future

Where did you get those pointed ears, big nose and long toes from? Features or traits that are inherited are passed from one generation to the next in the form of a genetic code. This code is written in a molecule called **deoxyribonucleic acid (DNA)** and is located within the **nucleus** of your cells.

Most of your body cells contain all of the DNA instructions needed to make another you. Your DNA, however, is more than just a genetic blueprint of instructions; it is also an 'ID tag' and a very special ancient 'book' that holds secrets both from your ancestral past and for your possible futures.

An animal cell

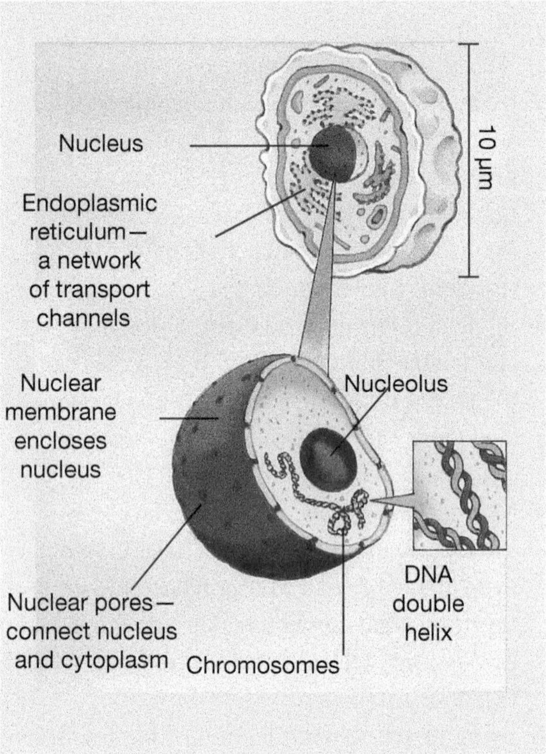

Suggest words to describe each link.

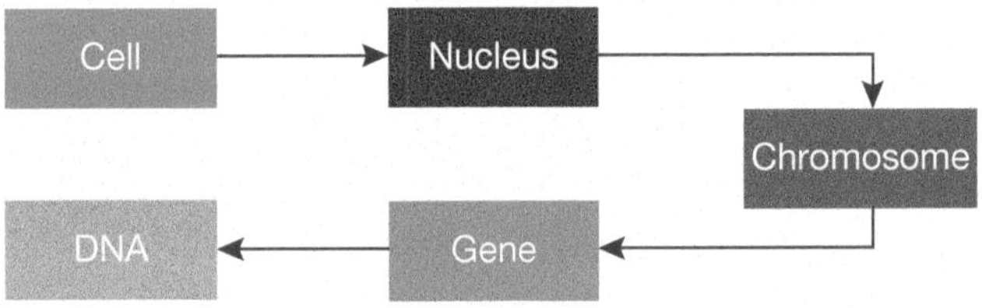

2.2.2 Genes

Each genetic instruction that codes for a particular trait (for example, shape of ear lobe, blood group or eye colour) is called a **gene**. Genes are made up of DNA and are organised into larger structures called **chromosomes**, which are located within the **nucleus** of the cell.

Scanning electron micrograph showing double-stranded chromosomes

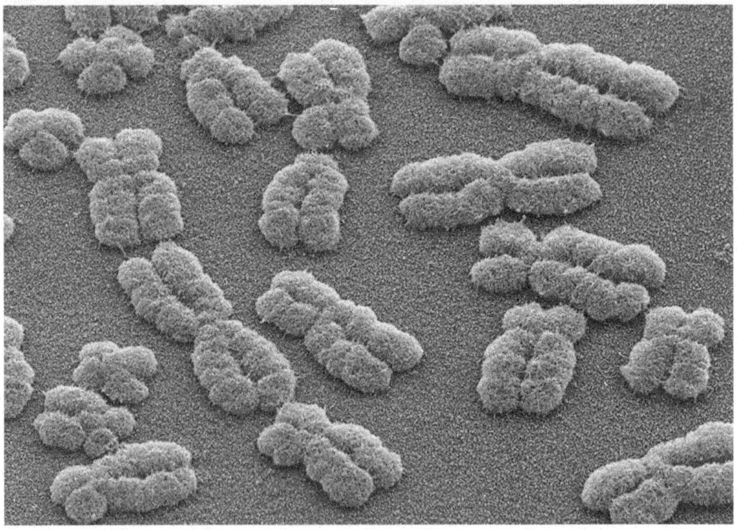

2.2.3 Chromosomes

Your body is constantly making new cells for replacement, growth and repair. It achieves this by a process called **mitosis**, which is a type of **cell division**.

Prior to cell division your DNA replicates itself, and this long molecule (2–3 metres) bunches itself up into 46 little packages called chromosomes. They are called chromosomes (*chromo* = 'coloured' + *some* = 'body') because scientists often stain them with various dyes so that they are easier to see.

Chromosomes are only visible when a cell is about to divide or is in the process of dividing. When your cells are not dividing, chromosomes are not visible as the coils are unwound and the DNA is spread throughout the nucleus.

Suggest how these terms are linked.

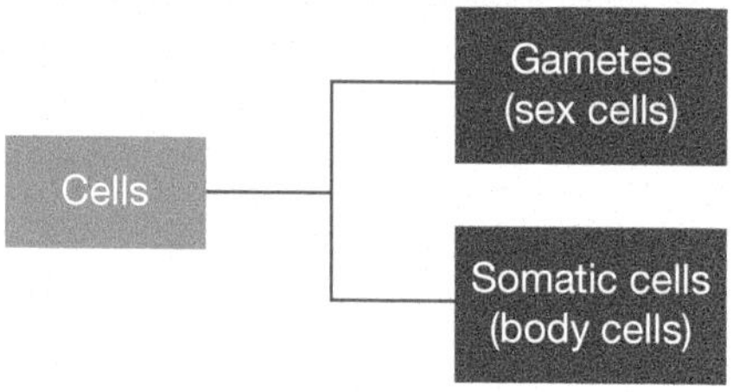

2.2.4 Sex cells

Another type of cell division called **meiosis** is used in the production of sex cells or **gametes**: **ova** (ovum is the singular form) and **sperm**. Meiosis results in the chromosome number being halved, so instead of pairs of chromosomes in each resulting cell, there is only one chromosome from each pair.

The genetic information that you received from your mother was packaged into 23 chromosomes in the nucleus of her egg cell (ovum), and the genetic information that you received from your father was packaged into 23 chromosomes in the nucleus of the sperm that fertilised your mother's egg cell. When these gametes fused together at **fertilisation**, the resulting **zygote** contained 23 pairs of chromosomes (one pair from each parent) — a total of 46 chromosomes.

Which words could be used to describe each link?

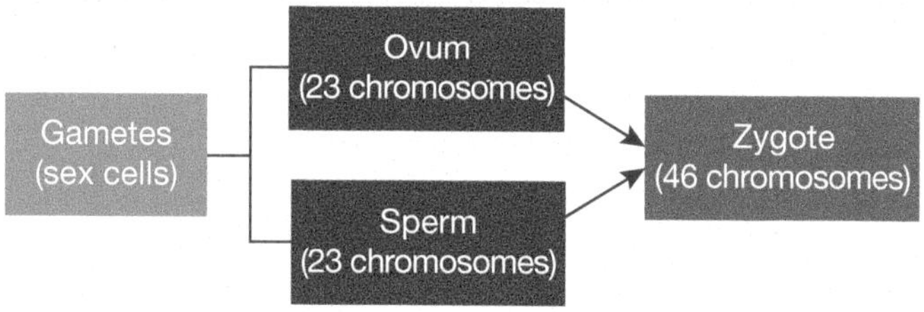

Suggest words to describe each link.

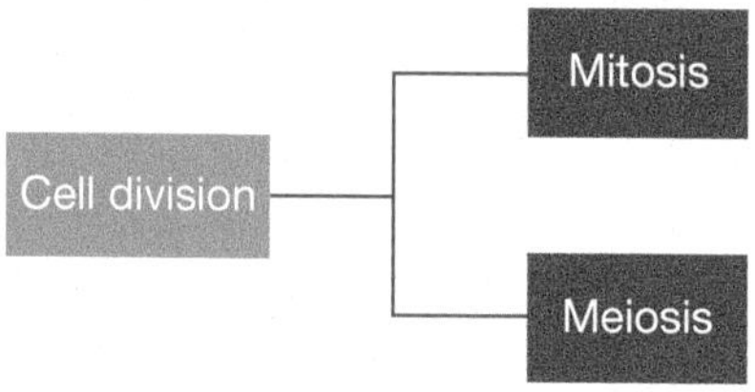

2.2.5 Somatic cells

Cells of your body that are not your sex cells are often referred to as body cells or **somatic cells**. With the exception of your red blood cells (that lose their nucleus when mature so they can carry more oxygen), all of your somatic cells contain chromosomes in pairs within their nucleus. This double set of genetic instructions (one set from each parent) makes up your **genotype**. The visible expression of the genotype as a particular trait or feature is called the **phenotype**. The phenotype may also be influenced by your **environment**.

Suggest how these terms are linked.

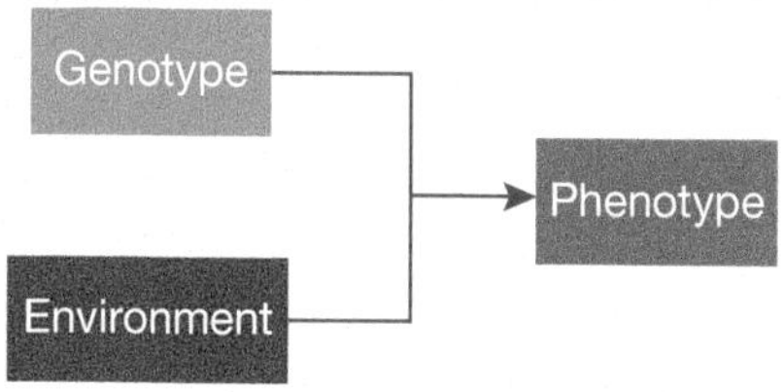

2.2.6 Chromosomes — more than one type

Chromosomes can be divided into two main types: autosomes and sex chromosomes.

2.2.7 Autosomes

Of the 46 chromosomes in your somatic cells, 44 are present in both males and females and can be matched into 22 pairs on the basis of their relative size, position of centromere (refer to the figure on the following page) and stained banding patterns. These are called **autosomes**. They are numbered from 1 to 22 on the basis of their size: chromosome 1 being the largest of the autosomes and chromosome 22 the smallest.

Which words could be used to describe each link?

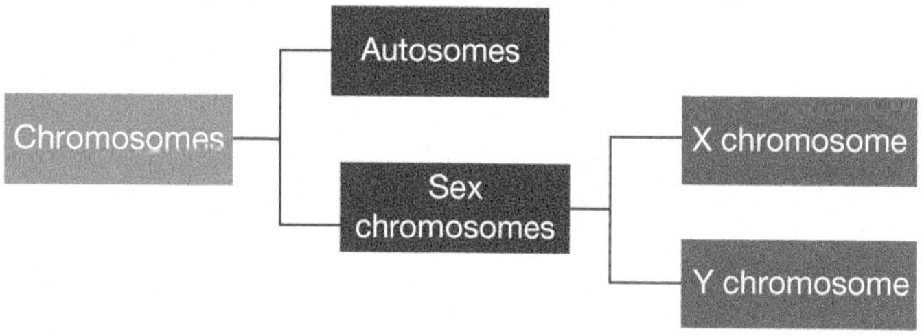

The members of each matching pair of chromosomes are described as being **homologous**. Those that don't match are called **non-homologous**. For example, two number 21 chromosomes would be referred to as homologous, and a number 21 chromosome and a number 11 chromosome would be non-homologous.

Homologous chromosomes have the same relative size, position of centromere and stained banding patterns.

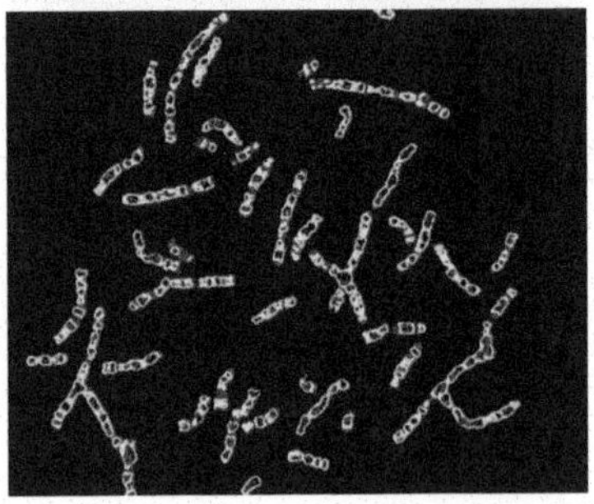

Fluorescent-dyed chromosomes showing stained centromere

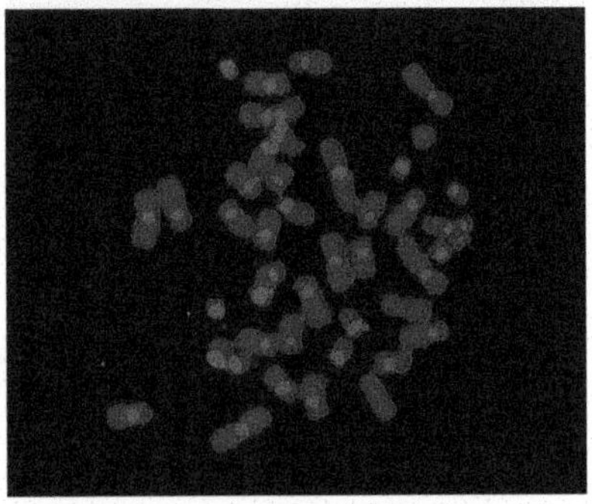

2.2.8 Sex chromosomes

The two remaining chromosomes are the **sex chromosomes**. In humans, these differ between males and females. Females possess a pair of X chromosomes (XX) and males have an X chromosome and a Y chromosome (XY). The sex chromosomes are important in determining an individual's gender (whether they are male or a female).

2.2.9 Too many or too few

Sometimes a genetic mistake or mutation can occur that results in more or less of a particular type of chromosome. Down syndrome is an example of a **trisomy** mutation in which there are three number 21 chromosomes instead of two. Turner syndrome is an example of a **monosomy** mutation that results in only one sex chromosome (XO).

Human chromosomes in order of size with banding patterns and centromere

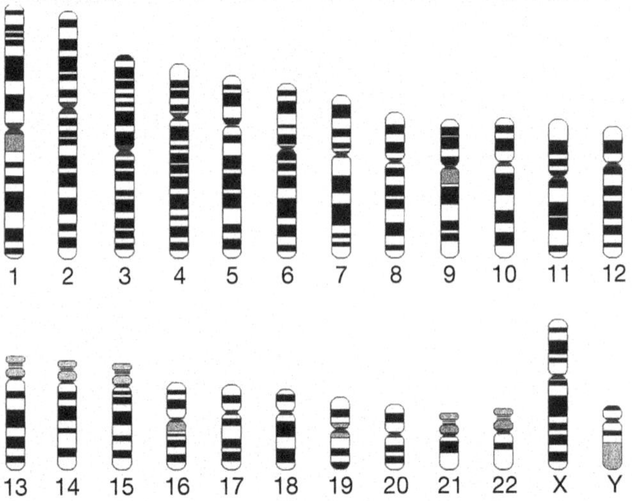

2.2.10 Your kind of karyotype

Differences between the chromosome pair size, shape and banding can be used to distinguish them from each other. Scientists use these differences to construct a **karyotype**. Cells about to divide are treated and stained, mounted on slides for viewing, and photographed. These photographs are cut up and rearranged into pictures that show the chromosomes in matching pairs in order of size from largest to smallest. Karyotyping can reveal a variety of chromosomal disorders such as Down syndrome and Turner syndrome.

The gender of an individual can also be determined using karyotyping. In humans, females possess two similar-sized X sex chromosomes. In males, however, their sex chromosomes are not matching — they possess an X chromosome and a smaller Y chromosome.

A human karyotype

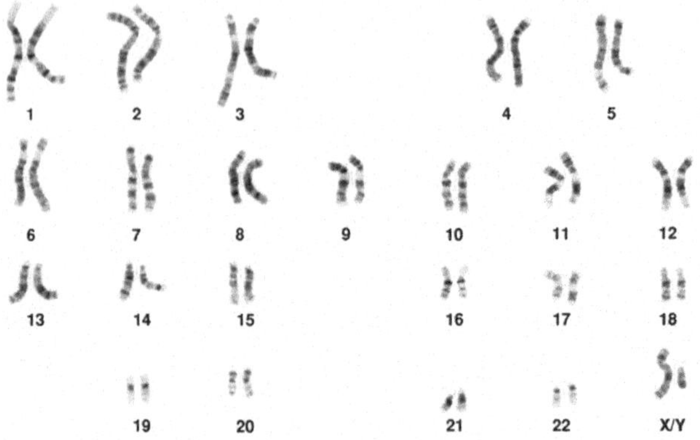

Some examples of chromosome changes and approximate incidence rates. Which syndrome is an example of a trisomy? A monosomy?

Chromosome change	Resulting syndrome	Approximate incidence rate
Addition: whole chromosome		
Extra number 21 (47, +21)	Down syndrome	1/700 live births
Extra number 18 (47, +18)	Edwards syndrome	1/3000 live births
Extra number 13 (47, +13)	Patau syndrome	1/5000 live births
Extra sex chromosome (47, XXY)	Klinefelter syndrome	1/1000 male births
Extra Y chromosome (47, XYY)	N/A	1/1000 male births
Deletion: whole chromosome		
Missing sex chromosome (46, XO)	Turner syndrome	1/5000 female births
Deletion: part chromosome		
Missing part of number 4	Wolf–Hirschhorn syndrome	1/50 000 live births
Missing part of number 5	Cri-du-chat syndrome	1/10 000 live births

2.2.11 Has the secret of age reversal been discovered?

In the 1970s, Tasmanian-born scientist Dr Elizabeth Blackburn made a discovery that contributed to our understanding of how cells age and die. She showed how the presence of a cap of DNA called a **telomere** on the tip of the chromosome enabled DNA to be replicated safely without losing valuable information. Each time the cell divides, however, these telomeres shorten. When the telomeres drop below a certain length, the cell stops dividing and dies. This is a normal part of ageing. Blackburn and her colleagues later discovered an enzyme, **telomerase**, that was involved in maintaining and repairing telomeres. In 2009, Blackburn and her colleagues were awarded the Nobel Prize in Physiology and Medicine for their work on how chromosomes are protected by telomeres and the enzyme telomerase.

Other scientists are now involved in finding out more about the exciting possibilities that our understanding of this process may open up. In 2010, for example, Mariela Jaskelioff and her colleagues in America genetically engineered mice with short telomeres and inactive telomerase to see what would happen when they turned this enzyme back on. Their results showed that after four weeks new brain cells were developing and tissue in several organs had regenerated — and the mice were living longer. If this happens in mice, what might future research suggest for humans?

Dr Elizabeth Blackburn

INVESTIGATION 2.1

Extracting DNA

AIM: To extract DNA from ground wheatgerm

Materials:

1 teaspoon of finely ground wheatgerm
14 mL of isopropyl alcohol (or equivalent)
1 mL of liquid detergent
20 mL of hot tap water (50–60 °C)
test tube
measuring cylinders
rubber stopper
test-tube rack
Pasteur pipette and bulb
glass stirring rod

Method and results

1. Draw a table in your book, allowing room for observations in the form of a diagram: immediately after adding the alcohol; at 3- and 15-minute intervals; and after you have collected and removed the DNA.
   - Add the wheatgerm and hot water to a test tube. Twist the stopper in and shake for 3 minutes.
   - Add 1 mL of detergent and mix gently with the glass rod for about 5 minutes. Do not create foam.
   - If you do create foam, suck it out with the Pasteur pipette.
   - Tilt the tube at an angle and slowly pour in the alcohol so that it sits at the bottom.

CAUTION

Do not mix!

2. Note your observations in your table.
3. Fill in your observations after 3 minutes and again after 15 minutes.
4. Collect the DNA with the glass rod. Feel it with your fingers and make your final observations.

Discuss

5. What colour did you expect DNA to be? Why do you think it was the colour that you observed?
6. How could you confirm that it really was DNA?
7. Suggest improvements to the experimental design.

2.2 Exercises: Understanding and inquiring

To answer questions online and to receive **immediate feedback** and **sample responses** for every question, go to your learnON title at www.jacplus.com.au. *Note:* Question numbers may vary slightly.

Remember

1. State the name of the:
   (a) molecule that DNA is an abbreviation for
   (b) location of DNA in a human cell
   (c) structures that genes are organised into
   (d) type of cell division used to produce gametes
   (e) male sex gamete
   (f) female sex gamete
   (g) process in which sex cells fuse together
   (h) cells of your body that are not sex cells
   (i) double set of genetic instructions
   (j) particular trait or feature that results from your genotype and environment
   (k) chromosomes that are not sex chromosomes
   (l) protective cap of DNA on the tip of chromosomes.
2. Suggest why chromosomes are stained with dyes.

3. Are chromosomes always visible in a cell? Explain.
4. State how many chromosomes you would expect to find in a human:
   (a) somatic cell
   (b) gamete.
5. Distinguish between the following pairs of terms.
   (a) Ovum and sperm
   (b) X chromosome and Y chromosome
   (c) Sex chromosomes and autosomes
   (d) Somatic cells and sex cells
   (e) Mitosis and meiosis
   (f) Homologous and non-homologous
   (g) Gene and DNA

Analyse, interpret and discuss

6. The following questions refer to figures A and B below.
   (a) Carefully observe the figures and suggest features that are useful in pairing the chromosomes.
   (b) On the basis of information in the karyotype, suggest the gender of A and B. Justify your responses.
   (c) Suggest why karyotyping can be carried out only on cells that are about to divide.

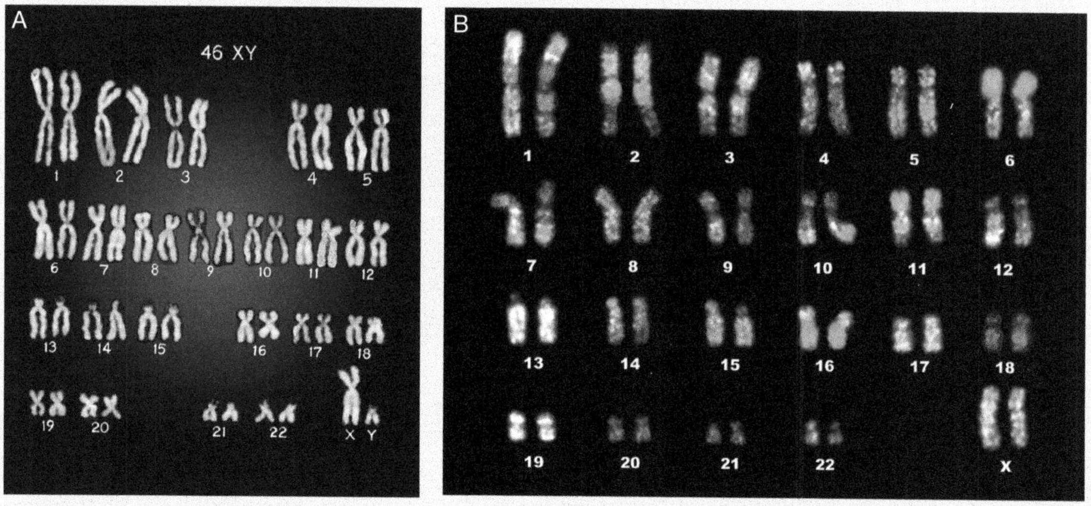

7. Each species has a particular number of chromosomes. The table at the top of the next page shows some examples of the number of chromosomes in the body cells of some organisms.
   (a) Using the data in the table, construct a column graph.
   (b) Identify the species with the:
      (i) highest total number of chromosomes
      (ii) lowest total number of chromosomes.
   (c) Carefully observe your graph, looking for any patterns. Discuss possible reasons for these.
   (d) Do you think that the number of chromosomes reflects the intelligence of an organism? Provide reasons for your response.
   (e) Suggest the number of chromosomes in the sex cells of a:
      (i) housefly
      (ii) sheep.

Number of chromosomes in body cells (non-sex cells) of some living things

Species of living thing	Number of chromosomes in each body cell	Species of living thing	Number of chromosomes in each body cell
Chimpanzee	48	Tomato	24
Euglena (unicellular organism)	90	Cabbage	18
Fruit fly	8	Frog	26
Human	46	Housefly	12
Koala	16	Pig	40
Onion	16	Platypus	52
Shrimp	254	Rice	24
Sugarcane	80	Sheep	54

8. Carefully observe the karyotypes A and B below.
   (a) Suggest the gender of the individual in A and in B. Justify your responses.
   (b) Use your observations and the chromosome change table in this subtopic to:
      (i) suggest which type of chromosome change is shown in each figure
      (ii) suggest the name of the resulting genetic disorders.
   (c) One of these disorders is also sometimes described as trisomy 21. Suggest a reason for this description.

A

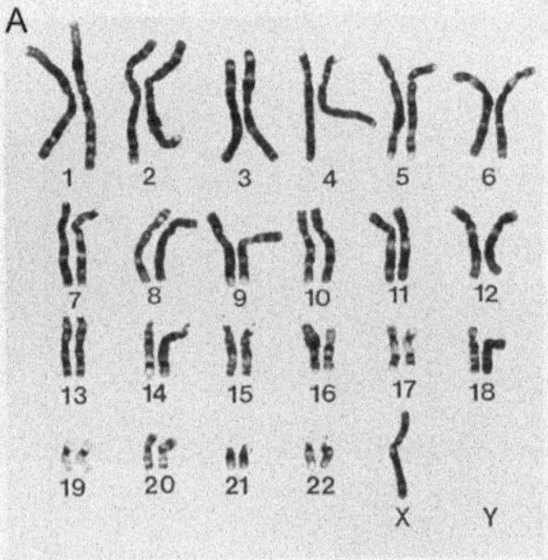

B

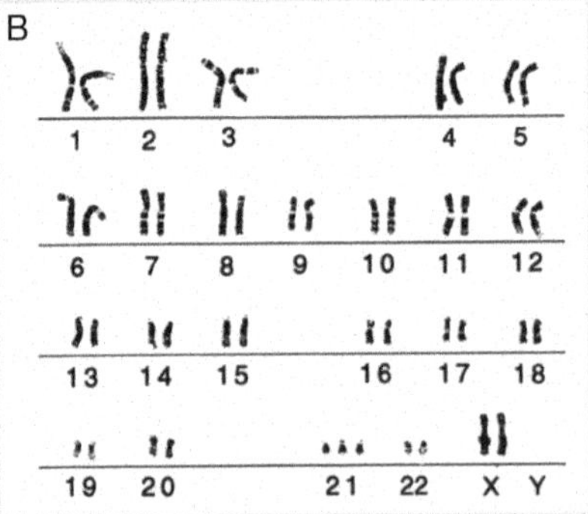

Analyse, interpret and investigate

9. Observe the figures at right that show the chromosomes belonging to four different types of organisms.
   (a) Suggest whether the chromosomes are from somatic cells or sex cells. Justify your response.
   (b) Suggest which organisms possess chromosomes:
      (i) most like humans
      (ii) least like humans.
   Justify your responses.
   (c) Do any observations in (b) surprise you? Why?
10. The graph on the next page shows the relationship between Down syndrome and maternal age.
   (a) Observe the graph and describe any patterns.
   (b) Suggest a hypothesis about Down syndrome and maternal age.
   (c) Research and report on types, causes and symptoms of Down syndrome.

Kangaroo (6 pairs)

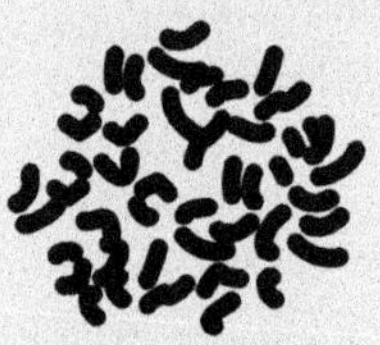
Human (23 pairs)

Domestic fowl (18 pairs)

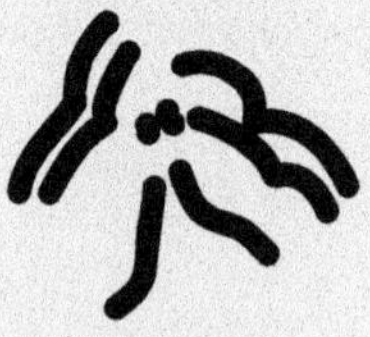
Fruit fly (4 pairs)

Investigate, discuss and report

11. (a) Research and report on Elizabeth Blackburn's:
    - (i) contribution to our understanding of DNA
    - (ii) stance on stem cell science that resulted in her losing her position on the President's Council on Bioethics.
   (b) (i) What is bioethics?
    - (iii) What is the Presidential Commission for the Study of Bioethical Issues, and what does it have to do with science? How does it differ from the President's Council on Bioethics?
    - (iv) Find out more about the types of issues that have been considered by the Commission.
12. If each cell nucleus has about a metre of DNA, how does it all fit in? Use the internet to find animations that show how DNA is organised so that it can fit into cells.

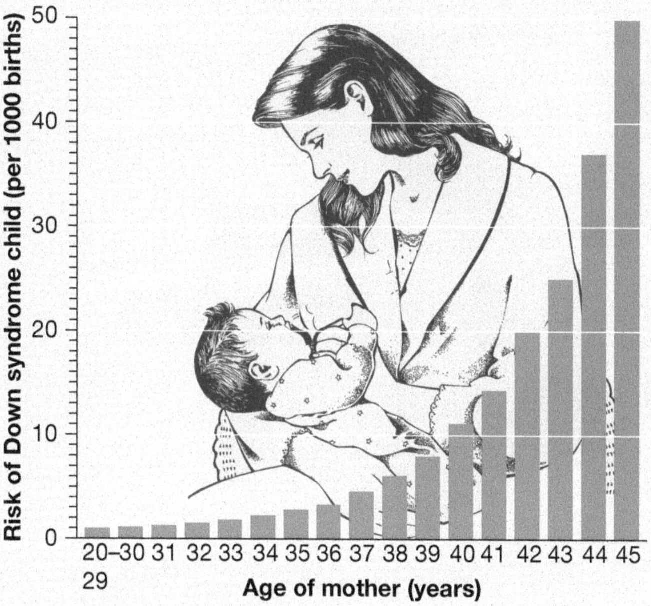

learnon RESOURCES — ONLINE ONLY

Complete this digital doc: Worksheet 2.1: Genes and chromosomes (doc-19405)

2.3 DNA — this is your life!

2.3.1 Greetings, DNA!

Even though DNA is as old as life itself, we have only recently been introduced to it. Like the story of the theory of evolution, DNA has its own story: a story of passion, imagination and determination that has involved the use of new technologies and the development of many more.

Friedrich Miescher

The abbreviation **DNA** is so well known, that it is often used as a word itself. DNA is the abbreviation for **deoxyribonucleic acid**. As the name suggests, it is a type of nucleic acid. It was not until 1869 that we were formally introduced to DNA, when it was discovered by Friedrich Miescher. Working in a laboratory located within a castle in Germany, Miescher — a young Swiss postgraduate student — isolated DNA from the nuclei of white blood cells from pus on bandages. Miescher named the compound he isolated from these cell nuclei 'nuclein'.

2.3.2 In parts of three

In 1929, over 50 years after its discovery, Phoebus Levene showed that DNA was made up of repeating units called **nucleotides**. Each of these nucleotides consisted of a sugar, a phosphate group and a nitrogenous base. In the nucleotides that make up DNA, the sugar is **deoxyribose** and the nitrogenous base in each nucleotide is one of four different types: **adenine** (A), **thymine** (T), **guanine** (G) or **cytosine** (C).

Phoebus Levene

Levene suggested that the nucleotides could be joined together to form chains. Although his theory was correct in terms of the chain formation, it was incorrect in other aspects of its structure. His tetranucleotide model, which proposed that DNA had equal amounts of all four nitrogenous bases, contributed to scientists of the time favouring proteins, rather than DNA, as the carrier of genetic information.

A nucleotide is made up of a phosphate group, a sugar and a nitrogenous base.

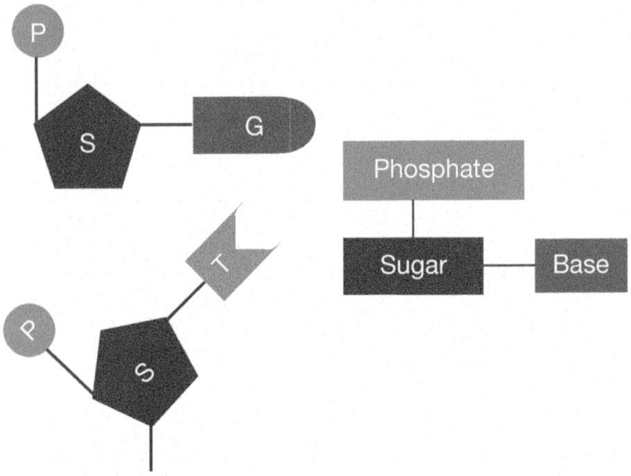

A nucleotide

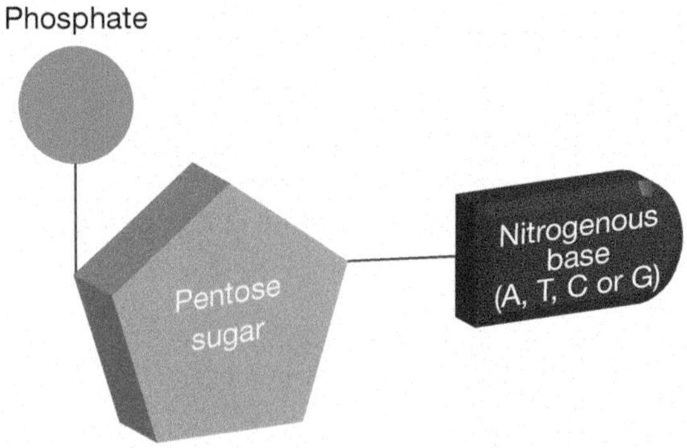

2.3.3 DNA carries messages from one generation to next

The experiments of Alfred Hershey and Martha Chase in 1953 supported those of Oswald Avery in 1943, suggesting that DNA was the molecule through which genetic information was carried between generations, not proteins.

Oswald Avery

2.3.4 A with T and G with C

In 1950, Erwin Chargaff contributed to our understanding of the structure of DNA by his careful and thorough analysis of the four different types of nucleotides and their ratios in DNA. His research led to the concept of **base pairing**. This concept states that in DNA every adenine (A) binds to a thymine (T), and every cytosine (C) binds to a guanine (G). This is now known as **Chargaff's rule**.

Erwin Chargaff

2.3.5 A key piece of the puzzle

The next piece of the puzzle to solve the structure of DNA was contributed (some say without her knowledge) by Rosalind Franklin. Rosalind Franklin and Maurice Wilkins had decided to crystallise DNA so that they could make an X-ray pattern of it. They were specialised in making X-ray diffraction images of biological molecules so that they could be analysed to find out information about their three-dimensional structures. Franklin's X-ray diffraction picture of a DNA molecule provided important clues about the shape of the molecule.

Examples of how base pairing using Chargaff's rule can be shown

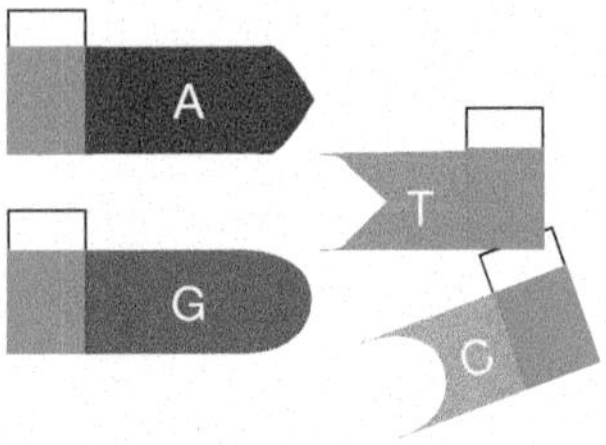

2.3.6 Double helix

James Watson and Francis Crick were building a DNA model to try to solve its structure. They were shown Franklin's X-ray diffraction image of DNA, which strongly suggested that DNA had a helical shape. They used this information, as well as that from Chargaff and other researchers (such as their American colleague Linus Pauling), to successfully solve the structure of DNA. At last the structure was identified!

Rosalind Franklin's X-ray diffraction picture provided important clues about the shape of the DNA molecule.

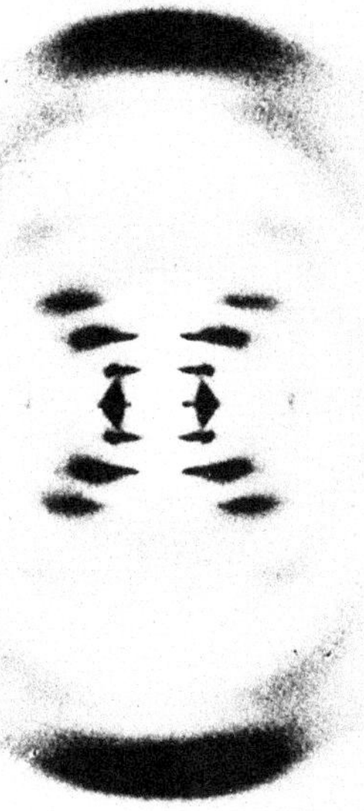

Rosalind Franklin provided a key clue to solve the structure of DNA.

2.3 Exercises: Understanding and inquiring

To answer questions online and to receive **immediate feedback** and **sample responses** for every question, go to your learnON title at www.jacplus.com.au. *Note:* Question numbers may vary slightly.

Remember

1. State what DNA is an abbreviation for.
2. Provide an example of a nucleic acid.
3. Identify the year and name of the scientist who first discovered DNA.
4. State the source of cells in which DNA was first isolated.
5. Use a diagram to show how the three sub-units that make up DNA are organised.
6. Outline what the research of Hershey, Chase and Avery suggested.
7. Describe what is meant by Chargaff's rule.
8. Describe Rosalind Franklin's contribution to the discovery of the structure of DNA.
9. Explain how Watson and Crick used information available to determine the structure of DNA.

Think and create

10. Use your own materials to construct a model of the double helix structure of DNA.
11. Use the information in this subtopic and other sources to construct a timeline on the development of our understanding of DNA.
12. Construct a paper model of DNA. Some suggested shapes you could use to represent the parts that make up a DNA molecule are shown at right.
13. Evaluate the model you made in question 12. Which aspects of the structure of DNA does your model show accurately? In what ways is your model different from an actual DNA molecule?

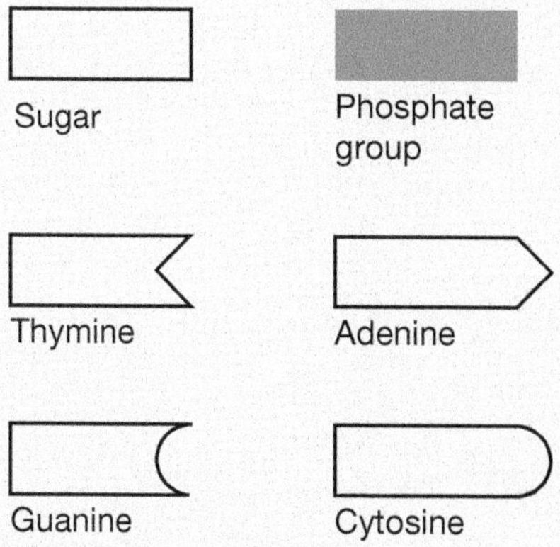

Investigate

14. Select one of the scientists discussed in this subtopic, research them and write a biography about their life and scientific contributions.

15. Find out more about DNA and how knowledge about its structure is being used in research and other applications. Present your findings as a documentary, animation or in a multimedia format.
16. Investigate more about the history of how we have obtained our genetic knowledge. Present your findings as a timeline.
17. Investigate the effect that our increased knowledge about the structure and function of DNA has had on:
    (a) our species
    (b) other species
    (c) our planet.
18. Investigate and report on the development of the Watson and Crick double helix model of the structure of DNA.
19. Use internet research to help you to identify three questions that could be investigated about DNA. Collate these questions as a class, and then select one to investigate and report on.
20. Discuss the following statement: 'Had Maurice Wilkins and Rosalind Franklin had a more harmonious working relationship it is likely that Franklin would have been involved in writing the scientific paper where the structure of DNA was first described, and that she would have been given the same credit for discovering the structure of DNA as Watson and Crick.'

Watson and Crick with their DNA molecule figure

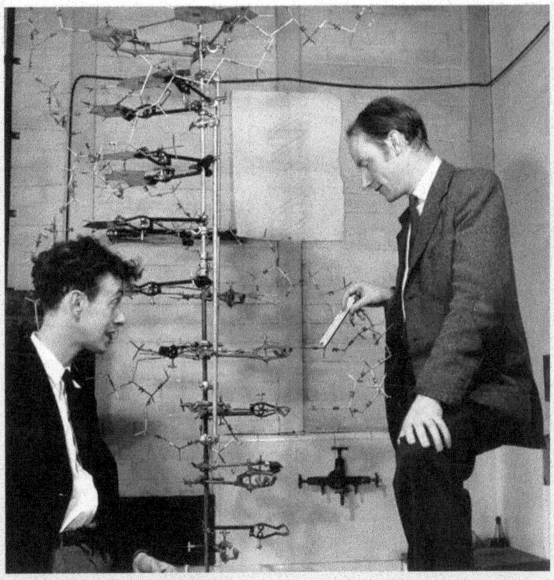

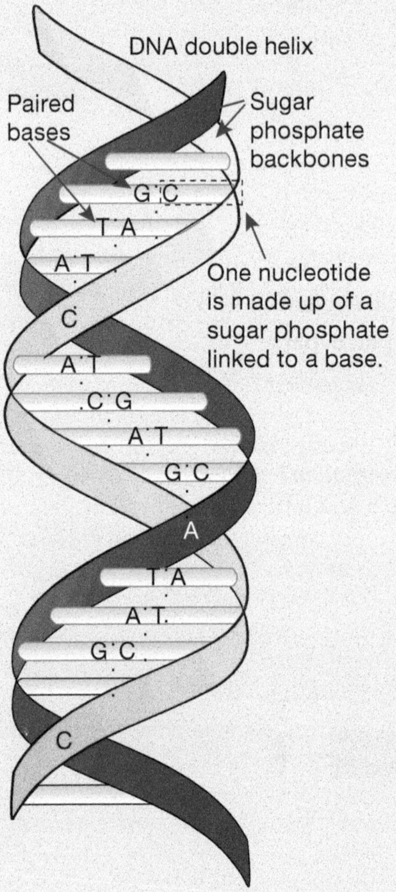

learnon RESOURCES — ONLINE ONLY

Watch this eLesson: Rosalind Franklin and Watson and Crick (eles-1782)

Try out this interactivity: Complementary DNA (int-0133)

2.4 Unlocking the DNA code

2.4.1 Key codes

Did you know that all living things share the same genetic letters? This universal genetic language provides strong evidence that all life on Earth evolved from one ancient cell line.

Suggest how these terms are linked.

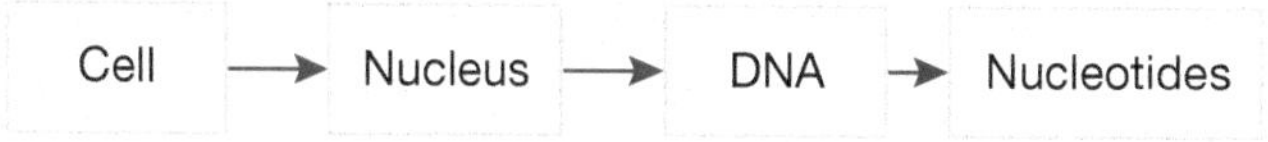

Like other eukaryotic organisms, DNA is located within the nucleus and mitochondria of your cells. While both types of DNA (nuclear and mitochondrial) share a number of features in common, they also differ. These differences will be considered later. In this subtopic, we will focus on your nuclear, rather than mitochondrial, DNA.

2.4.2 Stepping down the DNA ladder

Like other **nucleic acids**, DNA molecules are made up of building blocks called nucleotides. Each nucleotide is made up of three parts: a sugar part, a phosphate part and a **nitrogenous base**. The figures at right and below show some ways in which the components of nucleotides may be drawn.

While the sugar (for example, deoxyribose) and phosphate are the same for each nucleotide in DNA, the nitrogenous base may vary. The four possible nucleotides in DNA are adenine (A), thymine (T), cytosine (C) and guanine (G).

The nucleotides are joined together in a chain. The sugar and phosphate parts make up the outside frame and the nitrogenous bases are joined to the sugar parts.

DNA is made up of nucleotides.

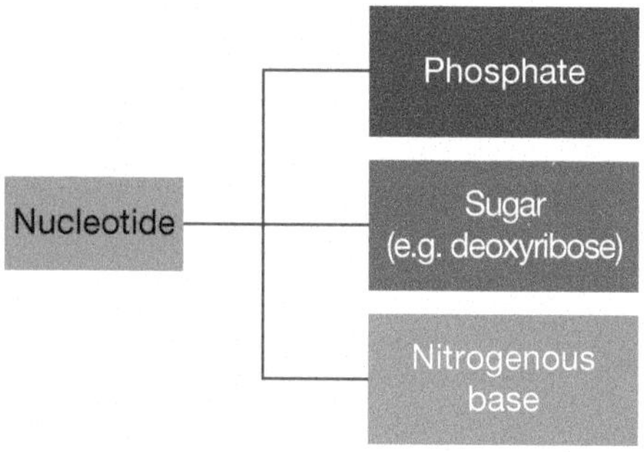

A nucleotide is made up of phosphate, sugar and a nitrogenous base.

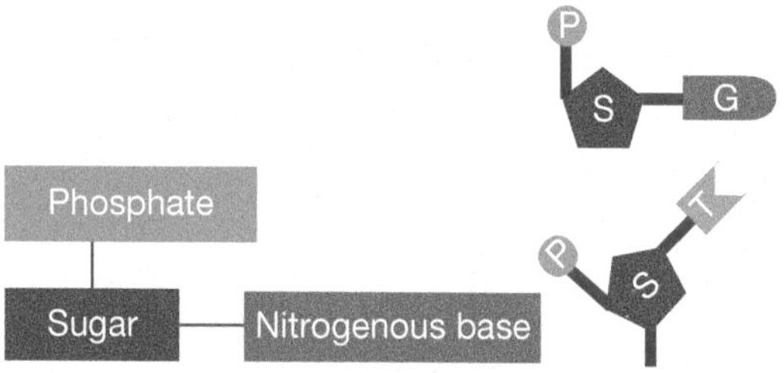

Each nucleotide in DNA may contain one of four nitrogenous bases.

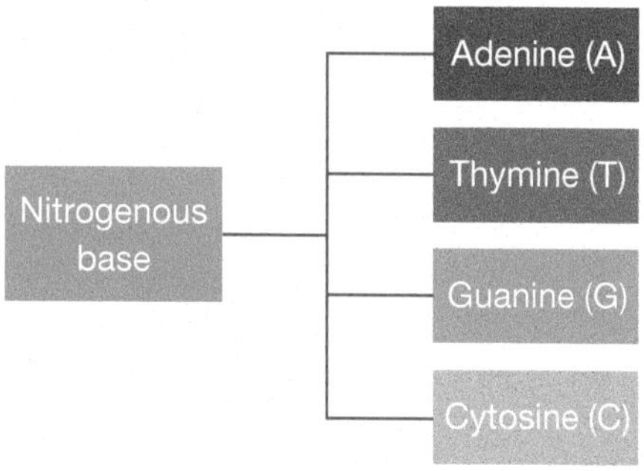

Nitrogenous bases are attached to the sugar part of the nucleotide.

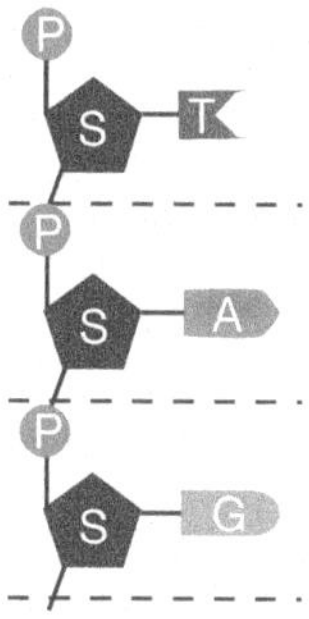

2.4.3 Nitrogenous bases in pairs

A DNA molecule is made up of two chains of nucleotides. Hydrogen bonds join them at their complementary (or matching) nitrogenous base pairs. Adenine binds to thymine and cytosine to guanine. This matching of the nitrogenous bases is often referred to as the **base-pairing rule**.

For example, a segment of DNA that has one strand with the code GATTACA would have a complementary strand of CTAATGT. In its double-stranded view it looks like this:

G A T T A C A
C T A A T G T

DNA molecules have the appearance of a **double helix** or spiral ladder. Using the spiral ladder metaphor, DNA could be considered as having a sugar–phosphate backbone or frame, and rungs or steps that are made up of **complementary base pairs** of nitrogenous bases joined together by hydrogen bonds.

2.4.4 Unlocking DNA codes

The sequence of nucleotides in DNA is often described in terms of the nitrogenous bases that they contain. For example, if the first nucleotide contains guanine, the second contains adenine and the third thymine, then this sequence would be described as GAT. This sequence of three nucleotides in DNA is referred to as a **triplet**. Although some of these DNA triplets code for a start (e.g. TAC) or stop (e.g. ATT, ATC or ACT) instruction, most code for a particular **amino acid**. The triplet GAT, for example, codes for the amino acid aspartine.

The sequence of these triplets in DNA contains the genetic information to make your body's proteins. This includes all of your hormones, enzymes, antibodies and many other proteins that are essential for your survival. If one of these triplets (or its bases) is incorrect or missing, it may result in a protein not being coded for or produced — which could result in death.

The DNA code is read three bases at a time.

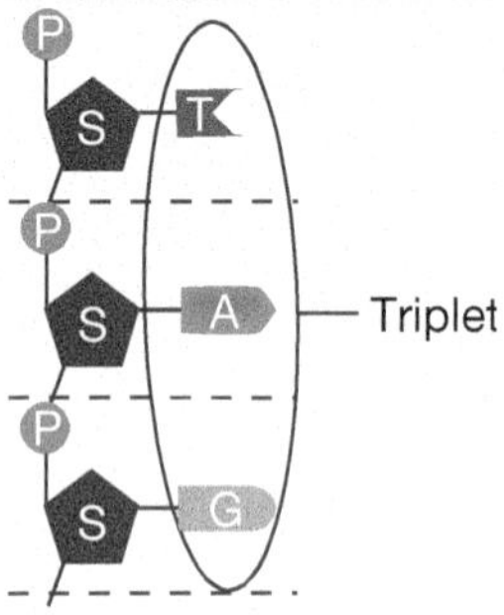

Can you suggest how ideas of patterns, order and organisation can be used to describe the structure and function of DNA?

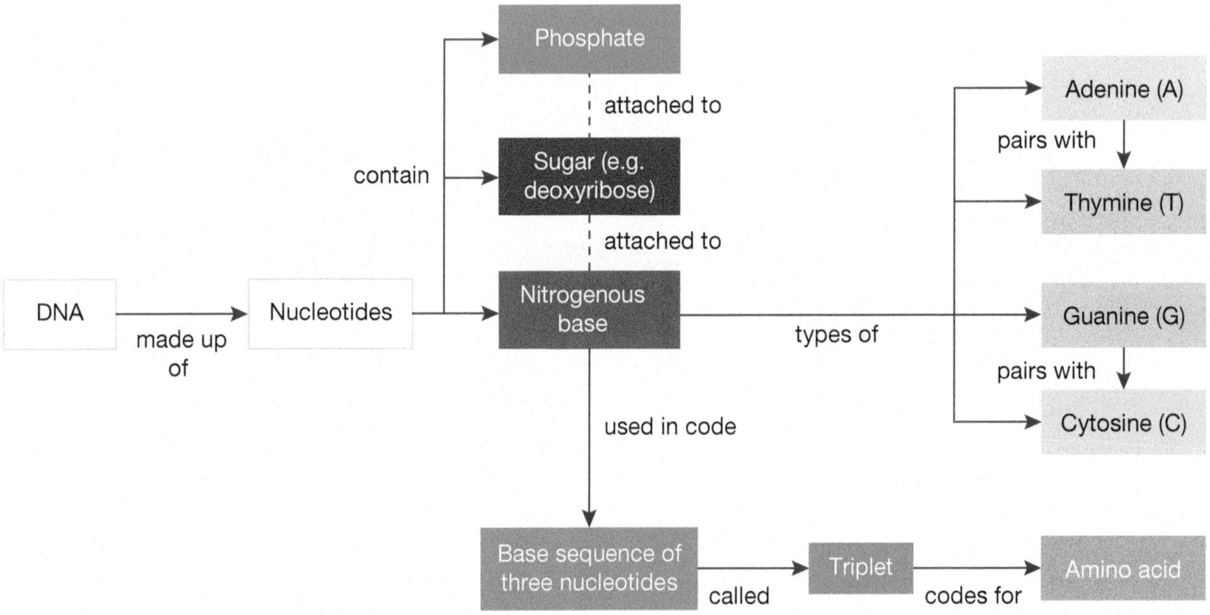

Double helix — the code for life. The DNA molecule is really very complex, but a model can help understanding.

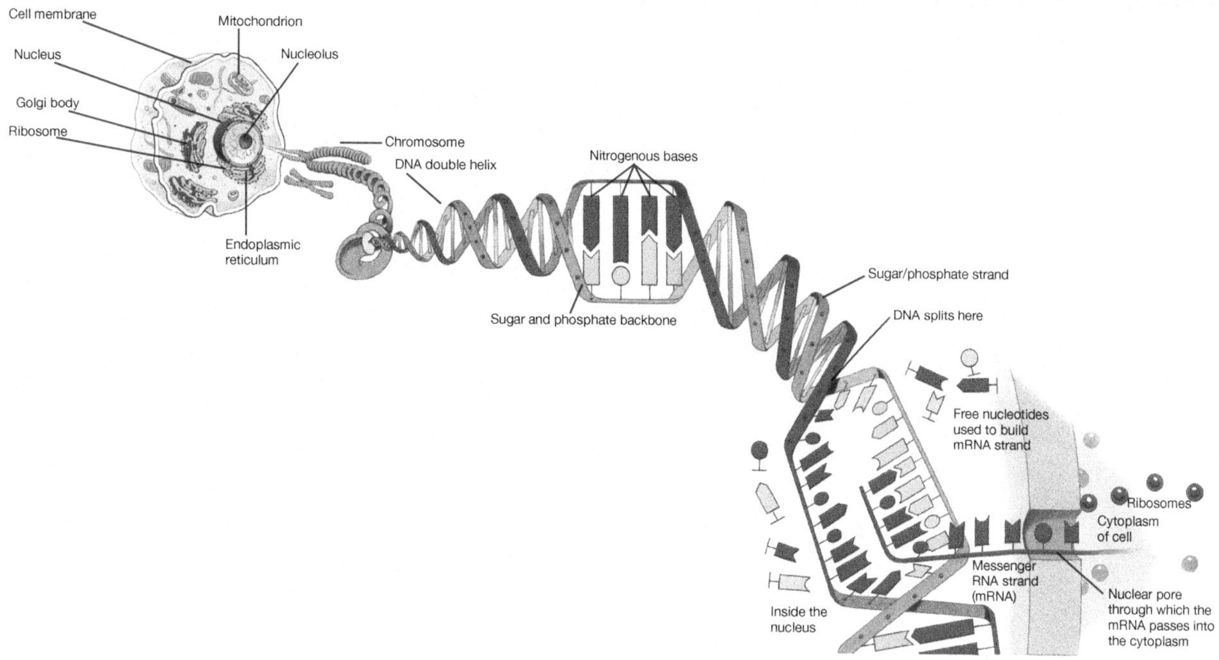

2.4.5 Protein synthesis: reading the code

DNA is in your genes, and it tells your cells how to make proteins. The instructions for making proteins are coded for in the sequence of nitrogenous bases in DNA. Within the nucleus, these instructions are transcribed into another type of nucleic acid called RNA in a process called transcription. This RNA copy then moves to a ribosome in the cytoplasm where the genetic message is translated into a protein.

The DNA message is transcribed into a mRNA message that is translated into a protein.

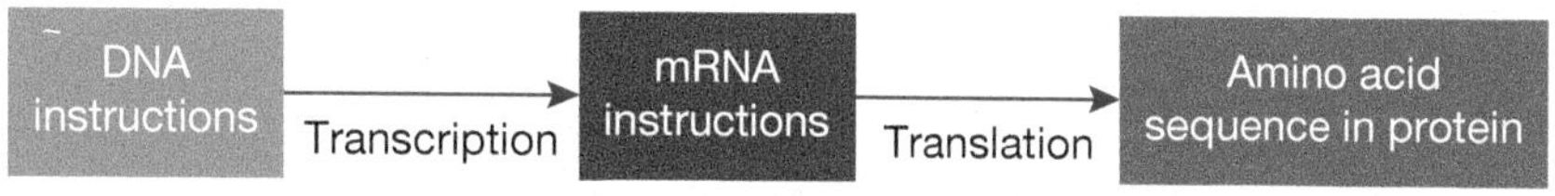

2.4.6 Introducing RNA

Like DNA, RNA is a type of nucleic acid and is made up of nucleotides. Its nucleotides, however, are different from those of DNA. RNA contains the sugar **ribose** (instead of deoxyribose), and **uracil** (instead of thymine) is one of its nitrogenous bases. It is also shorter and single-stranded.

Another difference is that the triplet code in mRNA is referred to as a **codon**. The complementary mRNA codon for the start triplet TAC in DNA, for example, would be AUG.

2.4.7 Transcription

The first step in making a protein involves the unzipping of the gene's DNA. When the relevant part of the DNA strand is exposed, a special copy of the sequence is produced in the form of **messenger RNA (mRNA)**. The process of making this complementary mRNA copy of the DNA message is called **transcription**.

How many differences can you identify between DNA and RNA?

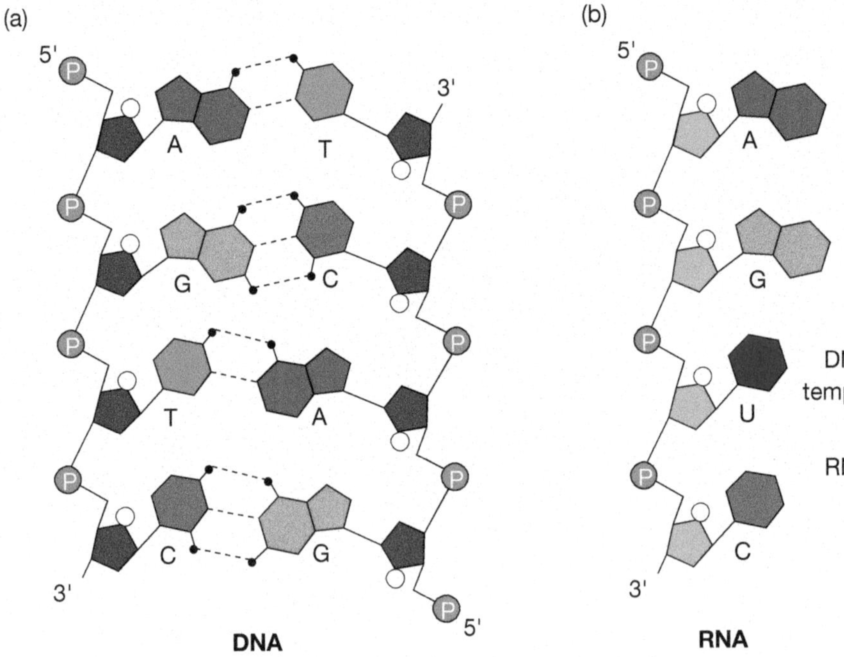

As its name suggests, messenger RNA (mRNA) passes through the pores of the nuclear membrane into the cytoplasm to take its genetic copy of the protein instruction message to **ribosomes**. These may be free floating in the cytosol or attached to the rough endoplasmic reticulum.

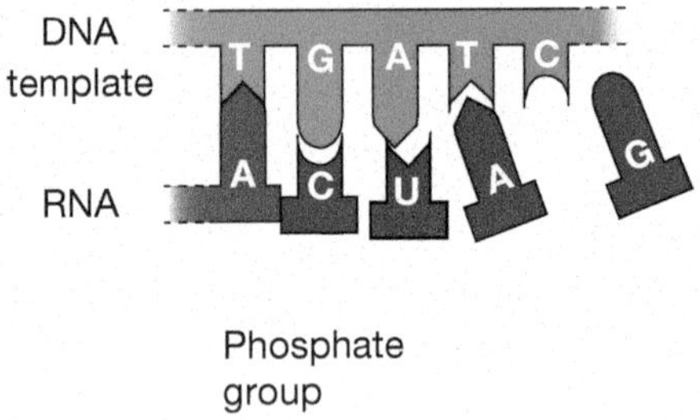

The complementary mRNA that had been transcribed from a segment of DNA such as TACATGCCA would be AUGUACGGU.

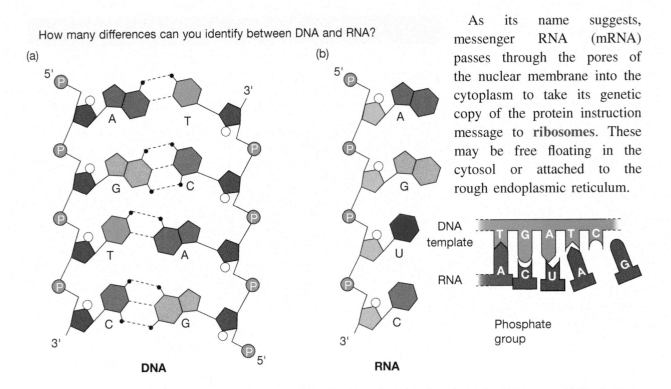

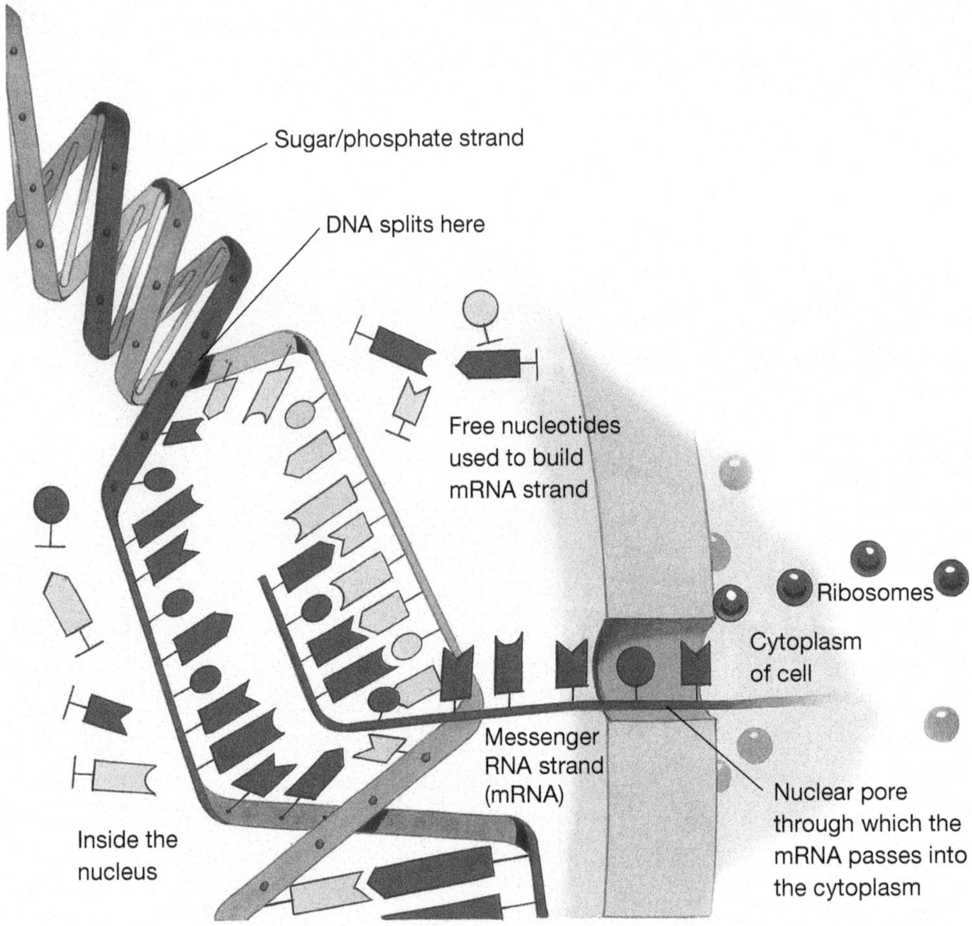

A section of the DNA unzips so that the mRNA copy can be made.

During transcription, an RNA molecule is formed with bases complementary to the DNA's base sequence.

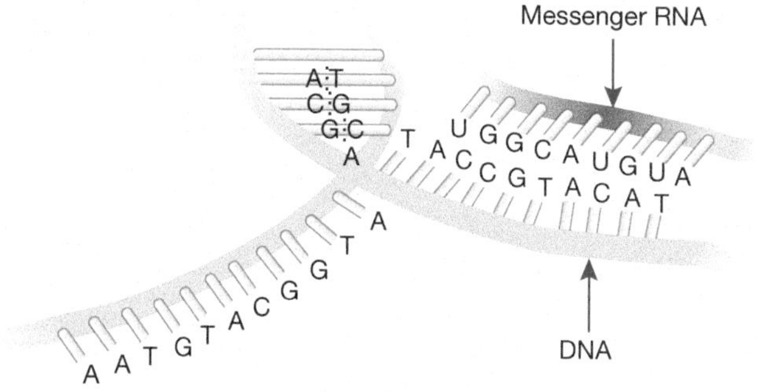

2.4.8 Translation

Ipsa scientia potestas est. Unless you speak Latin, you will need some help to translate this sentence! Once it is translated, you can then do something with it. This is similar to the meaning of the sentence: Knowledge itself is power.

Once the mRNA has reached the ribosome, its message needs to be **translated** into a protein. The ribosome and another type of molecule called **transfer RNA (tRNA)** are involved in this process. tRNA already located in the surrounding cytosol collects and transfers the appropriate amino acid to its matching code on the mRNA. These amino acids are joined together by peptidwe bonds to make a protein.

DNA triplet and corresponding mRNA codon and amino acid

DNA triplet	mRNA codon	Amino acid
AAT	UUA	Leucine (leu)
ACG	UGC	Cysteine (cys)
TAC	AUG	Start/methionine (met)
ATT	UAA	Stop
CGG	GCC	Alanine (ala)
CAT	GUA	Valine (val)
ATG	UAC	Tyrosine (tyr)
CCA	GGU	Glycine (gly)

2.4.9 Precious proteins

Why are proteins so important? Proteins form parts of cells, regulate many cell activities and even help defend against disease. Your heart muscle tissue contains special proteins that can contract, enabling blood containing haemoglobin and hormones to be pumped through your body. Haemoglobin is a protein that carries oxygen necessary for cellular respiration. Many hormones are proteins. Insulin, glucagon and adrenaline, for example, are hormones that influence activities of your cells. Enzymes are also made up of protein and can be involved in regulating metabolic activities such as those in chemical digestion and respiration. Antibodies are examples of proteins that play a key role in your immune system in its defence against disease.

Proteins are made up of amino acids.

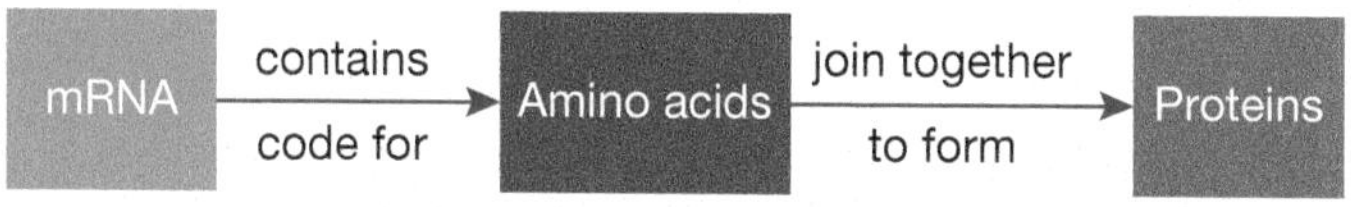

mRNA codons code for particular amino acids.

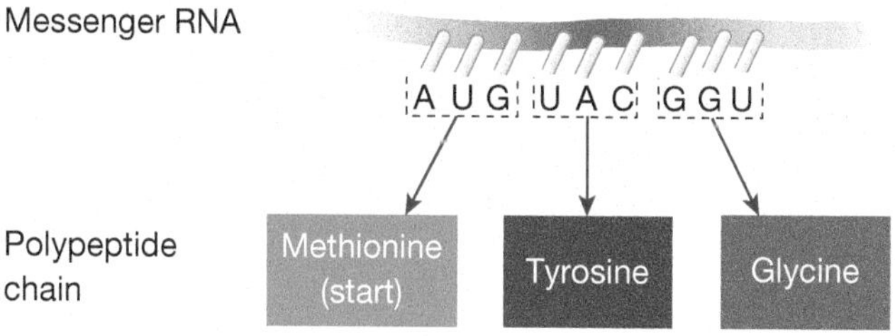

Plants also rely on proteins for their survival. Their growth and many other essential activities are regulated by hormones (such as auxins) and enzymes.

2.4.10 Switched on or off?

Different genes are responsible for different characteristics, such as the colour of flower petals, the markings on a snail shell, or a person's blood group or eye colour. Every body cell in an organism has the same set of genes called a genome, but not all genes are active. Some have to be switched on to act and some have to be switched off at different stages in the life of a cell. This is why hairs do not grow on the stomach lining and cheek cells do not grow on toenails.

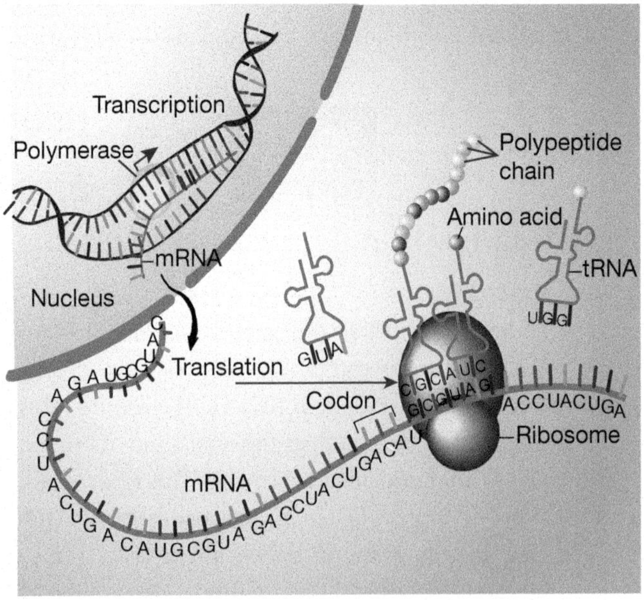

2.4 Exercises: Understanding and inquiring

To answer questions online and to receive **immediate feedback** and **sample responses** for every question, go to your learnON title at www.jacplus.com.au. *Note:* Question numbers may vary slightly.

Remember

1. Who am I? State the name(s) of the:
   (a) building blocks that make up DNA
   (b) three parts that make up nucleotides
   (c) four possible types of nitrogenous bases in DNA
   (d) four possible types of nitrogenous bases in RNA
   (e) complementary base that pairs with thymine in DNA
   (f) complementary base that pairs with adenine in RNA
   (g) sequence of three nucleotides in DNA that code for an amino acid
   (h) sequence of three nucleotides in mRNA that code for an amino acid
   (i) two steps in protein synthesis
   (j) site of protein synthesis
   (k) set of genes within a cell of an organism.
2. Construct a Venn diagram or matrix table to summarise the similarities and differences between:
   (a) DNA and RNA
   (b) transcription and translation
   (c) nucleic acids and amino acids
   (d) codons and triplets.
3. What is meant by the base-pairing rule? Use a diagram in your response.
4. Explain the importance of protein synthesis.

Analyse, interpret and think

5. Construct flowcharts, diagrams or concept maps to show connections or links between the following terms:
   (a) cells, DNA, nucleotides, nucleus
   (b) nitrogenous base, sugar, phosphate, nucleic acid, nucleotides
   (c) nitrogenous base, sugar, phosphate, deoxyribose, ribose, DNA, RNA, uracil, thymine, guanine, cytosine, adenine
   (d) DNA, mRNA, transcription, translation, amino acids, protein.
6. For the DNA code below, suggest the (a) corresponding mRNA strand and (b) amino acids it codes for.
   TAC CAT CGG CCA ATG ACG CGG CGG ATT

7. All cells of a particular living thing, such as a spider, have the same sets of genetic instructions, but not all of that organism's cells have the same structure and function. Suggest what causes this and why cell specialisation is so important.

Think and discuss

8. Copy and complete the Venn diagram at right using the following terms: thymine, uracil, deoxyribose, ribose, double-stranded, single-stranded, triplets, codons, adenine, guanine, cytosine, phosphate, nucleic acid.
9. Copy and complete the Venn diagram below right using the following terms: nucleus, protein, mRNA, ribosome, DNA, mRNA, protein, mRNA, mRNA.

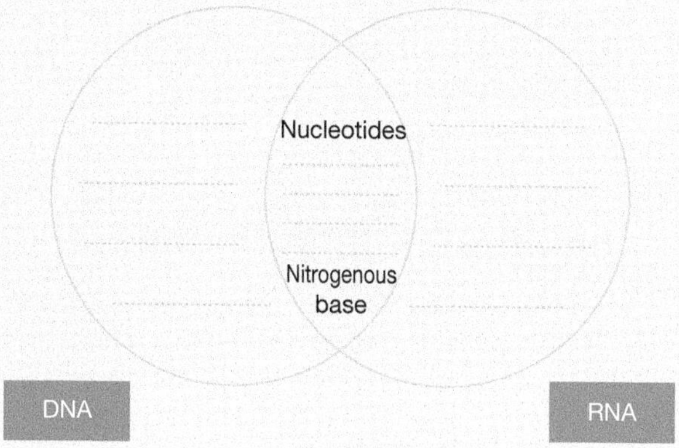

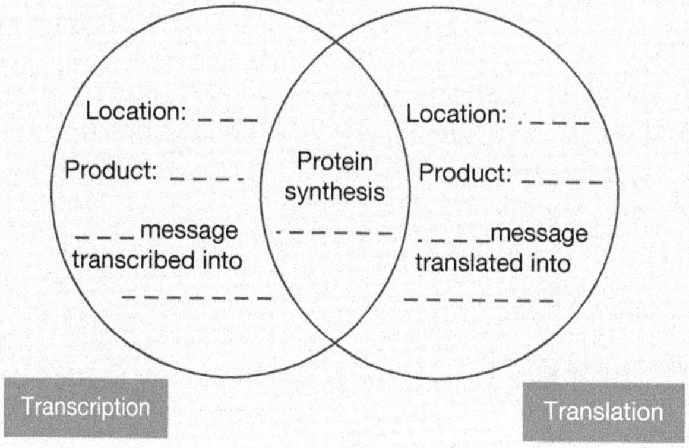

Create

10. Design and make a model showing a simplified structure of DNA. Decide whether you wish to make a 3D or a 2D (flat) representation.
    (a) What kinds of materials could you use in your construction?
    (b) Evaluate your model. What does it show or do well? What is it not able to show or do well?
11. Devise a role-play that demonstrates the way proteins are formed.
12. Rhymes such as the one below help us remember new information. Read or sing it, spelling out the triplets and codons with your fingers. Create your own rhyme about protein synthesis.

DNA is in my genes
Tells me how to make proteins
Got my genes from Mum and Dad
Mixed them up and made me glad
DNA is in my genes
Tells me how to make proteins.

DNA bases read times three
Always starting with TAC
mRNA codon would be AUG
DNA triplets tell the story of me
DNA bases read times three
Always starting with TAC.

ATT, ACT, ATC
Stop making proteins for me
mRNA codons for this would be
UAA, UGA, UAG
ATT, ACT, ATC
Stop making proteins for me.

Investigate

13. With increased knowledge and understanding, previous metaphors used to describe DNA are increasingly appearing to be less accurate in describing its complexities. The double helix, for example, describes its shape but not its function.
    (a) Find out more about two of the following metaphors and suggest reasons why each is becoming less useful.
    - Double helix
    - Computer code of life
    - Chemical building block
    - Symphony of life
    - Alphabet of life
    - Blueprint
    - Book of life

    (b) In six words or less, suggest a metaphor that could be used to communicate what DNA is all about — especially to those who do not have a background in biology. Provide reasons to support the use of your metaphor.
14. James Watson (co-discoverer of the structure of DNA) and Craig Venter were both involved in investigating the human genome. Find out more about science as a human endeavour by following their two different stories of genome exploration, what they have in common, and how and why they clash.
15. Scientists have discovered a gene switch that has restored youthful vigour to ageing, failing brains in rats. Results from investigations suggest an 'on switch' for genes involved in learning. Injection of an enzyme flips the switch on and improves the learning and memory performance of older rats. Find out more about this type of research or other research that involves switching on genes.
16. Draw a timeline to show the rate of identification of human genes. A computer database called OMIM (On-line Mendelian Inheritance in Man) keeps a regular update.
17. Use this subtopic of the textbook and other sources to research two scientists who contributed to the discovery of the double helix model of DNA. Write a brief account of their work.
18. Investigate further discoveries that have been made about DNA. Construct a timeline to share the who, what and when of your findings.

learnon RESOURCES — ONLINE ONLY

Complete this digital doc: Worksheet 2.2: DNA (doc-19406)

2.5 Who do you think you are?

2.5.1 Where are your genes?

You are very special. You have your very own unique DNA sequence. You have inherited this sequence from your ancestors. You are a human.

Much of who and what you are is determined by your genes. Genes determine many of the traits and characteristics that make you, you. A gene is a segment of double-stranded DNA that contains information that codes for the production of a particular protein or function. Located on specific chromosomes, humans possess around 20 000–24 000 genes. The position occupied by the gene on the chromosome is called its **locus**. Genes that are located on the same chromosome are described as being **linked**.

2.5.2 Genome maps

The total set of genes within an individual or cell is referred to as its **genome**. The study of genomes is called **genomics**. **Genome maps** describe the order of genes and the spacing between them on each chromosome. The genome size is often described in terms of the total number of base pairs (or bp). The genome size for organisms varies considerably: humans have about three billion base pairs, fruit flies about 160 million and brewer's yeast around 12 million.

2.5.3 The Human Genome Project

The Human Genome Project (HGP) was an international investigation to identify, sequence and study the genetic instructions within humans. Now that the information has been obtained, how can it be interpreted and further analysed? What are the potential benefits of this new knowledge? What ethical, social and political issues may arise?

2.5.4 Just the ingredients

It was anticipated that once we had the human genome sequenced, many mysteries would be unlocked and a new understanding of who we are would be unwrapped. Unfortunately, rather than an explosion of wonder and explanation, the sequencing only promoted more questions. Just like knowing the ingredients for a cake or the components of a car, we had the list, but not the delicious cake or speeding car.

The study of various genomes has revealed that the same genes that caused a fly to be a fly were also used to make a human a human. Parts of our genome are virtually interchangeable with those of our close primate 'cousins'. Rather than revealing the source of our diversity and uniqueness, our genome brought us closer to that of other life on Earth.

2.5.5 Epigenetics

While the HGP and its technologies provided us with information about the sequence of DNA, it is only part of the story. To understand more about its function, we may need to know more about the DNA of our ancestors. Maybe there are environmental triggers that switch on or off particular genes? If some of these involve lifestyle triggers, then could we be affected by the events our ancestors experienced? A new field called **epigenetics** suggests that this may be the case. This idea suggests that chemical changes can occur through environmental exposures and experiences that modify the DNA to a switched on or off form, and that these changes can be inherited.

This theory suggests that experiences of your great grandmother, for example, may have switched particular genes of hers on or off, and the modified gene(s) may have been passed on. Will you be involved in activities or events that change which of your genes are switched on, and then pass genes in this form to future generations?

2.5.6 Gene sequencing

Gene sequencing identifies of the order of nucleotides along a gene. DNA sequencers use four different-coloured fluorescent dyes (each binding to A, T, C or G in DNA) to identify the nucleotide sequence as it builds a complementary copy to the DNA template sample provided. An example of the output of a DNA sequencer is shown at right.

The locus for the cystic fibrosis gene is on chromosome 7. Polydactyly, cystic fibrosis and one form of colour blindness are linked genes — they are located on the same chromosome.

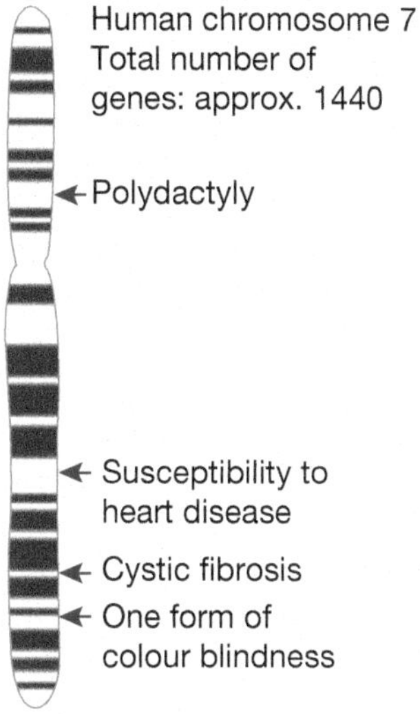

DNA sequencers identify the base sequence of sections of a DNA fragment.

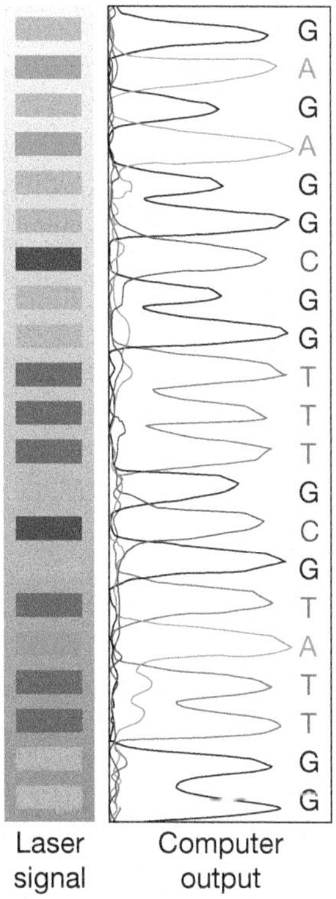

HOW ABOUT THAT!

- Since mice and humans diverged from a common ancestor millions of years ago, most of the DNA that codes for functional genes has remained similar, whereas the 'junk' DNA has mutated and is now extremely different.
- Is this 'junk' DNA really junk? Could it have a purpose? What have we found out about it?

2.5 Exercises: Understanding and inquiring

To answer questions online and to receive **immediate feedback** and **sample responses** for every question, go to your learnON title at www.jacplus.com.au. *Note:* Question numbers may vary slightly.

Remember

1. Define the following terms: gene, locus, linked, genome, gene sequencing, gene map, genomics.
2. Describe the relationship between the following terms.
   (a) Gene, chromosome, locus, linked
   (b) Gene, genome, genomics, genome map
   (c) Gene, gene sequencing, nucleotides, nitrogenous bases, DNA
3. Now that the human genome has been mapped, suggest three questions that could be asked.
4. Did the sequencing of the human genome answer our questions about why humans were unique? Explain.

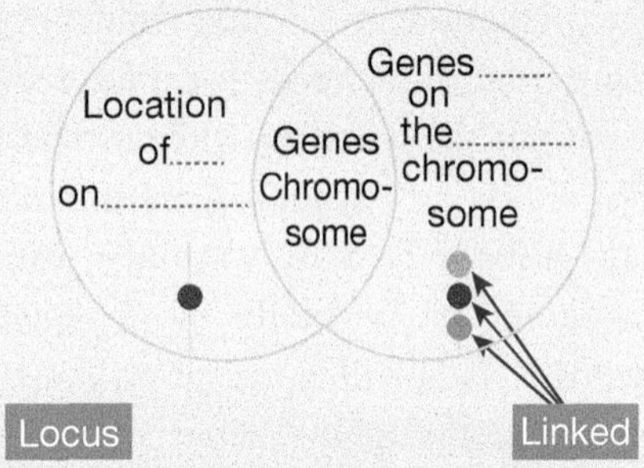

Think and discuss

5. Copy and complete the Venn diagram at right using the following terms: gene, located, chromosome, same.
6. Different genetic instructions within and between species are due to different nucleotide sequences in their genes. The table below shows part of the sequences of different genes from various organisms.

Type of organism	Section of gene sequence
Duck	TAG GGG TTG CAA TTC AGC ATA GGG ATC
Human	TTG TGG TTG CTT TTC ACC ATT GGG TTC
Bacteria	AAT GAA TGT AAC AGG GTT GAA TTA AAA

   (a) Suggest how they are similar.
   (b) Suggest how they are different.
   (c) Suggest a reason why they all use the same letters in their genetic coding system.
7. Who do you think should know if you had a higher risk of dying from a genetic disorder in 15 years' time?

Investigate and report

8. Research and report on Craig Venter and Francis Collins and their research on the human genome.
9. Guidelines have been developed for companies in the US that supply 'custom DNA' or DNA sequences to order. These guidelines have been introduced to make it harder for bioterrorists to build dangerous viruses as potential bioweapons. There is concern, however, that as these rules are voluntary and most custom DNA is made outside the US, they may have limited value. Find out more about custom DNA, bioterrorism and bioweapons, and how these relate to gene sequencing.
10. Personal genome scans can provide a lot of information about your genetic disposition for particular diseases and disorders. They do not, however, always guarantee that you will show the disease. Find out more about the relationship between genotype, phenotype and environmental factors and how these relate to the use, accuracy and effectiveness of personal genome scans.
11. If you had a personal genome scan that suggested you have a 25 per cent chance of developing a disease, and if you were told that environmental factors such as diet and exercise were more important than possession of the genes, how would this affect your future lifestyle? Why?

12. Is *bio* the buzzword of the twenty-first century? Research and report on at least two of the following.
   (a) Biotechnology
   (b) Biomedicine
   (c) Biomolecular scientist
   (d) Biochemist
   (e) Biophysicist

Investigate and create

13. Find out more about careers in genomics and genetic engineering, and research science fiction stories that include inheritance of interesting traits or genetic engineering. Based on your research, construct your own science fiction story about how knowledge of genetics may change our lives in the future. Share your work as a short story, animation or multimedia movie.

2.6 Dividing to multiply

2.6.1 All cells come from pre-existing cells

All cells arise from pre-existing cells. That's pretty amazing when you really think about it! This means that all organisms living today originated from cells from the past. The cells you are made up of come from an unbroken line of cells. Where, when and who did your original cell come from?

2.6.2 Cell division in eukaryotes

Scientists are still grappling with many questions about the origin of life. Maybe you will be the one to shed new light on some possible answers in the future? What we do know, however, is there are two key types of cell division. Mitosis is the type of cell division involved in growth, development and repair of tissues. Some eukaryotic organisms also use mitosis for **asexual reproduction**. Organisms involved in **sexual reproduction** use another type of cell division in their reproductive process called meiosis.

2.6.3 Nucleus, chromosomes and DNA

All eukaryotic cells have a nucleus, which contains genetic information with instructions that are necessary to keep the cell (and organism) alive. This information is contained in structures called chromosomes, which are made up of a chemical called deoxyribonucleic acid (DNA).

DNA is contained in the chromosomes, which are located in the nuclei of cells.

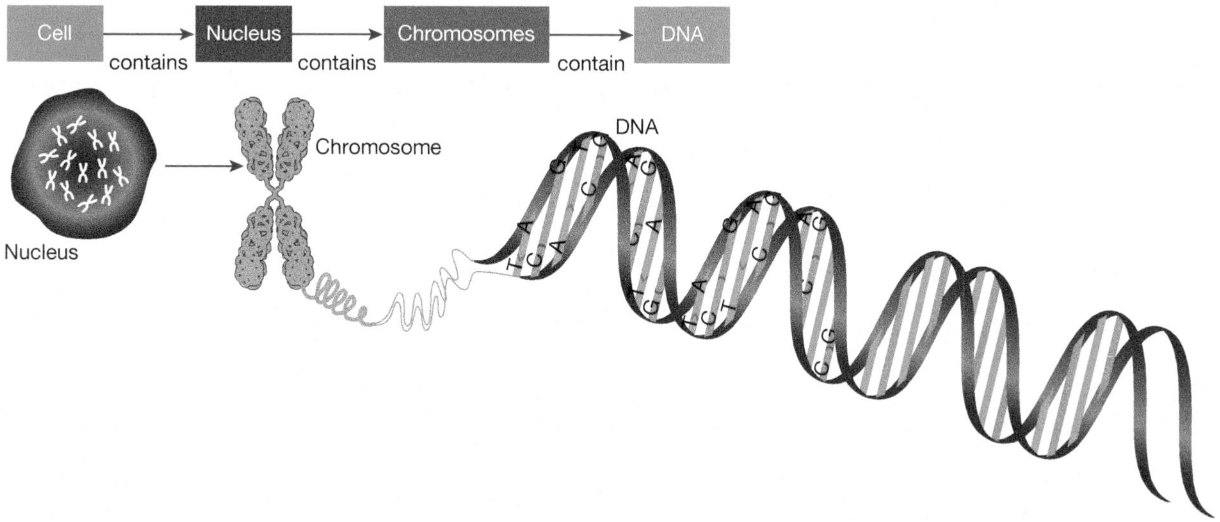

2.6.4 Mitosis

What happens when skin wears away and damaged tissues need repairing? How do seedlings grow into giant trees? How did you get to be so big? Throughout the life of multicellular organisms, mitosis is the type of cell division that is used for growth, development, repair and asexual reproduction.

The cells produced by mitosis are genetically identical to each other and to the original cell. They have the same number of chromosomes and DNA instructions. As they have identical genetic information, they are described as being **clones** of each other.

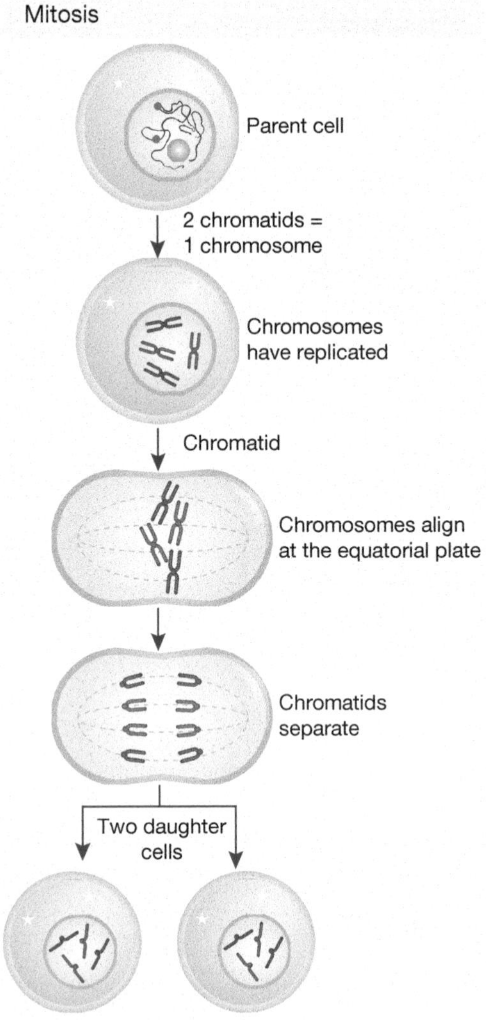

Each daughter cell contains the diploid ($2n = 4$) number of chromosomes

2.6.5 Cytokinesis

Mitosis is a process that involves division of the nucleus. Once a cell has undergone this process, the cell membrane pinches inwards so that a new membrane is formed, dividing the cell in two. The process of dividing the cytoplasm is called **cytokinesis**.

2.6.6 Counting chromosomes

Within the **somatic cells** (or body cells) of an organism, there is usually a particular number of chromosomes that is characteristic for their species. In humans, the total number of chromosomes in a somatic cell is 46. These chromosomes appear as 23 pairs in each body cell. The term used to describe chromosomes in pairs is **diploid**, because there are two sets of chromosomes.

Our gametes (or sex cells), however, contain only one set of chromosomes. They are referred to as being **haploid**. You may see the symbol n used to identify the haploid number. The diploid number would be identified as $2n$. How many sets of chromosomes do you think an organism would have if it was identified as $4n$ and **tetraploid**?

Eukaryotic unicellular organisms such as (a) *Amoeba* and (b) *Euglena* divide by binary fission involving mitosis. Unlike meiosis, mitosis produces identical cells.

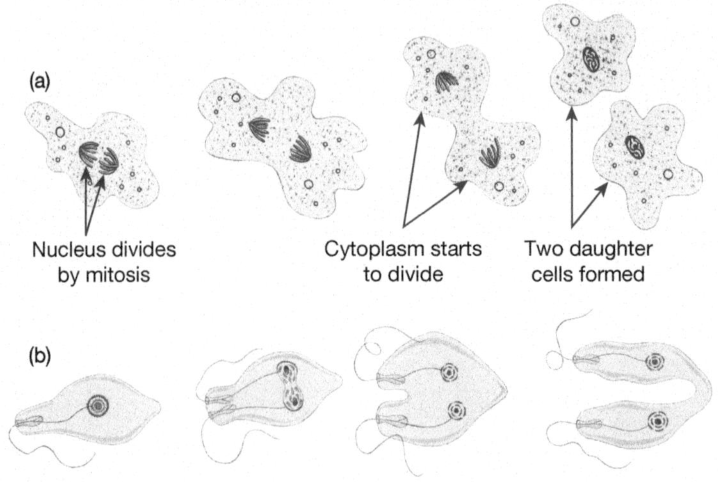

2.6.7 Meiosis

Why do gametes only have one set of chromosomes? If they didn't, then each time the egg and sperm nuclei combined during fertilisation, the number of chromosomes in the next generation of cells would double! For example, if each gamete had 46 chromosomes, the resulting cell after fertilisation would have 92 chromosomes.

Mitosis and meiosis are two types of cell division.

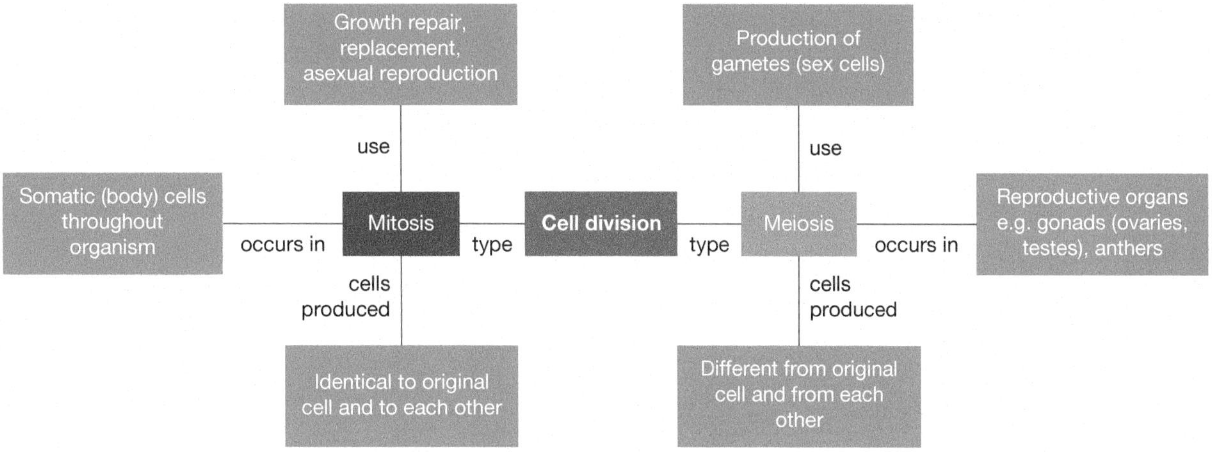

Meiosis is the kind of nuclear division that prevents the doubling of chromosomes at fertilisation. It is a process in which the chromosome number is halved. In humans, that means the parent cell that is to undergo meiosis would initially be diploid ($2n$) and the resulting daughter cells or gametes produced by meiosis would be haploid.

2.6.8 A key source of variation

Variation within a species can provide some individuals with an increased chance of surviving over others. Depending on the environment and selection pressures at a particular time, different variations may be advantageous. Lots of different variations among individuals will mean that there is more chance that some will survive to reproduce. This improves the chances of the species surviving.

2.6.9 Meiosis mix-up

Each parent produces gametes by the process of meiosis. Within each gamete are chromosomes from each parent. Chromosomes carried in the sperm are referred to as **paternal chromosomes**, and chromosomes from the ovum are referred to as **maternal chromosomes**.

You are a product of both meiosis and mitosis.

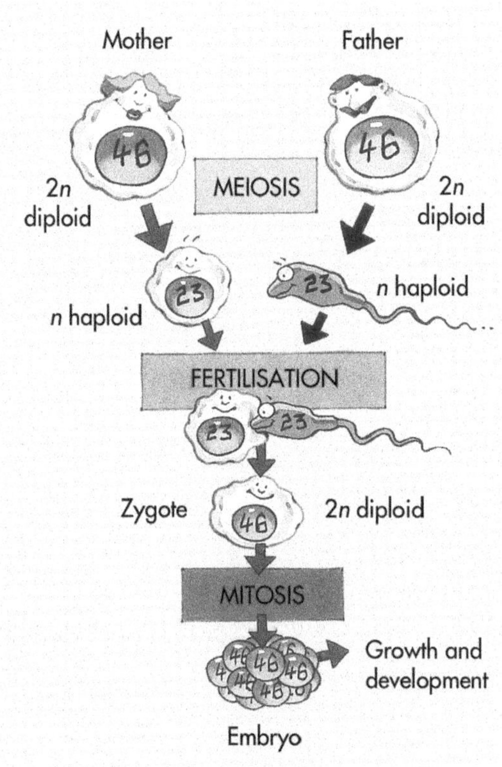

Meiosis provides sexually reproducing organisms with a source of variation. One way in which it increases variation is in terms of the number of combinations in which the chromosomes could be divided up into the gametes. For example, given that humans have 23 pairs of chromosomes, there are around 8 388 608 (2^{23}) different possible ways to divide up these chromosomes into each type of gamete.

Another source of variation in meiosis is **crossing over** between chromosomes of each pair. This results in a section of one chromosome swapping its genetic information with another. For example, genes that were once on a paternal chromosome can be transferred or crossed over onto a maternal chromosome and vice versa.

2.6.10 Fertilisation

In humans, fertilisation occurs when a **haploid gamete** from each parent fuse together to form a **diploid zygote**. But which sperm will fertilise the ovum? The identity of the lucky sperm that will contribute its genetic information to the next generation depends largely on chance. Depending on which sperm fertilises the egg, there are many different genetic combinations possible. This is another source of genetic variation that can give sexually reproducing organisms an increased chance of survival.

The zygote contains 23 paternal chromosomes from its father and 23 maternal chromosomes from its mother. Each pair of chromosomes will consist of a chromosome from each parent. The zygote divides rapidly by mitosis to form an embryo that will also use this type of cell division to develop and grow. Each time this process occurs, cells with this complete new set of chromosomes are produced.

Meiosis: Crossing over of genetic information between each pair of chromosomes is a source of variation in a species.

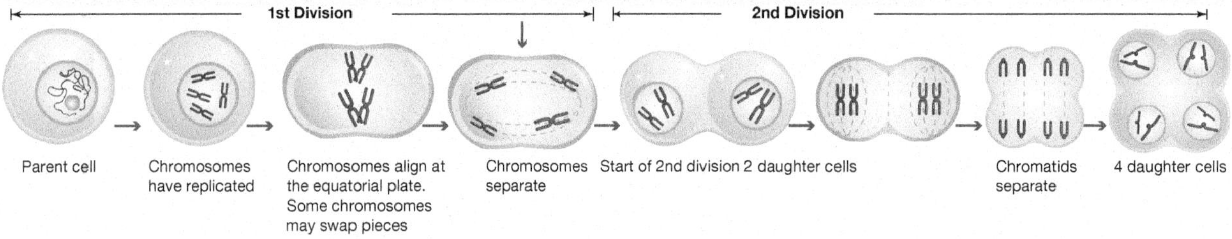

2.6.11 Boy or girl?

When a friend or family member is expecting a baby, one of the first questions people wonder or ask is whether it will be a boy or a girl. Probability suggests that the answer is that there is a 50 per cent chance either way. This can be predicted because it is determined by the sex chromosome combination that the child receives when the gametes from each parent fuse together at fertilisation.

Human somatic cells contain 22 pairs of autosomes and a pair of sex chromosomes. The sex chromosomes in the body cells of males and females differ. While females contain a pair of X sex chromosomes, males contain one X and one Y sex chromosome. Often this gender sex chromosome difference is abbreviated, so that females are described as being XX and males as being XY.

As a result of meiosis, gametes will contain only one sex chromosome. Human females (XX) can only produce gametes that contain an X chromosome. Human males (XY), however, will produce half of their gametes with an X chromosome and the other half with a Y chromosome. So, if a gamete containing a Y chromosome fuses with the ovum (which contains an X chromosome), the resulting zygote will be male (XY). Likewise, if the ovum is fertilised by an X-carrying gamete, then a female (XX) will result.

Is the mother or father the key determiner of the gender of the child?

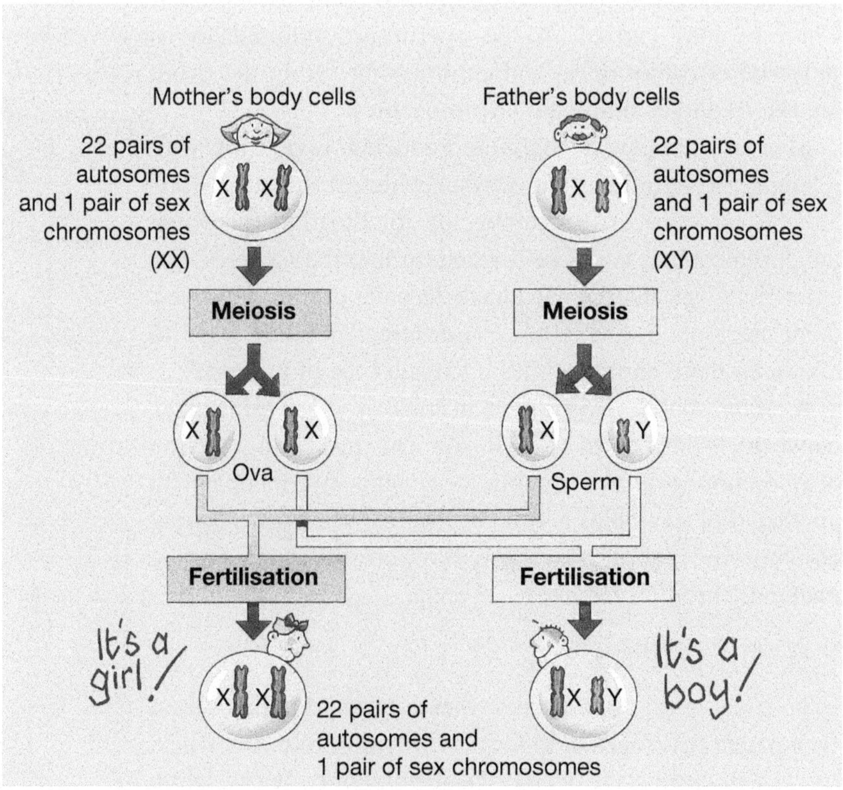

HOW ABOUT THAT!

The gender-determining factors of other animals can be quite different from those of humans. In birds, for example, it is the female that has different sex chromosomes, Z and W, and the male has two Z chromosomes. In some reptiles, gender is determined by the temperature at which the egg is kept rather than chromosomes. The temperature of the sand in which some crocodiles and turtles bury their eggs can determine whether the offspring will be male or female. The gender of brushturkey chicks is also determined partly by temperature.

2.6.12 Twins — or more!

Sometimes in the very early stages of division following fertilisation, clusters of a few cells develop into two separate individuals. If this happens, identical twins result as each cluster has the same genetic make-up as the other.

Usually, only one ovum is released at a time. However, if several are released, twins can result from fertilisation by different sperm. In this case, the babies are not identical because they have different genetic make-ups.

Identical or fraternal twins — one sperm or more?

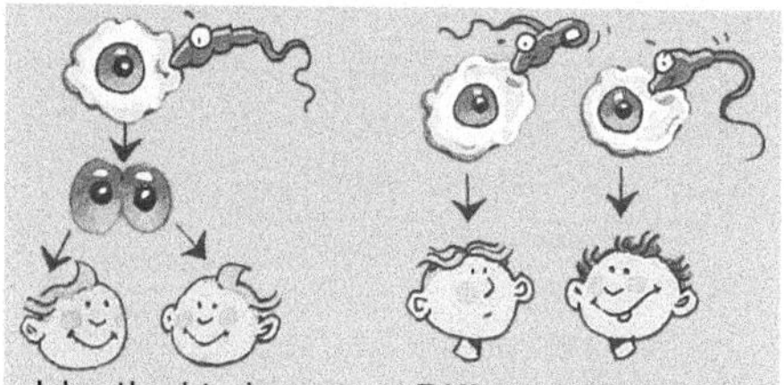

INVESTIGATION 2.2

What's the chance?

AIM: To simulate the chance of a male or female being conceived at fertilisation

Materials:

20-cent coin

Method and results

1. After reading the instructions and before you carry out the experiment, predict the number of times you will toss heads and the number of times you will toss tails. Give a reason for your prediction.
2. Toss a coin 50 times. Count the number of heads and tails and record the data in a table like the one below.

	Number of heads	Percentage of heads	Number of tails	Percentage of tails
Individual tosses				
Combined class result				

3. Calculate the percentage chance of obtaining heads and the percentage chance of obtaining tails.
4. Combine the results of the whole class and calculate the percentage chance of obtaining heads and tails.

Discuss and explain

5. Draw a graph of your results.
6. Analyse your data.
   (a) Was your prediction supported or not?
   (b) Were the percentage results obtained for 50 tosses the same as or different from the total class results? Suggest reasons for the similarities or differences.
7. If you tossed a coin 1000 times, would you obtain similar results?
8. What is the chance of obtaining heads each time you toss the coin?
9. If heads represented a sperm carrying an X chromosome and tails represented a sperm carrying a Y chromosome, suggest how this activity could link to the chances of a male or female baby being conceived.
10. Suggest a strength, a limitation and an improvement for this investigation.

2.6 Exercises: Understanding and inquiring

To answer questions online and to receive **immediate feedback** and **sample responses** for every question, go to your learnON title at www.jacplus.com.au. *Note:* Question numbers may vary slightly.

Remember

1. Where does the cell theory suggest that cells come from?
2. State the names of the two main types of cell division.
3. List three functions of mitosis.
4. What is DNA an abbreviation of?
5. Use a flowchart to show the link between the terms: DNA, cell, nucleus and chromosome.
6. Describe the features of the offspring cells produced by mitosis.
7. Distinguish between the following pairs of terms.
   (a) Cytokinesis and mitosis
   (b) Mitosis and meiosis
   (c) Diploid and haploid
   (d) Ovum and sperm
   (e) Maternal chromosomes and paternal chromosomes
   (f) Gamete and zygote
   (g) Fertilisation and meiosis
   (h) Autosomes and sex chromosomes
   (i) XY and XX
   (j) Somatic cells and gametes
   (k) Identical twins and fraternal twins

Think and discuss

8. Who am I?
   (a) My other name is sex cell.
   (b) I reduce the number of chromosomes in the daughter cells by half that of the parent cell.
   (c) I describe the number of chromosomes in normal human somatic cells.
   (d) I describe the fusion of gametes.
   (e) I describe the number of chromosomes in human gametes.
   (f) I am a type of cell division important for growth, repair and replacement.
9. Copy and complete the table below.

Type of cell division	Why use it?	Where does it occur?	Features of cells produced
Mitosis			
Meiosis			

10. With the use of a diagram, explain how the sex of a human baby is determined.
11. If a woman has already given birth to three boys, what are the chances of her next child being a girl?
12. In many cultures throughout history, a woman has been blamed for not producing sons and has been divorced. From a biological point of view, could this be justified? Explain your answer.
13. A few genetic traits, such as hairiness in ears, are due to genes carried on the Y chromosome. Would males and females have the same chance of having the trait?
14. Copy and complete the Venn diagram on the right, choosing from the following terms: somatic, only, body, gonads, gametes, anywhere, different, identical, chromosomes, cell division, eukaryotes.

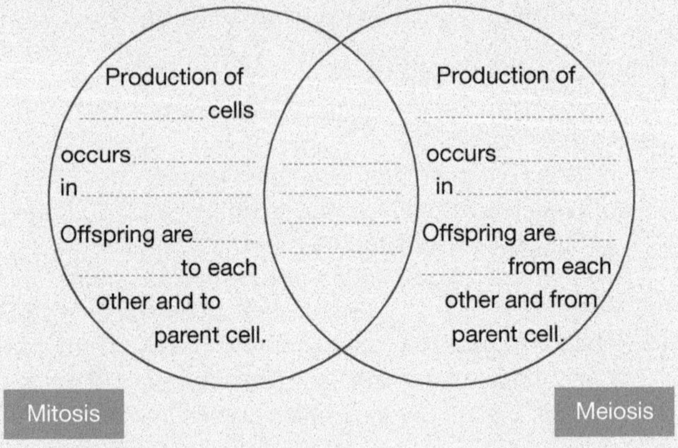

15. The Y chromosomes of human males are shorter than the X chromosomes. Would the same number of genes be carried by both chromosomes? Discuss your response.

Think, analyse and discuss

16. Figures (a)–(d) at right show bluebell cells in various stages of mitosis. Suggest which order they should be placed in.
17. Using the table below, suggest the possible effect of increasing global temperatures on turtles, crocodiles and lizards.

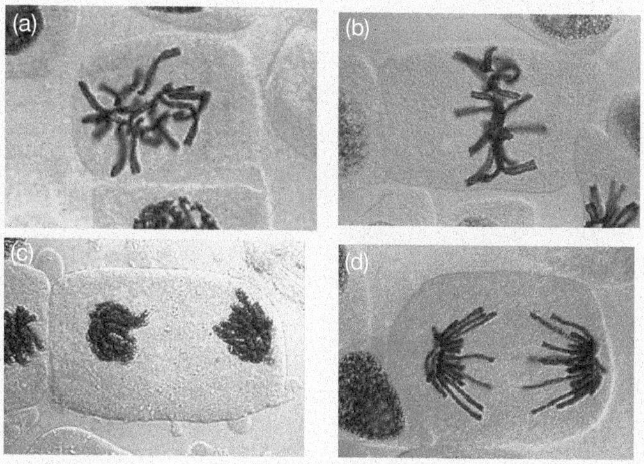

Temperature control of sex in some reptiles

Reptile	Cold 20–27 °C	Warm 28–29 °C	Hot > 30 °C
Turtle	Male	Male or female	Female
Crocodile	Female	Male	Female
Lizard	Female	Male or female	Male

Investigate

18. The kind of job a man does can affect whether he produces more or less Y sperm or any sperm at all. Chemicals and hormones washed into waterways or used in producing food can affect fertility. Research an example of an environmental impact on fertility and report your findings. Make sure you quote the sources of your information.

Remember

19. Complete the **Mitosis and meiosis** interactivity in your Resources tab to test your knowledge of the different processes of cell division, and challenge yourself to see if you can differentiate between mitosis and meiosis.

learnon RESOURCES — ONLINE ONLY

Complete this digital doc: Worksheet 2.3: Mitosis (doc-19408)

Complete this digital doc: Worksheet 2.4: Meiosis (doc-19409)

2.7 The next generation

2.7.1 From one generation to the next

Have you ever browsed through the family photo album and looked at family members at different ages? Did any look like you? Which features do you share with them? Do certain characteristics seem to appear and disappear from one generation to the next? How could this happen?

The passing on of characteristics from one generation to the next is called **inheritance**. The study of inheritance involves a branch of science called **genetics**. These characteristics or features are examples of your **phenotype**. Your phenotype is determined by both your **genotype** and your environment. Your genotype is determined by genetic information in the chromosomes that you received in the gametes of your parents.

2.7.2 It's not all about your genes!

Environmental factors contribute to characteristics that make up your phenotype. Your weight, for example, although influenced by genetic factors, is also influenced by what you eat and how active you are. Exposure to and use of chemicals in your environment (such as pollution, hair dyes, tanning lotions and make-up), stress, intensity of sunlight and temperature ranges are other examples of environmental factors that can contribute to your phenotype.

2.7.3 A product of chance

The similarities and differences in how you look compared to your relatives are partly due to chance. Chance was involved in which of the many sperm produced by your father fertilised your mother's ovum.

When fertilisation takes place, the zygote receives a pair of each set of chromosomes, the maternal and paternal chromosomes. Located within these chromosomes are the genes for particular characteristics. In the family generations diagram on the next page, the inheritance of the gene for eye colour is illustrated. There are two different eye colours shown. These alternative forms or expressions of a gene are called **alleles**.

INVESTIGATION 2.3

How does the environment affect phenotype?

AIM

1. Write an aim for this experiment.

Materials:

10 seedlings grown from cuttings of the same plant
potting mix in two small pots

Method and results

2. Copy and complete the following table:

	Pot A	Pot B
Number of seedlings that are still alive		
Colour of leaves		
Average height of seedlings		
Average number of leaves per seedling		

- Plant five of the seedlings in pot A and five in pot B.
- Place pot A in a dark cupboard and pot B near a window.
- Leave the plants undisturbed for two weeks. Water both pots when necessary. Ensure you use the same amount of water for both plants. After two weeks compare the plants in both pots.

Discuss

3. Explain how you calculated the average number of leaves and the average height of the seedlings.
4. In this experiment:
   (a) what is the independent variable
   (b) what is the dependent variable
   (c) which environmental factors were controlled?
5. Why is it important to use seedlings grown from cuttings of the same plant for this experiment?
6. Why were five seedlings planted in each pot?
7. Construct graphs of your data.
8. Comment on observed patterns in your data.
9. Explain why this experiment demonstrates that environmental factors play a part in determining the phenotype of an organism.

2.7.4 Hide and seek

The mixing of your parents' chromosomes at fertilisation resulted in two alleles for each gene coming together. Each of these alleles can be described using a letter. In the family generations diagram below, the expression of the gene for eye colour is shown. The allele for brown eyes is denoted as a capital letter *B*, because it is the **dominant** trait. The allele for blue eyes has been denoted by a lower case letter *b*, because this trait is **recessive** to the brown eye trait. If the allele for a dominant trait is present, it will always be expressed. The recessive trait is hidden in the presence of the dominant trait and can be expressed only if the allele for the dominant trait is not present.

You have a particular combination of alleles in your genotype.

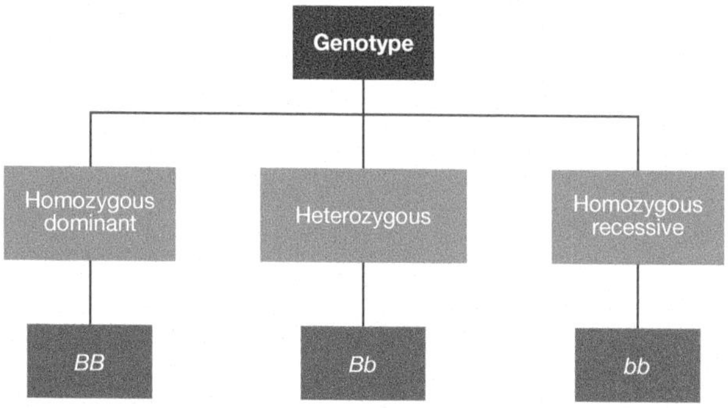

2.7.5 Mix and match

The combination of the alleles that you have for a particular gene is called your genotype. If your alleles for that gene are the same (e.g. *BB* or *bb*), then you are described as **homozygous** (or **pure breeding**) and

The letters *B* and *b* can be used to represent the genetic code for eye colour in the Davis and Swift families.

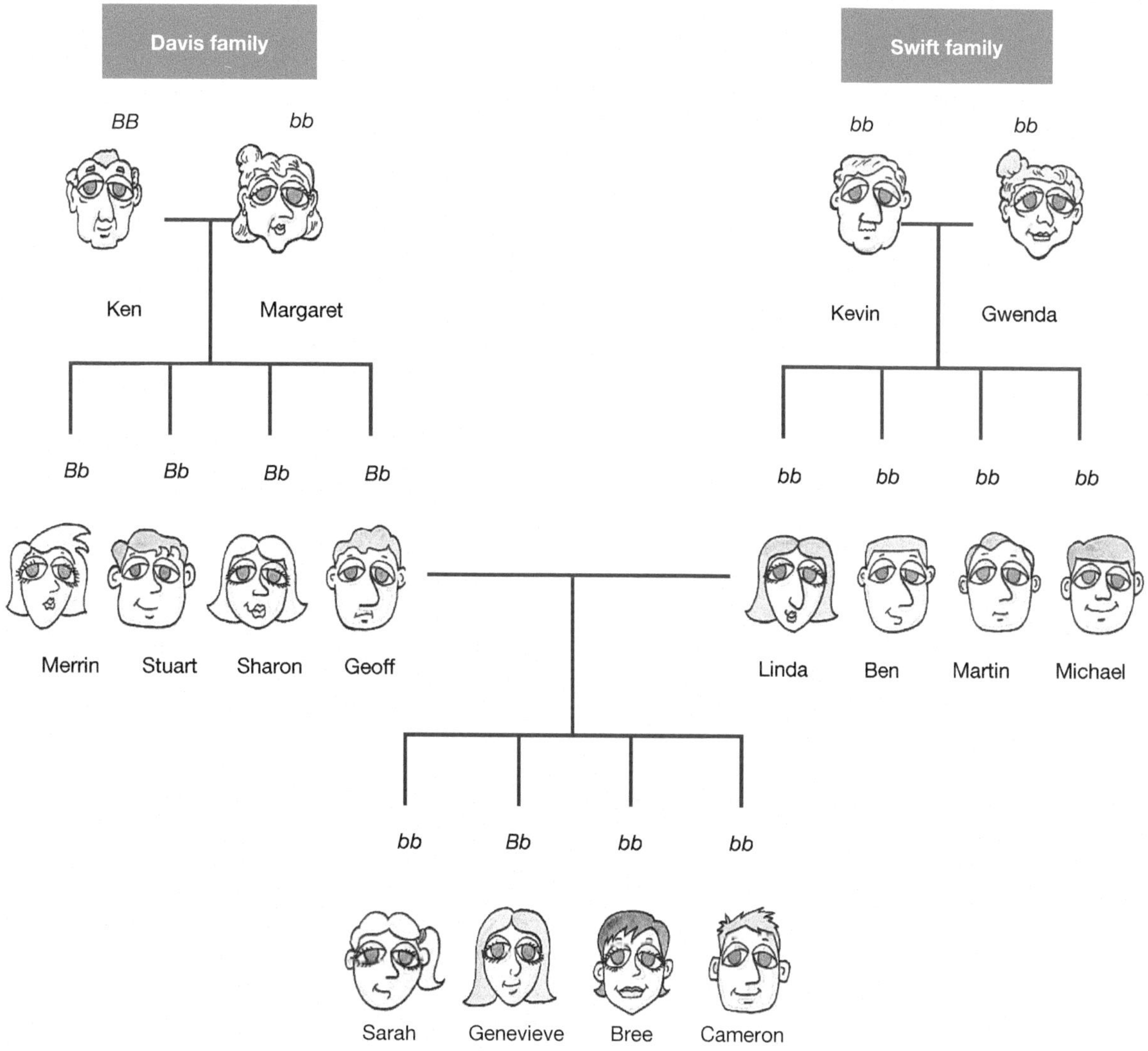

if they are different (e.g. *Bb*) then you are **heterozygous** (or **hybrid**) for that trait. A genotype of *BB* can be described as **homozygous dominant**, while a genotype of *bb* can be described as **homozygous recessive**. So the genotype has to do with the combination of alleles present; the phenotype describes the expression of the trait (e.g. brown or blue eyes).

INVESTIGATION 2.4

Genetics database

AIM: To increase awareness of a variety of inherited traits

Method and results

1. Copy and complete the table below. Enter data for 10 students in the table. You may need to refer to the pictures below to work out what each characteristic means.

Name of student				
Widow's peak?				
Can roll tongue?				
Right thumb over left when clasping hands?				
Cleft chin?				
Right handed?				
Ear lobes attached?				
Freckles?				
Gap between front teeth?				
Hair naturally straight?				
Colour blind?				

Do you have a widow's peak (left) or a straight hairline (right)?

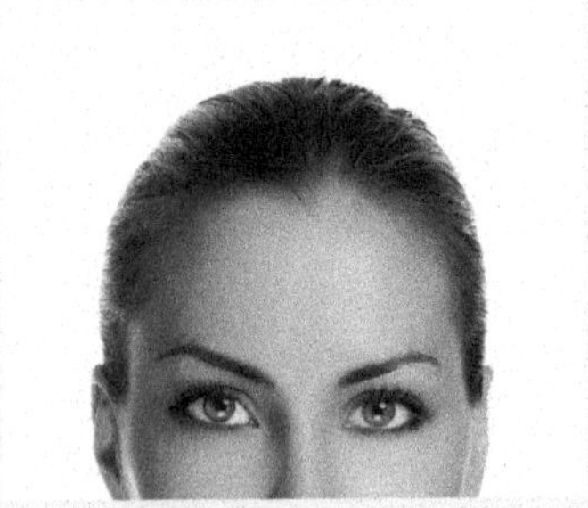

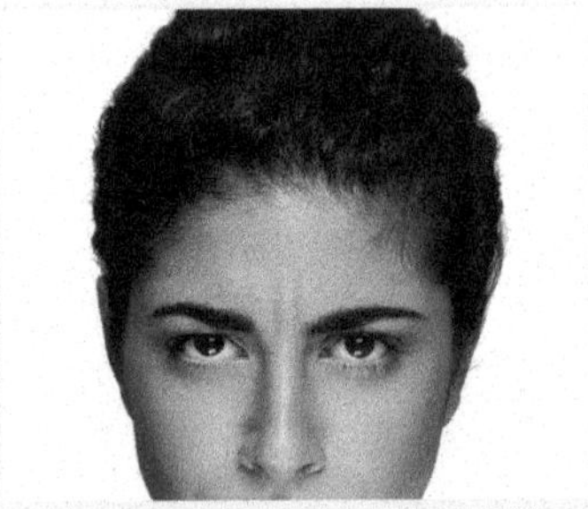

Do you have a gap between your front teeth?

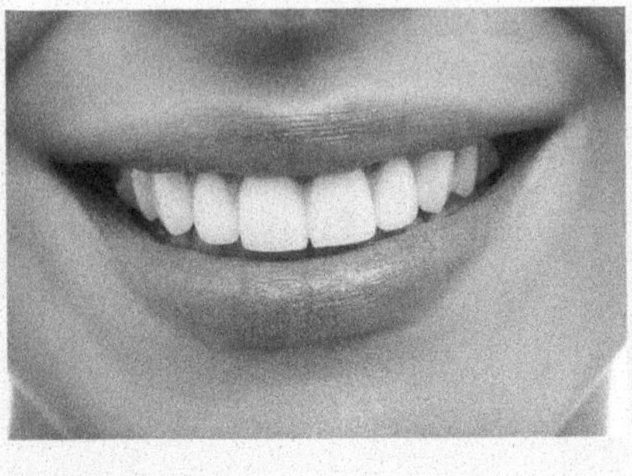

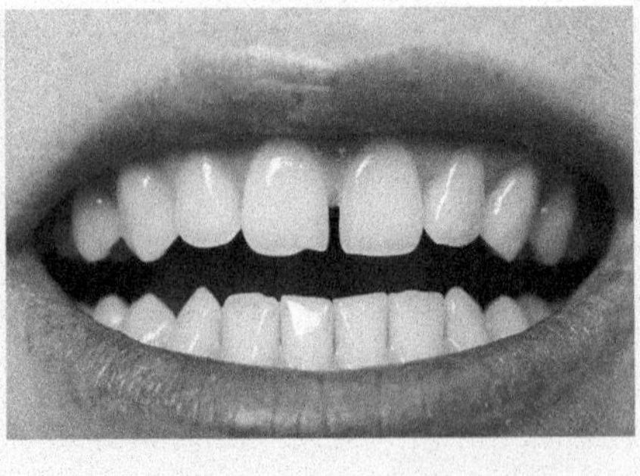

2. Use the instructions provided in your Resources tab to create an **Access** database where you will enter the data you collected and run a query on the database.

Discuss and explain

3. The database you created contains only a small amount of data so using a query to search for particular data did not save time. (It probably took you more time to set up the query than it would have taken to look through the data manually!) Can you think of examples of databases that contain so much information that it would take days to search the data manually?
4. Does your school keep a computerised database of student details? What type of information is kept in the database?

When you clasp your hands, is your right or left thumb on top?

Do you have a smooth or cleft chin (shown below)?

Are your ear lobes detached (left) or attached (right)?

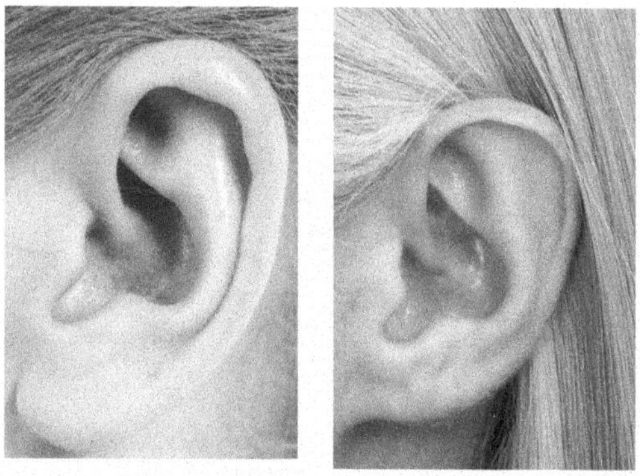

Can you roll your tongue?

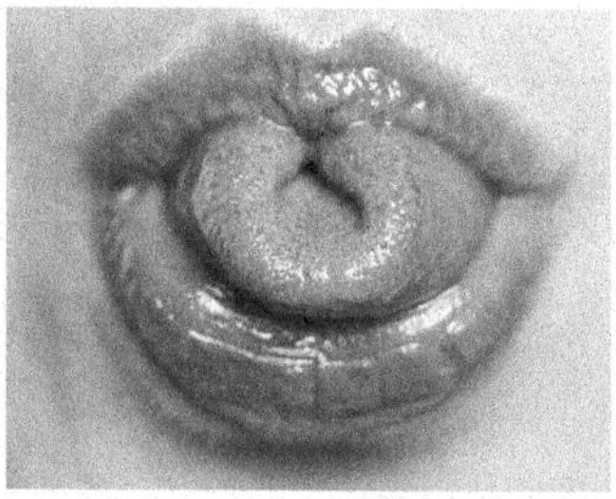

If you cannot see the number 47 in the picture below, you could be colour blind.

2.7.6 Are you a carrier?

The term **carrier** refers to someone who is heterozygous for a particular trait and carries the allele for the recessive trait (such as the alleles for blue eyes or red hair). Generally people are not aware of being a carrier because it does not show in their phenotype. They may, however, have children that show the recessive trait. Can you suggest how two brown-eyed parents (dominant trait) could have a child who has blue eyes (recessive trait)?

Alleles on chromosomes inherited from each of your parents contribute to your genotype.

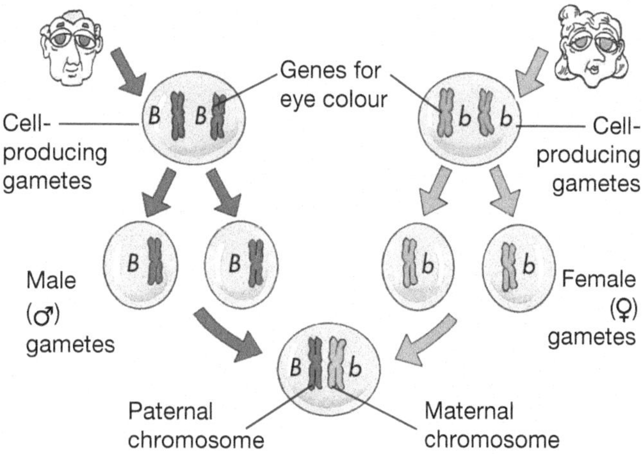

2.7.7 Degrees of dominance

In **complete dominance**, the expression of one trait is dominant over the other. This results in both the homozygous dominant and heterozygous genotypes being expressed as the same phenotype. There are two other types of inheritance in which neither allele is dominant over the other. In **codominance**, the heterozygote has the characteristics of both parents. An example of this type of inheritance is seen in the human blood groups. In **incomplete dominance**, the heterozygotes show a phenotype that is intermediate between the phenotypes of the homozygotes. An example of this type of inheritance is seen in the flower colour of snapdragons.

Incomplete confusion

You will find that there may be variations in the definitions of the terms recessive, dominant, codominance and incomplete dominance in various resources. New technologies and new knowledge can modify how we see, understand and communicate our knowledge. This eventually results in the creation, modification or replacement of terminology and theories that are used by a majority or enforced by those with the highest authority or persuasion.

The phenotype of the heterozygote can indicate the type of inheritance.

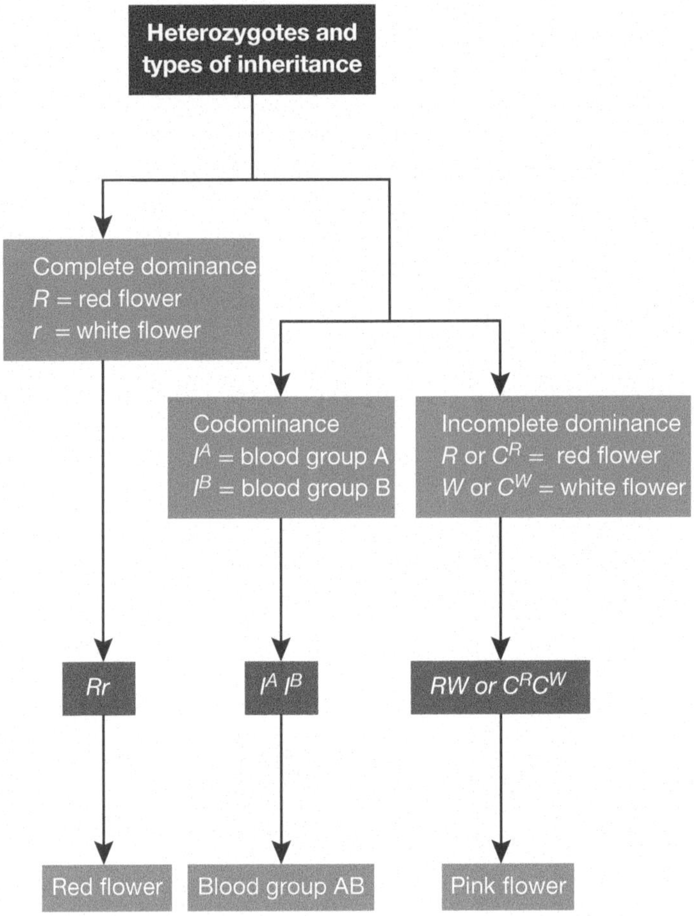

The inheritance of the human ABO blood groups is by codominance. The type of blood group you have determines who you can donate to or receive blood from. Which blood type are you? Are you the same blood type as either of your parents?

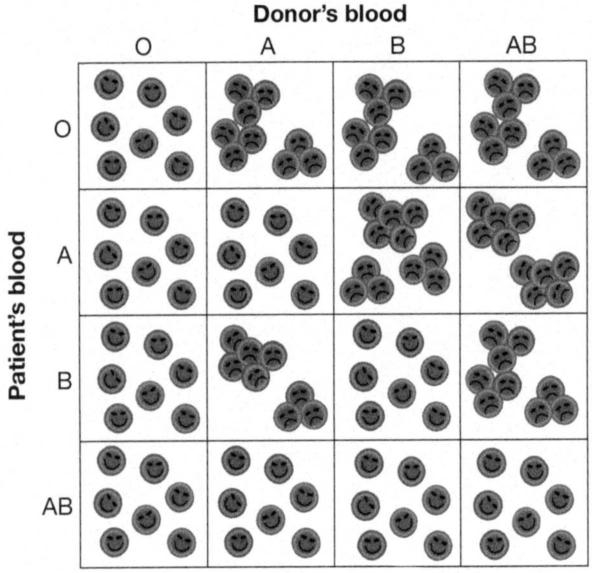

Suggest how red-, white- and pink-flowered offspring can result from pink-flowered parents.

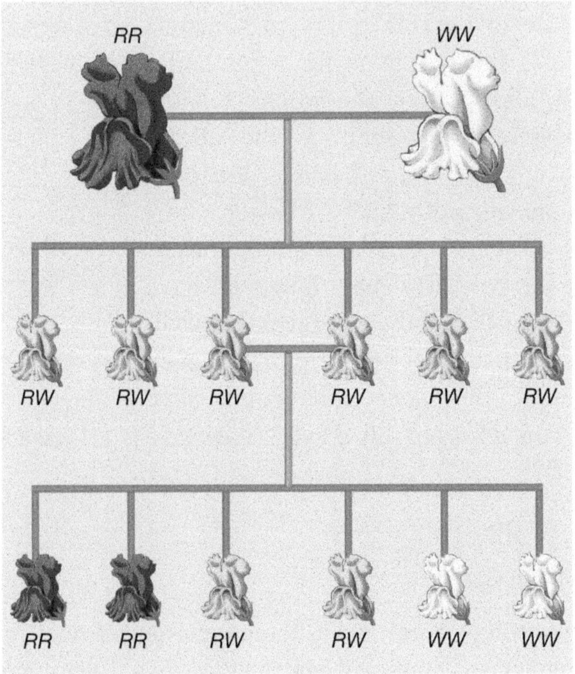

Both codominance and incomplete dominance can be considered examples of **partial dominance**. The common feature of these types of inheritance is that the **heterozygote** will show or express a phenotype that is different from the phenotype of an individual with either homozygous genotype.

Another area of confusion is with the terms recessive and dominant. Some resources abbreviate 'the allele for the recessive and dominant trait or phenotype' as 'recessive allele and dominant allele'. In terms of biology, it is increasingly accepted that the expression of the genotype as a particular phenotype is what is dominant or recessive, rather than the allele itself. If you continue on with senior Biology, it is important for you to check which of these definitions your authorities assess by.

Incomplete dominance can result in offspring that express a phenotype not observed in either parent.

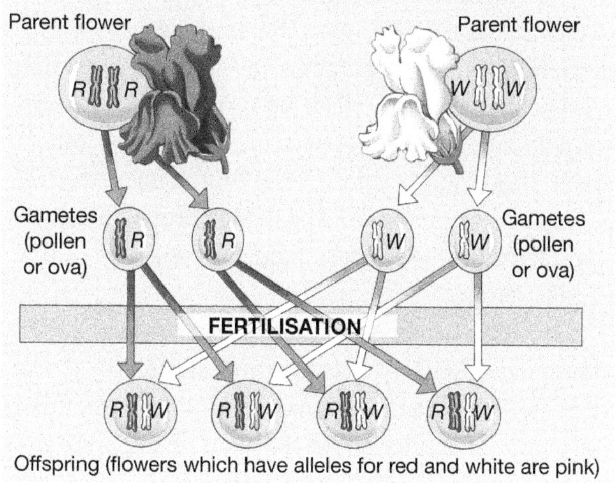

Offspring (flowers which have alleles for red and white are pink)

Gregor Mendel

2.7.8 Mendel's memos

Gregor Mendel (1822–1884), an Austrian monk, carried out experiments on pea plants in a monastery garden for 17 years. His work was unknown for about 35 years. When it was discovered in 1900, he became known as the 'father of genetics'. From his experiments, Mendel was able to explain patterns of inheritance of certain characteristics. Why did Mendel use pea plants and not cabbages? Pea plants are easily grown in large numbers and have easily identifiable characteristics that have either/or alternatives. Mendel could control their breeding by taking pollen from a particular pea plant and putting it on the stigma of another. Pea plants can also be self-pollinated.

Mendel's experiments were well designed and his record-keeping was meticulous.

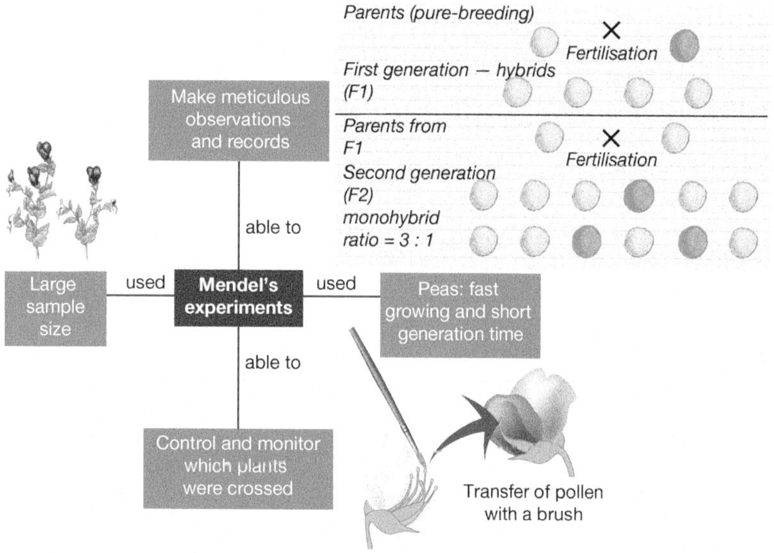

Mendel crossed a pure-breeding tall plant with a pure-breeding short plant. A plant is pure breeding for a characteristic if it has not shown the alternative characteristic for many generations.

Mendel showed the factor for shortness had not disappeared because when he crossed the tall offspring (called the F_1 generation) with each other, about a quarter of those offspring (called the F_2 generation) were short. He called shortness a recessive factor because it was hidden or masked in the F_1 generation.

A plant is a hybrid if it has parents with both alternatives, such as tallness and shortness, for a characteristic. We now know that Mendel's 'factors' are genes. The alternative forms of the factors are alleles.

Mendel bred plants for single characteristics such as height. He worked out that if many pure-breeding tall and short plants were crossed and then the first generation (F_1 generation) was also crossed, the ratio of tall to short plants would be about 3 : 1. He repeated these experiments many times using the other characteristics of the pea plants and came up with similar ratios. This is called the **monohybrid ratio**.

Pea plants showing the characteristics Mendel used in his experiments

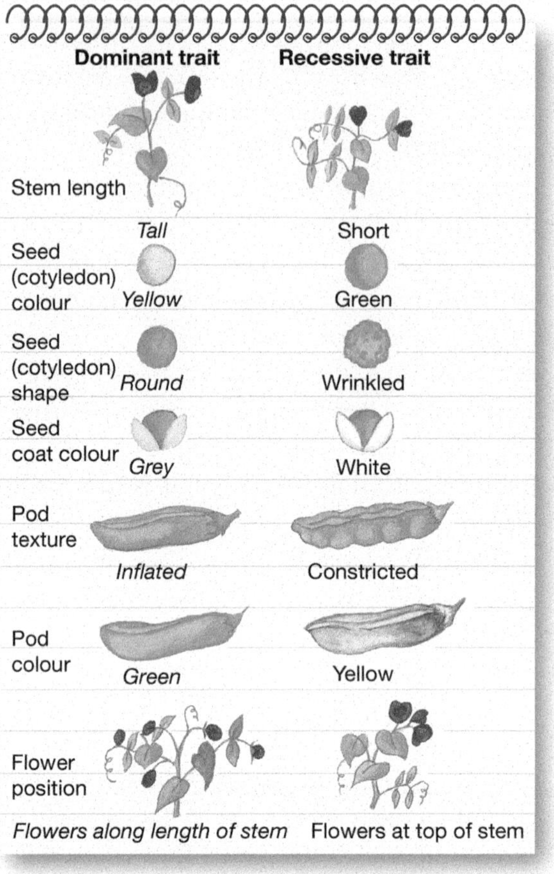

2.7 Exercises: Understanding and inquiring

To answer questions online and to receive **immediate feedback** and **sample responses** for every question, go to your learnON title at www.jacplus.com.au. *Note:* Question numbers may vary slightly.

Remember

1. Distinguish between the following pairs of terms. You may wish to use symbols, visual thinking tools or diagrams in your responses.
   (a) Inheritance and genetics
   (b) Genotype and phenotype
   (c) Chromosomes and gametes
   (d) Fertilisation and zygote
   (e) Genes and alleles
   (f) Dominant trait and recessive trait
   (g) Homozygous and heterozygous
   (h) Homozygous dominant and homozygous recessive
   (i) Carrier and recessive trait
   (j) Dominance and codominance
   (k) Maternal and paternal chromosomes
   (l) Pure breeding and hybrid
2. Describe the difference between the phenotypes of a heterozygous individual for a trait that shows complete dominance and for a trait that shows codominance.
3. State how many alleles there are on a homologous pair of chromosomes for a particular trait. Provide a reason for your response.
4. Suggest four strengths in the design of Mendel's experiments.

Think and discuss

5. Suggest why the ability to self-pollinate and cross-pollinate was an advantage for the pea plants in Mendel's experiments.
6. (a) Suggest the phenotype of a pea plant that showed:
   (i) dominant traits for seed colour, shape and coat colour
   (ii) recessive traits for stem length and flower position
   (iii) a dominant trait for pod texture, but a recessive trait for pod colour.
   (b) Devise a simple table to include the phenotype and genotype for the trait of each plant. Use an appropriate letter to match each of the characteristics Mendel studied.
7. Mendel obtained a ratio of 3 tall to 1 short plant in the offspring when he crossed pure-breeding tall and short plants. Convert this monohybrid ratio of 3 : 1 into a:
   (a) fraction
   (b) percentage.
8. Suggest the colour(s) of snapdragon flowers that you would expect in the offspring of a red-flowered plant and a pink-flowered plant.
9. Refer to the Davis family tree diagram top right to answer the following questions.
   (a) Could the parents of the Davis family, Ken and Margaret, ever have offspring with blue eyes? Explain your answer.
   (b) Suggest why all of Geoff and Linda's children do not have blue eyes.
10. Copy and complete the Venn diagram at right using the following terms: dominant phenotype, brown eyes, both phenotypes, blood type A, blood type B, blue eyes.
11. Construct a Venn diagram with the headings 'Determined by genetics' and 'Determined by environmental factors', and then place the following terms in the most appropriate category: eye colour, tattoo, skin colour, cleft chin, freckles, colour blind, hair colour, scar, widow's peak.

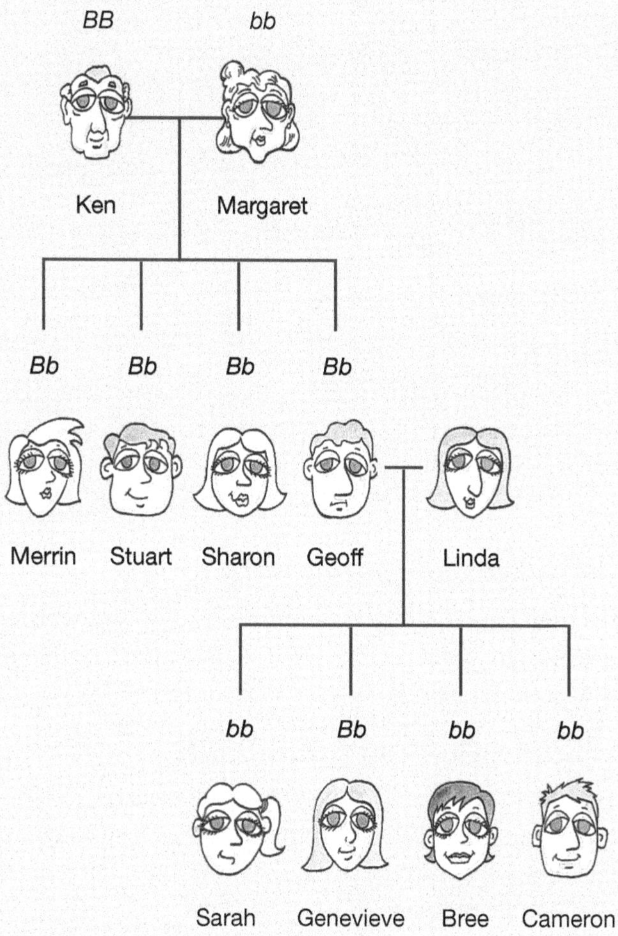

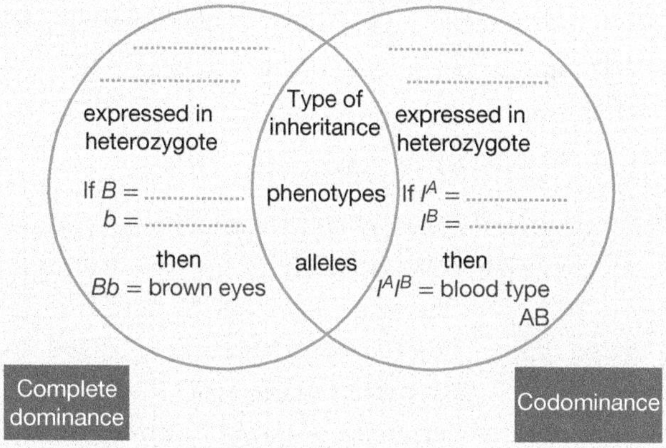

Think, discuss and create

12. On your own, in pairs or in teams, create a rhyme, song or poem that effectively uses as many of the key terms in this subtopic as possible. An example is shown below. Add movements or actions for each line and share it with your class.

Alleles are alternative forms of genes
Sometimes showing, sometimes behind the scenes
Genotypes are made up of two of them
Homozygotes have two the same
Heterozygotes have one of each kind
From each parent, alleles you will find.

Investigate

13. Investigate one of these genetic conditions: Huntington's disease, Tay–Sachs, cystic fibrosis, fragile X syndrome, PKU. Briefly describe the disease. Is it dominant or recessive?

learnon RESOURCES — ONLINE ONLY

Complete this digital doc: Worksheet 2.5: Dominant and recessive (doc-19412)

Complete this digital doc: Worksheet 2.6: Mendel's experiments (doc-19413)

2.8 What are the chances?

2.8.1 Mixing your genes?

Selecting a mate can be one of the most crucial decisions in your life. This selection process involves both conscious and unconscious choices. Next time you look at that special person, take a *really* good look. One day you might be mixing your genes together! What might the result be?

2.8.2 Predicting possibilities

Reginald Punnett (1875–1967) was a geneticist who supported Mendel's ideas. He repeated Mendel's experiments with peas and also did his own genetic experiments on poultry. Punnett is responsible for designing a special type of diagram, which is named after him. A **Punnett square** is a diagram that is used to predict the outcome of a genetic cross.

A Punnett square shows which alleles for a particular trait are present in the gametes of each parent. It then shows possible ways in which these can be combined. The alleles in each of the parent's genotypes for that trait are put in the outside squares and then multiplied together to show the possible genotypes of the offspring.

2.8.3 Punnett rules

When using a Punnett square for a dominant/recessive inheritance, you use a capital letter for the allele of the dominant trait (e.g. *B*) and a lower-case version of the same letter for the allele for the recessive trait (e.g. *b*). If the type of inheritance is incomplete or codominant, then different letters are used to represent them (e.g. *R* and *W* or I^A and I^B). The sex chromosomes are included when an **X-linked trait** is involved (e.g. X^B X^b and X^B Y).

In a Punnett square, alleles from each parent's genotype are used to determine the possible genotypes and phenotypes of the offspring.

Punnett square for *Bb* × *Bb*

B = allele for brown eyes

b = allele for blue eyes

Father

Possible gametes (Mother)	*B*	*b*
B	*BB*	*Bb*
b	*Bb*	*bb*

Offspring probabilities

Genotype: $\frac{1}{4}$*BB*: $\frac{1}{2}$*Bb*: $\frac{1}{4}$*bb*

Phenotype: $\frac{3}{4}$brown eyes: $\frac{1}{4}$blue eyes

2.8.4 What is the chance?

The chances of having offspring that show a particular trait depends on the trait's type of inheritance; that is, whether it is inherited by complete dominance, codominance, incomplete dominance or **sex-linked inheritance**.

2.8.5 Complete dominance

Remember Linda Swift and Geoff Davis from subtopic 2.7? The inheritance of eye colour was shown in their family. Inheritance of brown eyes was dominant to the inheritance of blue eyes. The diagrams below show that Geoff has brown eyes (*Bb*), Linda has blue eyes (*bb*) and their children have either brown or blue eyes.

The inheritance of blue eyes is recessive to the inheritance of brown eyes.

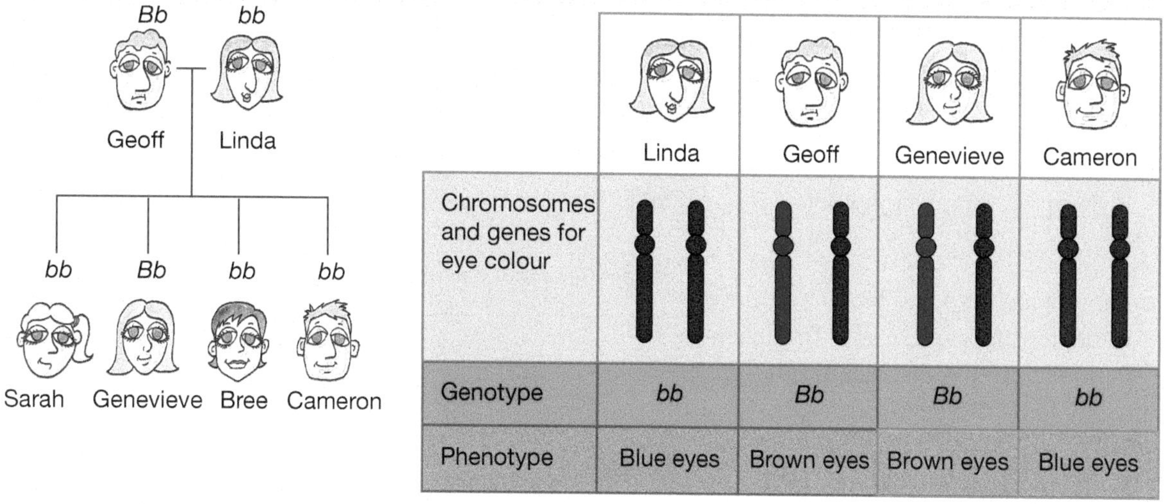

But how were the alleles from each parent inherited? The mix of alleles that Linda and Geoff contributed to Genevieve and Cameron's genetic make-up can also be shown below.

This can be more simply written as a Punnett square.

Each parent contributes alleles to the genotype of their offspring.

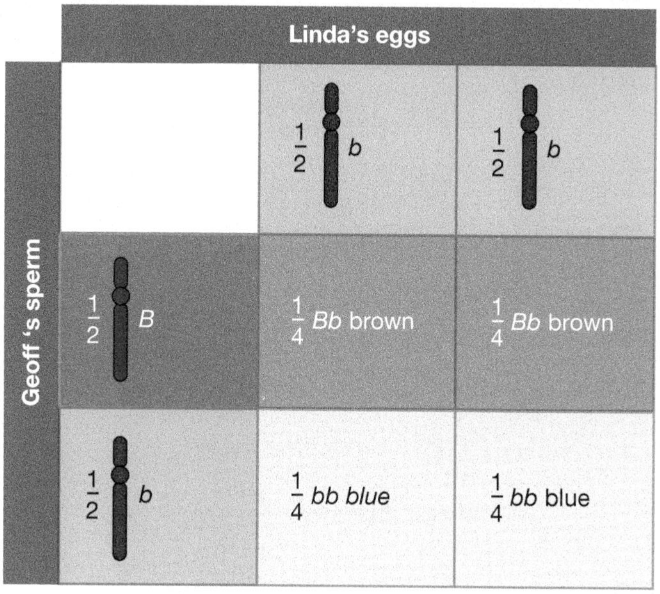

Punnett squares show us the chance of offspring inheriting particular combinations.

Punnett square *Bb* × *bb*

B = allele for brown eyes

b = allele for blue eyes

Geoff

Linda

Possible gametes	*B*	*b*
b	*Bb*	*bb*
b	*Bb*	*bb*

Offspring probabilities

Genotype $\frac{1}{2}$ *Bb*: $\frac{1}{2}$ *bb*

Phenotype $\frac{1}{2}$ brown eyes $\frac{1}{2}$ blue eyes

This tells us that the chance of producing a child with the combination *Bb* (heterozygote) is 2 out of 4 or $\frac{1}{2}$, and the chance of producing the combination *bb* (homozygous recessive) is 2 out of 4 or $\frac{1}{2}$. Hence, each of Linda and Geoff's children have a 50 per cent chance of having blue eyes and a 50 per cent chance of having brown eyes. All of their children could have had blue eyes or all brown eyes. It is important to note that the chance of inheritance calculated for one child is not dependent on the inheritance of another.

2.8.6 Pedigree charts

A diagram that shows a family's relationships and how characteristics are passed on from one generation to the next is a **pedigree chart**. A pedigree chart for Linda and Geoff's family is shown at right. Instructions on how to draw your own pedigree chart are provided below right.

2.8.7 What is your blood type?

Do you know which type of blood you have flowing through your capillaries? The inheritance of blood types A, B, AB and O are determined by the ABO gene.

Multiple alleles

There are three different alleles for the ABO gene. Two of these carry instructions to make a particular type of protein called an **antigen**; the other does not. The types of antigens coded for by the alleles are different. One allele codes for antigen A and the other codes for antigen B. If you possess both of these alleles, then you have the instructions to produce both antigen A and antigen B. This is an example of **codominant inheritance** because both blood types are expressed in the heterozygote.

If you refer to the ABO gene as I, then the allele that codes for:

- antigen A could be referred to as I^A
- antigen B could be referred to as I^B
- neither antigen could be referred to as i.

The ability to make antigen A or B is shown as a capital letter because it is dominant to the inability to make either antigen (which is recessive and shown as a lowercase letter).

There is an element of chance in why you are you!

Linda   Geoff

Genevieve   Cameron

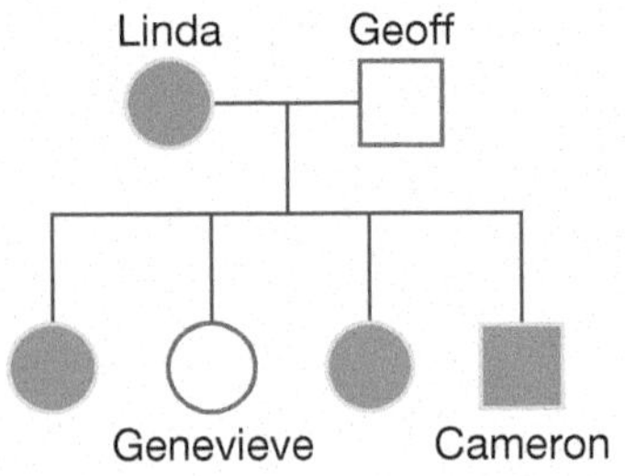

Key

⬤	= Female with blue eyes	◯	= Female without trait
⬛	= Male with blue eyes	☐	= Male without trait

SOME RULES FOR DRAWING PEDIGREE CHARTS

1. To show the gender of an individual:

☐ a square is used to represent a male

◯ a circle is used to represent a female.

2. To show the marriage or breeding relationship between individuals:

☐—◯ a line connecting the male and female is used to represent a breeding couple or marriage.

3. To show the offspring relationships:

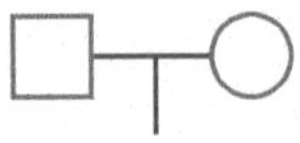

☐—◯ a line from the breeding couple/marriage line indicates children.

For example,

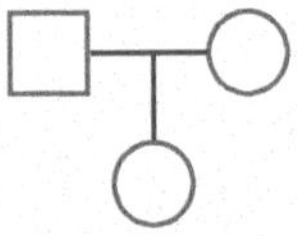

an only child (in this case, a daughter)

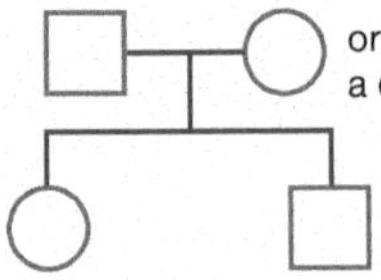

or two children (in this case, a daughter and son).

4. To show carriers of traits, the symbol may have a dot.

⊙ Female carrier   ▪ Male carrier

It is important to note, however, that carriers' symbols are not always dotted and may appear blank.

5. To show which individuals show a particular trait, an individual's symbol is shaded and this information is shown in a key next to the pedigree chart.

⬤ Female with trait   ⬛ Male with trait

◯ Female without trait   ☐ Male without trait

Genotype and phenotype of blood groups

Genotype	Phenotype
$I^A I^A$	Blood type A
$I^A i$	Blood type A
$I^B I^B$	Blood type B
$I^B i$	Blood type B
$I^A I^B$	Blood type AB
ii	Blood type O

Family blood

Can you have a blood type different from both of your parents? The answer is yes! The pedigree chart at right shows Tom (blood type O) and his wife Mallory (blood type AB) and their four children. A Punnett square can be used to predict the blood types that are possible for their children. This calculates that each child has a 50 per cent chance of inheriting blood type A or blood type B — blood types that neither parent possesses. What blood types do their children show in the pedigree chart at right? Can you suggest why $\frac{3}{4}$ of the children have blood type A, when their chance of inheriting it was $\frac{1}{2}$?

Depending on their inheritance of particular alleles, children can have different blood types from their parents.

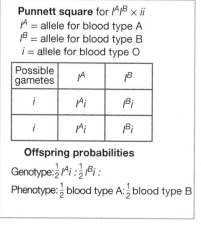

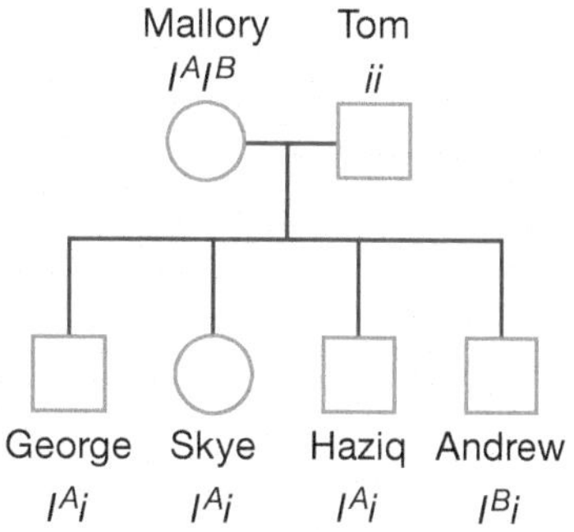

Punnett square for $I^A I^B \times ii$

I^A = allele for blood type A

I^B = allele for blood type B

i = allele for blood type O

Possible gametes	I^A	I^B
i	$I^A i$	$I^B i$
i	$I^A i$	$I^B i$

Offspring probabilities

Genotype: $\frac{1}{2} I^A i : \frac{1}{2} I^B i$:

Phenotype: $\frac{1}{2}$ blood type A: $\frac{1}{2}$ blood type B

2.8.8 Sex-linked inheritance

The genes that have been considered so far have been those on autosomes. These types of genes show **autosomal inheritance**. The genes for some traits, however, are located on sex chromosomes. These traits are referred to as being sex-linked, and the type of inheritance is called X-linked if they are located on the X chromosome and Y-linked if they are located on the Y chromosome. Because of the small size of the Y chromosome it does not contain many genes, and most examples of sex-linkage that you will come across will be those of X-linkage.

2.8.9 The X-files

Haemophilia and some forms of colour blindness are examples of X-linked recessive traits. This means that females need to receive two alleles for the recessive trait, whereas males need to receive only one. This is why there is a greater chance of males showing these traits than females.

The genotype for X-linked traits includes the sex chromosomes in its description. For example, females may be heterozygous, $X^B X^b$, or homozygous, $X^b X^b$ or $X^B X^B$. Males, who possess only one X chromosome, are hemizygous and would have the genotypes $X^B Y$ or $X^b Y$. When stating the phenotypes for X-linked traits it is important to also specify the person's gender (e.g. colour blind male).

2.8.10 Colour blindness

In the pedigree chart on the next page, Chris is colour blind and, although Heather carries the colour blindness allele, she also has the allele for normal vision and so does not show this X-linked recessive trait. The Punnett square shows the probabilities of their children inheriting colour blindness. Which children are colour blind? What were their chances of inheriting colour blindness?

Being colour blind is an X-linked recessive trait.

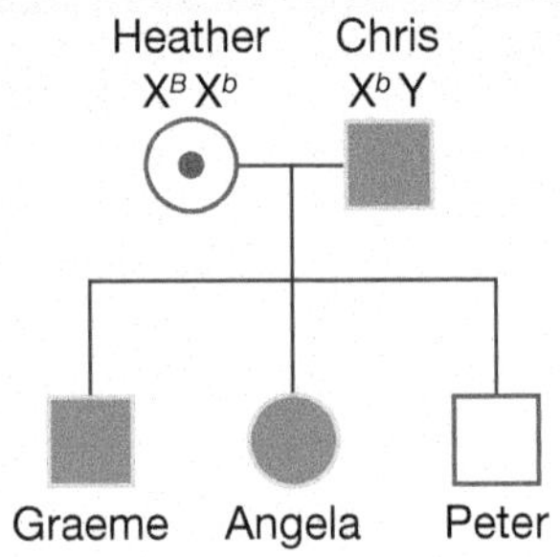

Key

● = Colour blind female	■ = Colour blind male

When writing out the phenotype of an X-linked trait, it is important to also show the gender of the individual.

Punnett square $X^BX^b \times X^bY$

X^B = allele for normal vision

X^b = allele for colour blindness

Possible gametes	X^B	X^b
X^b	X^BX^b	X^bX^b
Y	X^BY	X^bY

Offspring probabilities

Genotype: $\frac{1}{4}$ X^BX^b: $\frac{1}{4}$ X^bX^b: $\frac{1}{4}$ X^BY: $\frac{1}{4}$ X^bY

Phenotype: $\frac{1}{4}$ normal vision female: $\frac{1}{4}$ colour blind female: $\frac{1}{4}$ normal vision male: $\frac{1}{4}$ colour blind male

2.8 Exercises: Understanding and inquiring

To answer questions online and to receive **immediate feedback** and **sample responses** for every question, go to your learnON title at www.jacplus.com.au. *Note:* Question numbers may vary slightly.

Remember

1. Describe the function of a Punnett square.
2. Provide an example of a Punnett square.
3. Outline the differences between the symbols used to identify the alleles in dominant/recessive inheritance, codominance and sex-linked inheritance.
4. Describe the function of a pedigree chart.
5. In a pedigree chart, what do the circles and squares represent?
6. With regard to human blood type inheritance, identify:
   (a) the gene involved
   (b) four possible blood types
   (c) three possible alleles
   (d) the type of inheritance
   (e) the genotype of an individual with blood type O
   (f) the genotype of an individual with blood type AB
   (g) the phenotype of an individual with genotype I^AI^A
   (h) the phenotype of an individual with genotype I^Bi.
7. Distinguish between:
   (a) autosomal inheritance and sex-linked inheritance
   (b) X-linked and Y-linked inheritance
   (c) X-linked recessive and X-linked dominant traits.
8. Suggest why males have a greater chance of showing an X-linked recessive trait than females.

Analyse, think and discuss

9. Predict the probabilities of the phenotypes and genotypes of the offspring of:
   (a) a homozygous brown-eyed parent and a blue-eyed parent
   (b) two parents heterozygous for brown eyes.
10. Refer to Chris and Heather's family pedigree chart and information on their inheritance of colour blindness on this page.
   (a) State the genotype for:
      (i) Chris
      (ii) Heather.

(b) State the phenotype for:
  (i) Chris
  (ii) Heather.
(c) What is the chance of:
  (i) Graeme having colour blindness
  (ii) Peter having colour blindness
  (iii) Angela having colour blindness?
(d) Is it possible for Peter to have a child who is colour blind? Explain.

11. State the genotype of the following individuals.
  (a) Heterozygous for blood type A
  (b) Homozygous for blood type B
  (c) Blood type O
  (d) Blood type AB

Punnett square for *BB* × bb
B = allele for brown eyes
b = allele for blue eyes

	B	*B*
b		
b		

Offspring probabilities
Genotype: ....
Phenotype: ....
....

Punnett square for *Bb* × *Bb*
B = allele for brown eyes
b = allele for blue eyes

	B	*b*
B		
b		

Offspring probabilities
Genotype: ....
Phenotype: ....
....

12. If a man who was homozygous for blood type A ($I^A I^A$) had a child with a woman who had blood type O (*ii*), what is the chance that the child would have:
  (a) blood type A
  (b) blood type B
  (c) blood type O?
13. If a child had blood type AB, suggest the possible combinations of genotypes of the parents.
14. Determine the chance of a couple with blood types AB and A having a child with:
  (a) blood type A
  (b) blood type B
  (c) blood type AB
  (d) blood type O.
15. Can a father with blood type A and a mother with blood type B have a child with blood type O? Explain.
16. What is the chance of Linda and Geoff, described on page 43, producing a child with the homozygous dominant combination (*BB*)?
17. Refer to the pedigree of the family in the diagram at right. The inheritance of broad lips (*B*; unshaded individuals) is dominant to the inheritance of thin lips (*b*; shaded individuals).
  (a) How many females are shown in the pedigree chart?
  (b) How many males are shown in the pedigree chart?
  (c) How many females have the thin lips trait?
  (d) Suggest the genotype of Maggy's parents.
  (e) Suggest how Maggy inherited thin lips, when her parents did not.
  (f) Suggest the genotypes of (i) Peter, (ii) Kurt, (iii) George and (iv) Rebecca.

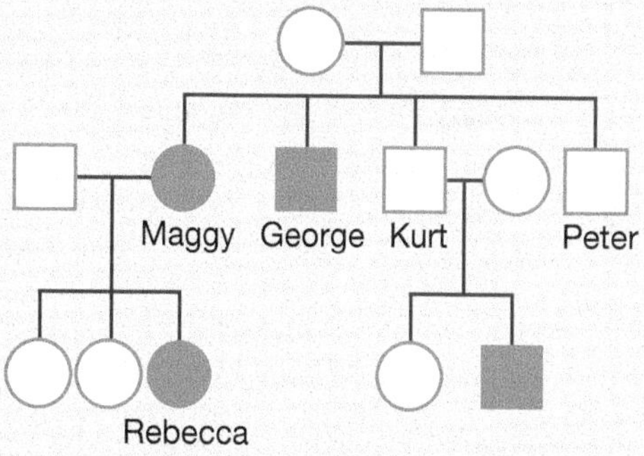

18. The pedigree at right traces the recessive trait of albinism in a family. The shaded individuals lack pigmentation and are described as being albinos.
  (a) List any observations from the pedigree that support albinism being a recessive trait.
  (b) If the albinism allele was represented as *n* and normal skin pigmentation as *N*, state the possible genotypes for each of the individuals in the pedigree.

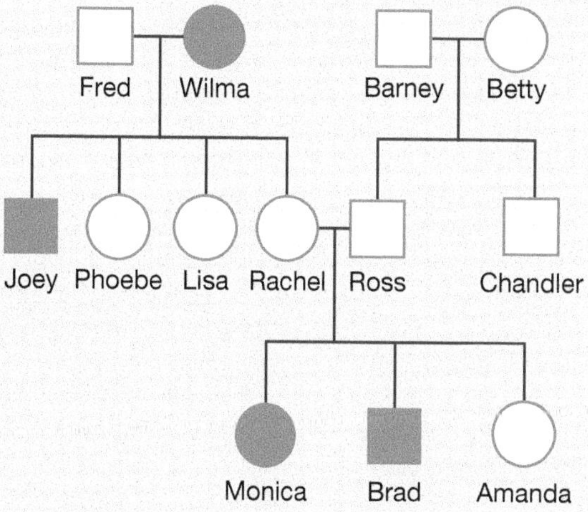

19. The pedigree at right traces the dominant trait, a widow's peak, in a family.
    (a) List any observations from the pedigree that support the widow's peak being a dominant trait.
    (b) If the widow's peak allele is represented as *W* and the straight hairline as *w*, state the possible genotypes for each of the individuals in the pedigree.
    (c) If Jo and Mark were to have another child, what is the chance of it having a widow's peak?
    (d) If Ronnie were to have a child with a man who did not have a widow's peak, what is the probability that their child would have a widow's peak?
    (e) If Norm and Rona were to have another child, what is the probability that they would have a child without a widow's peak?

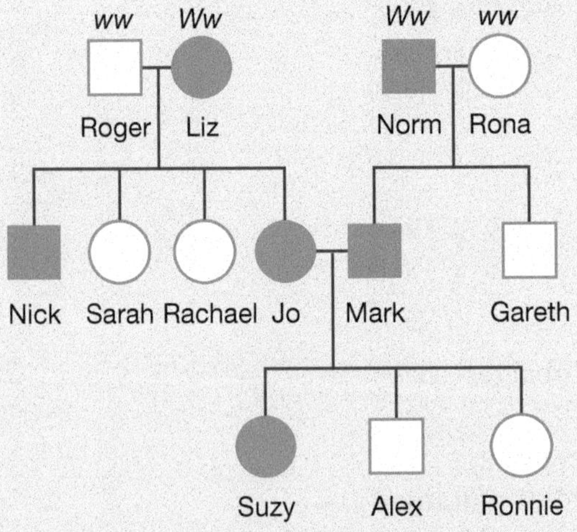

20. Use the dominant and recessive table below and Punnett squares to assist you in answering the following questions.

Dominant trait	Recessive trait
Free ear lobes	Attached ear lobes
Mid-digital hair present	Mid-digital hair absent
Normal skin pigmentation	Pigmentation lacking (albinism)
Non-red hair	Red hair
Rhesus-positive (Rh +ve) blood	Rhesus-negative (Rh –ve) blood
Dwarf stature (achondroplasia)	Average stature
Widow's peak	Straight hairline

(a) Find the probability of Sally (who is homozygous for dwarf stature) and Tom (who has average stature) having a child with dwarf stature.
(b) Find the probability of Fred (who is heterozygous for dwarf stature) and Susy (who has average stature) having a child with dwarf stature.
(c) What is the chance of two parents who are both heterozygous for free ear lobes having a child with attached ear lobes?
(d) Michael is heterozygous for mid-digital hair, whereas Debbie does not have mid-digital hair. What is the chance of their children having mid-digital hair?

Investigate, think and discuss

21. What are some physical attributes of males that suggest sexual potency and good genes?
22. Suggest what the major histocompatibility complex has to do with mate selection.
23. (a) What is sexual selection? Give two examples.
    (b) How is sexual selection different from natural selection?
    (c) Suggest implications of sexual selection for our species.
    (d) Suggest the possible impact of sexual selection on your future reproductive life.
24. While the science of love is still in its infancy, advances in molecular biology and technology have increasingly allowed us to peer through its window.
    (a) Find out examples of research on the chemistry of love or love potions.
    (b) Do you believe that this research should be continued? Give reasons.
    (c) Suggest possible issues that may arise with the knowledge obtained and its possible applications.
    (d) Discuss if, how and who should regulate or control this type of research.
25. Increasing numbers of people are finding love and their partners on the internet.
    (a) What is your opinion on this?
    (b) Discuss your opinion with others in your team.
    (c) Discuss the use of internet dating from biological, cultural, social and ethical viewpoints, and construct a PMI chart to summarise your discussion.

learnon RESOURCES — ONLINE ONLY

Complete this digital doc: Worksheet 2.7: Pedigrees (doc-19414)

2.9 Changing the code

2.9.1 Oops!

Errors or changes in DNA, genes or chromosomes can have a variety of consequences. These genetic mistakes are called **mutations**.

Polydactyly (having more than 10 fingers and toes) is usually due to a DNA mutation.

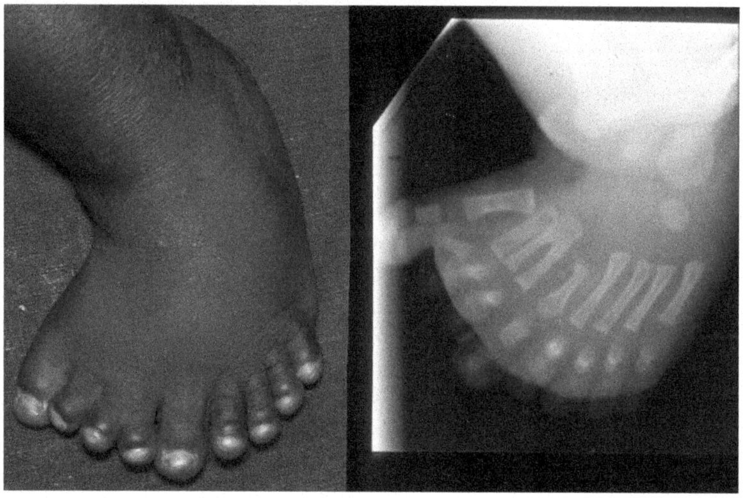

2.9.2 DNA replication

DNA is very stable and can be replicated into exact copies of itself. This process is called **DNA replication** and enables genetic material to be passed on unchanged from one generation to the next. DNA replication begins with the 'unzipping' of the paired strands. A new complementary DNA strand is made for each original DNA strand. This results in the formation of two new double-stranded DNA molecules, each containing one new DNA strand and one original DNA strand. This model of DNA replication is called the **semi-conservative model** because it has conserved one of the old DNA strands in each new double-stranded DNA molecule.

The process of DNA replication has a number of checkpoints to test for any mistakes that may be made, so that they can be corrected or destroyed. Sometimes, however, the mistakes get through this screening process. When this happens, we say that a mutation has occurred.

DNA replication is semi-conservative. The new DNA has one old and one new DNA strand.

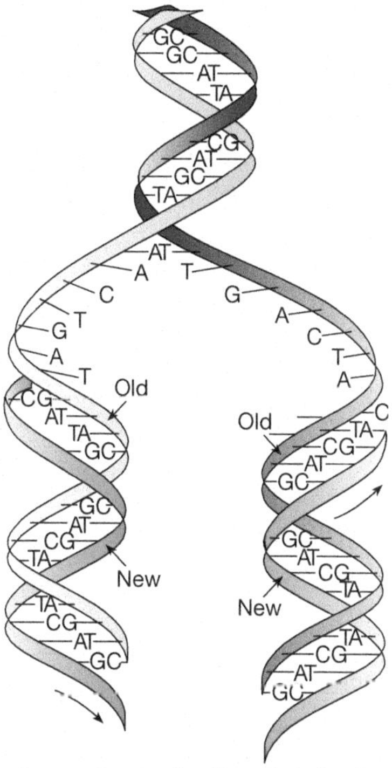

Arrows denote direction of synthesis.

2.9.3 Mutagenic agents

Mutations can happen by chance or have a particular cause. When the cause of the mutation cannot be identified it is called a **spontaneous mutation**, and when the cause can be identified it is referred to as an **induced mutation**.

A factor that triggers mutations in cells is called a **mutagen** or **mutagenic agent**. Examples of mutagenic agents include radiation (e.g. ultraviolet radiation, nuclear radiation and X-rays) and some chemical substances such as formalin and benzene (which used to be common in pesticides).

As a result of the thinning of the ozone layer in the atmosphere, we are exposed to increasing amounts of **UVB radiation** that can damage (or mutate) our DNA. This can lead to the development of skin cancers. Protective clothing and sunscreens can help reduce our exposure to this dangerous, potentially mutagenic environmental radiation.

learn on RESOURCES — ONLINE ONLY

Watch this eLesson: DNA and Hiroshima (eles-1781)

2.9.4 Errors in the code

Many important hormones and enzymes are made up of protein. Changes in the genetic code due to mutations may result in a particular protein not being made or a faulty version being produced. In some cases, the production of an essential enzyme may be impaired, disrupting chemical reactions and resulting in the deficiencies or accumulation of other substances. This may cause the death of the cell and, eventually, the organism.

2.9.5 Point mutations

Occasionally, errors can occur during DNA replication as DNA is being copied. This means that the instructions carried by the code are not followed exactly. This may be the result of an incorrect pairing of bases; the **substitution** of a different nucleotide; or the **deletion** or **insertion** of a nucleotide. Such a **point mutation** can change the genetic message by coding for a different amino acid, leading to the production of a different or non-functional protein. This can have consequences to the phenotype of the organism.

Just like changing letters in a word can change its meaning, changes in the DNA sequence can change the meaning of the genetic code.

Deletion:
I will send a *fr* iend to collect the jewellery.
Insertion:
She said, 'Talk, *s*talk, that's all you do'.
EXPRESS POST
Inversion:
The guerrillas are sending *ar*ms to the rioters.

Sickle-cell anaemia

Sickle-cell anaemia is a disease that is usually associated with a mutation in the gene that codes for one of the polypeptides that make up **haemoglobin** in red blood cells. In this mutation, an adenine base is substituted by a thymine base. The result is a phenotype of misshapen red blood cells that can clump together and block blood vessels.

2.9.6 Chromosome mutations

Point mutations relate to changes in the genetic information in genes; however, mutations can also involve chromosomes. These may involve the addition or deletion of entire chromosomes, or the deletion, addition or mixing of genetic information from segments of chromosomes. Some examples of disorders that result from chromosome mutations are shown in the table below.

Examples of human chromosome abnormalities (mutations)

Chromosome abnormality	Resulting disorder	Incidence (per live births)
Extra chromosome number 21	Down syndrome	1 in 700; risk rises with increase in maternal age
Missing sex chromosome (XO)	Turner syndrome	1 in 5000 (90% of these conceptions are aborted)
Extra sex chromosome (XXY)	Klinefelter syndrome	1 in 1000; risk rises with increase in maternal age

2.9.7 Mutants unite!

Not all mutations are harmful. Some mutations can increase the survival chances of individuals within a population, and hence the survival of their species.

Spray resistance

Pesticides kill the majority of insects sprayed. Some insects within the population, however, may survive because they possess slight variations or mutations in their genes that give them **resistance** to the pesticide. The mutated gene in the surviving insects is passed on to their offspring, who gain that resistance too. While the insects without the resistance die out, those with resistance increase in numbers.

Good for you, but not for me

When we look at natural selection as a mechanism for evolution in chapter 3, we see how mutations can be a very important source of new genetic material. While such mutations can be beneficial for the survival of the species under threat, they are not necessarily beneficial to humans. The resistance of bacteria to antibiotics, for example, has resulted in selection for antibiotic-resistant bacteria. This means we are unable to use these antibiotics to treat diseases caused by these resistant bacteria, as the drugs are no longer effective.

Malaria and sickle-cell mutation

Malaria is a disease that is very common in many parts of Africa, Asia and South America. It is caused by a parasite that uses a species of mosquito as a vector and grows in red blood cells of its human host. This disease is one of the main global causes of human disease-related deaths.

	Normal red blood cell	Sickle-cell red blood cell
DNA sequence	GAC TGA GGA CTC	GAC TGA GGA CAC
Complementary RNA sequence	CUG ACU CCU GAG	CUG ACU CCU GUG
Amino acid sequence	leu — thr — pro — glu	leu — thr — pro — val
Phenotype of red blood cell	Normal doughnut-shaped blood cell 	Sickle-shaped blood cell 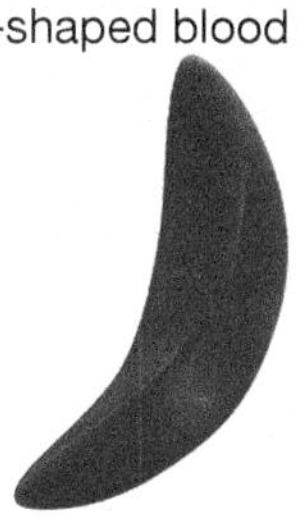

The mutation that results in sickle-cell anaemia can increase your resistance to malaria. If you are heterozygous for this trait (you have one copy of the sickle-cell allele), the parasite cannot grow as effectively in your red blood cells; hence you are less likely to die from malaria than people in the population without the allele. This is an example of what is known as **heterozygote advantage**.

Mutations can be caused by chemicals in your environment and may result in cancerous growths (tumours) within your body show as foggy areas here.

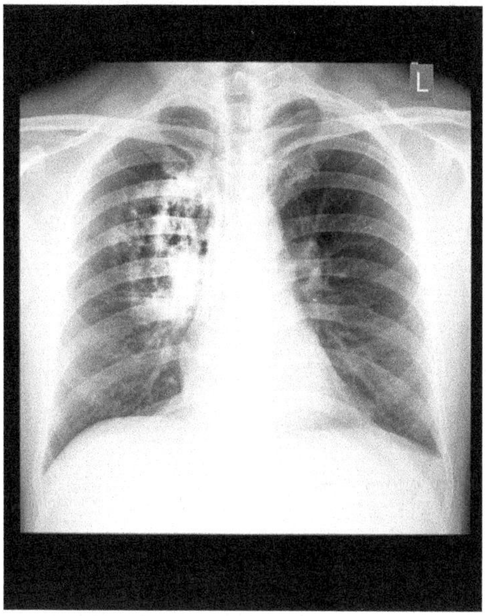

2.9.8 Not all mutations are inherited

Only mutations that have occurred in the germline cells such as the sex cells or gametes (sperm and ova) are inherited. In sexually reproducing organisms, mutations that occur in somatic cells are not passed on to the next generation.

2.9 Exercises: Understanding and inquiring

To answer questions online and to receive **immediate feedback** and **sample responses** for every question, go to your learnON title at www.jacplus.com.au. *Note:* Question numbers may vary slightly.

Remember

1. Name the process by which DNA makes copies of itself.
2. Explain why the model used to describe the process identified in question 1 is called semi-conservative. Include a diagram in your response.
3. Describe what is meant by the term *mutagenic agent*. Provide an example.
4. Distinguish between the terms *spontaneous mutation* and *induced mutation*.
5. Suggest the relationship between the thinning of the ozone layer and the increased incidence of skin cancer.
6. Outline the relationship between sickle-cell anaemia and mutated DNA.
7. Identify two disorders associated with chromosome mutations.
8. Are mutations always detrimental? Provide an example to justify your response.

Investigate, think and discuss

9. Suggest why radiographers wear special protective clothing and use remote controls for taking X-rays.
10. Suggest mutations that increase chances of survival.
11. Examine the graph at right.
    (a) Describe the pattern or trend. Incorporate the axis labels in your description.
    (b) Suggest an interpretation of the data.

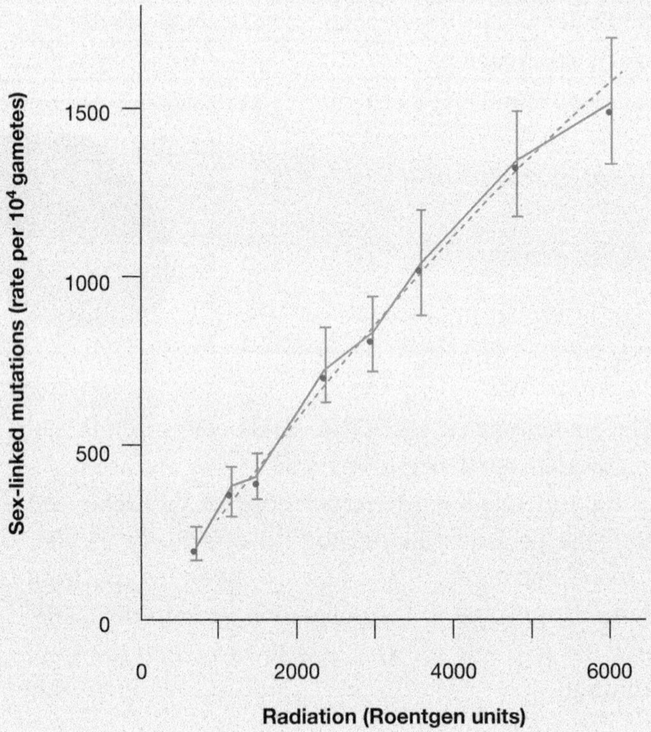

Analyse, interpret and think

12. Search online for Down syndrome research. Use your own knowledge and information found to answer the following questions.
    (a) How many chromosome 21 copies are in the somatic cells of a person with Down syndrome?
    (b) Is this the same number of chromosome 21 copies that are in the somatic cells of a person who doesn't have Down syndrome? Explain.
    (c) Suggest why the *DSCR1* gene is of importance.
    (d) On which chromosome is the *DSCR1* gene located?
    (e) Outline the advantage suggested by the research of possessing an extra copy of the *DSCR1* gene.

2.10 Predicting with pedigree charts

2.10.1 Inherited gene and chromosome abnormalities

You are the combined result of your parents' gametes and your environment. If someone's sperm or ovum carries a DNA abnormality, there is a chance that their child will be affected. Inherited gene and chromosome abnormalities may result in genetic disorders. These can be slight, such as red–green colour blindness, or more severe, such as haemophilia, a disorder in which the blood does not clot.

The image at right shows a potential inheritance pattern for haemophilia. Haemophilia is a rare genetic disorder characterised by an inability for blood to clot, meaning even minor cuts and bruises can be life threatening. It is passed on by recessive X-linked inheritance, which means that a mother may carry the disease without knowing and may pass it on to some of her sons.

Inheritance of haemophilia

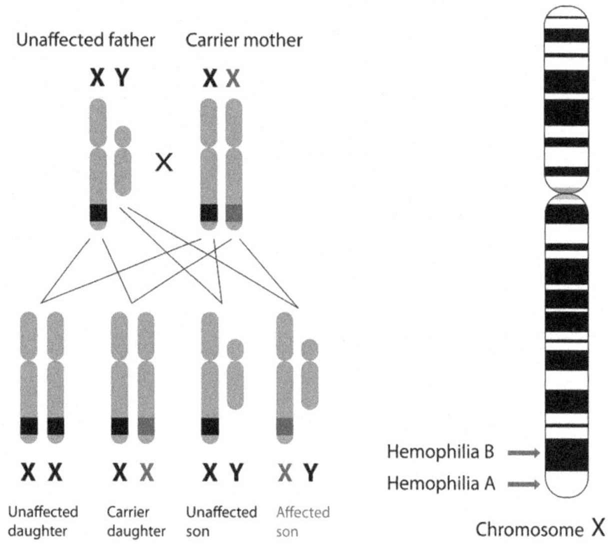

2.10.2 Who's who in a pedigree chart?

Pedigree charts can be used to observe patterns and to predict the inheritance of traits within families. Patterns in the inheritance of these traits can also show whether the trait is dominant or recessive and whether it is carried on an autosome or sex chromosome.

The pedigree chart below right shows how individuals and generations can be identified so that interpretation of patterns can be more effectively communicated. The shaded individual at the top of the chart is identified as I-1 (individual 1 in the first generation) and the shaded individual in the bottom row is identified as III-3 (individual 3 in the third generation). The daughters of individual I-1 are identified as II-3 and II-4.

Pedigree charts can show patterns of inheritance in families and enable identification of individuals within the family. Which individual do you think could be described as II-3?

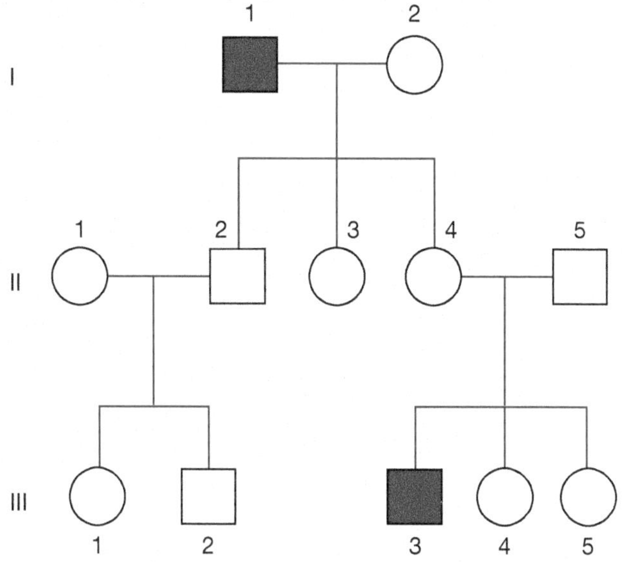

2.10.3 Naming inheritance

Within the nucleus of each human somatic (body) cell are 46 chromosomes. There are two sex chromosomes (either XX or XY) and 22 pairs of autosomes. The autosomes are numbered, based on their size and shape, from 1 to 22.

The inheritance of various traits can be described in terms of the location of the gene responsible and whether the inheritance is dominant or recessive. For traits located on the sex chromosomes, the trait is considered to be sex-linked. For traits located on the autosomes, the trait is considered to be autosomal.

A trait that is inherited recessively and caused by a gene on an autosome (e.g. chromosome 21) is described as **autosomal recessive**. Likewise, a trait located on the X chromosome and inherited recessively is described as **X-linked recessive**. The table on the next page provides examples of some inherited diseases and how they can be inherited.

How do you read a pedigree chart?

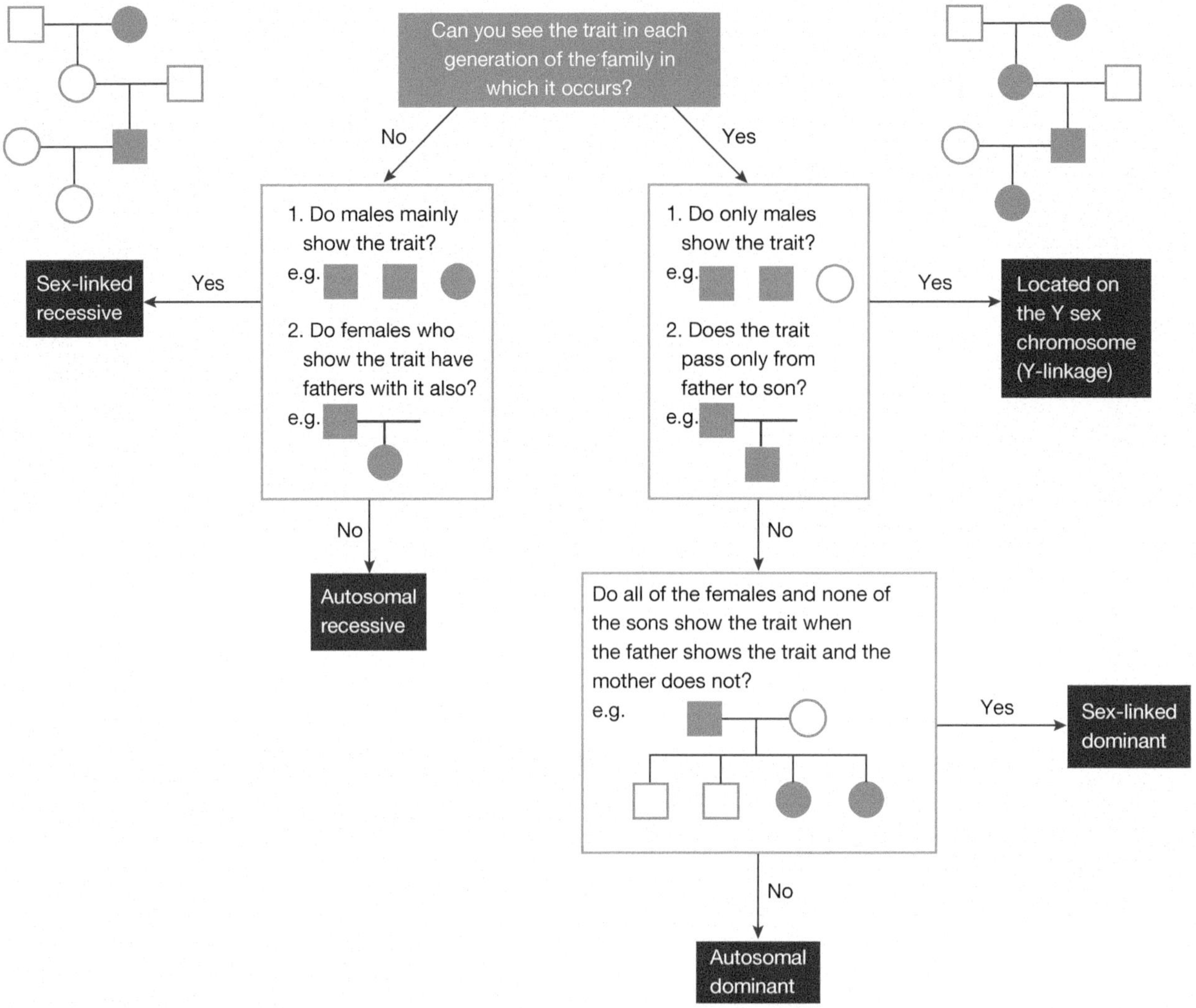

Some diseases that can be inherited

Inherited disorder	Type of inheritance	Symptoms of disorder
Fragile X syndrome (FRAX)	Sex-linked	Leading cause of inherited mental retardation
Haemophilia A and B (HEMA, HEMB)	Sex-linked recessive	Bleeding disorders
Huntington's disease (HD)	Autosomal dominant	Usually mid-life onset; progressive, lethal degenerative neurological disease
Intestinal polyposis	Autosomal dominant	Many small bulges in the colon form; may lead to colon cancer
Dwarfism	Autosomal dominant	Inhibited growth
Sickle-cell disease	Autosomal recessive	Red blood cells become deformed into a sickle shape when oxygen levels are low, which can lead to impaired mental function, paralysis and organ damage
Thalassaemia (THAL)	Autosomal recessive	Reduced red blood cell levels

2.10.4 Cystic fibrosis

About 1 in 2500 people suffer from an autosomal recessive genetic disorder called cystic fibrosis (CF). The CF allele is located on chromosome number 7. One amino acid in a chain of 1480 amino acids is not produced, causing a faulty protein to be synthesised. This results in the production of large amounts of thick mucus by cells lining the lungs and in the pancreas where digestive juices are secreted. The mucus interferes with the working of the respiratory and digestive systems. Infection readily occurs and sufferers tend to have a shortened life span. Pedigree analysis can show the likelihood of a child suffering from cystic fibrosis.

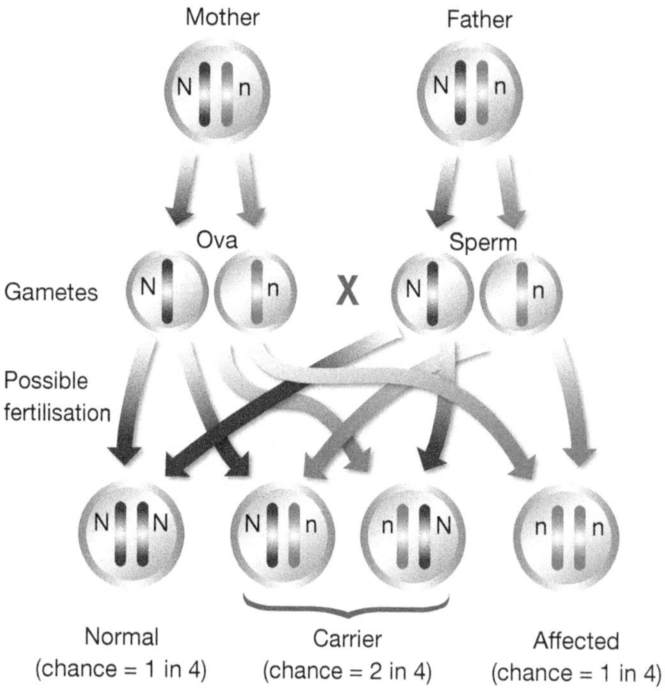

Checking to see if you are a CF carrier

Since the identification of the defective allele in 1989, the DNA of parents-to-be can be analysed to find out if they are one of the 1 in 25 people who carry the allele. This is useful information because, although they may not have cystic fibrosis themselves, they may be a carrier. This means that there is a chance they will have a child with cystic fibrosis. For example, if both parents are carriers, there is a 1 in 4 chance that they will have a child with cystic fibrosis.

A chance event?

Genetic counselling can help parents-to-be who are both carriers of the CF allele with their decision about whether to have a child. If they decide to go ahead, genetic tests can be used to determine the genotype of the embryo. If both parents are heterozygous for cystic fibrosis, each child they have has a 25 per cent chance of having two CF allelles, and hence having cystic fibrosis, and a 75 per cent chance of not having the disease. It is important to note that the chance is independent for each child. If the parents already had one child with cystic fibrosis, the next child would still have a 25 per cent chance of also having it. There is a 75 per cent chance that the child will not have the disorder, but only a third of these children will not have the CF allele. There is a 50 per cent chance that the child will not have the disorder but will be a carrier with one CF allele. You can see how these chances are calculated in the Punnett square at right.

If the genetic test shows that the child will have cystic fibrosis, the parents need to make an important decision — will they keep the baby? Genetic counselling may also help the parents with this very difficult decision. What would you do? If it was a more severe genetic disease, would that change your response?

If these two heterozygous parents had five children, would it be possible for none of them to have the disease?

Punnett square for *Nn* × *Nn*
N = normal allele
n = cystic fibrosis allele

	N	*n*
N	*NN*	*Nn*
n	*Nn*	*nn*

Offspring probabilities

Genotype: $\frac{1}{4}$*NN*: $\frac{1}{2}$ *Nn*: $\frac{1}{4}$*nn*

Phenotype: $\frac{3}{4}$ normal: $\frac{1}{4}$ cystic fibrosis

HOW ABOUT THAT!

Many genes, such as those controlling the production of enzymes necessary for respiration, are active throughout the life span of a person. Some are switched on only at particular times and in specific tissues. This regulates development. Late onset genetic disorders, such as Huntington's disease and a form of Alzheimer's disease, result from particular defective genes becoming activated later in life. Duchenne muscular dystrophy or muscle deterioration is another disease that gradually develops from late childhood.

2.10 Exercises: Understanding and inquiring

To answer questions online and to receive **immediate feedback** and **sample responses** for every question, go to your learnON title at www.jacplus.com.au. *Note:* Question numbers may vary slightly.

Remember

1. State the type of inheritance by which haemophilia is passed between generations.
2. Outline two uses of pedigree charts.
3. Outline how you could describe the identity of individuals with the following notation in a pedigree chart.
   (a) I-1
   (b) III-3
   (c) II-4
4. State the key difference between autosomal inheritance and sex-linked inheritance.
5. Identify an inherited disorder that is:
   (a) X-linked recessive
   (b) X-linked dominant
   (c) autosomal dominant
   (d) autosomal recessive.
6. On which chromosome is the cystic fibrosis allele located?
7. What are the symptoms of cystic fibrosis and what causes these symptoms?
8. Suggest why people may be tested for the cystic fibrosis allele.
9. Suggest how genetic counselling can be helpful in decision making.
10. What are the chances of two CF carrier parents having a child:
    (a) who has cystic fibrosis
    (b) who does not have cystic fibrosis
    (c) who is a carrier for cystic fibrosis
    (d) who does not have cystic fibrosis and is not a carrier?

Analyse, interpret and think

11. Jacob has hypertrichosis or werewolf syndrome, as does his mother. His father, however, does not. Hypertrichosis is an X-linked dominant trait that is characterised by increased hair growth on the face and upper body. Jacob is shown in the pedigree chart at right as individual II-4. For the following questions, assume that X^H = hypertrichosis and X^h = normal hair growth.

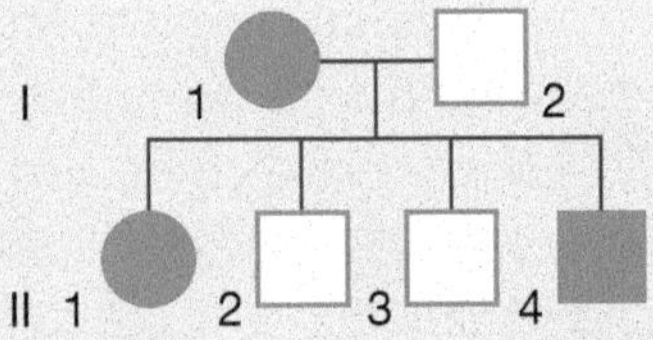

    (a) Use the Punnett square at right to determine the chances of Jacob's parents having children with the syndrome and state the chance of their having a:
        (i) daughter with hypertrichosis
        (ii) son with hypertrichosis
        (iii) child with hypertrichosis.

	X^H	X^h
X^h	$X^H X^h$	$X^h X^h$
Y	X^H Y	X^h Y

    (b) If Jacob had children with Bella (who does not have hypertrichosis), use a Punnett square to determine the chances of their child being:
        (i) a daughter who has hypertrichosis
        (ii) a son who has hypertrichosis
        (iii) a daughter who does not have hypertrichosis
        (iv) a son who does not have hypertrichosis.
    (c) Since Jacob is affected with hypertrichosis, do his sons and daughters have the same chance of inheriting the condition? Explain.

(d) How does this compare to a father who has an autosomal recessive trait? If he shows the trait and his wife does not, what are the chances that his daughters will show the trait? Explain.

(e) How does this compare to a father who has an autosomal dominant trait? If he shows the trait and his wife does not, what are the chances that his daughters will show the trait? Explain.

12. Huntington's disease is an autosomal dominant condition. Refer to the diagram below to answer the following.

(a) If H = Huntington's disease and h = normal, state the genotype(s) and phenotypes of:

(i) I-1

(ii) I-2

(iii) II-1

(iv) II-4

(v) II-5.

(b) Use a Punnett square to predict the chances of:

(i) I-1 and I-2 having a child with Huntington's disease

(ii) II-4 and I-5 having a child with Huntington's disease.

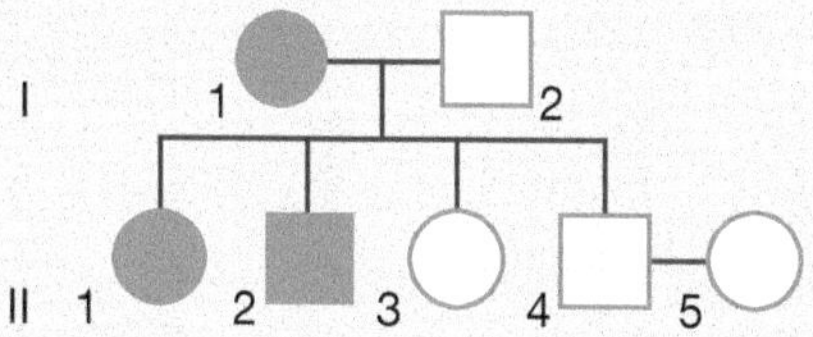

13. Queen Victoria was a carrier of the X-linked recessive trait haemophilia. This trait affects blood clotting. Use the figures provided below to answer the following.

(a) If X^H = normal trait and X^h = haemophilia, state the genotype of:

(i) Queen Victoria

(ii) her husband

(iii) her daughter Beatrice

(iv) her son Leopold (Duke of Albany)

(v) her granddaughter Alexandra

(vi) her great grandson Alexis.

(b) If Queen Victoria and her husband had had another child, what was the chance that their child would have:

(i) had haemophilia

(ii) been a carrier for haemophilia?

(c) If Alexandra and her husband, Tsar Nikolas II of Russia, had had another child, would they have had the same chance of having a haemophilic son as her mother and father? Explain.

(d) Suggest why our current Queen Elizabeth doesn't have haemophilia and why none of her children are haemophiliacs.

Queen Victoria carried the allele for haemophilia on one of her X chromosomes. This germline mutation was passed on to other members in her family.

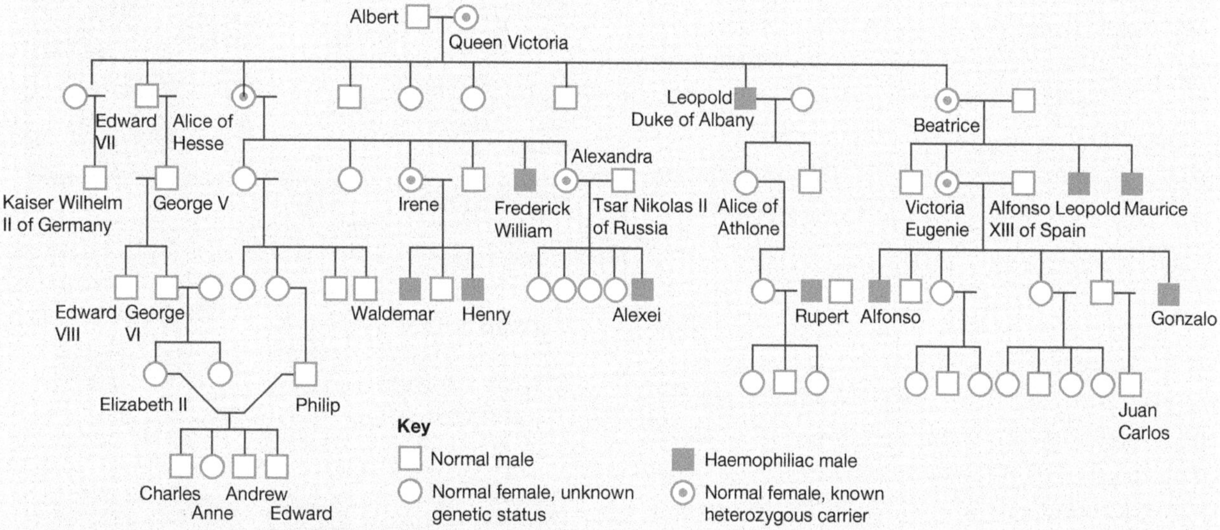

Think, investigate and discuss

14. (a) Find out what types of genetic testing occur in Australia.

(b) Are there any laws, rules or regulations associated with genetic testing? If so, what are they?

(c) List examples of different views and perspectives on genetic testing.
(d) Suggest why there are differing views.
(e) Construct a PMI chart on genetic testing.
(f) What is your opinion on genetic testing?

15. In a group, make a list of human genetic disorders. Each person is to write a report on one. Your report could take the form of a poster or brochure.
    (a) Include which gene or chromosomal abnormality is responsible for the disorder and some of the characteristics the affected person would show.
    (b) Find out whether there are organisations available to support people who have the disorder and their families.

Create

16. Create a rhyme, song or poem about pedigree analysis or the types of inheritance that effectively uses as many of the key terms in this subtopic as possible. An example is given at right.

> Dominant traits always with an affected parent
> While in recessive sometimes there aren't
> X-linked dominant — dad to his daughters
> No skipping seen, always loiters
> X-linked recessive — mum to her sons
> Can skip generations and hide in carrier ones.

2.11 Exposing your genes

Science as a human endeavour

2.11.1 Exposing your deepest secrets

Do you think that you can hide your identity? Maybe today you can, but that probably won't be the case in the future. DNA technology is rapidly providing techniques that will bring your deepest secrets out into the open.

How much do you want to know about your genes? How much do you need to know about the genes of a potential partner or members of your own family? Who should have access to your genetic information, and what should they be allowed to do with it? Who owns your genes?

2.11.2 Genetic tests

There are various tests that can be used to find out about your genetic information. Among these are gene tests and DNA-based tests that involve direct examination of the DNA molecule itself. Other tests include biochemical tests for various gene products (for example, enzymes and other proteins) or the microscopic examination of stained or fluorescent chromosomes.

Over 6000 single-gene disorders have been identified. Many other inherited diseases are

Already companies around the world are offering DNA profiling. Will you soon be required to carry your DNA profile around with you — and show it on request?

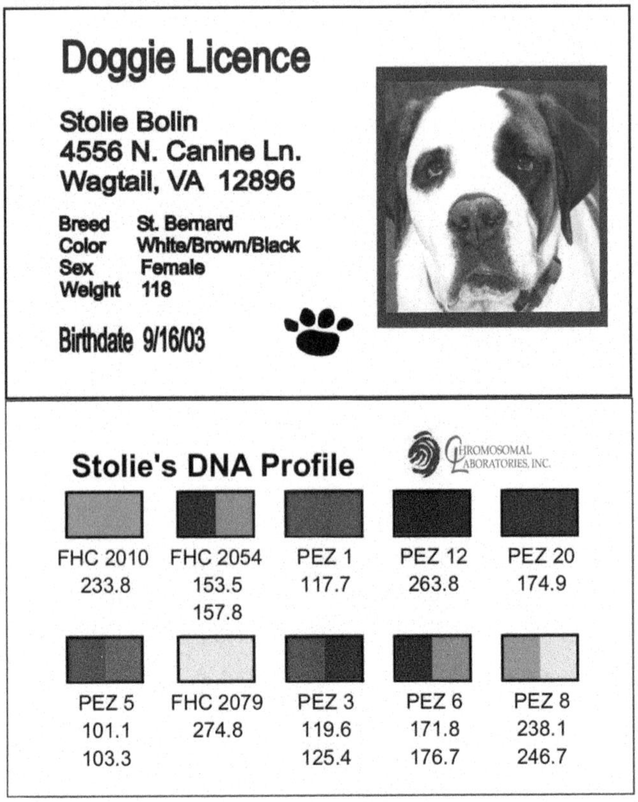

considered multifactorial because they may be caused by a combined effect of the interaction of a number of different genes with each other and the environment (for example, Alzheimer's disease, diabetes and asthma).

2.11.3 Why use genetic tests?

Genetic tests can provide information that can be used in gender determination; carrier screening for genetic mutations; or in the diagnosis, prediction or predisposition to particular genetic diseases or other inherited traits. These tests may be performed prenatally or on newborns, children or adults. Trying to control the characteristics of human populations by selective breeding or by genetic engineering is called **eugenics**.

2.11.4 Testing at birth

Australian state screening laboratories carry out a series of tests on a newborn baby's blood (see the table below). Early testing allows doctors to start any necessary treatment.

Screening tests

Genetic disorder	Symptoms	Incidence
Cystic fibrosis	Respiratory and digestive problems; shortened life expectancy	1 in 2500
Phenylketonuria (PKU)	Brain damage due to excessive levels of an amino acid in the blood	1 in 12 000
Hypothyroidism	Slowed growth and mental development owing to a poorly developed thyroid gland	1 in 3400

2.11.5 Counting chromosomes

The presence of chromosomal abnormalities such as Down syndrome can be determined by analysing cells of the developing fetus. These cells can be obtained by a technique called chorionic villus sampling (CVS), which collects actual cells of a fetus that is 10–12 weeks old. Another technique called amniocentesis can be used to collect samples of fluid from the uterus that contains cells shed by a fetus that is 14–16 weeks old.

These techniques can also be used to obtain cells that can be analysed for the presence of particular alleles. However, both of these techniques of cell sample collection are accompanied by some risk of miscarriage or damage to the fetus.

2.11.6 Counting DNA stutters

Did you know that the function of a large percentage of your DNA is still unknown? These supposedly non-functional or non-coding parts vary in length and can consist of patterns of repetitive base sequences called **microsatellites**.

2.11.7 DNA fingerprints

Patterns of variations in these repeated base sequences form the basis for **DNA fingerprinting**. This technique produces a kind of barcode of the natural variations found in every person's DNA. It is this barcode or DNA fingerprint that enables the identification of an individual to be made.

DNA fingerprinting is based on variations in the patterns of repeating base sequences in DNA between individuals.

DNA fingerprinting involves analysing DNA fragments. After the extraction of DNA fragments from biological material, **restriction enzymes** are used to cut the DNA into specific fragment lengths. The technique of **electrophoresis** is used to separate the fragments on the basis of their size and charge. DNA probes are then used so that the DNA patterns can be observed.

Amniocentesis and chorionic villus sampling can be used in the identification of chromosomal abnormalities.

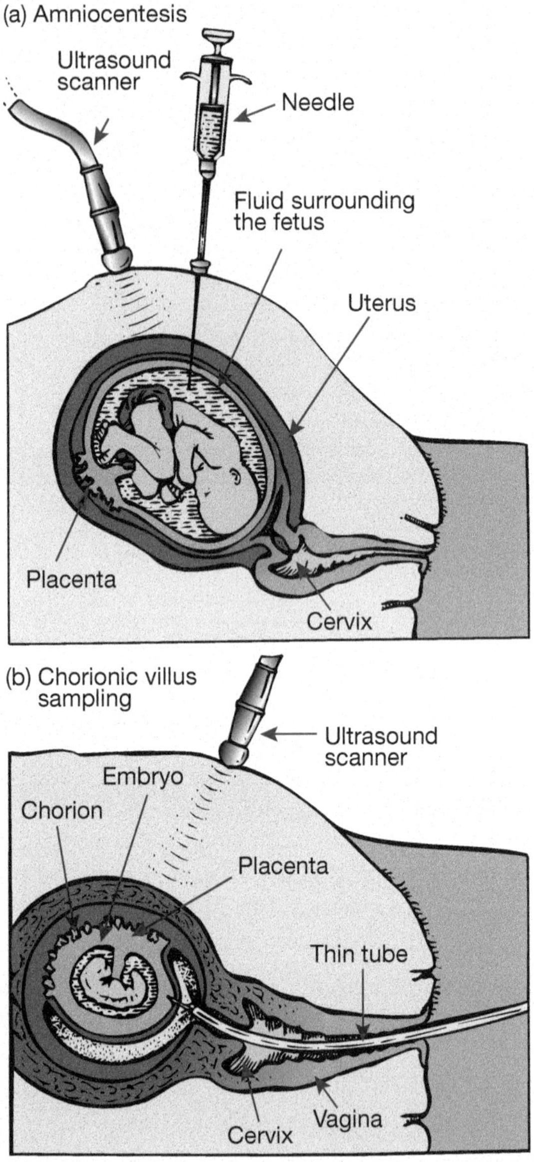

What's the use of DNA fingerprints?

DNA fingerprints can be useful in forensic investigations, paternity tests and evolutionary studies (to determine the relatedness of different organisms), and to search for the presence of a particular gene.

2.11.8 Fast computers and statistical genetics

Statistical methods have been used to establish gene linkage and estimate recombination fractions (due to crossing over in meiosis) since the 1930s. British scientists Bell and Haldane were the first to establish linkage between haemophilia and colour blindness with X-linked genes in 1937, and Mohr found linkage between blood group types on an autosome in 1954.

It was not until around 1980 that DNA sequence differences were used as **molecular markers**. The combination of these new markers with the use of **restriction fragment length polymorphisms (RFLPs)**, new multilocus mapping methods, suitable algorithms, and the affordability and availability of fast computers revolutionised human genetic mapping. In the late 1980s, the **polymerase chain reaction (PCR)** technique was beginning to revolutionise **molecular genetics**. This technique enabled amplification of small amounts of DNA, increasing the amount and hence the depth to which it could be studied. During this time, new types of genetic markers (such as RAPD, STRP and SSCP) were developed using PCR. The increase in the number of markers available enabled genome-wide scans to be performed. These scans could search the entire genome for linkage beween a trait and markers, identifying the location of genes that contributed to a wide range of phenotypes — including those associated with inherited diseases.

More recent research has focused on **single nucleotide polymorphisms (SNPs)** in the human genome. A current map of these in our genome contains more than 10 million SNPs. These SNP markers can be used in **genotyping** — a process that determines the alleles at various SNP markers within the human genome. Current technologies enable the genotyping of around 1 million SNPs per person within 24 hours for a cost of around $1000!

Linkage analysis

A team of scientists at the Walter and Eliza Hall Institute in Melbourne are using statistical models and fast computers to identify possible locations of particular genes within genomes. Information from families in the investigation is collected so that pedigrees can be constructed. They then use markers to scan the genome and perform a **linkage analysis** in their attempt to map the gene.

DNA fingerprinting uses a variety of different genetic engineering tools and techniques.

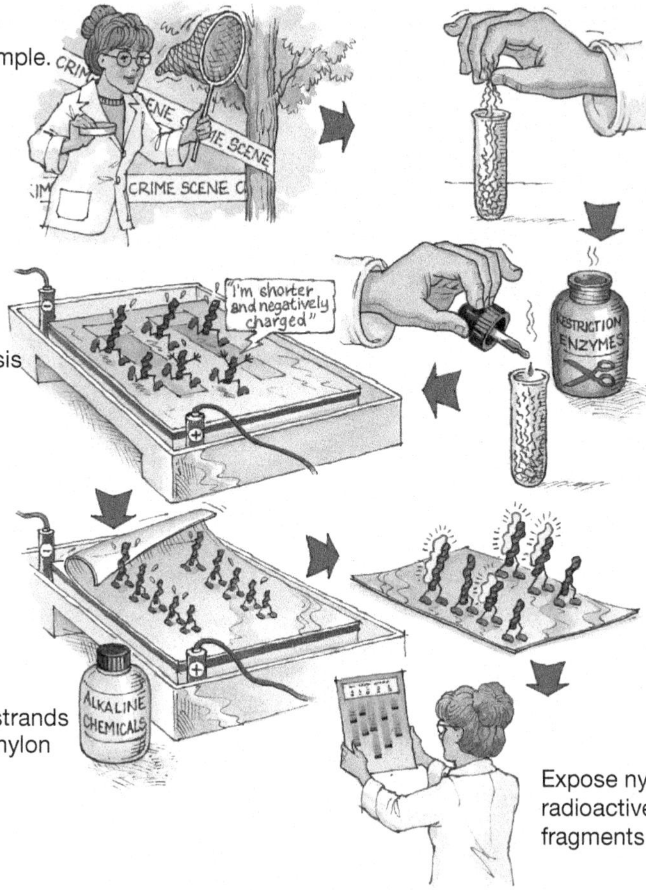

Collect the sample.

Extract the DNA from the sample.

Cut the DNA into fragments using restriction enzymes, which cut at certain base sequences they recognise.

Separate the DNA fragments on the basis of their length, using electrophoresis.

Split the DNA into single strands and transfer them onto a nylon sheet.

Immerse nylon sheet in bath with DNA radioactive probes. Probes bond to the core sequences of the sample DNA fragments.

Expose nylon sheet to X-ray film. The radioactive probes attached to DNA fragments show up as dark bands.

Who are the parents of individual 4?
Are persons 1 and 2 related?

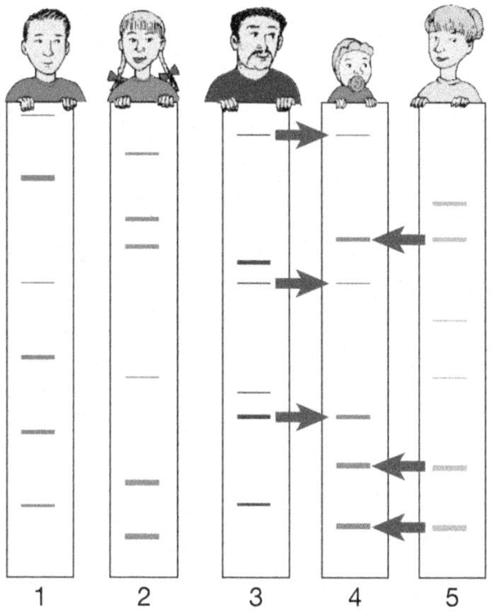

Australian scientists are involved in increasing our knowledge about our genes and inheritance.

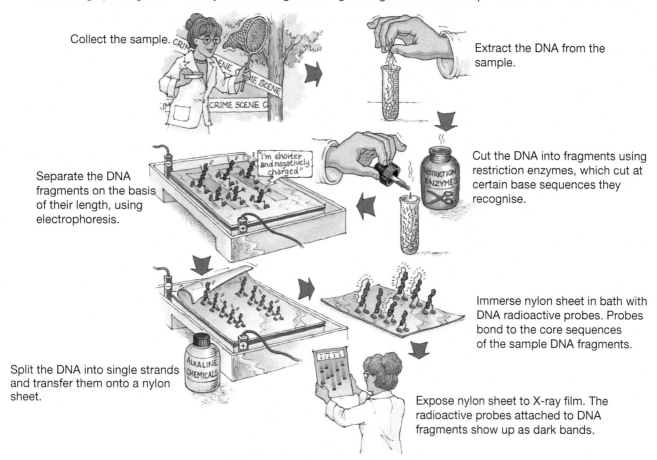

The team analyse the pedigree, trait and genotyping information using probability models that measure the significance of the linkage. Linkage analysis has already proved successful in mapping the genes for Huntington's disease and muscular dystrophy, and the breast cancer genes *BRCA1* and *BRCA2*.

2.11.9 Bioinformatics

Bioinformatics involves the use of computer technology to manage and analyse biological data. It has implications for a variety of fields, one of which is biotechnology.

2.11.10 DNA chips

DNA chips are made up of probes consisting of short fragments of selected genes attached to a wafer. Adding a sample of an individual's DNA to a particular type of DNA chip will result in a 'light up' response in the presence of one of the genes on the chip. This positive response is caused by the matching DNA locking onto the relevant probe on the chip. For example, a chip that carries probes for cystic fibrosis will search for the cystic fibrosis gene (allele), ignoring all other genes.

Implications of genetic testing

DNA chips are critical for making some sense of the enormous amount of information that research has supplied us with about the human genome. Although the most obvious use of these chips (and similar future technologies) is for diagnosis, there are important implications related to the ease with which genetic information can be accessed. With increasing knowledge about the human genome and inheritance, probes for genes associated with particular phenotypes (both favourable and unfavourable) may be incorporated into these DNA chips. Which gene probes should be used? Could they be used without your permission (or awareness)?

The construction of DNA databases, applications of bioinformatics and increased availablity of other types of DNA profiling techniques will open up a database of new questions, problems and issues. Who owns the resulting genetic information and decides what is done with it? Who should make these decisions?

NO ROOM FOR ERROR

The beady eye of DNA regulators needs to fall on paternity testing. The genetics revolution has progressed at breakneck speed since the discovery of the structure of DNA, and regulators have often struggled to keep up. It has been a few years since the 'personal genomics' industry took off, and the US Food and Drug Administration is only now warning firms that genome scans are 'medical devices' that require approval.

New Scientist

DROP OF BLOOD REVEALS AGE OF PERPETRATOR

Blood left at a crime scene could be used to estimate the age of a perpetrator, thanks to a new DNA test. The test could narrow down the range of possible suspects.

New Scientist

GENETIC RIGHTS

After more than a decade, the US Senate has finally passed the *Genetic Information Nondiscrimination Act* (GINA).

New Scientist

2.11 Exercises: Understanding and inquiring

To answer questions online and to receive **immediate feedback** and **sample responses** for every question, go to your learnON title at www.jacplus.com.au. *Note:* Question numbers may vary slightly.

Remember

1. Suggest six reasons for using genetic tests.
2. State three genetic disorder screening tests performed on newborn babies.
3. Outline similarities and differences between chorionic villus sampling and amniocentesis.
4. Identify the function of:
   (a) restriction enzymes
   (b) electrophoresis
   (c) DNA probes
   (d) PCR (polymerase chain reaction).
5. List four reasons for using DNA fingerprinting.
6. Describe how the use of computers and technology has increased our knowledge about genetics.

Think, analyse and investigate

7. The symptoms of the autosomal dominant inherited disorder Huntington's disease (HD) don't usually appear until the affected person is over 30 years old. Predictive testing can be carried out to detemine an individual's genetic status. In the figure at right, *H* represents the faulty HD allele and *h* the normal allele. The mother, I-1, is currently showing the symptoms of HD and has the genotype *Hh*.

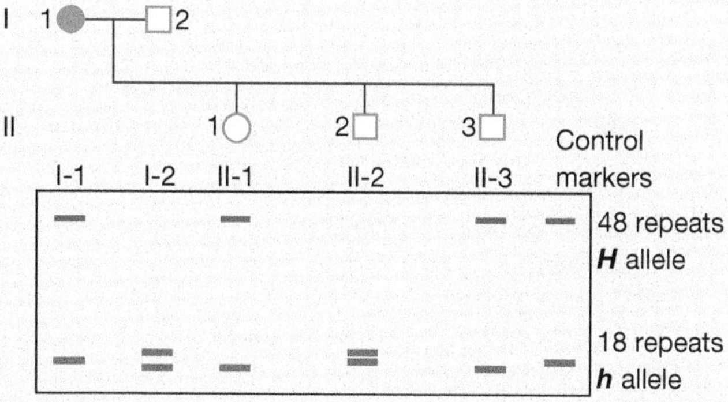

   (a) Suggest the genotype of:
      (i) individual II-3
      (ii) individual I-2.
   (b) Suggest whether the children are likely to develop HD. Justify your response.
   (c) Find out more about HD and current related research and issues.
8. Duchenne muscular dystrophy (DMD) is an inherited X-linked recessive disorder. The figure at right shows a pedigree and RFLP patterns that were obtained using a direct gene probe. Those individuals affected are shaded.

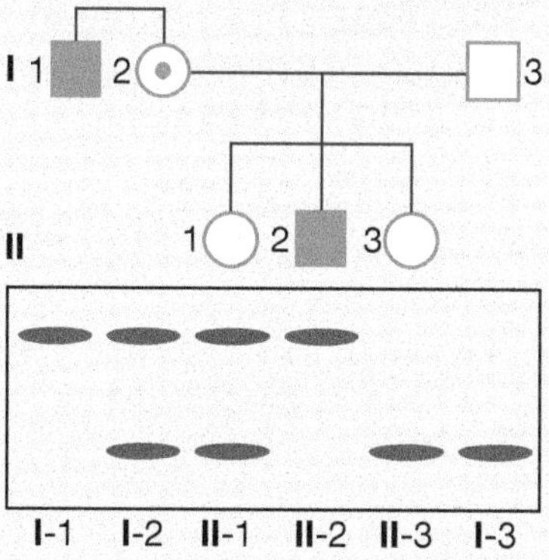

   (a) State the individuals with DMD.
   (b) Describe the RFLP pattern of those individuals with DMD.
   (c) If the mother, I-2, is a carrier of the DMD gene, which of her daughters is also a carrier?
   (d) Is the father, I-3, a carrier of the DMD gene? Explain.
   (e) Find out more about DMD and current related research.
9. The diagram at right shows the DNA fingerprint of a victim, the DNA fingerprint from evidence taken from her body after an attack, and the DNA fingerprints of three suspects.

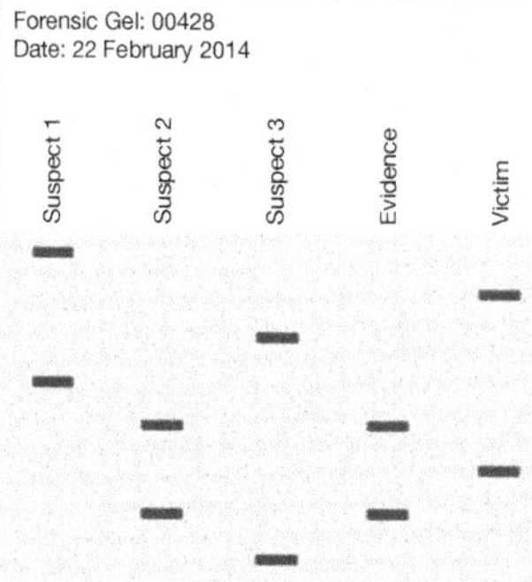

   (a) Using the information in the DNA fingerprints, which of the three suspects is most likely to be guilty of the crime against the victim?
   (b) Give reasons for your response to part (a).
   (c) Suggest why a sample was taken from the victim as well as the foreign DNA sample being collected from her body.
   (d) (i) State some other forensic diagnostic tools that exist to identify those guilty of crimes.
      (i) How do these compare with DNA fingerprinting?

10. (a) Consider and answer each of the following questions, justifying your responses.
    - Should the creation of a child from two different genetic mothers be encouraged?
    - Should a killer's jail sentence be reduced because they have a genetic disposition towards violence?
    - Can your genes absolve you of responsibility for a particular crime?
    - If you could 'engineer' your own child, would you?

    (b) Propose three of your own genetic issue questions for class discussion.

Investigate, think and discuss

11. Select two of the article headlines from in this subtopic and find out more about the topics. Write your own article on one of the topics for your class science magazine.
12. Find out more about the Australian Genome Research Facility (AGRF) and its involvement in genotyping many markers for many individuals in a single day.
13. What does a museum have to do with genetics? Find out about bioinformatics at a museum.
14. In 1993, American scientist Kary Mullis won the Nobel Prize in Chemistry for investigating PCR. Find out more about the discovery and applications of PCR.
15. What are bioethics and how do they relate to genetic testing? Research the use of a particular type of genetic testing (such as embryo selection, personal genomes, carrier status, predictive testing) and consider relevant bioethics issues.
16. Find out any implications of having a genetic disease for obtaining life and health insurance in Australia.
17. Suggest ways in which information from genetic tests may be used by organisations such as insurance companies, medical facilities and workplaces.
18. Suggest implications of patenting any of the following.
    (a) Genes
    (b) Gene products
    (c) Specific drugs that target a gene or gene products
19. Various countries and organisations are already developing DNA databases.
    (a) What is a DNA database?
    (b) Use a PMI chart to categorise possible applications of DNA databases.
    (c) Provide your own personal opinion on DNA databases. Include reasons for your opinion.
20. Research one of the following genetic careers: genetic statistician, genetic engineer, bioethicist, genetics counsellor, genetic researcher, genetic pathologist, molecular biologist, forensic scientist, sequencing specialist, bioinformatics/functional genomics officer.
21. Research one of the following Australian research institutes and find out more about their genetic research.
    - Ludwig Institute for Cancer Research
    - Howard Florey Institute of Experimental Physiology and Medicine
    - Walter and Eliza Hall Institute
22. What is thalassaemia? Find out more about screening and diagnostic tests for this genetic disorder. Find out more about the Thalassaemia Society and its involvement in genetic counselling.
23. Find out more about:
    (a) the Genetic Support Network in your state
    (b) companies that provide gene testing and screening.

When involved in a bioethical discussion, you need to consider these questions.

1 What is the issue?
2 Who will be affected by the issue?
3 What are the positive points of view? (Who or what benefits, and why and how?)
4 What are the negative points of view? (Who or what is disadvantaged, and why and how?)
5 What are some of the possible alternatives?
6 How may these alternatives affect those involved?
7 What is a possible solution that may be acceptable to those involved?
8 What is your opinion on the issue?

2.12 Domesticating biotechnology

2.12.1 Tomorrow, today?

Human genes in bacteria? Insect genes in plants? Cotton plants producing granules of plastic for ultra-warm fibre? These sound bizarre but are not in the realms of fantasy — they are happening now! What new creations will tomorrow bring?

You are living in the midst of a biological and technological revolution. Advances in biotechnology are gathering momentum so fast that your life will never be the same. We are speeding towards a future in which domesticated biotechnology may be a way of life.

2.12.2 Tools of the trade

Genetic engineering is one type of biotechnology that involves working with DNA, the genetic material located within cells. **Genetic engineers** use special tools to cut, join, copy and separate DNA.

2.12.3 Transferring the code

Genes from Arctic fish can be added to the genome of tomato plants so that they become frost resistant. This is an example of **recombinant DNA technology**. This technology uses specific enzymes called restriction enzymes to cut the DNA at specific points, so that a particular gene is removed. This DNA can then be pasted into the DNA of another organism using **DNA ligase**. If the organism belongs to another species, it is described as being **transgenic**. The feature coded for by the foreign gene is then expressed by its new host.

2.12.4 Cloning

If you saw the movie *Jurassic Park*, you may recall the scene in which scientists extract dinosaur DNA from mosquitoes that had been trapped in amber. They placed this prehistoric DNA (with a mix from some other living organisms to fill the gaps) into surrogate eggs. While the science in *Jurassic Park* has more than a few holes in it, we do have (and are still developing) technologies to clone single genes, some types of tissues and organs, and entire organisms.

Gene cloning involves the insertion of a specific gene into bacteria, so that the bacteria will act as microfactories and produce considerable quantities of desired proteins. This type of cloning has been used to produce insulin for diabetics and missing clotting factors required by haemophiliacs.

Therapeutic cloning and nuclear transfer cloning both involve the insertion of a nucleus from a somatic cell into a fertilised egg cell (which has had its own nucleus removed or destroyed) to create **totipotent stem cells**. The cells in therapeutic cloning are treated so that they will grow and divide into cells of a particular type or produce a specific type of tissue or organ. The cells in nuclear transfer cloning are transplanted into a surrogate host animal and result in the production of identical copies of the organism that supplied the donor DNA.

Reproductive cloning involves separating the cells of the developing embryo and implanting them into different surrogate mothers. The offspring from these surrogates will be identical to each other.

Bacteria with the human gene inserted into their DNA make human insulin.

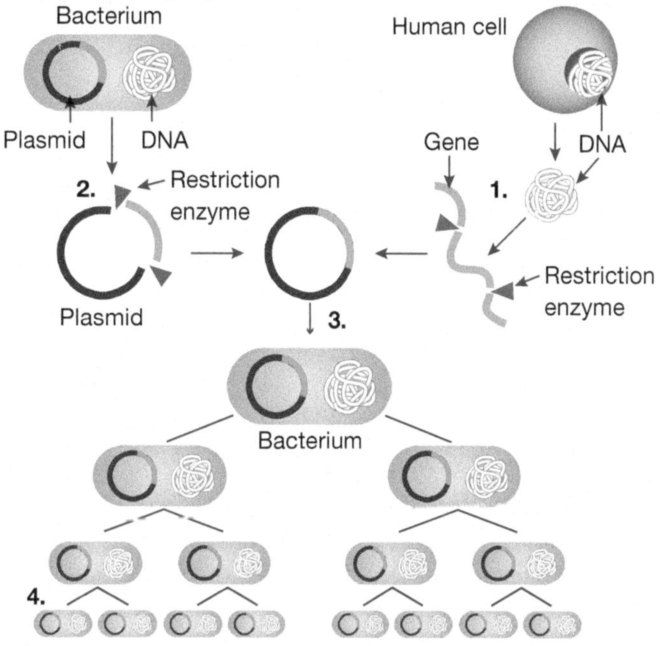

Are we in the midst of a molecular biology and biotechnology revolution? What new discoveries will the future bring and what implications will they have on our lives?

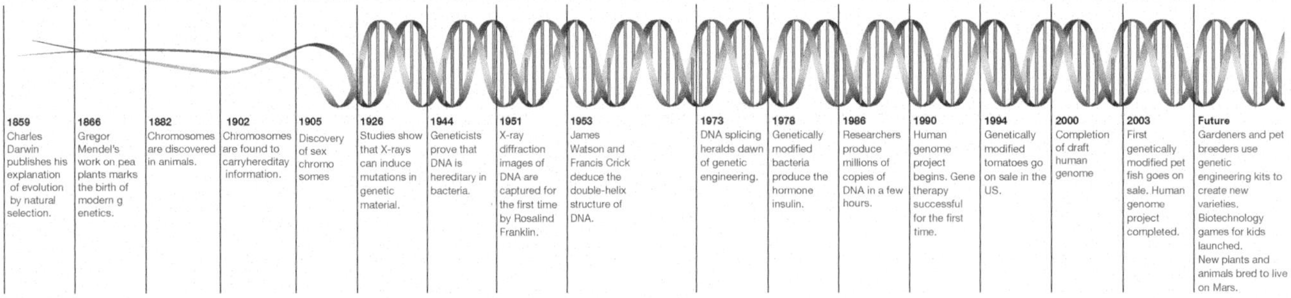
1859
Charles Darwin publishes his explanation of evolution by natural selection.
1866
Gregor Mendel's work on pea plants marks the birth of modern g enetics.
1882
Chromosomes are discovered in animals.
1902
Chromosomes are found to carryhereditay information.
1905
Discovery of sex chromo somes
1926
Studies show that X-rays can induce mutations in genetic material.
1944
Geneticists prove that DNA is hereditary in bacteria.
1951
X-ray diffraction images of DNA are captured for the first time by Rosalind Franklin.
1953
James Watson and Francis Crick deduce the double-helix structure of DNA.
1973
DNA splicing heralds dawn of genetic engineering.
1978
Genetically modified bacteria produce the hormone insulin.
1986
Researchers produce millions of copies of DNA in a few hours.
1990
Human genome project begins. Gene therapy successful for the first time.
1994
Genetically modified tomatoes go on sale in the US.
2000
Completion of draft human genome
2003
First genetically modified pet fish goes on sale. Human genome project completed.
Future
Gardeners and pet breeders use genetic engineering kits to create new varieties.
Biotechnology games for kids launched.
New plants and animals bred to live on Mars.

learnon RESOURCES — ONLINE ONLY

Watch this eLesson: Ancient resurrection: Do we have the right to resurrect ancient species? (eles-1070)

2.12.5 Human cloning

It can be argued that Australian scientists are leaders in the field of molecular biology and genetics. Around Australia science researchers are involved in cutting-edge investigations in genetics and molecular biology.

Australia made headlines in December 2006 when a ban on research on therapeutic cloning or **somatic cell nuclear transfer** was lifted in a national parliament vote, and Australia issued the first licence to clone human embryos.

Suggest why these surrogate mothers produce offspring that are identical to one another.

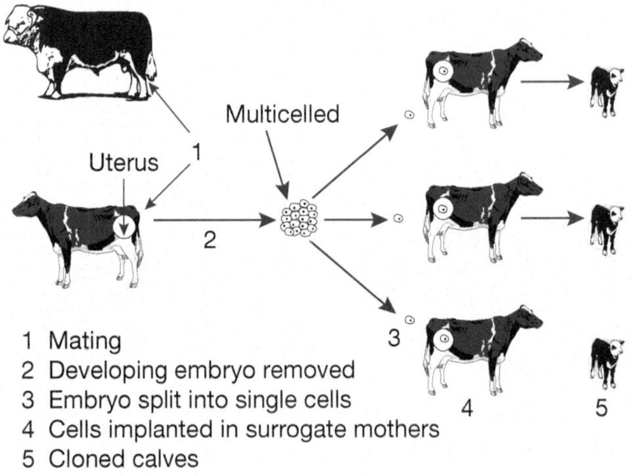
Multicelled
Uterus
1
2
3
4
5
1 Mating
2 Developing embryo removed
3 Embryo split into single cells
4 Cells implanted in surrogate mothers
5 Cloned calves

Therapeutic cloning can be used to produce stem cells.

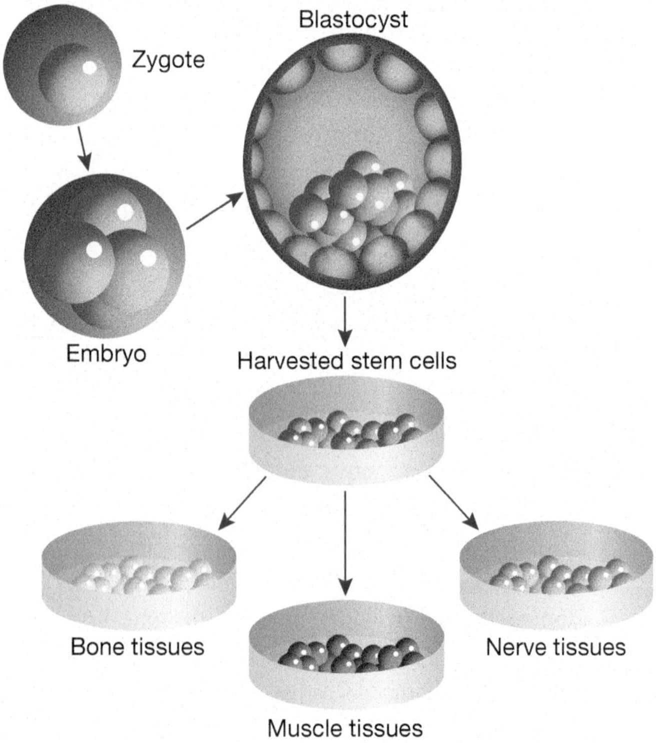
Blastocyst
Zygote
Embryo
Harvested stem cells
Bone tissues
Nerve tissues
Muscle tissues

2.12.6 The plant invader

Agrobacterium tumefaciens is a soil bacterium. It is able to get inside and infect many plants such as vines and fruit trees. In doing so, it transfers a tiny piece of its DNA into the host cell. This programs the host cell to make chemical compounds for the sneaky bacterium to feed on. Genetic engineers saw the possibility of using this bacterium as a **vector** to carry the genes they wanted from one plant into another.

Other kinds of bacteria and viruses act as vectors and carry genetic information from one organism (or synthesised to be like that organism) to another organism. Vectors can be used to carry genes for producing protein in soybean and sunflower plants, enzymes to control chemical processes, or compounds that keep insects or pathogenic viruses at bay.

HOW ABOUT THAT!

The story of Dolly the sheep began at the Roslin Institute in Scotland or, more specifically, as a single cell from the udder of a ewe. Follow her story in the diagram at right.

Dolly made history as the first mammal to be cloned from a single adult cell. Until then, biologists did not believe that once a cell had developed and become specialised, it could be reprogrammed to become different.

A group of cells that come from a single cell by repeated mitosis have the same genetic coding as each other. They are clones of each other. All of Dolly's cells came from the original fusion of an unfertilised egg and DNA from an udder cell. As there was no genetic input from another sheep, Dolly was a clone of the parent ewe from which the udder cell came.

How Dolly the sheep was created

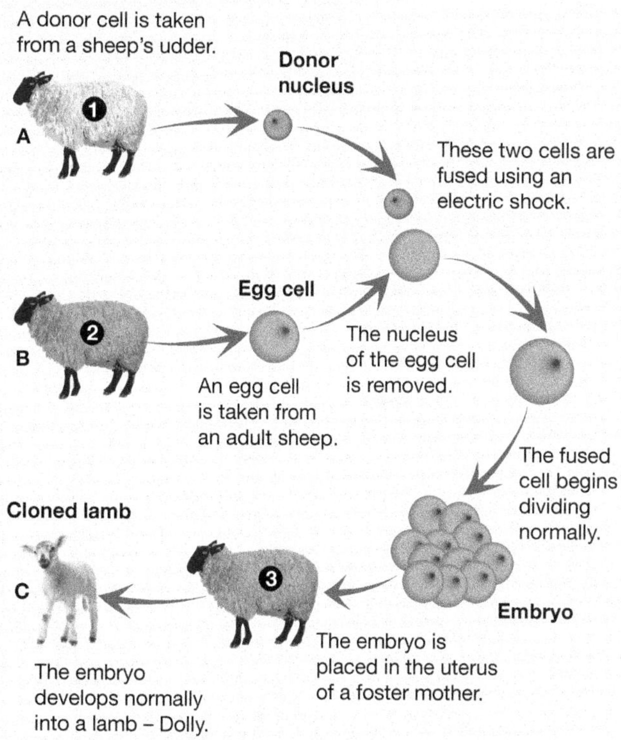

A donor cell is taken from a sheep's udder.
Donor nucleus
1
A
These two cells are fused using an electric shock.
Egg cell
2
B
The nucleus of the egg cell is removed.
An egg cell is taken from an adult sheep.
The fused cell begins dividing normally.
Cloned lamb
C
3
Embryo
The embryo is placed in the uterus of a foster mother.
The embryo develops normally into a lamb – Dolly.

2.12.7 Take aim … fire!

Recent developments have enabled foreign genes to be inserted into plant tissues by shooting them in with gas guns. Fine particles of gold are coated with the DNA and shot into the cells. Some cells are killed in the process but some survive, carry out mitosis and develop into complete plants with an altered genotype. Many plant crops such as maize and soybean have had favourable genes added to them in this way.

Biotechnology involving gene technology is a rapidly expanding branch of science. Already there have been trials of viral-vector nasal sprays to help treat people affected by cystic fibrosis. Some of the viruses carried by these vectors penetrate cells lining the respiratory tract and insert the normal gene into those cells. Vaccines against a number of diseases in humans and other animals are being investigated. Biotechnologists are trying to genetically engineer bananas to produce a vaccine against the hepatitis B virus. Will genetically engineered vaccines eventually prevent diseases such as AIDS?

Gene guns can be used to insert DNA into cells.

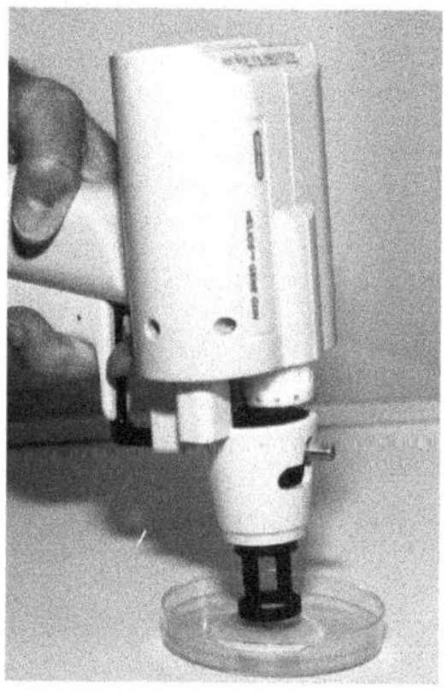

2.12.8 Do-it-yourself creation kits

Imagine if everyone was able to access genetic engineering tools to create and develop new forms of life. Imagine being able to create new plants for your garden, and pets that you could only dream about. Will designing genomes become a personal thing — a type of art form or expression of creativity?

What sorts of biotechnology games will be designed and produced? What sorts of lessons and learning may kindergarten children get from creating their own organisms to watch grow and interact with? Should there be rules and

and regulations for this new biotechnology? What sort of rules and regulations should there be? Who should make them? How can they be enforced?

2.12.9 Should we or shouldn't we?

Is transferring genes a wise use of gene technology? Some people argue that not enough is known about the way genes can jump the species barrier. Maybe they will end up where they shouldn't, such as in food chains. What could be the effect on other species in the environment? Could foreign genes interact with host genes and cause problems? Could viral vectors and genes mutate so that they would infect not only the target species but others too?

2.12.10 Gene therapy

Gene therapy is currently an experimental discipline and there is still considerable research required before it reaches its full potential. Gene therapy has a specific goal of targeting the gene that is responsible for a genetic disease. This type of therapy can be used to replace a faulty gene with a healthy version or insert a new gene that may cure or reduce the effects of a genetic fault.

The bacteria *Agrobacterium tumefaciens* has the ability to infect plants by inserting some of its DNA into the DNA of the plant. It has been used by geneticists to insert specific genes into plant DNA.

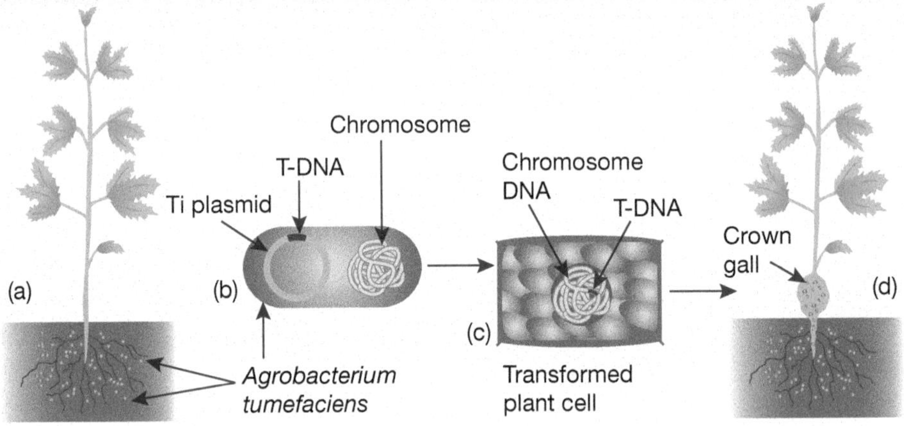

A delivery problem

There has been extensive research on the use of gene technology to treat or cure a variety of inherited diseases. If this is to be a viable alternative, the gene that causes the disease and the location of the affected cells need to be found. It also requires the availability of a healthy version of the gene and a way for it to be delivered to the cell. The delivery of the new genetic material has been one of the major stumbling blocks so far.

Viral transport

Early trials used a type of adenovirus with a healthy version of the cystic fibrosis gene inserted into its DNA. It was anticipated that this altered adenovirus would infect cells in the respiratory system, take over the cell's genetic machinery and make viruses that would make the required protein. While there was some success, there were also complications. This led to the development of different types of vectors that were less likely to mutate or cause adverse reactions within the hosts of these genetically engineered delivery vehicles.

Risks

While there are considerable potential benefits from the use of gene therapy, there are also risks. Some of these include the host's immune response to the foreign genetic material, the incorrect insertion of the new genes into the DNA or into an unintended cell, and the production of too much of the missing enzyme or protein. If viruses are used as vectors, the deactivated virus may target unintended cells or may be contagious, spreading to other organisms. The consequences of many of these risks remain unknown.

Issues

If and when these stumbling blocks in the delivery of the new genetic information are overcome, some new ethical and moral issues may arise in their place. Will gene therapy have the potential to create a more elite human being? Will there be attempts to alter characteristics such as height, intelligence or whatever the fashionable traits are at the time? Who will have access to this technology? Will gene therapy be available only to the rich and those in power? Who should be able to own and profit from this technology?

Although cloning produces identical offspring, there are a variety of ways of achieving it.

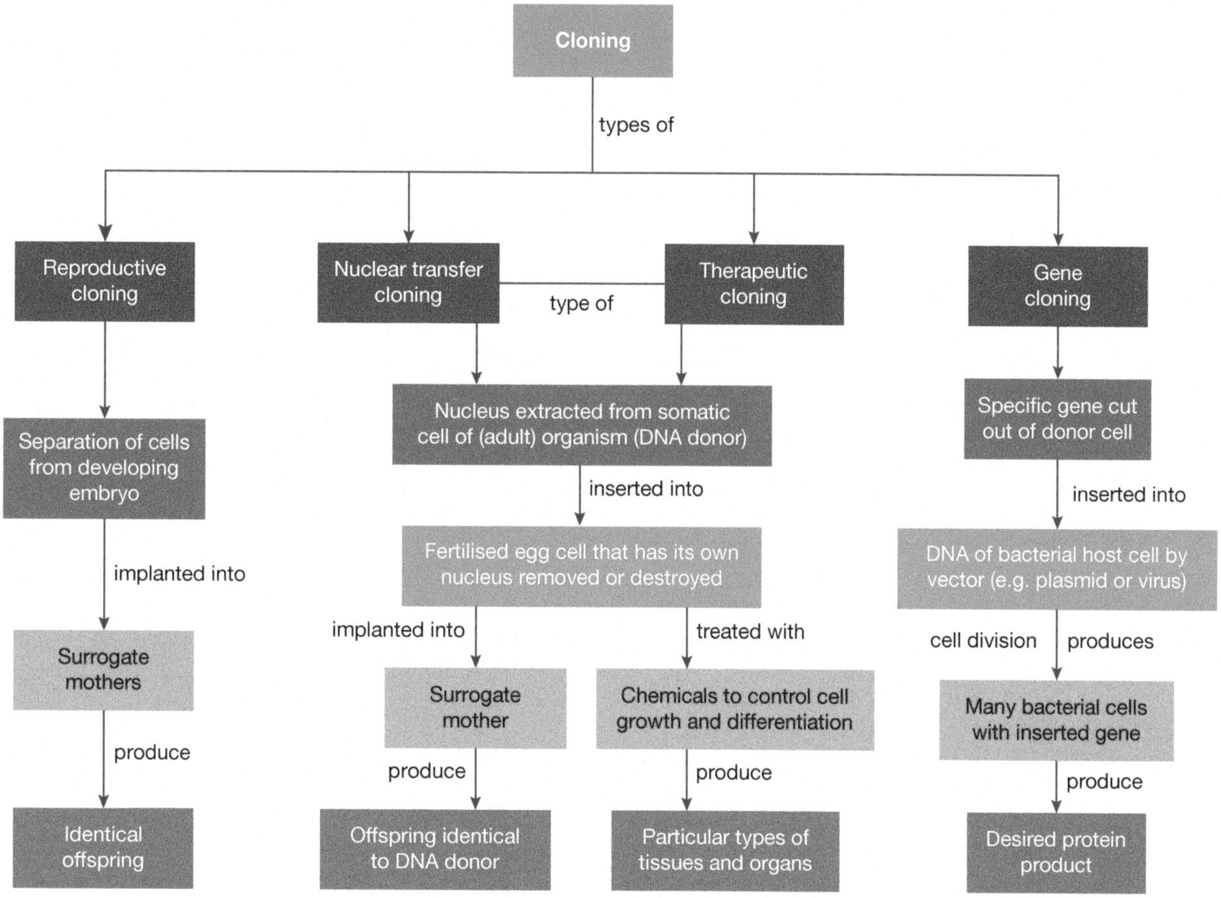

2.12 Exercises: Understanding and inquiring

To answer questions online and to receive **immediate feedback** and **sample responses** for every question, go to your learnON title at www.jacplus.com.au. *Note:* Question numbers may vary slightly.

Remember

1. What are clones?
2. Why is Dolly famous?

3. Refer to the diagram showing Dolly's creation.
   (a) What had to be done to the unfertilised egg taken from ewe number 1?
   (b) How did the donor cell DNA (from ewe 2) get inside the empty cell from ewe 1?
   (c) What was ewe 3's role?
4. Summarise the tools of genetic engineering into a mind, concept or cluster map. Include diagrams or figures to help you remember what each of the tools does.
5. What is gene technology? What are its potential benefits?
6. Draw a simple flow diagram to summarise how bacteria are used to produce insulin that people with diabetes can use.
7. Explain what transgenic organisms are and give an example.
8. What is a vector?
9. What methods are used to transfer foreign genes into other organisms?

Investigate, think and discuss

10. Why is Dolly called a clone?
11. Which ewe was the original Dolly?
12. Does Dolly have any genetic material from a ram?
13. There is now no theoretical scientific reason why cloning of humans, even dead ones, is not possible. Some people have suggested that cloning of humans should be permitted, while some countries, such as the United States, have outlawed it.
    (a) What reasons would people give for wanting to clone humans?
    (b) Research media stories that raise issues related to cloning, particularly of humans. Draw an issue map to identify some of the implications. Keep a record of your references.
14. Since Dolly, a number of other animals have also been cloned. Find out more about the cloning of at least one other animal. Report on:
    (a) the reasons for cloning that particular animal
    (b) how the animal was cloned
    (c) the advantages and disadvantages associated with the cloning of the animal.
15. Use the timeline in this subtopic to answer the following questions.
    (a) In which year were sex chromosomes discovered?
    (b) Find out when Dolly was born and add the date to the timeline.
    (c) Find two other biotechnological events to add to the timeline.
    (d) Research one of the events on the timeline in more detail and share your findings with your class.
16. (a) Outline how gene therapy could be used to treat and cure genetic diseases.
    (b) Suggest reasons why gene therapy is not currently being extensively used to treat genetic diseases.
    (c) Identify issues related to gene therapy.
    (d) Use a SWOT analysis to categorise points made in a discussion on gene therapy.
    (e) Identify different possible uses of gene therapy.
    (f) Construct a priority grid to map out different potential uses of gene therapy.
17. Design a biotechnology game of the future.
18. Find out the environmental conditions on Mars and think about what sort of life form could survive them. Design a plant or animal that you think would have a good chance of surviving on Mars.
19. What are some of the advantages and disadvantages of gene technology? Use specific examples to support your views.
20. Would you eat genetically manipulated food? Why or why not?
21. In a group, compile a folio of about 10 journal and other media articles relating to gene technology. Make sure you note the date and source of each one. Each person will choose an article to analyse. In your analysis include:
    (a) the kind of gene technology being reported on
    (b) a simple description or explanation of the particular example of technology
    (c) any issues relating to the example.

2.13 DNA — interwoven stories?

Science as a human endeavour

2.13.1 Hologenome theory of evolution

Do you think you are special? Of course you are! Your DNA contains a wonderful tapestry of not just your human ancestry but also that of other genomes.

Although we see ourselves as being purely human, we are not. Like other plants and animals, we contain DNA from a variety of sources. The **hologenome theory of evolution** emphasises the role that micro-organisms have within our evolution.

Micro-organisms (microbiota) within their hosts (such as plants or animals) can be described as being **symbionts**. The term *holobiont* is used to describe the host and all of their symbiont microbiota collectively. Some scientists have suggested this as a unit of selection in evolution.

The hologenome is made up of the combined genomes of the host and the microbes within it. Genetic variation within the holobiont may occur in both the host and the microbial symbiont genomes, and the variation can be inherited by the offspring. Some of this variation may occur within existing microbes; other variation may be due to microbes newly acquired from the environment. Some scientists consider this view as Lamarckian, as the inheritance of variation via microbes follows Lamarck's idea that traits acquired within the lifetime of the parent can be transmitted to the next generation.

2.13.2 Coding and non-coding DNA

Did you know that our genome contains viral DNA? Our DNA is made up of coding and non-coding sections. It was long thought that the non-coding DNA served no purpose. As this DNA contained highly repetitive sections and did not code for amino acids, it was often referred to as 'junk' DNA. New technologies and discoveries, however, are fast changing our views of this. Scientists have discovered that some of these non-coding regions contain genes that regulate many activities, and without them, protein synthesis using the coding DNA sections would not occur.

DNA and tissue types

Scientists have found that the complexity of an organism is not matched to the total amount of DNA that it contains. What they have discovered is that there is a relationship between non-coding and coding DNA. When testing people's tissue types, Malcolm Simons, a New Zealand-born immunologist, discovered that the pattern of the 'junk' DNA surrounding MHC genes was a very good predictor of tissue type. Some scientists suggest that this 'junk' DNA has played a key role in making us human, as it distinguishes primates from other mammals. It has been suggested that this 'junk' DNA is due to the invasion of a million copies of jumping genes!

2.13.3 Jumping genes

Barbara McClintock (1902–1992) was a cytogeneticist whose scientific theories (and possibly her gender) clashed with the scientific community of her early research years. McClintock investigated

DNA sequences known as transposons or 'jumping genes' can copy themselves into other sections of the genome.

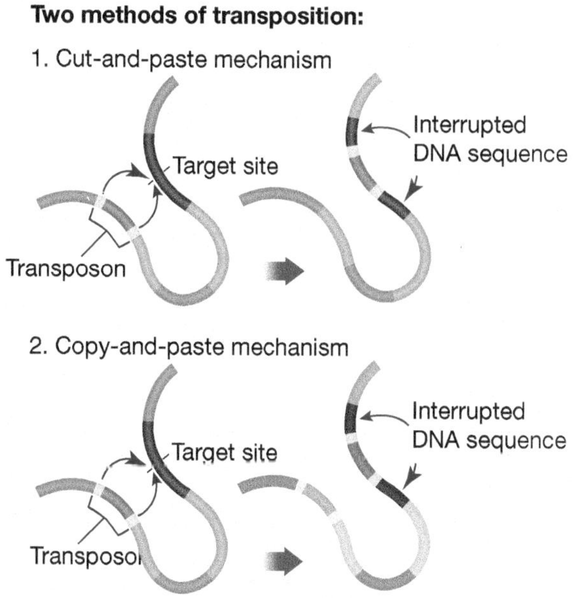

how chromosomes change during reproduction in maize (*Zea mays*). Maize proved to be the perfect organism for the study of transposable elements (TE) or 'jumping genes'. She contributed to our understanding of the mechanism of crossing over during meiosis and produced the first genetic map for maize. She linked regions of the chromosome with physical traits and demonstrated the roles of telomeres and centromeres. She also discovered **transposition** and outlined how it could be involved in turning on and off the expression of genes.

McClintock pioneered the study of **cytogenetics** in the 1930s. Before the structure of DNA was even discovered, McClintock was the first scientist to outline the basic concept of **epigenetics**, recognising that genes could be expressed and silenced. McClintock's theories were revolutionary because they suggested that an organism's genome was not a stationary entity, but something that could be altered and rearranged. This view was highly criticised by the scientific community at the time. She was eventually awarded the Nobel Prize in Physiology or Medicine in 1983, when she was over 80, for research she had done many years before.

Your jumping genes

Some of the repeating sequences within our non-coding DNA are known as **transposons** or 'jumping genes'. They may have originated from invading viruses. These sections have the ability to copy themselves independently of the rest of the genome and then randomly insert themselves in other sections of the genome. There have been suggestions that our evolution has been shaped by these transposons.

2.13.4 Genetic invasion

The human immunodeficiency virus (HIV) is an example of a retrovirus. Retroviruses convert their RNA genome into DNA before implanting it into host chromosomes. This process is called **endogenisation**. If the viral genome is incorporated into the chromosomes in the host's germline, it can become a part of the genome of future generations.

Are you aware that such germline endogenisation has happened repeatedly in our own lineage? This mechanism may help explain the varied sources of the DNA in your own genome. It provides an explanation as to why our genome may contain thousands of **human endogenous retroviruses (HERVs)**. Are these the legacies of viral invasions throughout our evolutionary history?

Viruses have an amazing ability to unite, genome to genome, with their hosts. This has a powerful evolutionary significance; it supplies the host with a new source of variation for evolution. If a virus happens to introduce a useful gene, natural selection will act on it. Like any other beneficial mutation, it will spread through that population.

If the human genome has evolved as a holobiontic union of vertebrate and virus, could plagues be considered a vital evolutionary survival tool for our descendants? What are the implications of this theory in terms of how we treat diseases and think about viral infections? Do viruses have a place not just in our present, but in shaping our future evolution?

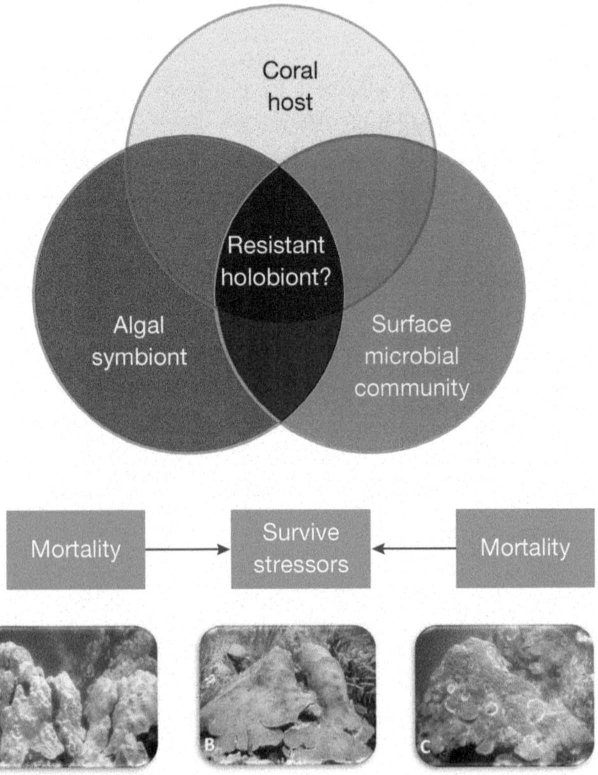

2.13.5 Coping with climate change

Some scientists have suggested that reef corals and possibly some other multicellular organisms may alter the microbial communities within their bodies to cope with environmental stresses such as those caused by climate change. More studies on this may provide some strategies we can use to help reduce the loss of biodiversity in ecosystems threatened by environmental changes related to climate change.

2.13.6 A career in Caribbean coral holobionts

Emmanuel Buschiazzo began his academic career as a scientist studying applied marine biology in Edinburgh, Scotland. He worked at the Marine Environment Laboratory in Monaco in marine ecotoxicology, then did his PhD in New Zealand studying the evolution of microsatellite DNA. His next adventure sent him to Canada for a postdoctoral fellowship in conifer genomics. He is currently involved in blending genomic approaches and population genetics to unravel the biology of two Caribbean coral holobionts, *Montastraea faveolata* and *Acropora palmata*.

(a) Dr Emmanuel Buschiazzo (b) *Montastraea faveolata* coral (c) *Acropora palmata* coral

2.13.7 Australian marsupials and jumping genes

Scientists have argued about how marsupials such as kangaroos, opossums and Tasmanian devils evolved in South America and Australia. DNA sequencing and the fossil record tell two different stories. Do jumping genes hold the answer to the mystery?

A German evolutionary biologist, Maria Nilsson, has been investigating this mystery by looking at strange bits of DNA called **retroposons**. Retroposons have the ability to break off chromosomal DNA and then copy and paste themselves elsewhere in the genome. Once they copy and paste themselves their locations are stable, making them a reliable marker for determining evolutionary relationships.

Nilsson's retroposon data suggest that the Australian and South American marsupials could be divided into two distinct groups that had little contact as they evolved. This supports the DNA sequencing data that they share a single South American ancestor that travelled to Australia before the continents drifted apart, and that they evolved separately afterwards.

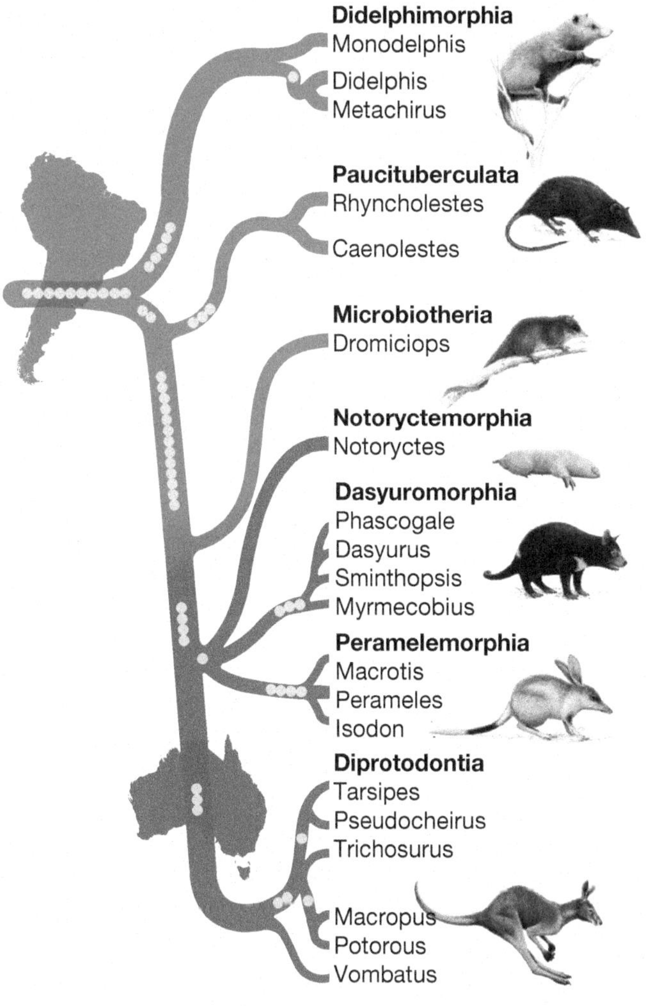

Our protein assassin

Australian and British scientists have identified the process through which our natural killer cells (part of our immune system) puncture and destroy virus-infected or cancerous cells. A protein called **perforin** is responsible for forming a pore in the diseased cell. The natural killer cells can then inject toxins into the diseased cell to kill it from within.

By using powerful technologies such as the Australian Synchrotron and cryo-electron microscopy, scientists have determined the structure of perforin and how it creates pores. This protein resembles cellular weaponry used by bacteria, and it is possible that our immune system may have incorporated the genetic information from bacteria within our evolutionary past.

Where did perforin, our protein assassin, come from?

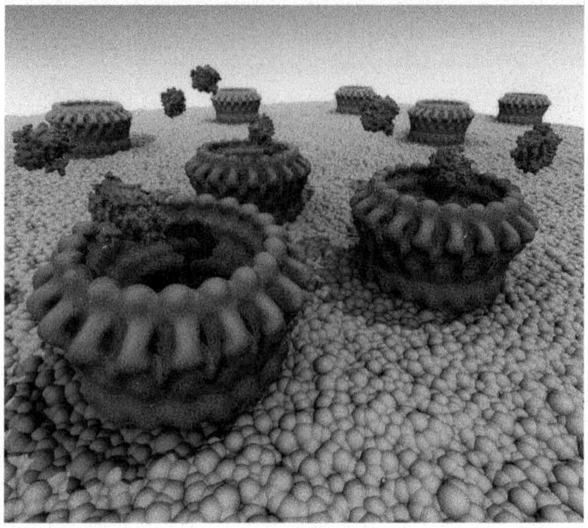

2.13 Exercises: Understanding and inquiring

To answer questions online and to receive **immediate feedback** and **sample responses** for every question, go to your learnON title at www.jacplus.com.au. *Note:* Question numbers may vary slightly.

Remember

1. Suggest why we should not think of ourselves as purely human.
2. Suggest what the hologenome theory of evolution emphasises.
3. Suggest what is meant by the following terms.
   (a) Symbiont
   (b) Holobiont
   (c) Hologenome
   (d) Endogenisation
   (e) Transposon
   (f) Retroposon
4. Suggest why we no longer call regions in our DNA that do not code for proteins 'junk' DNA.
5. Who was Barbara McClintock? List examples of ways in which she contributed to our understanding of genetics.
6. State another name for 'jumping genes'.
7. Provide an example of a retrovirus.
8. Suggest how we know that germline endogenisation has occurred within our own lineage.
9. Suggest how research on retroposons has contributed to our knowledge about the evolution of Australian marsupials.

Investigate, think, discuss and report

10. Research and report on one of the following.
    (a) The hologenome theory of evolution
    (b) Symbionts
    (c) Holobionts
    (d) Hologenome
    (e) Barbara McClintock
    (f) Transposons (or 'jumping genes')
    (g) Endogenisation
    (h) Microbiota and climate change
    (i) Caribbean coral holobionts
    (j) Australian marsupials and jumping genes
    (k) Retroposons
    (l) Epigenetics
    (m) Perforin
    (n) Astrobiologists
    (o) Artificial life
    (p) Mitochondrial DNA
    (q) Chromosome painting
11. If viruses kill about 20 per cent of all living material in the oceans every day, releasing their contents for other organisms to grow, does that mean that they drive ocean ecosystems?
12. Yellowstone Park in America is a place that attracts scientists. Suggest why many scientists view this park as a voyage back in time.

13. Discuss the following statement: If it weren't for viruses, ocean ecosystems would stop.
14. How do you feel about the possibility of having viral DNA in your DNA? Construct a PMI chart for a discussion on this question with your peers.
15. If you had a question to ask about the human genome, what would it be?
16. If you could investigate an aspect of human evolution, what would your research question or hypothesis be?
17. Some people suggest that there was a second genesis on Earth and that it is living among us undetected. It could even be extraterrestrial in its origin. Discuss this possibility and propose a number of questions that could be investigated.
18. Do you think that life (as we know it) could have occurred on planets other than Earth? Justify your response and discuss it with your peers.
19. Do you think that all life on Earth descended from a common origin? Discuss this and justify your opinion.
20. Suggest how astrobiologists could detect a life form on Mars.
21. The genetic code consists of 64 possible triplet DNA combinations that code for one or more of the 20 different natural amino acids; all species on Earth use the same code. Suggest why this might be used as evidence that there has only been one genesis on Earth.
22. Due to research by Carl Woese in the 1970s, many scientists now accept that prokaryotes can be divided into two distinct groups and that those in the archaea group are older than bacteria. These ancient ancestors of life on our planet were riddled with viruses. Does that mean that life on Earth originated from viruses? Discuss this question in groups and record comments made during your discussion.
23. If living things did not share an ancestor that shared ribosomes, ATP and the triplet code, then why are these found universally among all living things? Discuss and justify your reasons.
24. Suggest how we might identify life as we don't know it.
25. Given that chance plays a large part in the evolution of life, it is unlikely that life from a very separate origin would have the same biochemistry. Discuss this statement. Do you agree with it? Justify your response.
26. If you were an astrobiologist, you might refer to known organisms as 'standard' life and alternative forms as 'weird' life. You might not know what you are looking for, or where to look to find it. But how weird is 'weird'? Which criteria do you assign to life, and how will you know when you have found it?
27. *Deinococcus radiodurans* is a halophile and an example of an extremophile that can survive high doses of radiation. It can be found living in waste pools of nuclear reactors. Find out more about this microbe and others that can survive high levels of radiation.
28. The unusual Murchison meteorite fell north of Melbourne just two months after the lunar landing in 1969. It belonged to a rare class of carbonaceous chondrites. Find out more about this meteorite and its evolutionary implications.

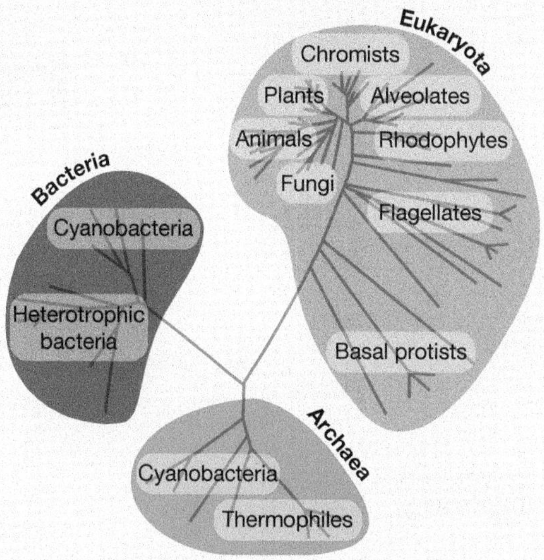

Carl Woese relied on RNA sequences rather than structural features to determine evolutionary relationships among prokaryotes. He discovered that prokaryotes were actually composed of two very different groups: bacteria (cyanobacteria and heterotrophic bacteria) and archaea (halophiles and thermophiles).

learnON RESOURCES — ONLINE ONLY

Complete this digital doc: Worksheet 2.8: Big picture science (doc-19415)

2.14 SWOT analyses and priority grids

2.14.1 SWOT analyses and priority grids

how to ...?

1. Draw up a square and divide it into four quarters. In the centre of the diagram write down the topic or issue that you are going to analyse.
2. Think about or brainstorm the positive features and behaviours, and record them in the Strengths section.
3. Think about or brainstorm the negative features and behaviours, and record them in the Weaknesses section.
4. Think about or brainstorm possible opportunities and record them in the Opportunities section.
5. Think about or brainstorm possible threats and record them in the Threats section.

question

What are the strengths and weaknesses of your project?

why use?

Allows you to prepare a plan of action and to consider possible 'blockers' to your project.

also called

No other names

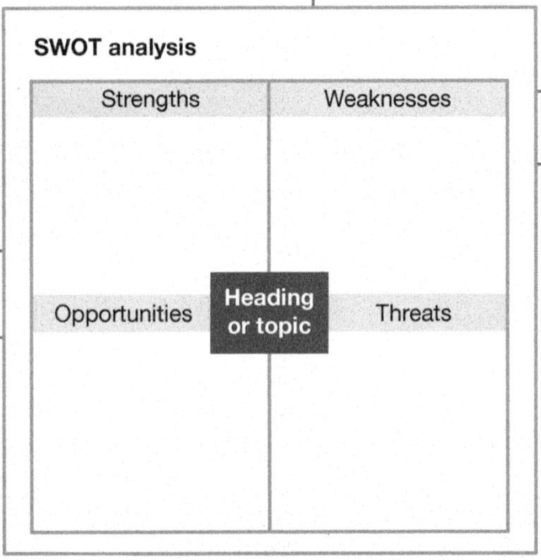

comparison

Priority grids

Similarity

Both can help you to consider various perspectives for decision making.

Difference

Priority grids compare two aspects of the idea; SWOT analysis considers a variety of positive and negative aspects.

example

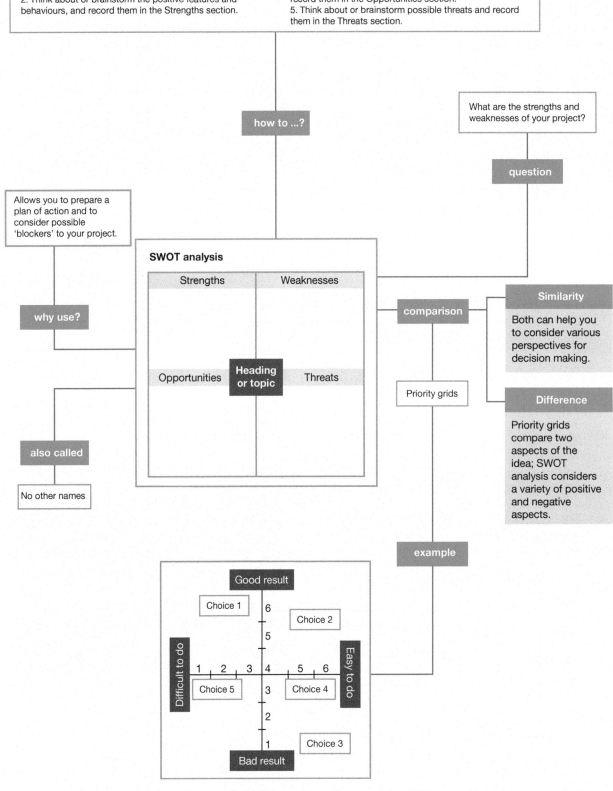

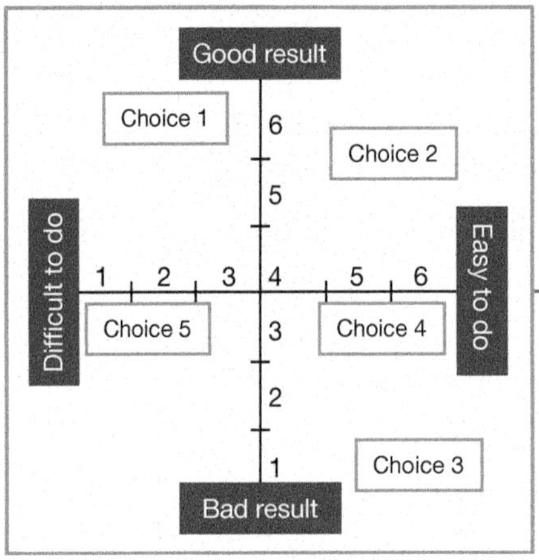

2.15 Project: The gene lab

2.15.1 Scenario

Think of how different dog breeds such as chihuahuas, great danes, dachshunds, blue cattle dogs and dobermans are from each other. Yet all of our pet dog breeds — regardless of size, colour, coat and intelligence — are still members of the same species. All dogs are descended from a long-gone species of wolf. Over the thousands of years that they have been humankind's companions, we have selectively bred dogs together so that particular characteristics became more pronounced while others faded out. For example, greyhounds with their long graceful legs were bred for speed while bull-mastiffs were bred for their size and strength. Over time, these characteristics became fixed in that breed. The breeding process is continuous, with new breeds being registered with the International Federation of Dog Breeders every few years.

Dog breeders try to produce dogs that are the ideal examples of their breed, and do so by carefully selecting which dogs to mate. Unfortunately, in their quest to establish these perfect examples, the dogs produced may inherit genetic disorders as a result of unfortunate genetic combinations or inbreeding. Purebred labradors, for example, may develop hip dysplasia, knee problems and eye problems such as progressive retinal atrophy which — as well as preventing the dog from being shown in competitions — have serious effects on the dog's quality and length of life. Now that genetics and DNA are more fully understood, it is not uncommon for dog breeders to consult with genetic scientists to ensure that the puppies they produce have the smallest risk of developing these disorders.

2.15.2 Your task

You are part of a team of vets that works for the Dog Breeders Association of Australia as genetic counsellors. Your client has a labrador bitch that has a family history of progressive retinal atrophy — a condition that causes gradual blindness. The client would like to breed her to produce for show as many puppies as possible that do not carry the gene for the disorder. There are three available stud dogs that the bitch can be mated with. Given the pedigree of each of these dogs, you must determine which of them should be selected to sire the litter. You will give your recommendations to the client in the form of a genetic report explaining your decision — including family trees, phenotype and genotype identification, and final breeding recommendations.

2.16 Review

2.16.1 Study checklist

DNA, genes and chromosomes

- describe the structure of DNA
- outline the process and importance of DNA replication
- relate the structure of DNA to its function
- distinguish between alleles and genes
- state the relationship between DNA, genes and proteins
- define 'karyotype' and describe its use
- explain how the gender of a baby is determined
- compare processes of mitosis and meiosis
- define the terms 'mutation' and 'mutagen'

- explain how mutations can reduce an organism's chance of survival
- identify examples of how mutations can be beneficial
- identify three different types of mutation

Inheritance

- outline the role that DNA plays in inheritance
- distinguish between genotypes and phenotypes
- define the following terms: dominant, recessive, heterozygous, homozygous
- explain how both genetic and environmental factors determine phenotypes
- predict the outcome of genetic crosses using Punnett squares
- interpret pedigree diagrams

Genetic applications

- identify applications of DNA fingerprinting
- outline the procedure to create transgenic species
- discuss issues relating to the positive and negative impacts of gene technology and cloning

Science as a human endeavour

- investigate the history and impact of developments in genetic knowledge
- suggest how values and needs of contemporary society can influence the focus of scientific research
- describe how science is used in the media to justify people's actions
- use knowledge of science to evaluate claims, explanations and predictions
- recognise that financial backing from governments or commercial organisations is required for scientific developments and that this can determine what research is carried out

Individual pathways

Activity 2.1	Activity 2.2	Activity 2.3
Revising genetics	Investigating genetics	Investigating genetics further
doc-8464	doc-8465	doc-8466

learnON ONLINE ONLY

2.16 Review 1: Looking back

To answer questions online and to receive **immediate feedback** and **sample responses** for every question, go to your learnON title at www.jacplus.com.au. *Note:* Question numbers may vary slightly.

1. Arrange the sentence fragments below to complete the sentence that has been started for you.

is made up of	which are made up of
cells	chromosomes
DNA	which contain
which contain in the nucleus	genes

A living organism ____
____.

2. Suggest the missing sex chromosome labels for the figure at right.
3. Copy and complete the following linked figures using the terms in the box below.

homozygous dominant	phosphate
cytosine	heterozygous
sugar	thymine
nitrogenous base	homozygous recessive
adenine	guanine

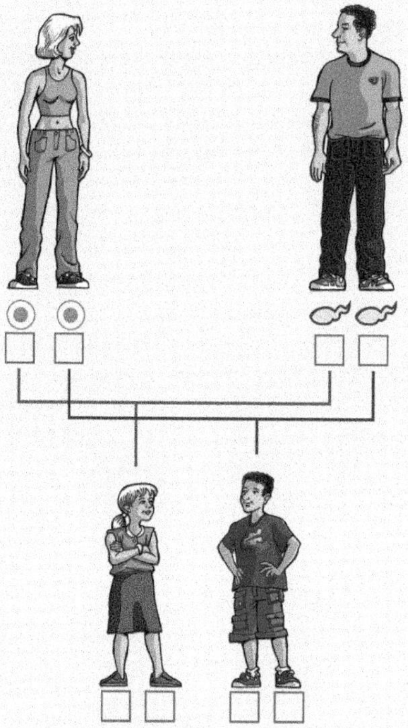

(a)

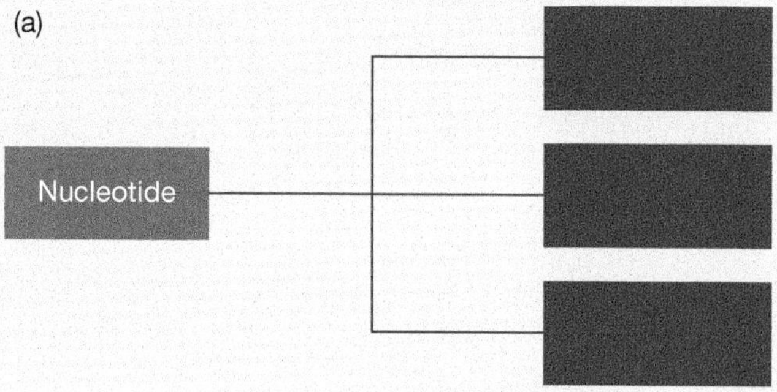

(b)

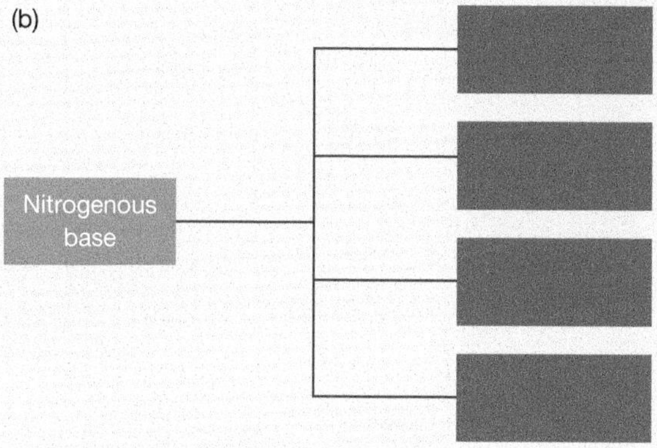

(c)

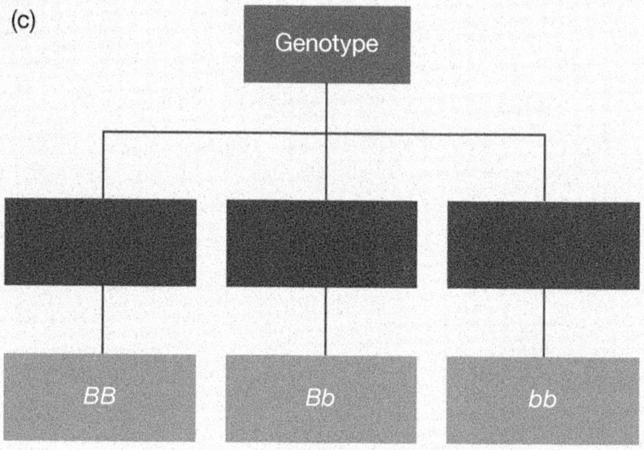

4. Use the *Some rules for drawing pedigree charts* diagram in subtopic 2.7 and the cystic fibrosis pedigree chart at right to determine:
   (a) which type of inheritance is responsible for cystic fibrosis
   (b) if individual II-3 is a male or a female
   (c) for each individual I-1, I-2, II-4, III-3 in the CF pedigree, whether you think they have cystic fibrosis or are carriers of cystic fibrosis. Give a reason for your suggestion.
5. A diagram representing a DNA molecule is shown on the following page. Which row in the following table shows the correct names for the structures labelled 1–4 in the diagram?

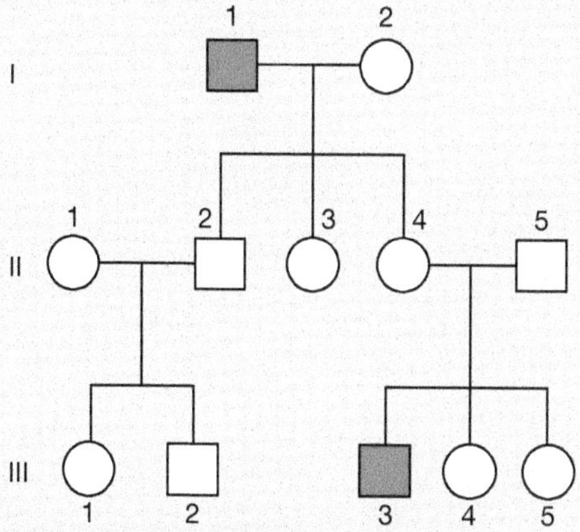

	Part 1	Part 2	Part 3	Part 4
A	Nucleotide	Nitrogenous base	Phosphate group	Sugar
B	Chromosome	Nucleotide	Adenine	Cytosine
C	Nitrogenous base	Sugar	Phosphate group	Polypeptide
D	Nucleotide	Nitrogenous base	Sugar	Phosphate group

6. (a) Using the letters *B* or *b* to represent the gene for coat colour, predict the results of the following crosses. Draw diagrams or Punnett squares to show your predictions.
   (i) A pure-breeding (homozygous) black mouse with a hybrid (heterozygous) black mouse.
   (ii) A pure-breeding black mouse with a pure-breeding white mouse.
   (b) Is black dominant to white or white to black? Support your answer.
7. The pedigree chart below right shows the inheritance pattern of Huntington's disease. This disease is due to a dominant HD gene on chromosome 4.
   (a) How many generations are shown?
   (b) How many females are in the pedigree?
   (c) How many males are in the second generation?
   (d) Identify three individuals who have Huntington's disease.
   (e) If *H* represents the allele for Huntington's disease, state the genotypes of:
      (i) individuals 1 and 2 in the first generation
      (ii) individuals 2 and 4 in the second generation.
   (f) How would the pattern in the pedigree be different for a recessively inherited trait?
8. Construct Venn diagrams and add shading to pedigrees like those shown in the figures below to illustrate the similarities and differences between each of the following types of inheritance.

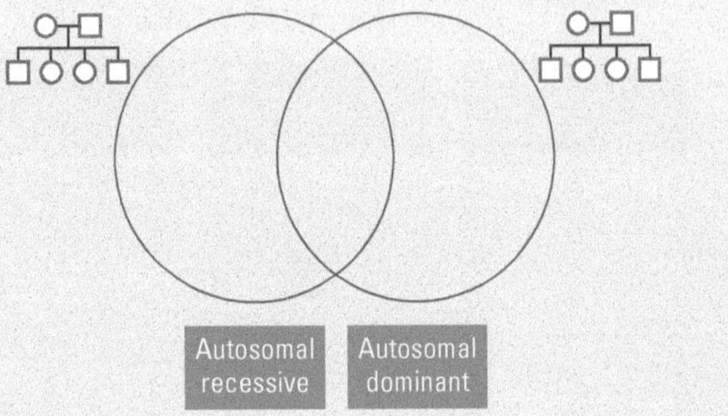

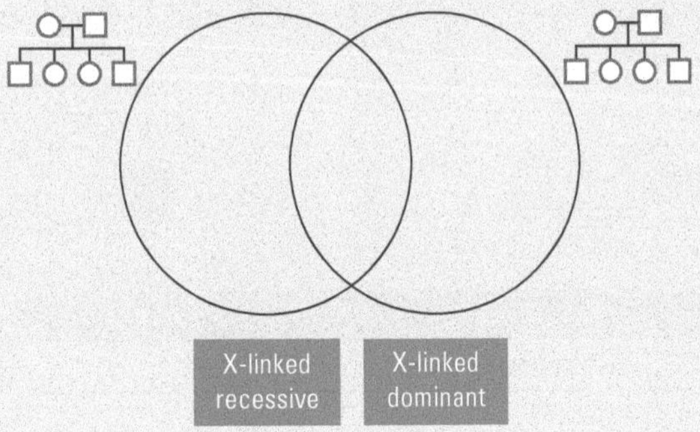

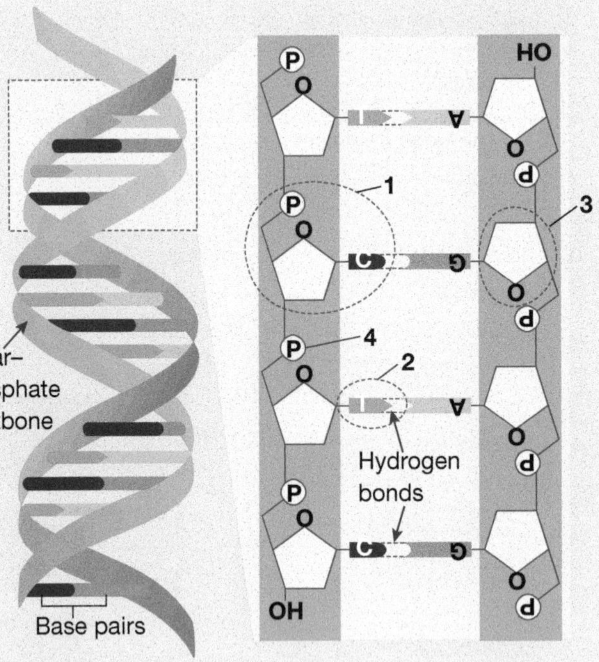

Pedigree chart showing the inheritance pattern of Huntington's disease

Key

□	Normal male	●	Affected female
■	Affected male	○	Normal female
(twin symbol)	Identical twin	1, 2, 3, etc.	Sequence of individuals

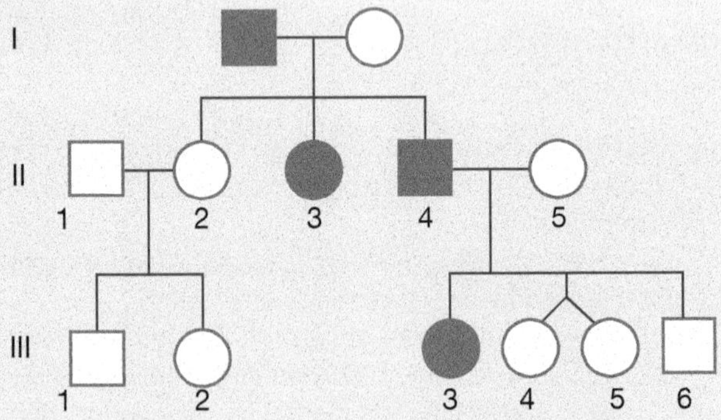

9. Examine the pedigree chart at right. Let the dark hair allele be represented by *B* and the red hair allele by *b*.
   (a) Write the genotypes for the individuals B, G, H and K.
   (b) Write the phenotypes for the individuals F and D.
   (c) If individuals G and H had another child, what is the chance that it would have red hair?
   (d) If individuals F and I had a child, do you think it might be possible for it to have red hair? Explain your reasoning.

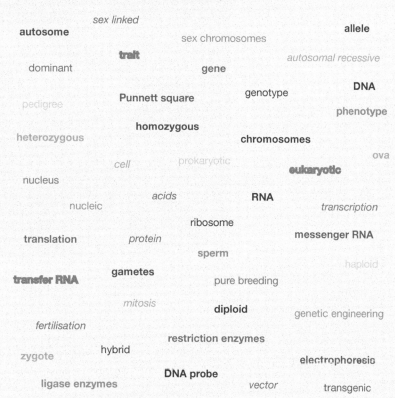

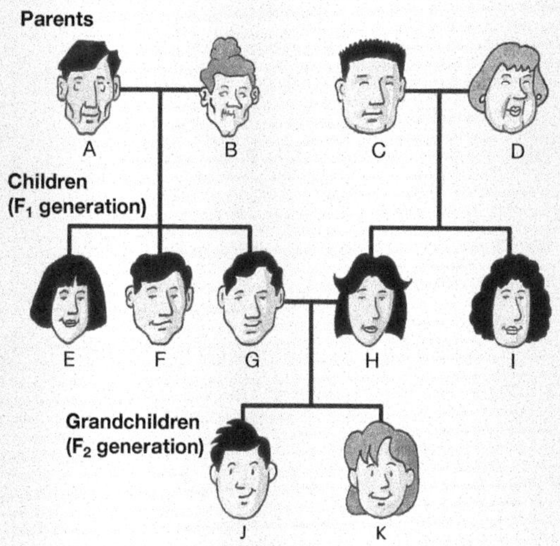

0	1	2	3
(Strongly disagree)	**(Disagree)**	**(Agree)**	**(Strongly agree)**

10. (a) For each of the following statements, rate your opinion on the scale shown below:
      (i) Insurance companies and employers should have access to your genetic information.
      (ii) Once the tools for genetic engineering are available, everyone should have access to them.
      (iii) You should be able to use gene technology to select specific characteristics of your future children.
      (iv) Before dating, you should exchange all of your genetic information with your potential partner.
    (b) Select and discuss one of the statements with a partner, constructing a PMI chart to summarise the key points.
    (c) Share your PMI chart with that of another pair or team. Add any points that you consider to be worthwhile and appropriate.
    (d) On the basis of the discussions with your peers, would you change your initial scaled opinion or is it exactly the same? Comment.
11. (a) Create a concept map that uses as many of the terms below as you can.
    (b) Select at least five terms and create a poem or song to help teach others how the terms are linked or connected. Share your creative piece with others in your class.

sex linked
autosome
allele
sex chromosomes
trait
autosomal recessive
dominant
gene
DNA
genotype
Punnett square
pedigree
phenotype
homozygous
heterozygous
chromosomes
ova
cell
prokaryotic
eukaryotic
nucleus
acids
RNA
nucleic
transcription
ribosome
translation
protein
messenger RNA
sperm
haploid
transfer RNA
gametes
pure breeding
mitosis
diploid
genetic engineering
fertilisation
restriction enzymes
hybrid
zygote
electrophoresis
DNA probe
ligase enzymes
vector
transgenic

12. Use the diagram at right to decide which of the statements below is correct.
   (a) Pig A and pig B are genetically identical.
   (b) Pig D and pig C are genetically identical.
   (c) Pig B is the surrogate mother of pig C.
   (d) None of the pigs are genetically identical as the environment in which the pigs grow can affect their genotype.

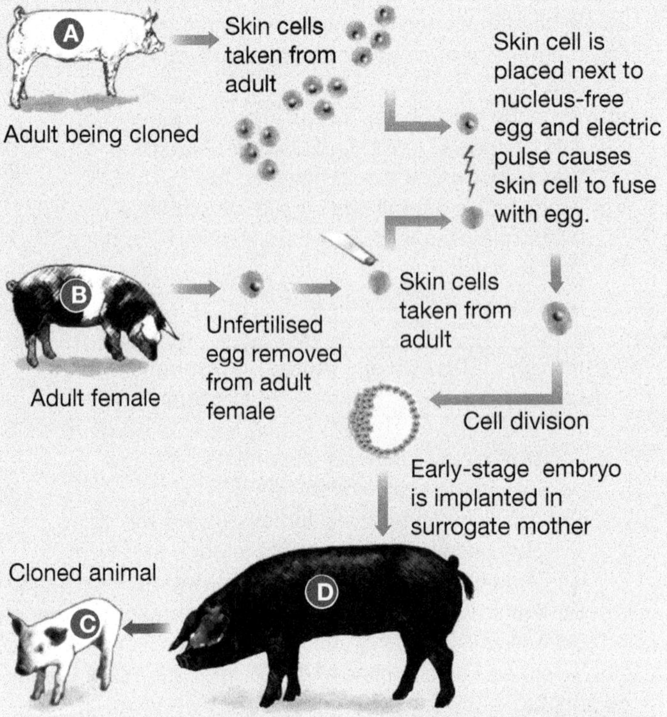

13. Search online and read the article *New genetic testing technology for IVF embryos* and answer questions 13–16.
   (a) What does PGD stand for?
   (b) What does the PGD test identify in embryos? Include four specific examples in your response.
   (c) Outline the opportunity that this test offers families with histories of genetic disorders.
   (d) Outline a negative aspect of the PGD test.
   (e) At which stage is the embryo when a single cell is removed from it?
   (f) Are you aware of any bias in the article? How many different perspectives were included? If you were to write the article, what other information or details might you include?

14. (a) In a team, construct a SWOT analysis on pre-implantation genetic diagnosis.
   (b) Share your SWOT with another team and discuss any similarities and differences between your SWOTs.
   (c) On the basis of your discussions, construct your own SWOT on the article.
15. In your team, brainstorm statements or choices related to genetic testing of embryos. Select five of these statements and position them on a priority grid with the following labels.

   Horizontal
   Left-hand side — difficult decision
   Right-hand side — easy decision

   Vertical
   Top — good result
   Bottom — bad result
16. Find out more about two of the genetic disorders listed in the article. Share and discuss your findings with your team or class members.
17. Genetic tests can be ordered on the internet!
   (a) Find out about examples of tests that are available.
   (b) Suggest some implications of being able to buy genetic tests in this way.
   (c) Is the unregulated sale of genetic tests potentially misleading or unethical? How can the accuracy and privacy of these tests be ensured?
   (d) Who owns the information of genetic tests ordered on the internet? Comment on your findings.
   (e) Construct your own SWOT analysis of your findings, and then share it with others in your class.

18. Search online and read the article *Australia issues first licence to clone human embryos*, and then answer the following questions.
   (a) State the number of human eggs that Sydney IVF was granted licence to access.
   (b) Is human cloning for reproductive purposes allowed? Explain.
   (c) State the stage of development to which the embryos are allowed to develop.
   (d) Find out more about therapeutic cloning or somatic cell nuclear transfer.
   (e) Identify issues related to therapeutic cloning.
   (f) Use a SWOT analysis to categorise points made on a discussion on therapeutic cloning.
   (g) Identify possible uses of different types of cloning.
   (h) Construct a priority grid to map out different potential uses of different types of cloning.

SWOT analysis

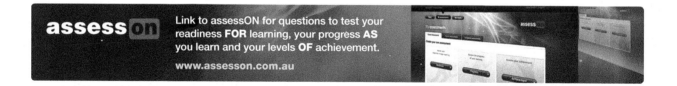

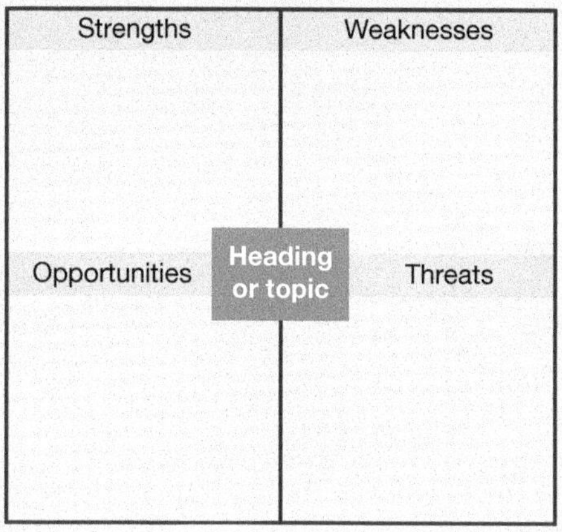

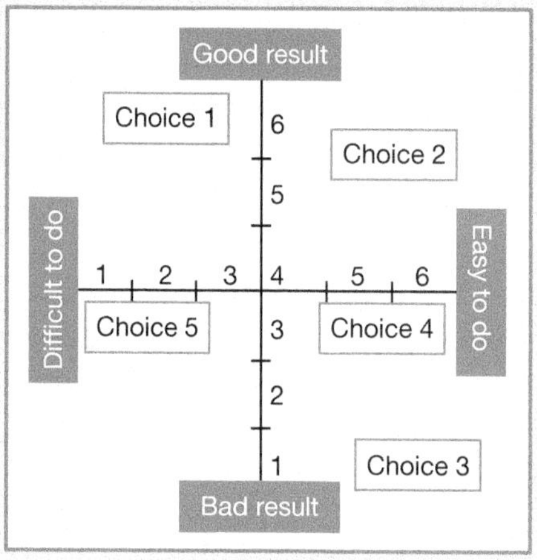

learnON RESOURCES — ONLINE ONLY

Complete this digital doc: Worksheet 2.9: Getting into genes: Puzzle (doc-19416)

Complete this digital doc: Worksheet 2.10: Getting into genes: Summary (doc-19471)

TOPIC 3
Evolution

3.1 Overview

3.1.1 Why learn this?

The great diversity of living things may be explained by the theory of evolution by natural selection. Variations upon which natural selection acts may be determined by both genetic and environmental factors. The selection of some variations over others is related to their possible effects on increasing the chances of survival and reproduction of individuals that possess them. In this way, favourable variations may be passed from one generation to the next. But what is the evidence for this theory and how can it be evaluated and interpreted?

3.1.2 Think about evolution

- How can crossing over increase variation?
- Why isn't it a good idea to have all of your eggs in one basket?
- What has dating got to do with rocks?
- How much Neanderthal DNA do you think you have in your genome?
- How can one species become two?
- What does a clock have to do with your ancestors?
- Why should we celebrate our differences?

LEARNING SEQUENCE

Numerous **videos** and **interactivities** are embedded just where you need them, at the point of learning, in your learnON title at www.jacplus.com.au. They will help you to learn the content and concepts covered in this topic.

3.1.3 Your quest

LUCA — your ancestor

Every living thing on Earth is thought to have descended from one single entity. This was a sort of primitive cell that floated around in the primordial soup over three billion years ago. It has been named the **last universal common ancestor LUCA** or **LUCA**. There is considerable controversy surrounding this ancestor, as it has left no fossil remains or any other physical clues of its identity. Researchers, however, are comparing genes from all forms of life and have put together a portrait of this cell that could be the ancestor of us all.

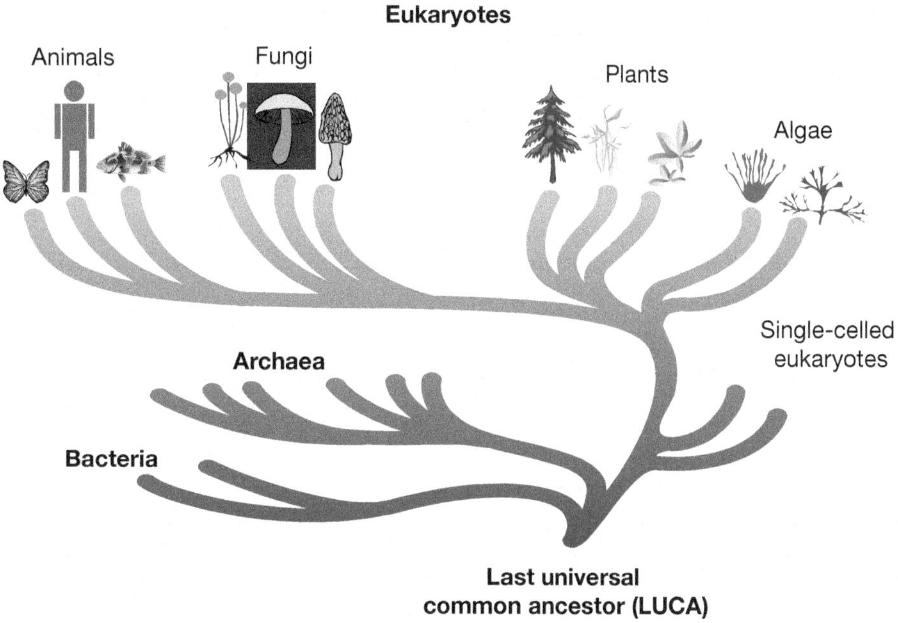

Think

1. (a) Suggest why all forms of life are coded by nucleic acids (DNA or RNA).
   (b) As nucleic acids are all made up of nucleotides, how can they differ?
2. What do you think Earth was like three billion years ago? Find out whether your hypothesis agrees with current evidence.
3. (a) Suggest the features of organisms that could survive on Earth three billion years ago.
   (b) Suggest processes or features that may have increased the chance of organisms passing on their traits to their offspring at this time.
   (c) Design an organism that could survive these conditions. Describe its features, including those that are present in organisms living today.

3.1.4 Five minutes to midnight

If we compressed Earth's 4.5 billion year history into a single year, Earth would have formed on 1 January and the present time would be represented by the stroke of midnight on 31 December. Using this timescale, the first primitive microbial life forms appeared in late March, followed by more complex photosynthetic micro-organisms towards the end of May. Land plants and animals emerged from the seas in mid-November. Dinosaurs arrived early on the morning of 13 December and then disappeared forever in the evening of 25 December. Although human-like creatures appeared in Africa during the evening of 31 December, it was not until about five minutes before the New Year that our species, *Homo sapiens*, appeared on Earth.

INVESTIGATION 3.1

The common ancestor

- Carefully observe the features of the possums in the figure.
  1. Make lists of how the possums are similar and how they are different.
  2. Suggest reasons for the differences.
  3. Suggest how the possums may have become different.

3.2. Classification

3.2.1 Classification can change

Classification is not fixed. With the wonders of new knowledge and understanding comes the excitement — and the frustration — of new theories and terminology.

There have been shifts from a model of two kingdoms, plants and animals, to a five-kingdom model and then a number of further variations. Initially, the main characteristics used to classify organisms into these groups were structures visible to the human eye, but with the development of microscopes, cell structure could be used as well. Now, due to new technologies, the chemical composition of organisms can be analysed at a molecular level.

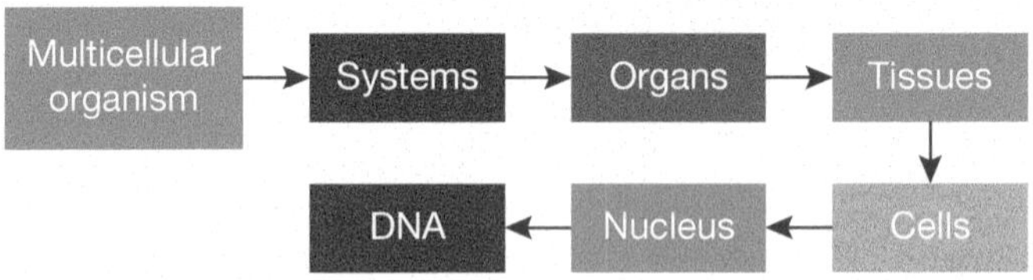

3.2.2 Changing tides of classification

The five-kingdom classification was proposed by Whittaker in 1969. In 1990, Woese proposed a model that focused on genetic rather than physical characteristics to divide organisms into groups. This new grouping added broader levels of classification (domains) that were then divided into kingdoms. With new technologies, what other types of classification systems might be suggested?

Although classification systems are not fixed and can change when new information is discovered, they are very useful for categorising organisms into groups. Classification systems help us to see patterns and order, so we can make sense and meaning of the natural world in which we live.

3.2.3 Binomial nomenclature

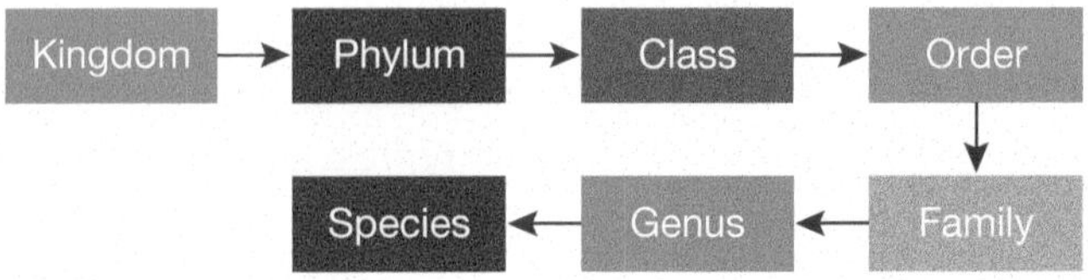

Classifying organisms into groups provides a framework that uses specific criteria and terminology and improves our communication about organisms. The Swedish naturalist Carolus Linnaeus (also known as Carl von Linné) developed a naming system that could be used for all living organisms. It involved placing them into groupings based on their similarities. He called the smallest grouping *species*.

Linnaeus' naming system was called the **binomial system of nomenclature** because it involved giving each species a particular name made up of two words. The scientific names given to organisms were often Latinised. In this system, the species name is made up of a genus name as the first word and a descriptive or specific name as the second word. A capital letter is used for the genus name and lower case for the descriptive name. If handwritten, the species name should be underlined; if typed, it should be in *italics*.

A painting of Carolus Linnaeus as a young man

WHAT DOES IT MEAN?

The word *binomial* comes from the Latin terms *bi-*, meaning 'two', and *nomen*, meaning 'name'.

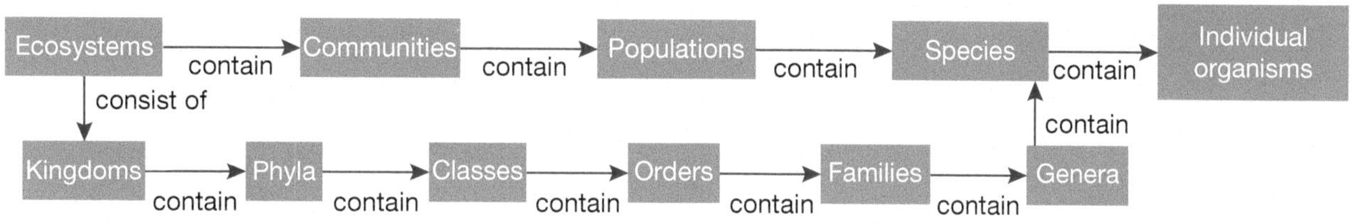

3.2.4 Species

There have also been changes in our definition of the term *species*. Although genetic technologies have blurred the lines of our classification system, our usual definition of species refers to individuals that can interbreed to produce fertile offspring.

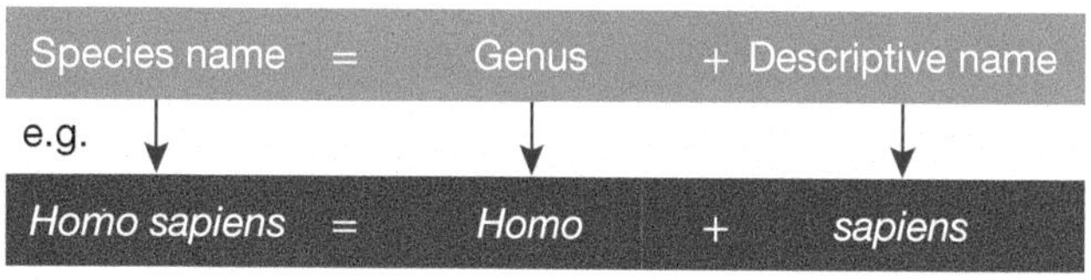

Species also fit into another grouping in terms of where they belong within an ecosystem. Ecosystems consist of a number of different communities. Within these communities are populations of individual organisms of a species living together in a particular place at a particular time.

3.2 Exercises: Understanding and inquiring

To answer questions online and to receive **immediate feedback** and **sample responses** for every question, go to your learnON title at www.jacplus.com.au. *Note:* Question numbers may vary slightly.

Remember

1. Suggest why classification of organisms is not fixed.
2. Select appropriate terms from the list at right and use flowcharts to show their connections.
3. Suggest why scientists classify organisms into groups.
4. State the name of the person who developed the binomial system of nomenclature.
5. Outline the naming system used in the binomial system of nomenclature.
6. Outline a definition of species.
7. Outline the relationship between species and ecosystems.

Populations DNA Multicellular
Ecosystems Organism Systems
Tissues Species Cells
Classes Communities Kingdoms
Families Phyla Orders
Genera Nucleus Organs

Think and discuss

8. Suggest criteria that are used to divide organisms into the five-kingdom classification system.
9. Explain why the number of kingdoms has changed from two to at least five.
10. State the relationships between the two terms in each of the following pairs.
    (a) Genus and genera
    (b) Phyla and phylum
    (c) *Homo and sapiens*
    (d) Genus and species
    (e) Species and populations

Investigate

11. Use the internet to find out the latest classification systems and reasons for their differences from previous systems.
12. Research and report on Carolus Linnaeus.

Create

13. Using magazines, the internet and other resources, find pictures of living things and classify them into kingdoms. Make a poster showing members of the five kingdoms.
14. Construct a table showing the key similarities and differences between the five kingdoms at a cellular level, and then construct a model that demonstrates these for each kingdom.

3.3. Biodiversity

3.3.1 I am a dog!

Look at the dogs in the photograph at right. What differences can you see? How did this variation come about when all the individuals belong to the same species?

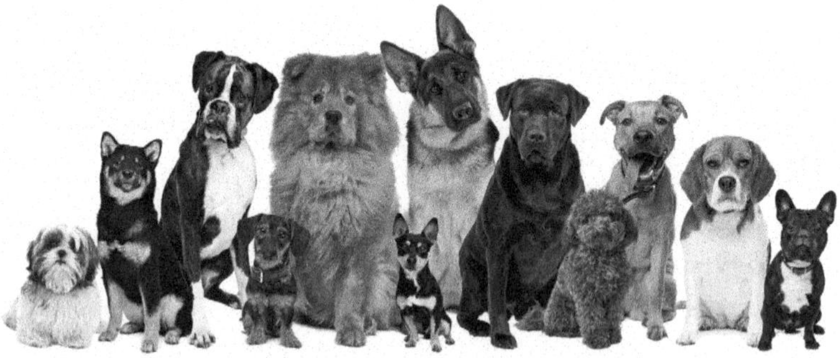

3.3.2 It's great to be different!

Have a look at the people around you. How many differences do you notice? How can you explain your observations? One part of your response might deal with genetics and inherited traits; another part might deal with the environment. The variation of characteristics or phenotypes within populations has contributed to the survival of our species.

3.3.3 Genetic diversity

Biodiversity (or biological diversity) has to do with variation within living things. It can be described in terms of an ecosystem, at the level of species, or even at the level of individual genes. **Species diversity** is the number of different species within an ecosystem. In contrast, genetic diversity is the range of genetic characteristics within a single species. The most important level in terms of evolution is that of the gene.

Genetic diversity is important because it codes for variations of phenotypes, some of which may better suit the individual organism to a particular environment than others, giving it an increased chance of survival. If this individual survives, there is an increased chance of it reproducing to pass the advantageous gene to its offspring, also giving them an increased chance of surviving. Overall, this genetic advantage will increase the survival of the species within that particular environment.

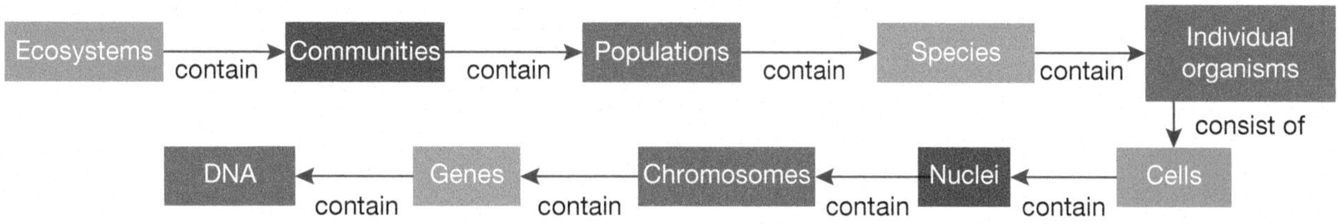

Mutation

Mutation can occur in all organisms and is the source of new genetic variation. A change in the genetic code in DNA can lead to a change in the protein that is coded for and produced by that segment of DNA. This can change the organism's characteristics. In the diagram on the right, for example, a change in DNA has led to the production of a protein that changes the colour of the mouse from white to black. Mutations that occur in germline cells (such as sperm and eggs) are the source of new alleles (alternative forms of genes) within populations.

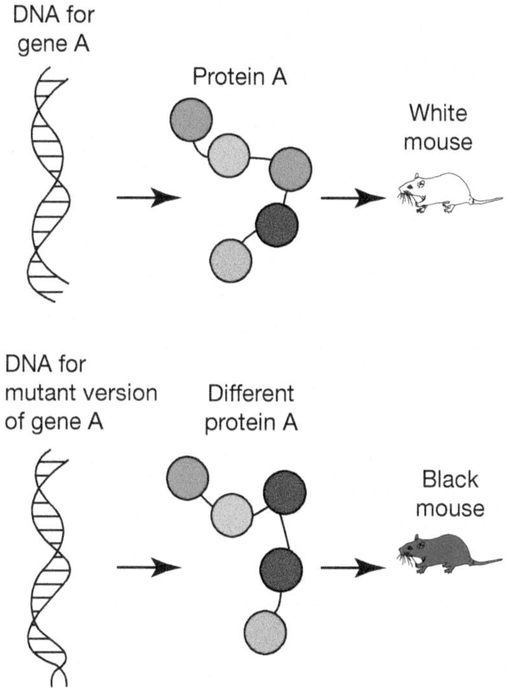

3.3.4 Variation between individuals

Variation between individuals that reproduce sexually may also be the result of several other factors besides mutation. Variation can occur during **meiosis** due to crossing over of sections of maternal and paternal chromosomes, and also due to the independent assortment of the chromosomes into the **gametes**. The combination of gametes that fuse together during **fertilisation** provides another source of variation, as does the selection of a particular mate.

Variation between individuals can be described in terms of **alleles** — the alternative forms of genes. The possible variation of alleles for a particular trait within an individual is called a **genotype**. Genotypes and the environment contribute to the variations in **phenotypes** or characteristics between organisms.

Independent assortment and crossing over during meiosis are two causes of variation between individuals.

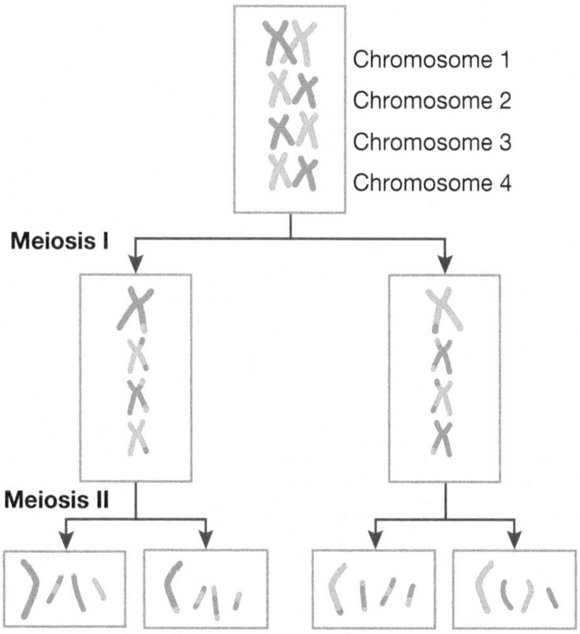

Adaptations

Variations that increase chances of survival may also be thought of as adaptations. An adaptation may be considered to be a special feature or characteristic that improves an organism's chance of survival in its environment. There are different types of adaptations; for example, structural adaptations (e.g. hair to keep warm), behavioural adaptations (e.g. courtship display to attract a mate) and physiological adaptations (e.g. ability to produce concentrated urine to conserve water). Can you think of adaptations that you possess that increase your chances of survival in your current environment?

3.3.5 Variation within populations

Variation can also be described within populations. Genetic variation within populations can be referred to in terms of the frequency of particular alleles within the population. While the genotype describes the variation of alleles for a particular trait within an individual, a gene pool describes the alleles for a particular trait within a population.

Genetic drift — changes due to chance events — and **natural selection** can have an impact on allele frequencies and hence variation within populations.

Frogs and gene pools

When individuals of a species of frog mate, they recombine their genetic material to produce offspring that show a wide variety of characteristics. Such variability within a species is important because it may enhance the chances of survival of the individual's offspring in a changing environment. Some individuals may have genes (or alleles) that assist in their survival. They may then pass these favourable genes on to their offspring. On the other hand, if a large number of frogs emigrated or were removed from a particular habitat without mating, their genes (or alleles) would be removed from the gene pool. Once removed, they are gone forever.

A species of frog might show a wide variety of characteristics; this variability enhances the chances of survival.

Gene flow

Movement of individuals between populations provides another possible source of diversity. **Emigration** (moving out) may result in the loss of particular alleles; **immigration** (moving in) may result in the addition of new alleles into the population.

Before advances in technology provided humans with relatively easy long-distance travel, our species was split into small groups. The separate identities of these groups was maintained by geographical barriers such as mountains and oceans, and by attitudinal and social barriers. With the advent of faster and more accessible means of transport and improved communication technologies, these barriers are now starting to break down, and migration and interbreeding between human groups is widespread.

Sometimes the variation introduced into a population is not beneficial. An inherited anaemic disease, thalassaemia, is common among people living along the Mediterranean coast, particularly among people of Greek origin. As people from this part of the world migrated to Australia, they brought with them the thalassaemia allele. Although this trait is recessive and requires two heterozygous parents to contribute it for their offspring to express it, there is an increasing number of people within the Australian population with this disease.

In other cases, the introduction of a genetic trait into a population may increase the chances of survival of individuals with the trait. This new variation may contribute to increasing the fitness of the population to the current or future environment.

3.3.6 Environments change

All environments change over time. If they change too rapidly, the genes required for survival in the changed environment may not be present in the gene pool and that species may become extinct. Over the last 100 years, the natural habitats of many species have been changed so significantly that they have not possessed the genetic variation to be able to adapt to the new conditions. Many species have therefore died out.

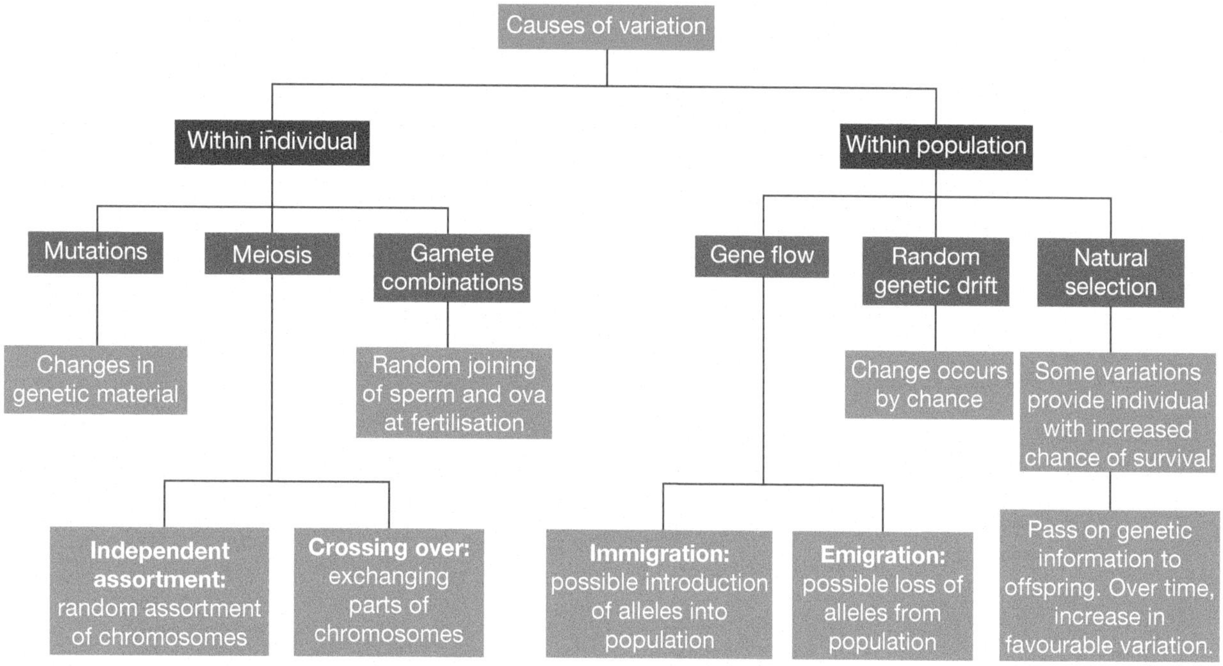

3.3.7 Reduced biodiversity

The use of reproductive technologies such as artificial selection, artificial insemination, IVF and cloning has the potential to unbalance natural levels of biodiversity. These technologies can be used in horticulture, agriculture and animal breeding to select which particular desired characteristics will be passed on to the next generation.

Artificial selection

For thousands of years, humans have used selective breeding techniques to breed domestic animals and plants. We have selected which animals will mate together based on their possession of particular features, to increase the chances that their offspring will have features that suit our needs. This type of selective breeding is called artificial selection.

Because fewer individuals are selected for breeding, genetic diversity is reduced and inbreeding may result. As well as decreasing variation in the traits of offspring, inbreeding can increase the chances of inherited diseases.

Artificial insemination

The sperm of a prize-winning racehorse may be used to inseminate many mares, increasing the chance of offspring that also possess the race-winning features of its father. This leads to a larger contribution of alleles from this horse than would naturally be possible. It can also lead to reduced genetic diversity within the populations of horses in which this occurs.

IVF and embryo screening

In-vitro fertilisation (IVF) techniques allow the testing and selection of embryos for particular characteristics prior to their implantation. This can also have an impact on genetic diversity. Imagine the effect of implanting only female embryos or only those with a particular recessive trait.

Cloning

Imagine the production of a population of genetically identical individuals. Although they may be well suited to a particular environment, what might happen when the environment changes to one that they are not suited to?

Consequences of reduced biodiversity

Reducing variation in genetic diversity can lead to the eradication of populations or entire species. If the population or species is exposed to an environmental change or threat (for example, disease, climate change or lack of a particular resource), the reduced variation may mean that there is less chance that some of the species will survive to reproduce.

3.3 Exercises: Understanding and inquiring

To answer questions online and to receive **immediate feedback** and **sample responses** for every question, go to your learnON title at www.jacplus.com.au. *Note:* Question numbers may vary slightly.

Remember

1. Suggest why genetic diversity is important to the survival of a species.
2. Select appropriate terms from the list at right and use flowcharts to show their connections.
3. Distinguish between:
   (a) genetic diversity and species diversity
   (b) gene and allele
   (c) immigration and emigration
   (d) genotype and gene pool.
4. Outline the relationship between mutation and genetic variation.
5. Explain why mutations are important to asexually reproducing organisms.
6. Are all mutations bad? Justify your response.
7. Suggest causes or sources of genetic variation for:
   (a) an individual
   (b) a population.
8. (a) Outline how humans have achieved artificial selection.
   (b) State the desired outcome of artificial selection.
   (c) Suggest possible consequences of artificial selection on genetic diversity.
9. Outline the advantages and disadvantages of artificial insemination.
10. Suggest how IVF and related technologies could reduce human genetic diversity.
11. Outline the consequences of reduced genetic diversity to the survival of a species.

Populations DNA Mutation Meiosis
Gamete combination Gene flow
Chromosomes Cells Ecosystems
Organisms Genes Species
Genetic drift Gene flow Natural selection
Communities Nucleus Variation
Crossing over Emigration Immigration
Fertilisation Alleles
Independent assortment

Think and discuss

12. (a) List the differences you can see among the dogs (*Canis familiaris*) at the beginning of this subtopic.
    (b) Suggest how these variations came about.
    (c) Find out more about artificial selection and how it is currently used in Australia.
13. Discuss the implications on human populations of allowing the selective implantation of embryos that:
    (a) are male
    (b) possess recessive traits such as red hair or blue eyes
    (c) have a potential for a higher IQ
    (d) have a potential for paler skin.
14. Unscramble the following terms.
    (a) vteydioirsib
    (b) gmtimiranio
    (c) egsatem
    (d) seosimi
15. Suggest what will happen if a species does not possess the genes that allow it to adapt to new environmental conditions.

16. *Brassica oleracea* is a common ancestor to a number of vegetables.
    (a) Identify differences between *Brassica oleracea* and each of the species in the figure at right.
    (b) Suggest how these differences came about.

All of these vegetables were produced by artificial selection and share a common ancestor. Could this have happened by natural selection also?

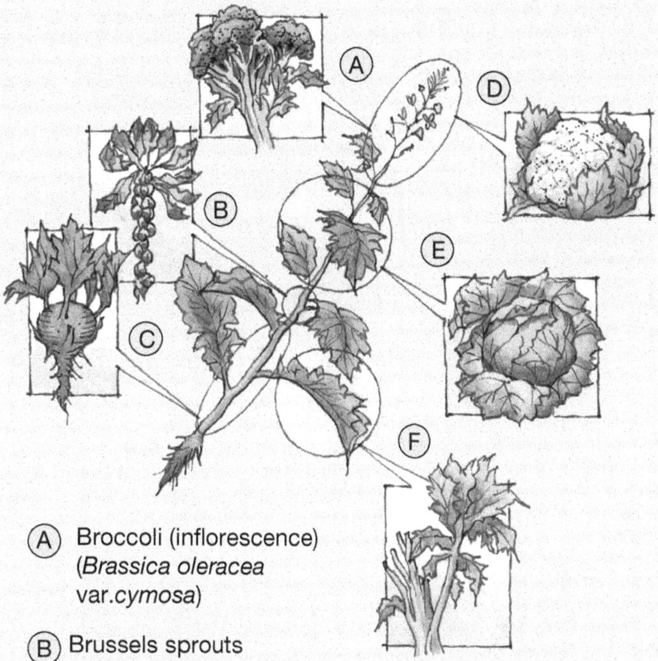

(A) Broccoli (inflorescence) (*Brassica oleracea* var.*cymosa*)

(B) Brussels sprouts (lateral buds) (*Brassica oleracea* var.*gemmifera*)

(C) Kohlrabi (stem) (*Brassica oleracea* var.*gonglyoides*)

(D) Cauliflower (flower) (*Brassica oleracea* var.*botrytis*)

(E) Cabbage (terminal buds) (*Brassica oleracea* var.*capitata*)

(F) Kale (leaf) (*Brassica oleracea* var.*acephala*)

Investigate and create

17. Choose a species. Using the internet, magazines or other resources, collect pictures that show variation within the species. Paste these onto a poster and label the types of variations.
18. Investigate meiosis and its involvement in generating genetic variation. Construct a model that shows independent assortment and crossing over.
19. Construct a model that shows possible variations of outcome from the fertilisation of gametes.
20. Find out more about genotypes and gene pools, and create your own animation or cartoon to teach other students the difference between the two.

Investigate, think and report

21. Read the article below and answer the following questions:
    (a) What question were the researchers trying to answer?
    (b) Describe the experiment they set up.
    (c) What were their results? What were their conclusions?
    (d) What implications do these conclusions have for life on Earth if the ozone layer continues to break down?
    (e) Use internet research to identify relevant questions that could be investigated scientifically.
    (f) Research and report on Australian research on the ozone layer.

BACTERIA DON'T SUNBAKE!

The ozone layer helps to filter out many harmful UV rays. In Precambrian times, over 540 million years ago, there was, however, no ozone layer. How then did life survive? To answer this question, a team of biologists from NASA's Ames Research Centre in California has been studying microbial mats of bacteria and blue–green algae, the types of organisms that lived in Precambrian times.

The natural production of DNA in each organism was studied by placing some of the mat into a plastic bag that was transparent to UV light. Some radioactive phosphate was added into the bag. (Phosphate is used by cells to produce DNA.) Every few hours the amount of phosphate in the DNA of some of the cells was measured. The results for both bacteria and blue–green algae showed the same pattern of DNA production. At sunrise, the amount of phosphate in the DNA was high. DNA production then ceased at noon for three to six hours, resuming just before sunset. Photosynthesis occurred throughout the whole day.

Head of the research team Lynn Rothschild concluded that the cells cease DNA production at noon because of the harmful UV light. The cells use this time to repair any DNA damage before they begin to divide again. This mechanism might give some unicellular organisms a natural advantage if the Earth's ozone layer continues to be destroyed.

3.4 The evolution revolution

3.4.1 The theory of evolution — bigger than one man

Do you and the apes that you see at the zoo share a common ancestor? This concept caused much controversy between religion and science.

The concept of organisms sharing common ancestors contributed to the development of the **theory of evolution**. Although this theory is usually credited to one man — Charles Darwin (1809–1882) — it is really a culmination of the ideas of many individuals both in Darwin's time and before it. With technological advances and new knowledge, refinements have been made to Darwin's theory and continue to be made. The theory of evolution itself is evolving.

Darwin points out the similarities between humans and apes, by an unnamed artist in *The London Sketch Book*, 1874

3.4.2 Evolution ownership

Like many other scientific theories, although one person may be credited as its sole creator, it is really a story of awareness, relationships, passion and wonder. The development of a theory usually requires an appreciation of and connection to what has been before and the transfer of this knowledge to new discoveries. It often involves seeing links, patterns or connections that can tie all of the knowledge together into a new framework of understanding.

3.4.3 Grandpa sows his own seeds

Erasmus Darwin (1731–1802) was not only a British physician and leading intellectual of his time, but also Charles Darwin's grandfather. He believed that all living organisms originated from a single common ancestor and in 1794 published *Zoonomia* — a book that sowed the seeds for later ideas regarding the theory of evolution.

Carolus Linnaeus (1708–1778), the 'father of taxonomy'

3.4.4 Classifying life

Carolus Linnaeus (1708–1778) is considered the founder of **taxonomy** — the branch of biology concerned with naming and classifying the diverse forms of life. He developed his classification system 'for the greater glory of God', rather than in the interest of scientific understanding. His ideas, however, were used as a basis for the development of the theory of evolution.

3.4.5 Hints in the ground

Without the contribution of geologists, the theory of evolution may still not have been developed. **Geology** bestowed a great gift upon Darwin's generation of scientific thinkers — the gift of time.

In the eighteenth century, many believed that the Earth was only around 6000 years old and that — other than changes brought about by sudden, dramatic catastrophes (like Noah's biblical flood) — it was unchanging. This was the theory of **catastrophism**.

3.4.6 An underground time machine

James Hutton (1726–1797) proposed the theory of **gradualism**, which suggested that Earth's geological features were due to the cumulative product of slow but continuous processes. He used money from farming and his invention of a process for manufacturing the chemical sal ammoniac to devote his life to his scientific quests. It was not until almost ten years after he presented his theories to the Royal Society of Edinburgh in 1785 that they were taken seriously enough to be vigorously attacked. Hutton's response was to publish *Theory of the Earth* in 1795. Hutton's geological theories were built upon by others who also observed evidence that contradicted the theological teachings of the time.

English surveyor William Smith (1769–1839) made great contributions to the development of geology and could be considered the 'father of English geology'. He is credited with creating the first nationwide geological map (more information can be found in his biography, *The Map That Changed the World* (2002)) and was the first person to make a systematic study of fossils.

Smith was the son of a blacksmith and his life was not an easy one. He published his first geological map of Britain in 1815. Unfortunately, the map was plagiarised and he became bankrupt and then served time in debtor's prison. He did not receive recognition for his contributions until many years later in 1831.

Smith's work as a surveyor took him down into mines where he observed different layers of rocks (strata) and the fossils that they contained. Smith noticed a regular pattern to the distribution of types of fossils in the particular rock layers in the different locations where he worked. This pattern suggested that the Earth must be very old and that successive strata had been laid down one on top of the other. His observations also suggested that different types of organisms had appeared, lived for a while, and then been replaced by others.

Born in the same year as William Smith, Baron Georges Cuvier (1769–1832) played a key role in the development of **palaeontology** (the study of fossils). He is credited with recognising that fossils in deeper strata were older than those in strata closer to the surface.

Geologist William Smith (1769–1839) is credited with creating 'the map that changed the world'.

Diagram showing characteristic fossils in the different layers of strata. These studies led to the development of stratigraphy.

Sir Charles Lyell (1797–1875) was born the year that James Hutton died. He incorporated Hutton's gradualism into **uniformitarianism theory** — the antithesis of catastrophism. Lyell also played a key role in Darwin's decision to finally publish, and formally presented Darwin's (and Wallace's) theory of evolution to the scientific community in 1858.

Baron Georges Cuvier (1769–1832)

3.4.7 Use it or lose it!

Jean-Baptiste de Lamarck (1744–1829) was one of the first scientists to suggest that populations of organisms changed over time and that old species died out and new species arose. He believed that if a particular feature was not used then it would eventually be lost over succeeding generations. He also suggested that changes acquired within the lifetime of an individual could be passed on to its offspring. The example often given to describe this theory relates to the long necks of giraffes. Lamarck's explanation would be that the giraffes had to reach the leaves high up in the trees, stretching their necks. He would then suggest that the lengthened necks that resulted from this stretching were passed on to their offspring.

3.4.8 Seeds of inheritance

Gregor Mendel (1822–1884), an Austrian monk, used peas of different colours and shapes in his experiments and is responsible for the development of the fundamentals of the genetic basis of inheritance. Although most of his work was destroyed, his gene idea was recognised 34 years after his death and provided a mechanism for natural selection.

Gregor Mendel (1822–1844) used peas of different shapes and colours to collect data that provided him with patterns of inheritance. Much of Mendel's work was destroyed by the church.

Jean-Baptiste de Lamarck (1744–1829) is often referred to in terms of the 'use and disuse' and acquired inheritance theories. Although his theories have been discredited in favour of Darwin's, they are making a comeback in new findings in the science of epigenetics.

Living around the same time as Mendel, Herbert Spencer (1820–1903) suggested the concept of survival of the fittest. Born twelve years after Mendel, August Weismann (1834–1914) demolished Lamarck's theory of the inheritance of acquired traits, and is well known for his experiment that cut the tails off mice to collect evidence that tail loss during the parent's lifetime was not inherited by their offspring. He later also suggested that **chromosomes** were the basis of heredity.

3.4.9 Darwin's journey of selection

In 1831, a 22-year-old Charles Darwin set sail on a five-year voyage on the *HMS Beagle*. It was a journey that would greatly change his views on life. He noted the similarities and differences in the flora and fauna inhabiting the different regions that he visited. His observations made him question the belief at the time that the Earth was only a few thousand years old and that its organisms were the unchanging work of a creator.

Darwin was particularly puzzled by the features of animals on the Galapagos Islands near South America. On these islands, he noticed a number of different species of finches that were similar in size and colour, but varied in the size and shapes of their beaks. He recorded that these **variations** suited them to particular types of foods.

Charles Darwin

Darwin's doubts doubled

By the time Darwin had sailed from Galapagos, his observations and awareness of the ideas of the geologist Sir Charles Lyell (who had been influenced by James Hutton) led him to doubt the church's position that the Earth was static and only a few thousand years old. He was particularly influenced by Lyell and Hutton's views that geological change resulted from slow, continuous actions rather than sudden events.

After returning home in 1837, Darwin began his notebooks on the origin of different species and in 1844 (at 35 years of age), wrote his essay *On the Origin of Species*. Aware of the controversy that such ideas may fuel, he left this essay unpublished for over ten years.

3.4.10 Why do some die and some live?

While Darwin continued to develop his theory, an English naturalist reached the same conclusion. His name was Alfred Wallace (1823–1913). Wallace was a school teacher with a passion for botany and collecting plants and insects. Like Darwin, Wallace had travelled extensively and made many detailed observations of variations in the species that he came across. In 1848, he began a series of expeditions, first to the Amazon and later to the Malay Archipelago where he stayed for eight years.

Alfred Wallace (1823–1913) sent his theory of evolution to Darwin.

In February 1858, while he was recovering from a bout of malaria, Wallace remembered reading a book titled *Essay on the Principle of Population* (1798). This book was written by the mathematician, economist and founder of demography Thomas Malthus (1766–1834), and had also influenced Darwin's thinking. Wallace connected what he had remembered from this book to his observations. It is documented that the idea of survival of the fittest then came to him in a flash. In his autobiography, *My Life: A Record of Events and Opinions* (1905), Wallace wrote:

> It occurred to me to ask the question, why do some die and some live? And the answer was clearly that, on the whole, the best fitted lived.

Within two evenings, Wallace had written an essay on his theory of evolution and sent it to Darwin in the next mail.

3.4.11 Publish or perish

Imagine the shock of seeing your life's work summarised in a letter, sent to you for comment on its possible publication. This is what must have happened when Darwin opened Wallace's letter describing his theory of evolution. This forced Darwin to reconsider publishing his previously unpublished work on his theory. Given his wife and family's religious connections, this must have been a difficult personal time for him and later for his family.

On the advice of Sir Charles Lyell and the eminent botanist Sir Joseph Hooker, Darwin decided to publish his work along with Wallace's essay in a joint paper. In July of 1858, Sir Charles Lyell presented Darwin's previously unpublished 1844 essay along with Wallace's work to the Linnean Society of London. Later, in 1859, Darwin finally published his book *On the Origin of Species by Means of Natural Selection*. Many people were outraged by the suggestion that humans could be related to apes. There were many debates and arguments about the theory of evolution. A young anatomist, Thomas Henry Huxley (1825–1895), fought the case for evolution in many public debates. He did this so fiercely that he became known as 'Darwin's bulldog'. Eventually, the scientific community came to accept Darwin's theory, some even expressing embarrassment at not having thought of such a simple explanation before.

Thomas Henry Huxley (1825–1895) was often described as 'Darwin's bulldog' and had his own adventures in his early twenties aboard the *HMS Rattlesnake*.

3.4.12 Natural selection

Darwin's theory was different from others in that it included a process by which evolution could occur. Although this process is often referred to as 'survival of the fittest', he called it **natural selection**. He believed that by this process a single species could have given rise to many new species, and that these new species were much better suited to the environment in which they lived.

Natural selection proposes the following:

1. There is variation of inherited characteristics in a species and some of these variations will increase the chances of surviving in a particular environment.
2. In the struggle to survive, those members with favourable traits will have an increased chance of survival over others.
3. Surviving members have an increased chance of reproducing and passing on their inherited traits to their offspring.
4. Over time and many generations, organisms will possess traits that are better suited to their environment and increase their chances of survival.

Some view this as the time when science broke away from religion. Would it still have occurred if Wallace had not sent his theory to Darwin? Would the theory of the less well-known Wallace have been taken seriously? Given the changing ideas and new knowledge being discovered at that time, would someone else have come up with the same idea?

3.4.13 Brave new world

There are two other Huxleys who have had an impact on how we see the world. Both of these are the grandsons of Thomas Henry Huxley. Sir Julian Huxley (1887–1975) was involved in the formulation of Darwinian evolution that incorporated developments in genetics and palaeontology. Aldous Huxley (1894–1963) was the author of novels that both inspired and caused fear, as well as increasing public awareness of the possible implications of science for our future humanity. The novels *Brave New World* (1932) and *Island* (1962) (and their subsequent movies) have caused many to pause, reflect and consider the potential ethical issues that new scientific discoveries and their applications may hold for our species.

Julian Huxley (left) with his brother Aldous

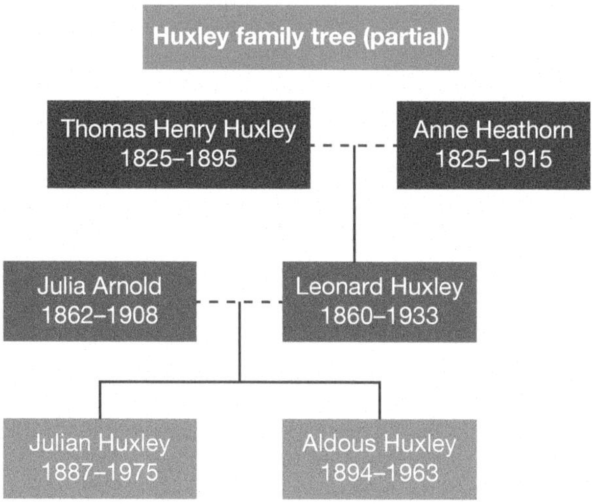

A young Aldous Huxley

The theory of evolution is a culmination of ideas from many different individuals.

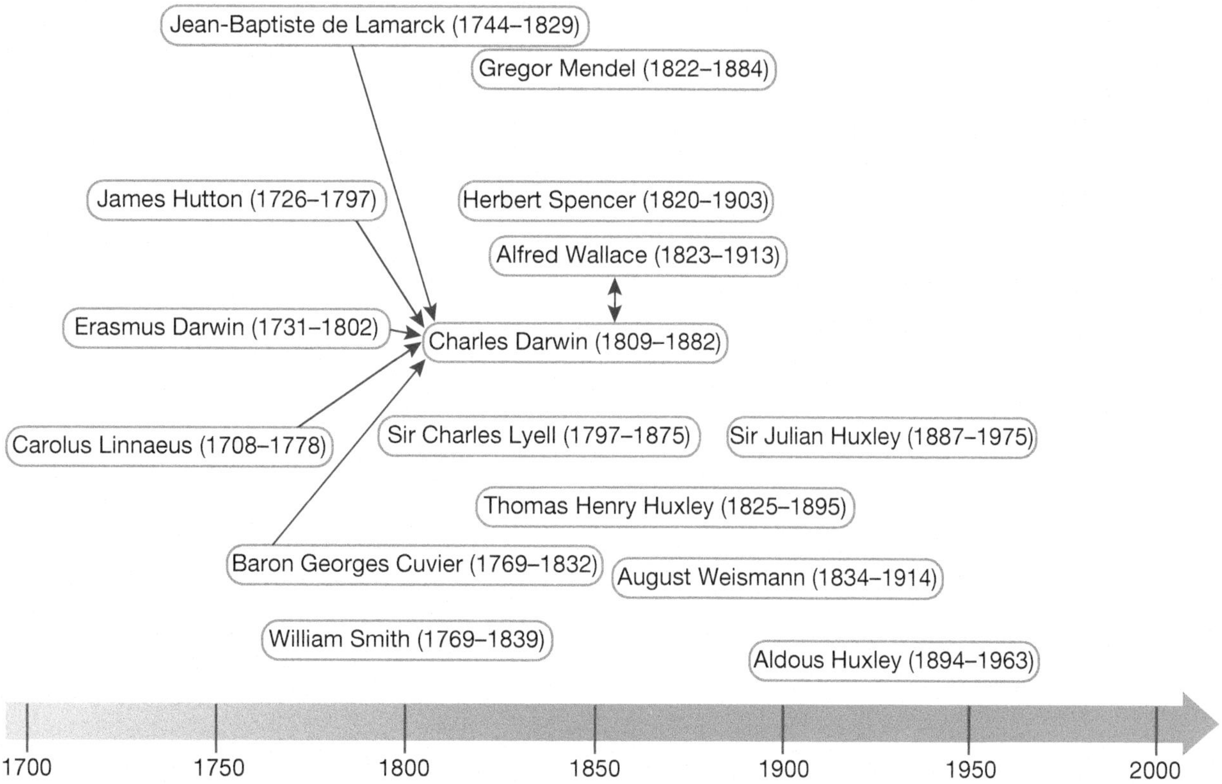

3.4 Exercises: Understanding and inquiring

To answer questions online and to receive **immediate feedback** and **sample responses** for every question, go to your learnON title at www.jacplus.com.au. *Note:* Question numbers may vary slightly.

Remember

1. Who are the two people jointly credited with developing the theory of evolution?
2. What did Darwin conclude from the observations he made during his voyage on the *HMS Beagle*?
3. Summarise the process of natural selection.
4. Outline the contributions of the following to the theory of evolution: Linnaeus, Darwin, Lyell, Hutton, Wallace, Mendel.

Investigate, think and discuss

5. (a) Use the timeline and information in this subtopic to answer the following questions.
   (i) Research the time period in which these scientists grew up.
   (ii) Imagine what life was like and the sorts of beliefs that were held by the majority in the society in which they lived. Collect resources and record notes to summarise your findings.
   (iii) As a class, in teams, select one of the scientists discussed in this subtopic. Rigorously research your scientist to find out as much as you can about their life (both personal and professional). Write a biography about your scientist.
   (iv) Research the other characters in this subtopic that may have influenced your selected scientist and construct a PMI chart about your findings.
   (v) As a class, discuss how your combined research could be organised into a novel about the development, presentation and final acceptance of Darwin and Wallace's theory of evolution.
   (vi) Write your own novel of the collective material and then create a screenplay or storyboard that enables you to tell the story of the development of the theory to others.
   (vii) Present your screenplay or storyboard to the class. Try to make your presentation as creative as possible.
   (viii) Find out more about the people and books mentioned (e.g. *Theory of the Earth* (1788), *Zoonomia* (1794), *An Essay on the Principle of Population* (1798), *Principles of Geology* (1830–33), *The Origin of Species* (1859), *The Map That Changed the World* (2002)) and their contributions to our scientific understanding of the world in which we live.

   (b) Add other social, cultural, religious, political or historical events to the timeline. Select one of the scientists and incorporate this information into what life must have been like for them. Comment on the influence that their contributions to the theory of evolution may have had on their personal lives. Do you think that some of today's scientists also face social pressures in regards to their work?

   (c) Suggest where the discovery of DNA would fit into the timeline. Discuss the effect that this has had on the evolution theory.

3.5 Natural selection

3.5.1 Best suited

Survival of the fittest is more than having muscles, being tough or working out at the gym. It's about being better suited to a particular environment and having an increased chance of surviving long enough to be able to have offspring that will take your genes into the next generation.

Can you see the butterfly in this image?

3.5.2 Vive la différence!

We are all different! Our differences increase the chance of survival of our species. If our environmental conditions were to change, some of us might have an increased chance of surviving over others. Those who survive might then pass on any genetically inherited advantage to their children, who would also have an increased chance of survival; at least unless the environment changed to their disadvantage. If this happened, then other variations may have increased chances of survival. This is what the theory of evolution and natural selection is all about.

3.5.3 The mechanism for evolution

Darwin and Wallace's theory of evolution included the suggestion that the mechanism for evolution was natural selection. The three key terms that will help you to understand this idea are variation, competition and selection.

Variation

The theory of natural selection starts with the observations that more individuals are produced than their environment can support and that individuals within populations are usually different from each other in some way — they show variations. (Some causes of these variations are outlined in subtopic 3.4.) According to this theory, some of these variations will provide an increased chance of survival over other variations within a particular environment. In other words, some variations will provide individuals with a competitive advantage.

Individuals that possess a favourable variation (or phenotype) will have an increased chance of reproducing and passing on this variation (through their genes) to their offspring. Inheritance of this variation (or phenotype) will also increase the chances of survival of the offspring and hence the possibility that they will contribute their genes into the next generation. Over time and many generations, if this variation continues to provide a selective advantage, the number of individuals within a population that show the favourable variation will increase.

Individuals that have less favourable variations or that are not as well suited to their environment will not be able to compete as effectively. They may die young or produce few or no offspring. Therefore they will have a limited contribution to the gene pool of the next generation. This will lead to a decrease in the number of individuals with that particular variation within the population.

Can you suggest features that provide the individuals in these photos with increased chances of survival?

3.5.4 Selection

Organisms live within ecosystems, which are made up of various living (biotic) and non-living (abiotic) factors. These factors contribute to selecting which variations provide the individual with an increased chance of surviving. It is for this reason that these factors may be referred to as **selective pressures** or **selective agents**. Biotic factors that may act as selective agents include predators, disease, competitors, prey and mating partners. Examples of abiotic factors include temperature, shelter, sunlight, water and nutrients.

3.5.5 Competition

Individuals within a population compete with each other for resources such as food, shelter or mates. Those with a selective advantage over other individuals are better able to compete for the resource.

There may be situations in which competing is not about resources, but about competing to not be eaten by a predator or killed by a particular disease. In this case, individuals with a particular variation that reduces their chance of being eaten or killed will have a higher chance of survival. Can you think of examples in which variations in phenotype might provide an individual with an increased chance of avoiding being eaten by a predator or dying from a disease?

3.5.6 Tales of resistance

Most mutations are harmful to the organism and decrease its chances of survival. Some, however, may actually increase chances of survival. If a mutation results in a characteristic that gives the organism an increased chance of survival, then it is more likely that the organism will survive long enough to reproduce.

If the organism's offspring inherit the genetic information for this new 'increased survival' trait, then over time an increased proportion of the population may possess this trait.

This is the way in which populations of organisms can become resistant to methods that humans have used to kill them or control their population sizes. Those individuals within the population with the mutation that confers resistance to the control method live long enough to produce offspring who also possess the resistant characteristic. Over time, future generations are likely to contain increased numbers of individuals with resistance against the control method, making it no longer effective.

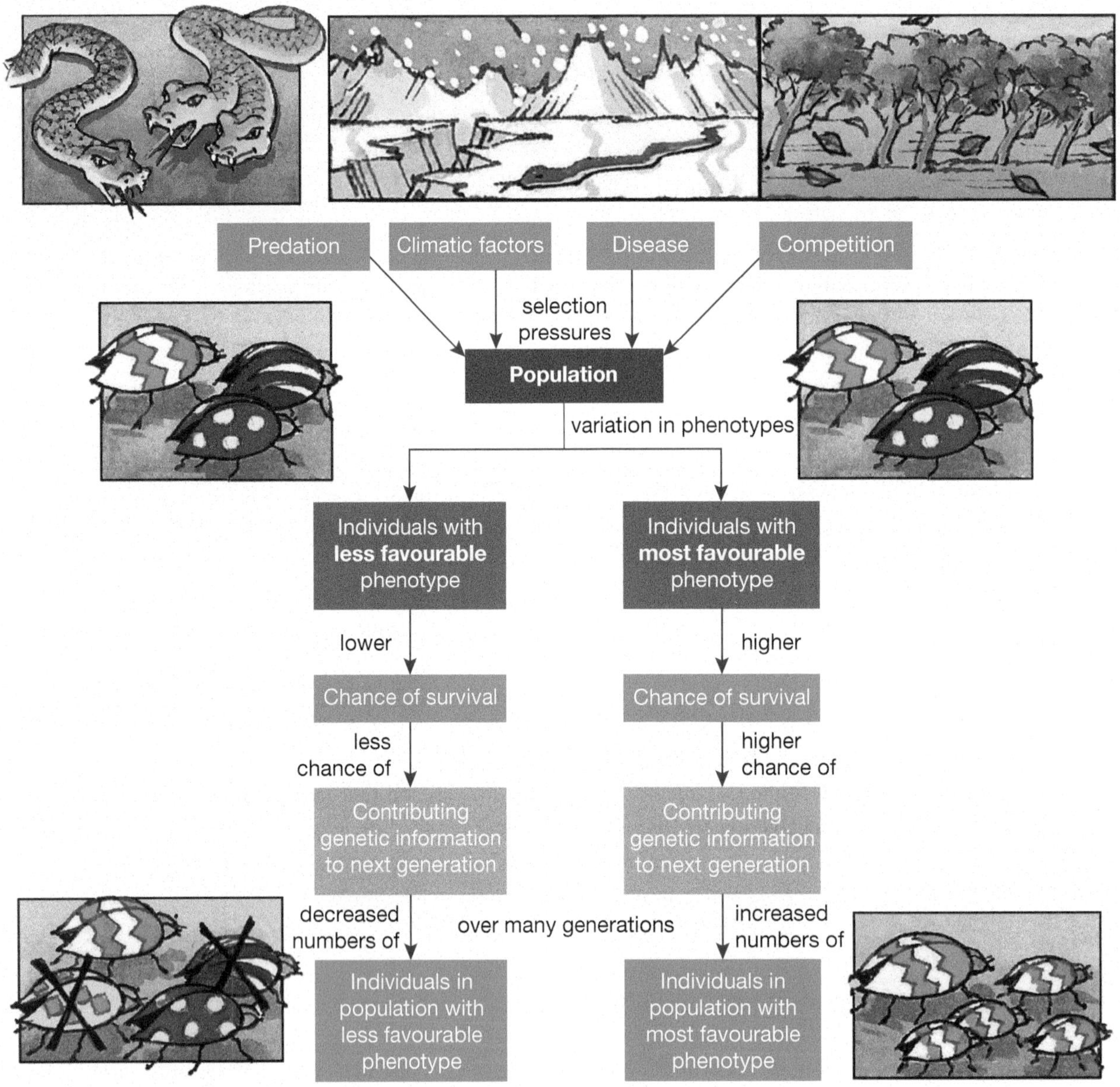

3.5.7 Resistant bacteria

Mutation is the key source of genetic variation in asexually reproducing organisms. Unless a mutation occurs, organisms that reproduce asexually produce clones of each other — they are genetically identical. Errors or changed sequences in their DNA during cell division can be the source of new alleles.

Prior to the discovery and use of penicillin, many people died from a variety of infections that can currently be treated with antibiotics. Penicillin and other antibiotics are drugs that can kill or slow the growth of bacteria. Referred to as 'magic bullets', these drugs revolutionised medicine. Although still widely used today, they are not as effective as they once were. Many bacteria have evolved to be resistant to them.

3.5.8 Resistant bugs and bunnies

Variations due to mutation can also be advantageous in sexually reproducing organisms. The mutation may lead to a resistance to a particular pesticide, such as insects developing resistance to the pesticide DDT or to a particular viral disease, as in the case of some rabbits being resistant to the myxomatosis virus that was used to try to control their populations. If some individuals within the population have a mutation that enables them to survive and reproduce, they may pass this trait on to their offspring, who will also be resistant. An increase in the number of organisms within the population that are resistant to the pesticide or virus reduces the effectiveness of the control method.

Mutations in bacteria can result in some individuals having resistance against antibiotics. When these bacteria reproduce, their offspring also show antibiotic resistance.

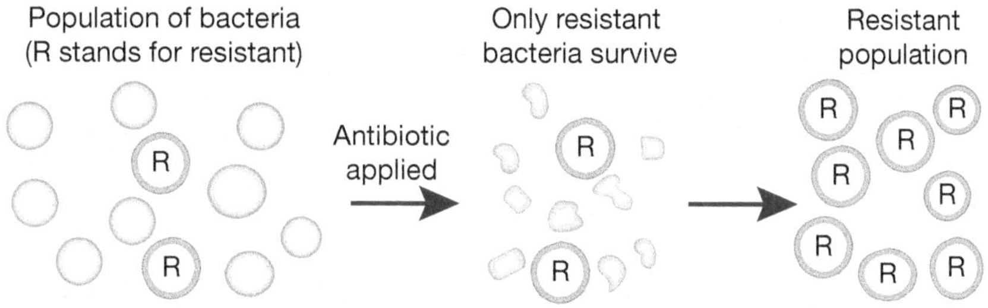

Mutations in flies can result in some individuals having resistance against a particular pesticide. When these flies reproduce, their offspring also show pesticide resistance.

INVESTIGATION 3.2

Modelling natural selection

AIM: To use a model to simulate natural selection

Materials:

100 green toothpicks (or rubber bands)
100 red toothpicks (or rubber bands)

Method and results

1. Copy and complete the table at right.
   - Scatter 50 green toothpicks and 50 red toothpicks over an area of grass measuring at least 10 m × 10 m. The toothpicks represent caterpillars.
   - One student will be the caterpillar-eating bird (CEB). Allow the CEB 15 seconds to 'eat' (pick up) as many of the caterpillars as she or he can.
2. Count how many caterpillars of each colour were eaten. That will tell you how many caterpillars of each colour are left in the grass. Record these figures in a result table similar to the one shown at right.
3. Allow the caterpillars to 'breed'. For every pair of caterpillars of a particular colour, add a third caterpillar of the same colour (e.g. if you have

Time	Number of caterpillars	
	Red	Green
Start	50	50
After first feeding frenzy		
After first mating		
After 2nd feeding frenzy		
After 2nd mating		
After 3rd feeding frenzy		
After 3rd mating		

15 green toothpicks [7 pairs] and 10 red toothpicks [5 pairs] left in the grass you should scatter an additional 7 green and 5 red toothpicks). Record the number of each type of caterpillar after breeding in your result table.

4. The CEB will now have two further 15-second feeding frenzies. After each feeding frenzy, record the number of each type of caterpillar left in the grass, and allow the caterpillars to 'breed'.

Discuss

5. By the end of the experiment were there more green or red caterpillars? Explain why.
6. Suggest why one colour of caterpillar may eventually disappear or become less abundant in the population.
7. Explain how this experiment models natural selection.

3.5.9 How about that!

Can adding variety to life increase your chances of survival? The ancestors of the eukaryotic cells that make up your body may have thought so! There is a hypothesis that, over a billion years ago, ancestors of complex cells like ours captured some little aerobic bacteria. Supplying the prehistoric cells with energy, these 'house guests' were fed and looked after. Over time, however, the independence and most of the genetic material and functions of these aerobic bacteria were lost. Their descendants are mitochondria, the organelles that supply our cells with energy using aerobic respiration. It is thought that chloroplasts, like those in plant cells, evolved in a similar way.

The origin of the eukaryotic cell? Some scientists also suggest that our nucleus may have come from a giant viral ancestor.

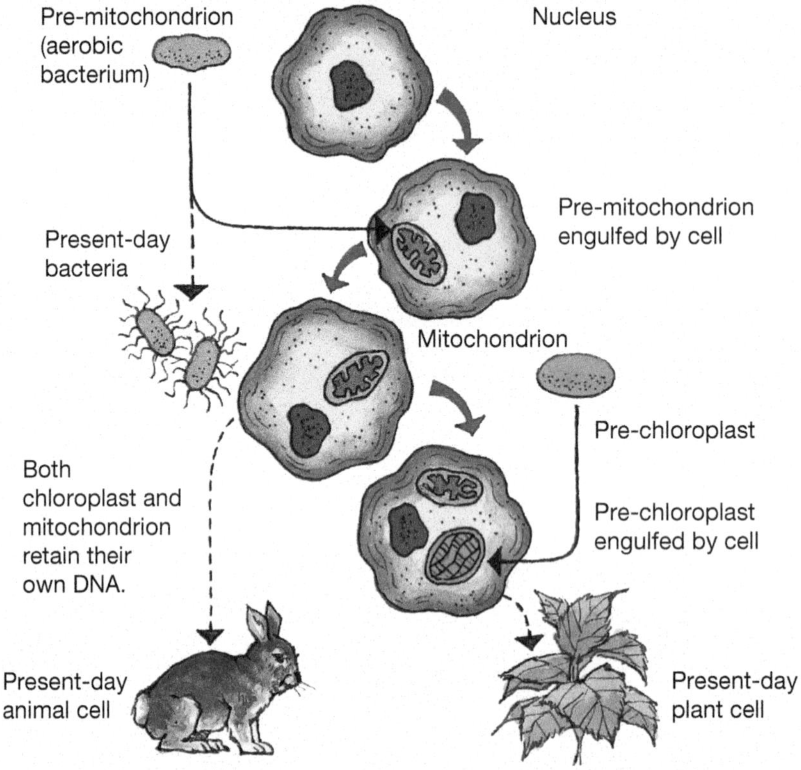

3.5 Exercises: Understanding and inquiring

To answer questions online and to receive **immediate feedback** and **sample responses** for every question, go to your learnON title at www.jacplus.com.au.
Note: Question numbers may vary slightly.

Remember

1. Suggest how variation within populations increases the chances of survival of the species.
2. Construct a flowchart to describe natural selection using the terms at right.

Predation Competition
Variation in phenotypes Selection pressures
Climatic factors Population
Less favourable Most favourable
Disease Lower chance Higher chance
Over many generations Survival Phenotype
Genetic contribution to next generation
Decreased numbers Individuals
Increased numbers

3. Suggest the link between natural selection and evolution.
4. Describe what is meant by the term *natural selection*.
5. Identify three examples of (a) biotic and (b) abiotic selective pressures or selective agents.
6. Identify three resources for which individuals within a population may compete.
7. Describe a link between mutation, variation and resistance to a pesticide or antibiotic.
8. Using diagrams, explain how bacteria may become resistant to a particular antibiotic.
9. Myxomatosis virus was used as a method to control rabbit populations in Australia. However, it is no longer effective. Suggest reasons for the ineffectiveness of a previously effective method of control.

Investigate, think and discuss

10. DDT was a pesticide used to kill mosquitoes. It is no longer effective and has caused some unexpected environmental and ecological issues.
    (a) Find out more about the pesticide DDT and its history and use.
    (b) Identify ecological and environmental concerns about the use of DDT.
    (c) Suggest reasons for the gradual decrease in DDT's effectiveness as a pesticide.
    (d) Link the story about DDT to natural selection.
11. Penicillin was a very effective antibiotic against a number of different types of bacteria.
    (a) Find out more about penicillin and its history and use.
    (b) Suggest reasons why some bacteria are now resistant to penicillin.
    (c) Link the story of penicillin to natural selection.

Investigate and create

12. Design and construct your own organism.
    (a) Give this organism a name and describe the environment in which it lives.
    (b) Use this organism as the common ancestor for four other variations of the organism.
    (c) Construct each variation, giving each a name and describing how the variation increases its chances of survival in its environment.

Investigate, think and report

Which moth is more easily seen?

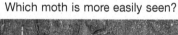

13. The English peppered moth, *Biston betularia*, rests on tree trunks during the day. Prior to 1850, this species had a speckled pale grey colour that effectively camouflaged it from predators as it rested on the pale lichen-covered trunks. In about 1850, a black version of this moth appeared. By 1895, these black moths made up about 98 per cent of the population.
    (a) Find out more about this species of moth.
    (b) Suggest the source of the new variation.
    (c) Find out what was happening in England between 1850 and 1895 that may have had an impact on the survival of these moths.
    (d) Suggest why and how the number of black moths in the population increased so dramatically.
14. Search the internet for cartoons and simulations on natural selection. Then use the best ideas to create your own cartoon, comic strip, picture story book or animation.

learnon RESOURCES — ONLINE ONLY

Complete this digital doc: Worksheet 3.1: Struggling to survive (doc-19421)

Complete this digital doc: Worksheet 3.2: Isolation and new species (doc-19422)

3.6 Evolution

3.6.1 The formation of new species

Variation, struggle for survival, selective advantage and inheritance of advantageous variations formed the basis for Charles Darwin's theory of evolution by natural selection. They also provided an explanation for how new species arise.

The formation of new species is called **speciation**. There are two ways in which speciation can occur. **Phyletic evolution** occurs when a population of a species progressively changes over time to become a new species. **Branching evolution** or **divergent evolution** is more common; in this case, a population is divided into two or more new populations that are prevented from interbreeding. When different selection pressures act on each population, different characteristics are selected for. Over generations, these new populations may become so different from each other that they can no longer interbreed and produce fertile offspring. At this point, they have become two different species.

learnon RESOURCES — ONLINE ONLY

Watch this eLesson: How a new species evolves (eles-0162)

Darwin's finches are examples of divergent evolution. They share a common ancestor, but over time and generations, different selective pressures led to the selection of different variations that are most suited to a particular environment or available niche.

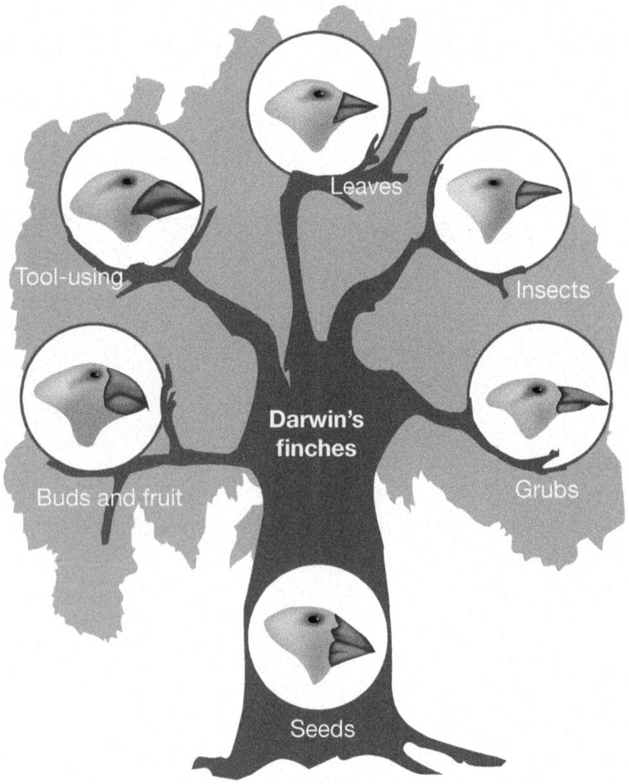

Speciation is a process by which a new species develops from another.

1 Variation of characteristics is present in a population.

2 The breeding population becomes isolated.

3 Different characteristics arise through random genetic drift, mutation and environmental pressures.

4 The environment changes. Because of selection, some characteristics are favoured over others. Those best suited to the environment survive.

5 Survivors reproduce and pass on favourable genes and features to offspring.

6 The frequency at which the genes for the new characteristics appear increases.

7 The isolated population is now quite different, producing a new species.

3.6.2 Divergent evolution

Divergent evolution is a type of evolution in which new species evolve from a shared ancestral species. That is, two or more new species share a common ancestor. At some point in history a barrier (such as a geographical barrier, for example a mountain or ocean) has divided the population into two or more populations and has also interfered with interbreeding between the populations.

Exposure of these populations to different selection pressures results in the selection of different variations or phenotypes. Over time, the populations may be so different that even if they were brought back together they would be unable to produce fertile offspring. It is at this point that they are referred to as different species. Speciation has occurred.

3.6.3 Adaptive radiation

Adaptive radiation is said to have occurred when divergent evolution of one species results in the formation of many species that are adapted to a variety of environments. Darwin's finches and Australian marsupials are two examples. Australian marsupials are thought to have evolved from a common possum-like ancestor. The photographs below show examples of species that have arisen from a common ancestor.

Australian marsupials show adaptive radiation as they have evolved from a common ancestor but, due to different selection pressures, different characteristics have been selected.

Divergent evolution

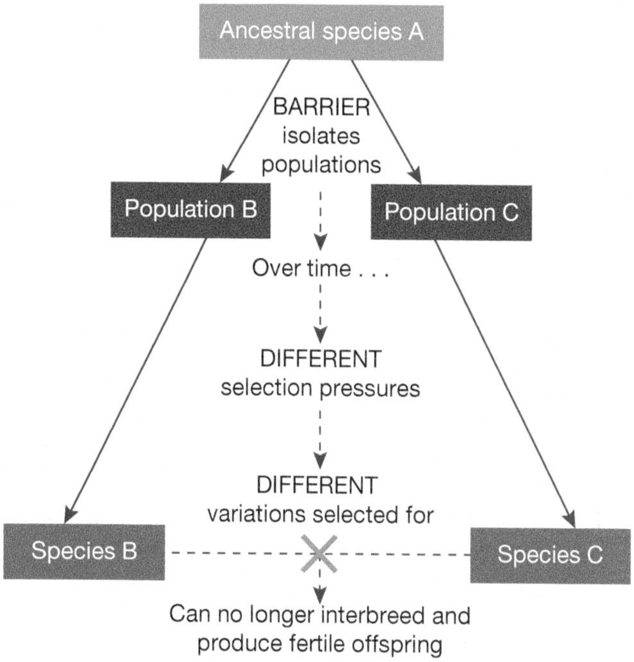

3.6.4 Convergent evolution

In divergent evolution, different selection pressures lead to the selection of different variations in evolution from a common ancestor. **Convergent evolution** is the opposite. Convergent evolution is the result of similar selection pressures in the environment selecting for similar features or adaptations. These adaptations have not been inherited from a common ancestor.

(a) The Australian echidna, (b) African aardvark, (c) South-East Asian pangolin and (d) South American anteater share similar features. These features were selected for because they gave a selective advantage in obtaining food supply their environment, rather than because of a recent common ancestry.

3.6.5 Coevolution

The evolution of one organism can sometimes be in response to another organism. Examples of this coevolution include parasites and their hosts, or birds and plants. If you look at the features of birds and the flowers that they pollinate, you may notice that some birds have evolved specialised features, such as beaks that are well suited for obtaining nectar from a flower with a particular shape. The plants have evolved flowers of a particular colour that may be attractive to its pollinator, and nectar that not only attracts but rewards the bird for its task of being involved in pollination.

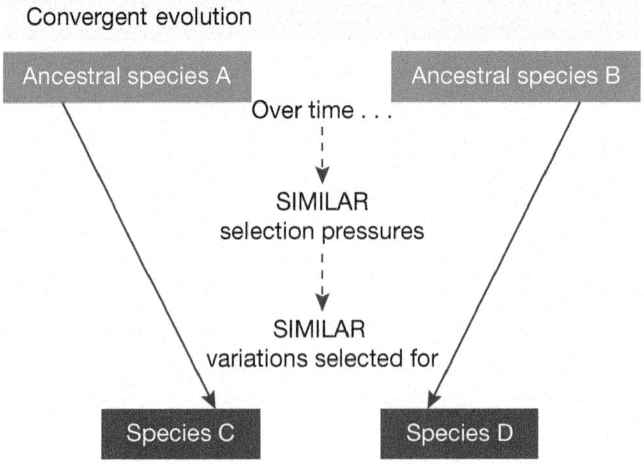

3.6.6 Extinction

Extinction is the loss or disappearance of a species on Earth. Extinction of a species may influence the evolution of another species, as it may provide the opportunity to move into the niche that the extinct species occupied. Extinctions and their effect on biological diversity are explored in subtopic 3.12.

3.6 Exercises: Understanding and inquiring

To answer questions online and to receive **immediate feedback** and **sample responses** for every question, go to your learnON title at www.jacplus.com.au. *Note:* Question numbers may vary slightly.

Remember

1. State four key ideas that formed the basis of Darwin's theory of evolution by natural selection.
2. What is meant by the term *speciation*?
3. Describe how a new species can be formed.
4. Distinguish between divergent evolution and convergent evolution.
5. Describe an example of divergent evolution.
6. Outline the relationship between adaptive radiation and divergent evolution.
7. Describe an example of adaptive radiation.
8. Identify examples of organisms that show convergent evolution.
9. What is meant by the term *extinction*?
10. Select appropriate terms from the following list and use flowcharts to describe:
    (a) divergent evolution
    (b) convergent evolution.

Ancestral species A   Similar selection pressures
Similar variations selected for   Species C
Population B   Different selection pressures
Over time   Barrier isolates populations
Different variations selected for   Species D
Population C   Ancestral species B   Species B
Can no longer interbreed and produce fertile offspring

Analyse, think and discuss

11. The figures on the next page show two species of North American hares that are closely related and share a common ancestor. The snowshoe hare, *Lepus americanus* (left), lives in northern parts of North America where it snows in winter. The black-tailed jack rabbit, *Lepus californicus* (right), lives in desert areas.

(a) Identify differences between these hares.
(b) Suggest reasons for these differences.
(c) Suggest how these differences came about.
(d) Is this an example of convergent or divergent evolution? Explain.

12. Identify where each of the figures shown at right belong in the convergent evolution table below.

Niche	Placental mammal	Australian marsupial
Anteater		
Cat		
Climber		
Glider		

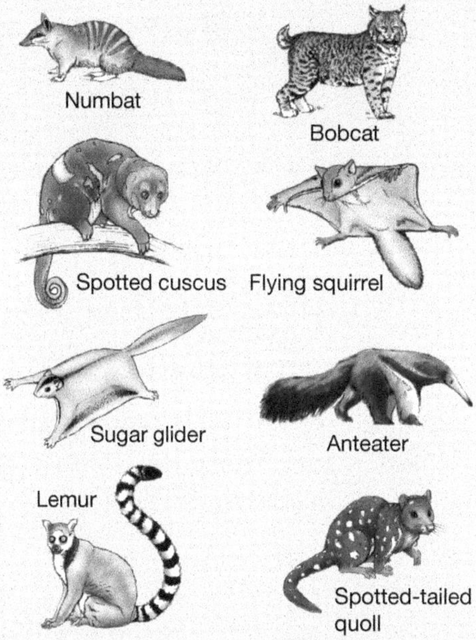

13. (a) Carefully examine each of the pairs of organisms shown bottom right. Identify whether each pair is an example of convergent or divergent evolution.
(i) Dolphin and shark
(ii) Numbat and anteater
(iii) Sea dragon and seahorse
(iv) European goldfinch and pine siskin (finch)
(b) Provide a reason for your response.
(c) Use the internet to check the accuracy of your response.

Investigate, think and create

14. The figure below shows how one ancestral species can undergo evolution and give rise to a number of new species.
(a) Select a species that is currently alive on Earth.
(b) Use the internet and other resources to find information to construct a figure similar to the one given, showing your selected species and other species to which it is related.
(c) Write a play, story or documentary to tell the tale of the evolution of your selected species from its ancestral species.
(d) Using puppets, animations or multimedia, develop your tale into a presentation that can be shared with others.

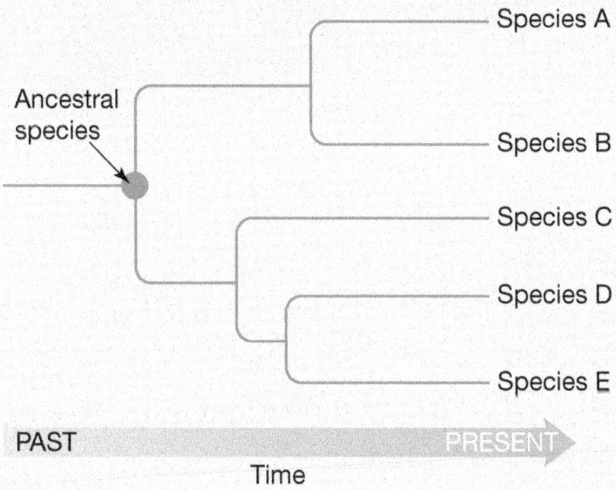

(i)

(ii)

(iii)

(iv)

15. Honeycreepers are found only in the Hawaiian Islands and share a common ancestry. Examples of four species of honeycreepers are shown below.
    (a) Suggest reasons for their different appearance.
    (b) Share your suggestions with others.
    (c) Create a story to explain how and why these honeycreepers look so different.
    (d) Collate the class's stories and read stories that others have written.

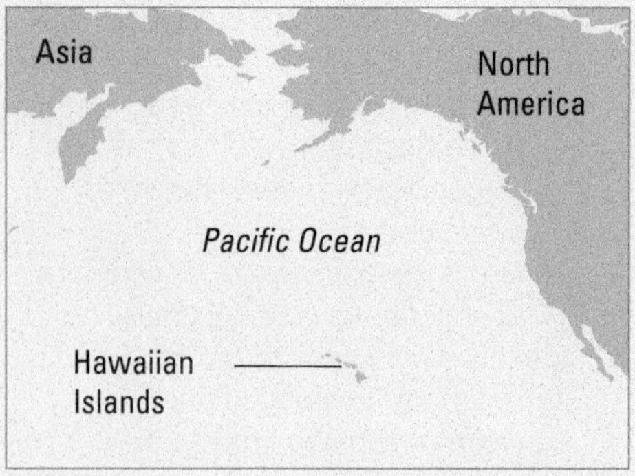

3.7 Long, long ago

3.7.1 Geographical jigsaw

A long time ago, long before humans inhabited the Earth, the continents were joined together. If you could travel back in time 10 million years, not only would the continents look different, but life on Earth would also be very different to what it is today.

learnon RESOURCES — ONLINE ONLY

Watch this eLesson: How did we get here? (eles-1776)

3.7.2 Moving plates

The theory of **plate tectonics** suggests that the Earth's crust is divided into about 30 plates, each about 120 kilometres thick. These plates move only several centimetres each year, sliding past, pushing against or moving away from each other.

3.7.3 From Pangaea to Gondwana

Over millions of years, some of these plates have moved further and further apart, separating what was once a single landmass into the continents that we know today. Scientists believe that Australia was once part of **Pangaea**, a giant landmass that comprised all the land on Earth. About 200 million years ago, Pangaea moved apart to form two supercontinents — **Laurasia** and **Gondwana**.

Laurasia in the Northern Hemisphere consisted of the plates that would eventually become North America, Greenland, Europe and Asia. In the Southern Hemisphere, Gondwana consisted of the plates that would become South America, Africa, India, Antarctica and Australia.

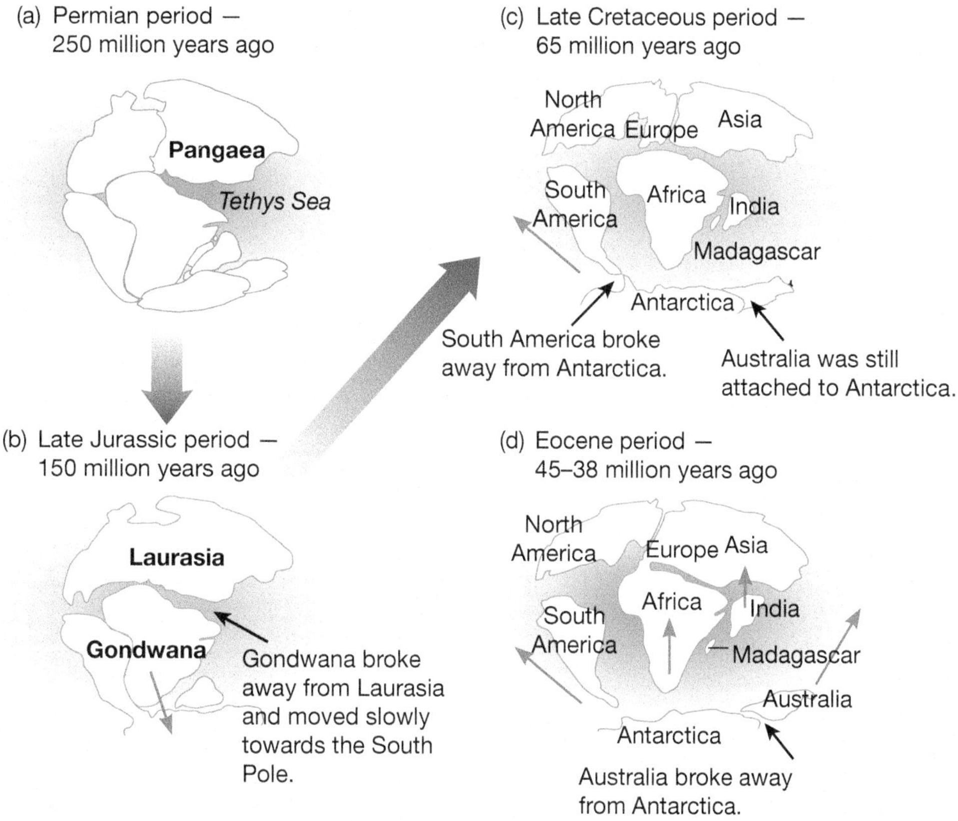

3.7.4 Plate tectonics — the evidence

The evidence that supports the theory of plate tectonics includes:

- data showing that continents are still moving apart. Australia is moving northwards at the rate of about 7 cm every year.
- the physical fit between the continents
- the remarkable similarities between rock and crystal structures at the edges of continents
- fossil evidence suggesting that many of Australia's marsupials originated in South America

Fossil evidence suggests that many of Australia's marsupials originated in South America.

- the discovery of fossils of the land-dwelling dinosaur *Mesosaurus* (which lived about 270 million years ago) in only two places in the world — the eastern side of South America and the western side of South Africa, which are now separated by 6600 kilometres of ocean
- the distribution of closely related animals and plants across the continents.

3.7.5 Biogeography

Biogeography refers to the geographical distribution of species. Observations by Charles Darwin and Alfred Russel Wallace of this distribution contributed to their development of the theory of evolution. For example, Darwin observed that islands with similar environments in different parts of the world were not populated by closely related species but with species related to those of the nearest mainland. He concluded that the species originated in one area and then dispersed outwards.

Divergent evolution can describe how isolated populations of a species can evolve into new species due to different selection pressures (see subtopic 3.6). Species A, initially living on a supercontinent, evolves into different species B–E as tectonic plates move apart. ***Source:*** Modified with permission from *Understanding Evolution* (www.evolution.berkeley.edu), University of California Museum of Paleontology.

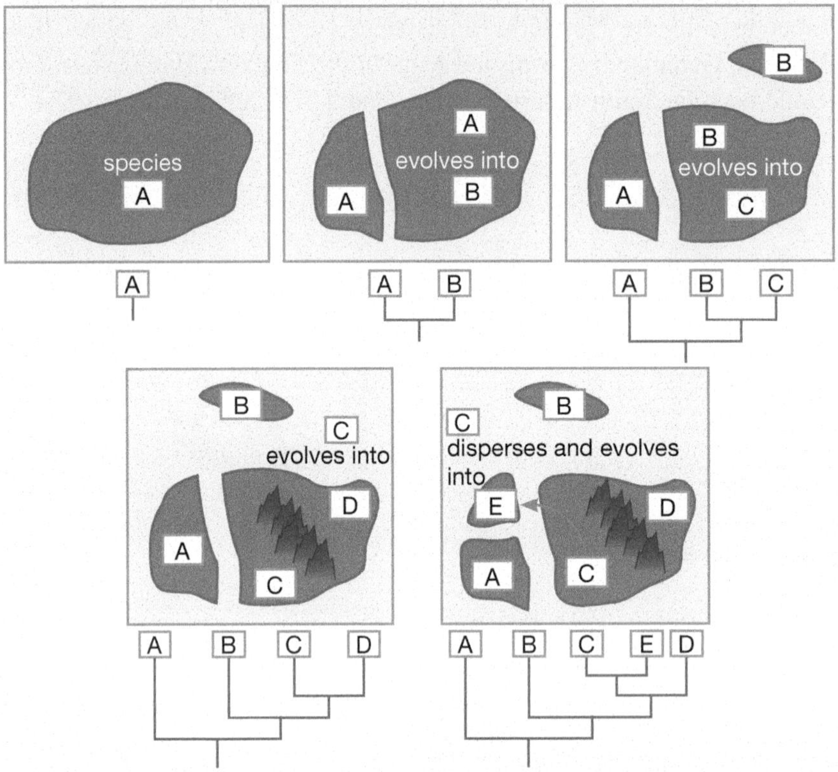

3.7.6 Analogous structures

Unrelated species living in very similar environments (with similar selection pressures) in different parts of the world have evolved similar structures. For example, the fins of a dolphin and a shark, or the wings of a bat and a butterfly, are the result of convergent evolution. Structures that perform the same role but have different evolutionary origins are called **analogous structures**.

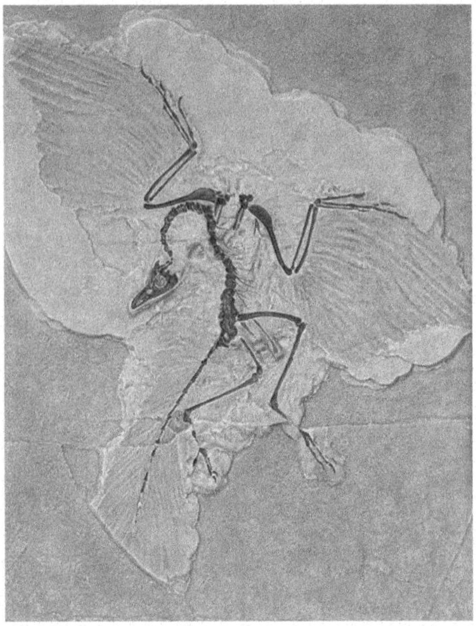

3.7.7 Geological time

Our Earth is old. Its current age is estimated to be around 4.6 billion years. Geologists have constructed a geological timeline that divides this time into five **eras**, some of which are further divided into **periods**. This timeline with its divisions and information from fossil records are shown in the figure on the next page.

The fossil record provides an incomplete picture of life as it has existed on Earth.

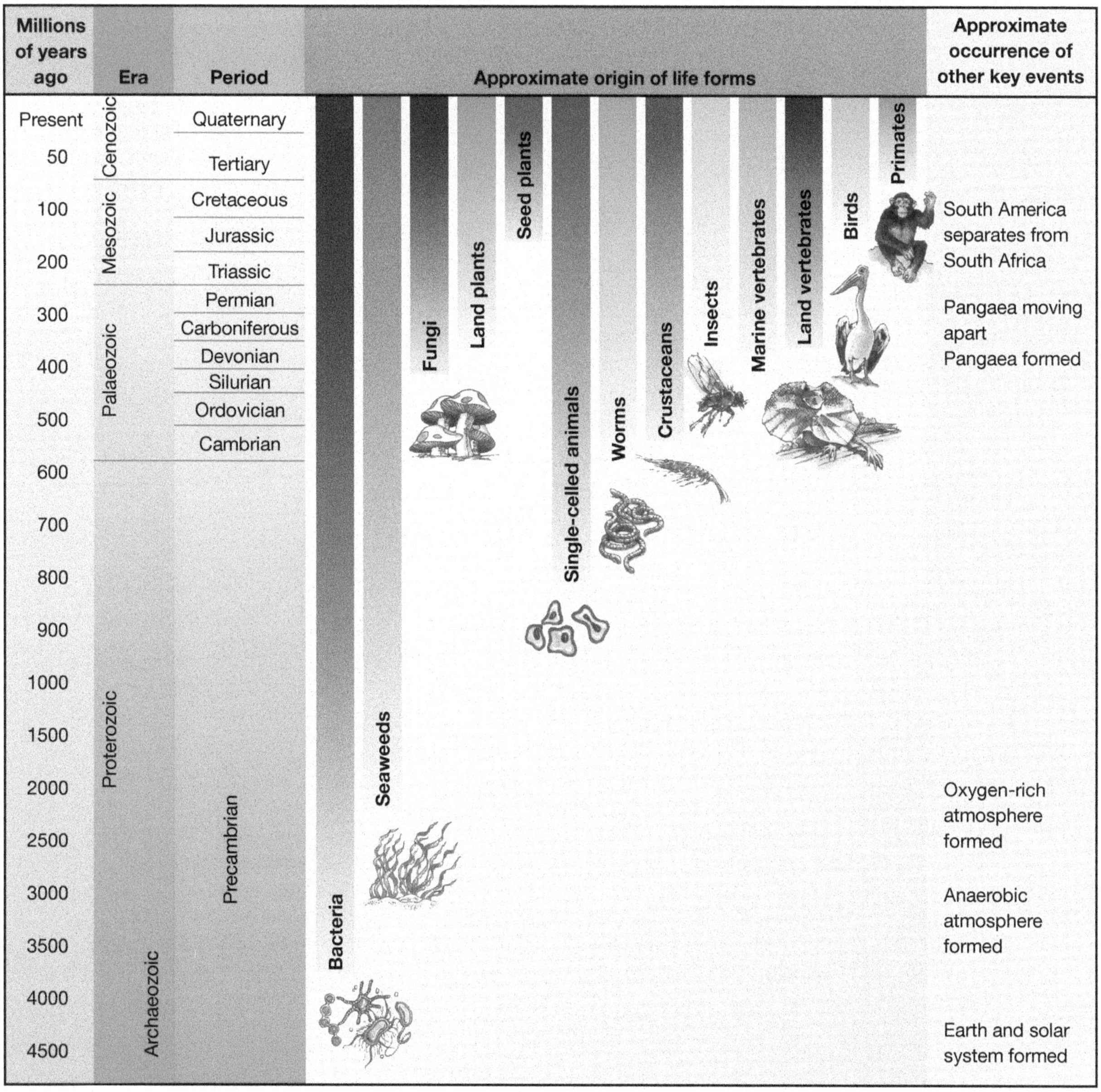

INVESTIGATION 3.3

4.6 billion years of history

AIM: To construct a timeline of the history of the Earth

Materials:

roll of toilet paper, cash register tape or similar

Method, results and discussion

- Use the roll of paper to create a timeline of the history of the Earth. Begin by choosing an appropriate scale to represent the 4.6 billion years of history.
- Indicate the events shown in the figure above on your timeline.

1. A student was describing the evolution of life on Earth and wrote 'for much of Earth's history not much happened'. Is this statement justified?
2. Explain why a long roll of paper is necessary to construct this timeline.

3.7 Exercises: Understanding and inquiring

To answer questions online and to receive **immediate feedback** and **sample responses** for every question, go to your learnON title at www.jacplus.com.au. *Note:* Question numbers may vary slightly.

Remember

1. About how many plates is the Earth's crust thought to be divided into?
2. Provide an example of evidence that the Earth's crust is still moving.
3. State the name of:
   (a) the giant landmass that once made up all of Earth's land surface
   (b) the two supercontinents
   (c) the supercontinent in which Australia was located
   (d) the other continents in the same supercontinent as Australia.
4. Outline five pieces of evidence that support the theory of plate tectonics.
5. What does biogeography refer to?
6. Suggest the relationship between biogeography and evolution.
7. What is meant by the term *analogous structures*?
8. Provide examples of analogous structures.
9. Approximately how old is Earth estimated to be?

Analyse, think and discuss

10. Read the article below and then answer the questions.
    (a) Using an atlas and the diagrams in this subtopic, locate Chile, Tasmania and Antarctica. Does their position on the supercontinent support the claims made by the researchers in the article? Explain.
    (b) Why aren't Fitzroya trees found on the Antarctic continent today?

FOSSIL FIND JOINS CONTINENTS

A team of researchers from the University of Tasmania has found fossils of a tree in the north-west of the state, estimated to be 35 million years old. At this time, Tasmania was supposedly moving away from Antarctica.

The tree, Fitzroya, is a giant conifer that can grow up to 50 metres high. Today, it is found only in Chile. The discovery is just one more piece of evidence to support the hypothesis that the continents in the Southern Hemisphere were once joined together as the supercontinent Gondwana.

Together with other discoveries made over the last decade, this find lends weight to the view that there were once forests growing in places of high latitude where today there is often nothing but pack-ice. It would seem that forests containing a large number of species once thrived in Gondwana.

11. Refer to the fossil record and geological timeline in this subtopic to answer the following questions.
    (a) List the eras from most recent to least recent.
    (b) List the periods in the Mesozoic era.
    (c) Which period came first, the Cambrian or the Permian?
    (d) In which period are we currently living?
    (e) Humans are primates. In which era did primates appear?
    (f) Dinosaurs became extinct about 65 million years ago. Identify the period and era of this time.
    (g) Humans have been blamed for causing the extinction of many other organisms. On the basis of data in the timeline, did they cause the extinction of the dinosaurs? Explain.
    (h) Suggest why humans could not have survived 4 billion (4000 million) years ago.
    (i) Identify the first life forms to appear.
    (j) Identify the most recent life forms to appear.
    (k) List the following life forms in order of their appearance: fungi, birds, worms, insects, primates, crustaceans.
    (l) Suggest the difference between land plants and seed plants.
    (m) Suggest a reason for the appearance of seed plants and birds around the same time.
    (n) Suggest why the term *Cambrian explosion* is often associated with the Cambrian period.

HOW ABOUT THAT!

Evidence that life may once have existed on Mars has been found inside a meteorite that landed in Antarctica. The meteorite has been dated at about 13 000 years old. Examination of thin slices of the meteorite under an electron microscope has suggested the presence of microfossils of single cells. This is significant because it is thought that life on Earth also started out as single cells.

learnon RESOURCES — ONLINE ONLY

Complete this digital doc: Worksheet 3.3: Geological time (doc-19424)

3.8 Yesterday's plants

3.8.1 First findings

Imagine walking along the shores of the primeval oceans and observing the first traces of life on Earth. What would you see?

If you were to observe the first traces of life on Earth, you would see a rich, slimy soup in the primeval oceans. The earliest known traces of life were primitive bacteria, the ancestors of modern-day organisms. Their fossil remains are found in the rock structures (stromatolites) they produced 1000 million years ago. These ancient **stromatolites**, like upside-down ice-cream cones up to 18 metres high and about 10 metres across, loomed above the silent seabed in greenish-white forests that stretched for hundreds of kilometres.

The earliest forms of animal life did not appear until plants were well established. In Australia, this was in the early Cambrian period, which was about 570 million years ago.

3.8.2 From forests to coal

Over 300 million years ago, during the Devonian and Carboniferous periods, plants had developed into a variety of complex forms. Close relatives of the horsetails, club mosses and ferns formed vast ancient forests. Thick layers of their rotting remains became solidified over time, forming the coal beds found today.

About 350 million years ago, the first seed-producing plants appeared. **Gymnosperms** were the dominant plants in the Permian, Triassic and Jurassic periods. Gymnosperms such as conifers, cycads and maidenhair trees are living descendants of the first pollen-producing plants.

Some of these ancient forests contained horsetails and club mosses 45 metres tall.

HOW ABOUT THAT!

Many conifers produce a sticky, aromatic, oily material called resin. This material gives them their characteristic smell. Although the resin is produced to stop the growth of microbes and to prevent insects from feeding on them, some insects have developed ways to put it to their own use. The female bark beetle converts the resinous chemicals into a substance that attracts males to her.

Blooming flowers

It was during the Cretaceous period, about 135 million years ago when dinosaurs still flourished, that flowering plants appeared. During this period, **angiosperms** or flowering plants became the dominant plants. These plants were closely related to those found today.

Pollen power

Fossilised pollen grains survive for millions of years. By studying ancient pollen, scientists can investigate vegetation that existed in the past. In Australia the oldest fossil pollen from a flowering plant is from the native holly genus *Ilex*. Millions of years ago, most of the surface of Australia was covered by forests. Over time, Australia gradually became drier. The change in climate resulted in fewer rainforests. Eucalypts, acacias and proteas, with their tough, hard leaves and often woody fruits, were well suited to these dry conditions. Pollen fossil evidence suggests that eucalyptus plants appeared about 30 million years ago.

Geological table of plant evolution

Era	Period	Millions of years ago	Plant evolution
Cenozoic	Pleistocene	2	
	Pliocene	7	
	Miocene	26	
	Oligocene	38	
	Eocene	54	
	Palaeocene	65	Age of flowering plants
Mesozoic	Cretaceous	136	
	Jurassic	193	
	Triassic	225	Age of gymnosperms
Palaeozoic	Permian	280	
	Carboniferous	345	Age of ferns and horsetails 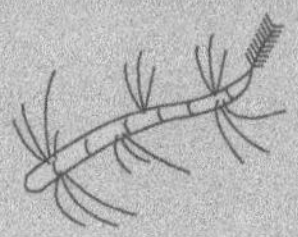
	Devonian	395	
	Silurian	435	Age of early land plants 
	Ordovician	500	
	Cambrian	590	Age of algae
	Precambrian	600+	New fossils

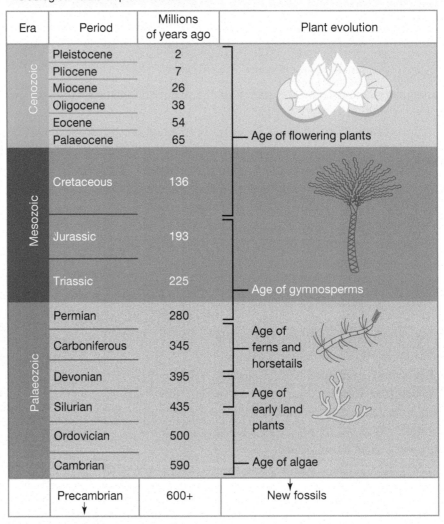

3.8.3 Plants tell tales of history

The Australian National Herbarium (ANH) contains over six million specimens of plants dating from the earliest days of European exploration. Each specimen has its own story and history documented. This has enabled the ANH to maintain a historical record of over two hundred years of changes to our vegetation.

The records that ANH have kept have also enabled monitoring of the changes in the names given to plants over the last 200 years in Australia. Their plant information system, which is based on 'scientifically verifiable voucher specimens', ensures the 'currency of names' as we continue to find out more about our Australian plants.

An example of a label from a plant specimen held at the Australian National Herbarium: each label includes the plant names, basic details about where, when and by whom it was collected, and information about the habitat and appearance of the plant. All these details are stored in a database so that they can be managed and made available for research and analysis.

Flora of AUSTRALIA

AUSTRALIAN NATIONAL HERBARIUM (CANB)

STERCULIACEAE

Brachychiton populneus (Schott & Endl.) R.Br. subsp. ***populneus***

AUSTRALIA: New South Wales

Spring Creek Crossing, on Bowning-Murrumburrah/Harden Road, ca 12.5km ESE of Murrumburrah/Harden.

34° 37' S 148° 28' E 370m

Slopes above creek. Growing among granite boulders. Soil gravelly, reddish-brown. Cleared *Eucalyptus melliodora* - *E. blakelyi* woodland, with no understorey. Groundcover of weed-infested pasture.

Tree to ca 10 m. Follicles collected from ground. Fruit separate.

One tree only.

B.J. Lepschi 231 **19 Nov 1989**

CANB 393422.1

Australia's Virtual Herbarium (AVH) is an online resource. It provides access to plant specimen data held by Australian herbaria.

ePlants

Data from Australian herbaria are maintained in a database called Australia's Virtual Herbarium (AVH) which you can search for information about many different plants online. Australia's Virtual Herbarium is a dynamic project that includes many plans for future developments. Perhaps you will be a part of this project in the future, providing images and descriptions or creating identification tools for future generations.

3.8 Exercises: Understanding and inquiring

To answer questions online and to receive **immediate feedback** and **sample responses** for every question, go to your learnON title at www.jacplus.com.au. *Note:* Question numbers may vary slightly.

Remember

1. What is the relationship between stromatolites and modern-day plants?
2. When did the angiosperms become the dominant plants?
3. Which characteristics make eucalypts well suited to the dry Australian environment?

Think and reason

4. Use the geological table to complete the following.
   (a) List the plant groups in order of their appearance, from oldest to most recent.
   (b) Draw a timeline showing the times when these plant groups dominated the Earth.

Think

5. Suggest why no new major plant groups have arisen over the past 130 million years.
6. Suggest why flower fossils are very scarce.
7. If a botanist studies plants and a palaeontologist studies fossils, what do you think a palaeobotanist studies?

Investigate

8. The ginkgo or maidenhair tree is often described as a living fossil. It is descended from trees that date back to the Triassic period, about 200 million years ago. Find out how it differs from the other groups of living gymnosperms, such as conifers and cycads. Suggest reasons for the differences.
9. Find out more about the following ancient plants:
   (a) giant club moss, *Archaeosigillaria*
   (b) horsetails
   (c) *Lepidodendron*
   (d) *Cooksonia*
   (e) *Baragwanathia.*
10. Investigate research on the history of Australian plants. Report your findings in a storyboard or as a PowerPoint or Photo Story presentation.

3.9 Fossils

3.9.1 Evidence of past life

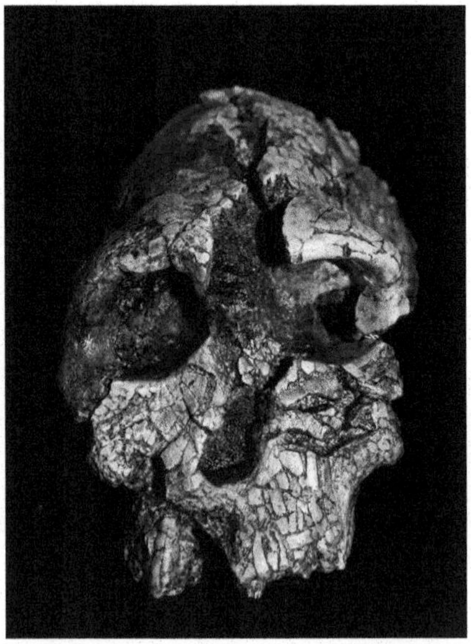

To gain insight into what life was like in the past, you need look no further than rocks. Within rocks you may find fossils — evidence of past life. The study of organisms by their fossil remains is called **palaeontology**.

3.9.2 Fossils

Fossils can be parts of an organism, such as its bones, teeth, feathers, scales, branches or leaves. They can also be footprints, burrows and other evidence that an organism existed in an area. For example, a dinosaur track has been discovered in the Otway Range in southern Victoria. By observing the footprints in the track, **palaeontologists** can work out the size, weight and speed of the dinosaur that made them.

3.9.3 How are fossils formed?

Fossilisation is a rare event. Usually when an organism dies, micro-organisms are involved in its decomposition so that eventually no part of it remains. However, if an organism is covered shortly after its death by dirt, mud, silt or lava (as can happen if it becomes trapped in a mudslide or in the silt at the bottom of the ocean), the micro-organisms responsible for decomposition cannot do their job because of the lack of oxygen. Over millions of years, the material covering the dead organism is compressed and turned into rock, preserving the fossil within it.

Fossil formation

1. A dinosaur dies and is quickly covered by sediment.

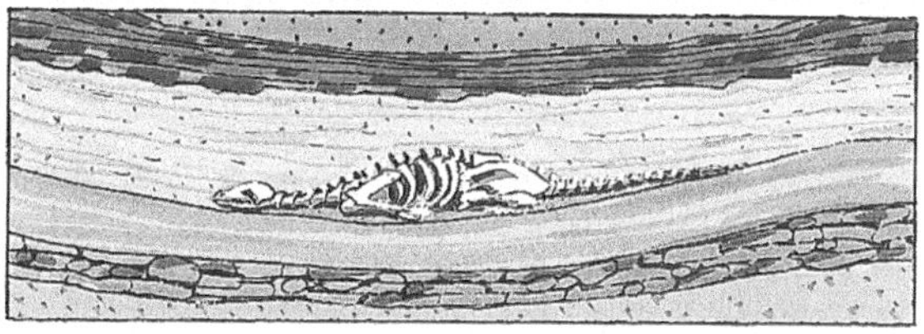

2. Over time, the sediment turns into rock. The remains of the dinosaur turn into a fossil.

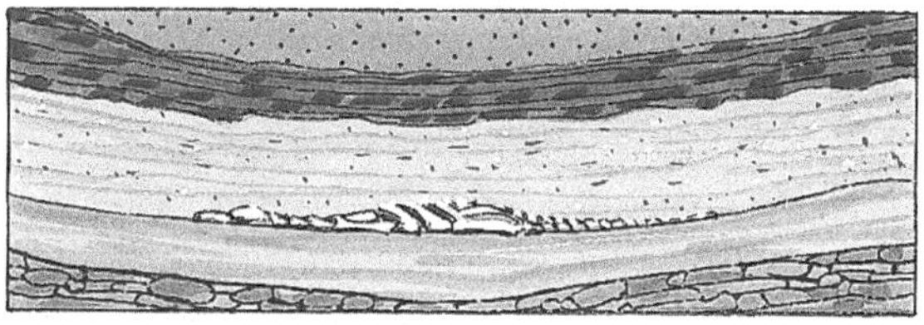

3. The fossil is flattened by the layers of rock.

4. The rock is folded and eroded and the fossil can be seen on the surface.

3.9.4 Dating fossils

There are two main ways in which the age of fossils is estimated. One is called **relative dating** and the other is called **absolute dating**. The key difference between these two types of dating can be outlined using the following analogy.

If you were to ask me 'What is your **relative age**?', I would reply, 'I am the eldest of three daughters'. If you were to ask me 'What is your **absolute age**?', I would tell you how old I am in years.

Fossils can be dated by observing their position in the rock layers.

Relative dating

Relative dating is used to determine the relative age of a fossil. As the layers of sedimentary rock are usually arranged in the order they were deposited, the most recent layers are near the surface and the older layers are further down. The position or location of a fossil in the strata, or layers, of rock gives an indication of the time in which the animal lived. Relative dating can also provide information about which other species were living at the same time and the order in which they appeared in the area.

Interpretation of the relative dating method requires considering the movement of tectonic plates in which the rocks lie. It is possible that a layer (or layers) containing fossils could have been thrust upwards by a sideways force to form a **fold**, or broken and moved apart in opposite vertical directions to form a **fault**.

The formation of folds and faults can cause changes in the rock layers.

(a) Folds

(b) Faults

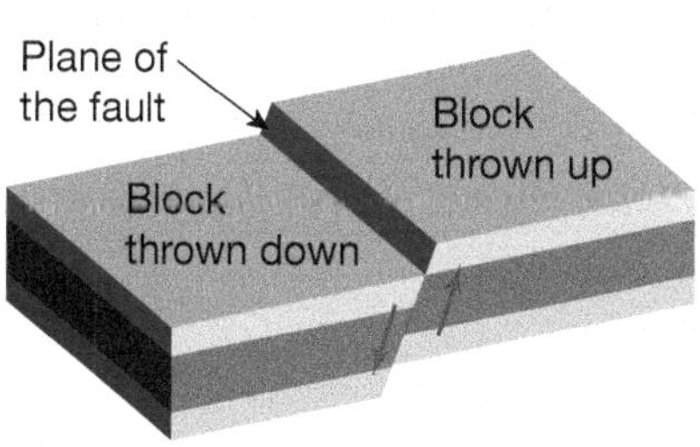

Absolute dating

Fossils or the rocks in which they were located can also be dated by various radiometric techniques, which are based on the rate of decay or **half-life** of particular **isotopes**. The half-life of an isotope is the amount of time it takes for its radioactivity to halve. The use of these techniques to determine the absolute age of rocks and fossils is called **radiometric dating**.

Carbon dating is a specific type of radiometric dating and can be used to date fossils up to about 60 000 years old. Most of the carbon contained in living things is carbon-12, but there is also a small amount of the radioactive isotope carbon-14. Organisms incorporate this into their bodies from the small amount of radioactive carbon dioxide that is naturally present in the air. When an organism dies, the unstable carbon-14 decays, but the carbon-12 does not. The ratio between carbon-12 and carbon-14 can be used to determine the absolute age of the fossil.

Radiometric dating can also be used to determine the age of inorganic materials (materials not containing carbon), such as the rocks surrounding fossils. **Potassium–argon dating** is commonly used to determine the absolute age of ancient rocks. Another example involves measurement of the ratios of decay of uranium-238 to lead-207 and uranium-235 to lead-206. The diagram on the right outlines how these ratios can change over time.

Over time the uranium present in rocks decays into lead. The half-life of uranium is the amount of time it takes for half the uranium initially present in the rock to decay into lead.

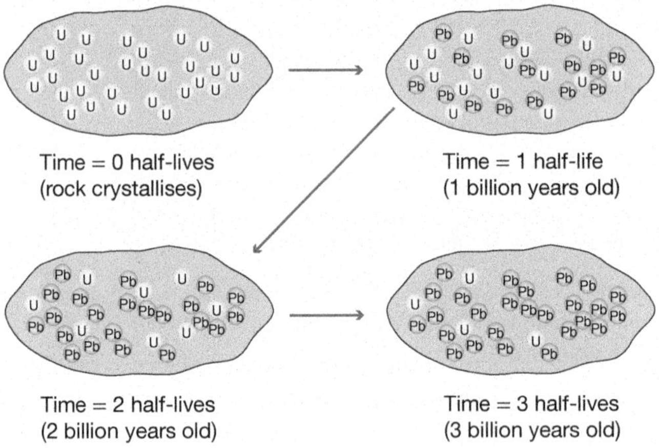

3.9.5 Fossils telling tales

The fossil record gives us evidence that species have changed over time. For example, a fossilised skeleton of a bird (*Archaeopteryx*) found in Bavaria has been dated at 150 million years old. It clearly shows feathers, which are a feature of all modern birds; however, it also has dinosaur characteristics such as teeth, claws on its wings and a long, jointed bony tail. From this, scientists have deduced that birds evolved from a dinosaur ancestor. The evolution, or change over time, of other species can also be followed by studying the fossil record.

The fossilised skeleton of *Archaeopteryx*

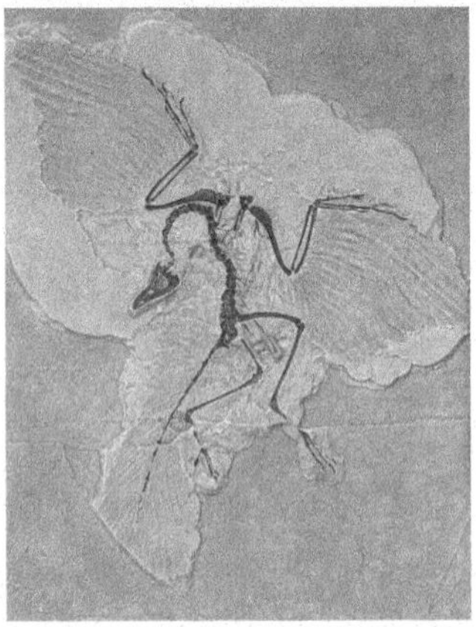

Some radioactive isotopes and their daughter products of decay

Radioactive parent isotope	Daughter product	Half-life (years)	Uses
Carbon-14	Nitrogen-14	5730	Used for dating organic (carbon-based) remains up to about 60 000 years old
Uranium-235	Lead-207	710 000 000	Used for dating igneous rocks containing uranium-based minerals in the range from about 1000 to 1 000 000 years
Potassium-40	Argon-40	1 300 000 000	Used for dating igneous rocks containing potassium-bearing minerals in the range from 500 000 years and older
Rubidium-87	Strontium-87	47 000 000 000	Used to date the most ancient igneous rocks

The longer the half-life of a radioactive isotope, the older the material that can be dated using a particular radiometric method.

Horsing around in time

The fossil record gives us evidence of gradual change occurring over time. An example can be seen in fossils of horse species from different times. Fossils indicate that horses have become taller, their teeth are now better suited to grazing than eating leaves and fruit, and their feet have a single hoof rather than spread-out toes. Over time, environmental changes have led to different variations having a selective advantage over others. Reduced availability of fruit and leaves but increased availability of tough grasses resulted in selection for teeth better suited to grazing. As forests were replaced by open plains, longer legs and hoofs may have been selected for to provide a better chance of escaping predators.

3.9.6 Types of fossils

There are many different types of fossils, including moulds, casts, imprints, petrified organisms, and whole organisms that have been frozen or trapped in sap or amber.

The evolution of the horse

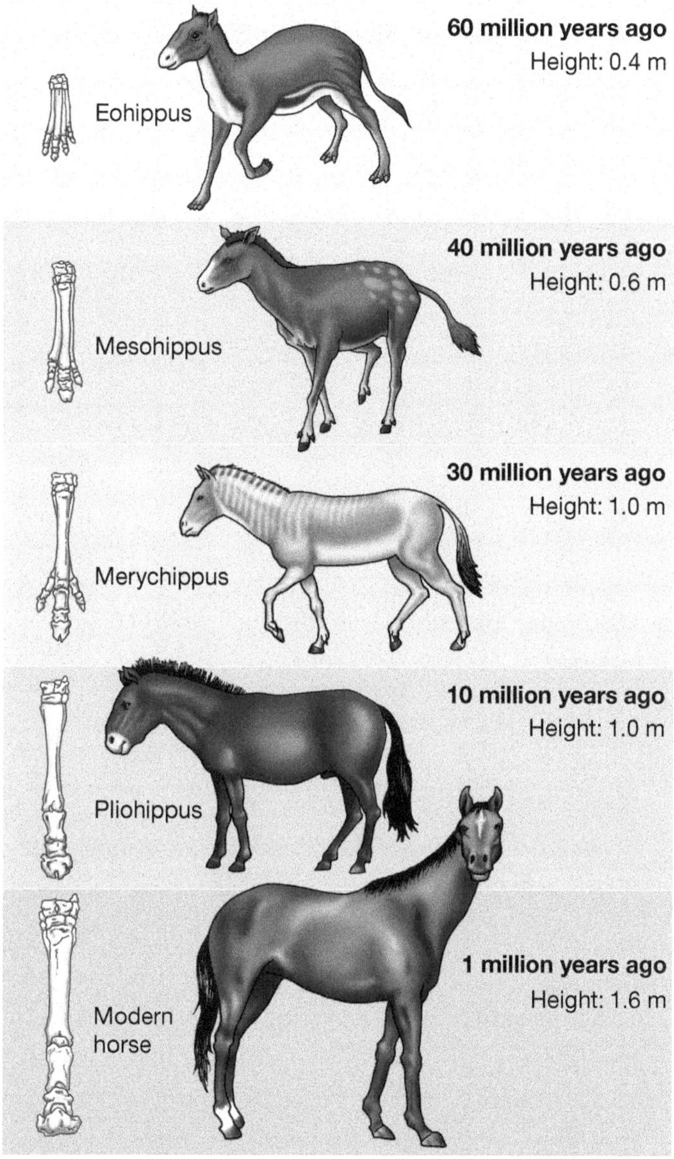

Cast: a rock with the shape of an organism protruding (sticking out) from it

Carbon imprint: the dark print of an organism that can be seen on a rock

Whole organism: larger organisms that have been preserved whole by being mummified or frozen, such as this baby mammoth found in 2007 in Siberia

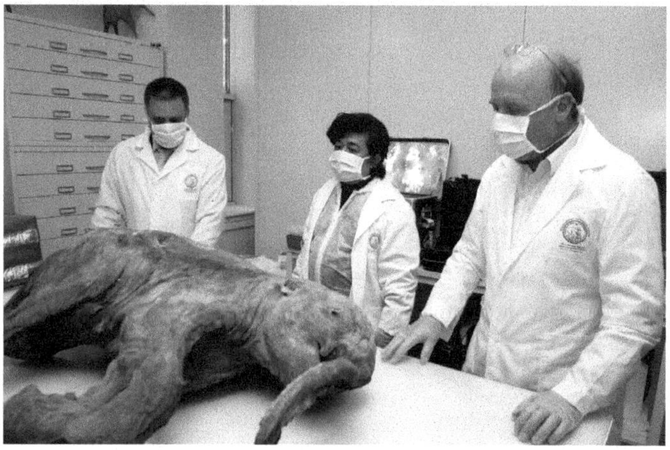

Mould: a rock that has an impression (hollow) of an organism

Amber fossils: parts of plants, insects or other small animals that have been trapped in a clear substance called amber

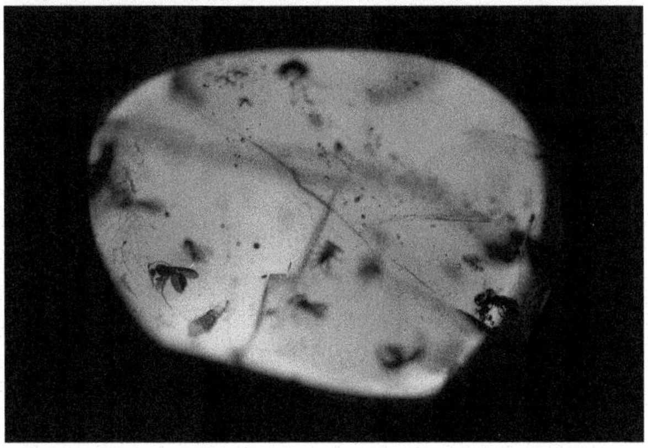

INVESTIGATION 3.4

Studying fossils

AIM: To describe and classify different types of fossil

Materials:

fossils, fossil casts or pictures of fossils

- Copy and complete the table for each type of fossil.

Name	Description	Type of fossil (cast, mould, imprint or other)

Petrified fossil: organic material of living things that has been replaced by minerals, such as petrified wood

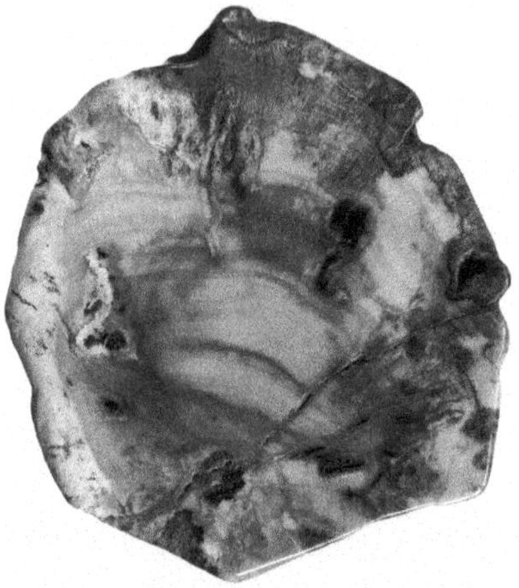

3.9 Exercises: Understanding and inquiring

To answer questions online and to receive **immediate feedback** and **sample responses** for every question, go to your learnON title at www.jacplus.com.au. *Note:* Question numbers may vary slightly.

Remember

1. Define the following terms.
   (a) Fossil
   (b) Palaeontology
   (c) Half-life of an isotope
2. Provide examples of different types of fossils.
3. Explain why fossilisation is a rare event.
4. Describe how fossilisation can occur.
5. Name the two main ways of dating fossils.
6. Outline the difference between the two main ways of dating fossils.
7. Suggest the connection between tectonic plates and dating fossils.
8. State the relationship between carbon dating and radiometric dating.
9. Describe how carbon dating is used to estimate the age of fossils.
10. Suggest why you might use potassium–argon dating rather than carbon dating.
11. Outline what the fossil record tells us about the evolution of horses.

Think and discuss

12. The Venn diagram on the right uses some analogies as well as key differences to distinguish the relative age from the absolute age of fossils. Copy and complete it using the terms top right.
13. Examine the diagram showing the evolution of the horse.
    (a) Describe how horses have evolved over the last 60 million years.
    (b) What type of horse would have been fittest (in terms of biological fitness) 60 million years ago? Explain your answer.
    (c) What type of horse would have been fittest one million years ago? Explain your answer.
    (d) Horse breeders pay large sums of money to have prize-winning racehorses breed with the mares in their stables. The fastest horses are flown around the world for breeding purposes. It is also possible to collect and freeze sperm from successful competition horses. This sperm can be used to impregnate many mares. Explain how this might affect the evolution of the horse. How might horses look in another million years?

Analyse and evaluate

14. Examine the picture of the fossilised dinosaur on the right. Write a brief description of the animal, including what it may have eaten and how it may have moved. Why do you think it had so many large openings in its skull?
15. Examine the dinosaur track on the right to decide:
    (a) which dinosaur walked along this track first, and which walked here last
    (b) which dinosaurs probably walked on two legs, and which probably walked on four
    (c) which dinosaur hopped
    (d) which dinosaur was the heaviest.
    (Identify the dinosaur(s) in your answers to the above questions as A, B, C or D.)
16. The layers of rock shown in the illustration bottom right have been disturbed by plate movements.
    (a) Was the plate movement caused by folding or faulting?
    (b) Which layer of rock is the youngest? Justify your answer.
    (c) Which layer of rock is the oldest? Justify your answer.

Order Potassium–argon dating
16 years old Most recent Strata
Eldest of three daughters
Carbon dating Oldest

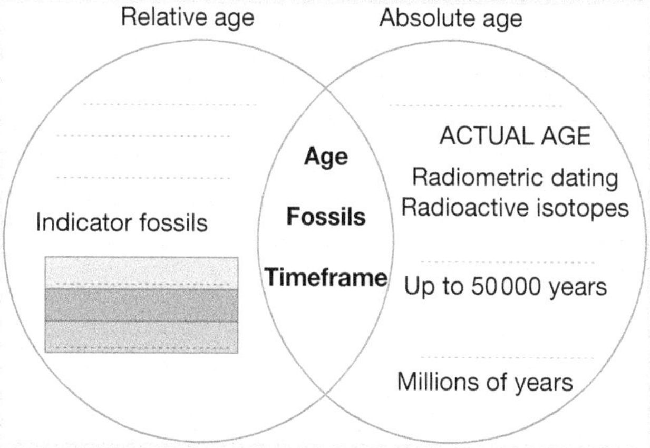

Fossilised skeleton of the Saurischian dinosaur

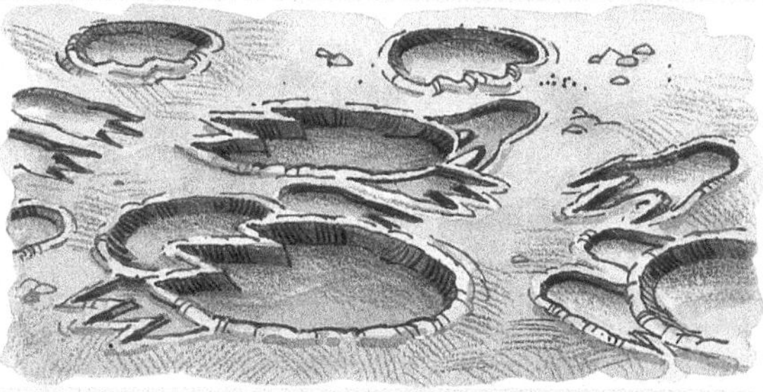

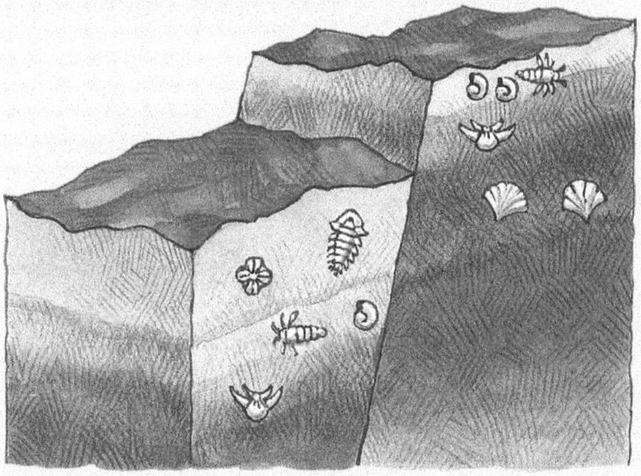

17. Carefully observe the graph on the right showing the decay of carbon-14 and answer the following questions.
   (a) Estimate the time taken for the radioactivity of carbon-14 to reduce by 50 per cent.
   (b) On the basis of this graph, what is the half-life of carbon-14?
   (c) Approximately how much radioactivity will be present at:
      (i) 10 000 years
      (ii) 30 000 years
      (iii) 80 000 years?

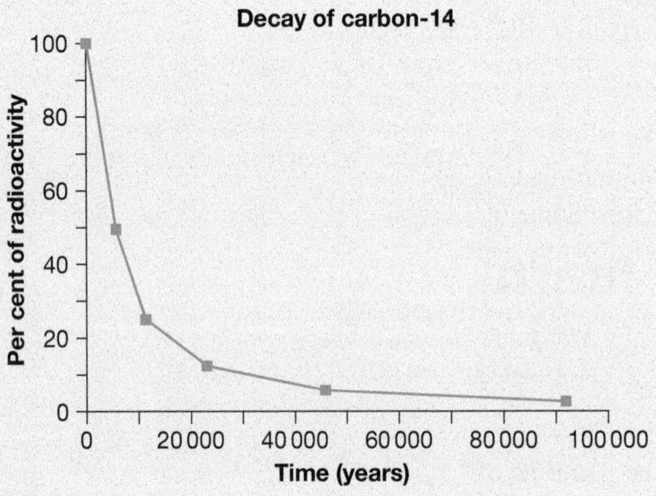

Investigate

18. What do geologists and palaeontologists do?
19. Investigate the claims of life on Mars (for example, information on the rock sample labelled ALH84001) through newspaper reports and websites. Are you convinced that life once existed on Mars?
20. Research and prepare a report on carbon dating.

Create

21. Create a dinosaur track using modelling clay.
22. Make a cast of a leaf fossil using modelling clay or plasticine. To do this, first roll out a rectangular piece of clay and cover it with petroleum jelly. Then press a ribbed leaf into the clay to make an impression. Remove the leaf and build some clay walls about 1 cm high at the edge of the rectangle. Cover the walls with petroleum jelly and pour in some mixed plaster. When the plaster has set, remove the clay and examine the cast for the leaf impression.
23. Make a model or draw a picture to show what a common ancestor of both mammals and birds may have looked like.
24. Use an image search engine to locate images of each of the types of fossils described under the heading *Types of fossils*. Cut and paste the pictures into a Word document. Write a caption for each image. The caption should include the name of the fossilised organism shown in the picture, the location where the fossil was found and the type of fossil (e.g. cast).
25. There are a number of great fossil sites in Australia. Use the internet and other sources to investigate one of the following fossil sites: Naracoote, Riversleigh, Bluff Downs, Murgon, Lightning Ridge.
   Summarise information about the fossil site you have chosen under the following headings:
   (a) Why the area is rich in fossils
   (b) Examples of fossils that have been found here
   (c) Age of the fossils found in this area
   (d) Important information revealed by the fossils found in this area.
26. Test your knowledge of all things old by completing the **Revelation: 'Fossils'** interactivity. Success rewards you with a video interview with a paleontologist where you can see some real fossils.

learnon RESOURCES — ONLINE ONLY

Complete this digital doc: Worksheet 3.4: Fossils (doc-19426)

Complete this digital doc: Worksheet 3.5: Ages of fossils (doc-19427)

3.10 More evidence for evolution

3.10.1 More evidence for evolution

The theory of evolution by natural selection was developed from the many observations that Darwin and Wallace made on their journeys. Since then, more evidence has been collected to further support their theory. Some of this evidence involved the use of new technologies.

learnon RESOURCES — ONLINE ONLY

Watch this eLesson: Ancient DNA (eles-1069)

3.10.2 Comparative anatomy

The forearms of mammals, amphibians, reptiles and birds are remarkably similar in structure. Each, however, is used for a different function, such as swimming, walking or flying. The structure of the forearm can be traced back to the fin of a fossilised fish from which amphibians are thought to have evolved.

Similarity in characteristics that result from common ancestry is known as **homology**. Anatomical signs of evolution such as the similar forearms of mammals are called **homologous structures**. For example, in the diagram at right, you can see that each limb has a similar number of bones that are arranged in the same basic pattern. Even though their functions are different, the similarity of basic structure still exists.

These structures have the same basic structure since they are all derived from a vertebrate forelimb. Do they have identical functions?

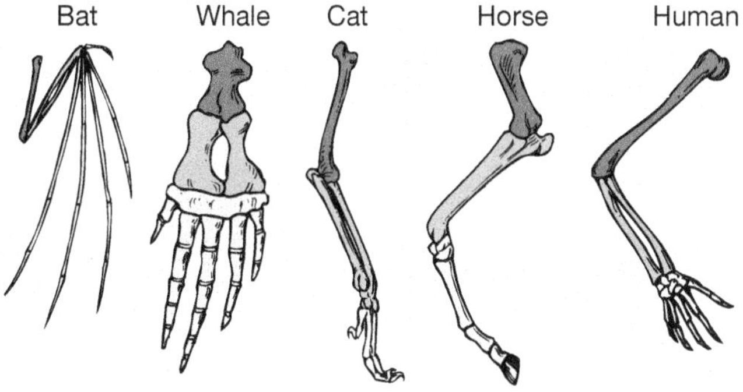

Gill slits are visible in sharks, but in bony fish they are concealed beneath a scaly panel known as an operculum.

3.10.3 Comparative embryology

Organisms that go through similar stages in their embryonic development are believed to be closely related. During the early stages of development, the human embryo and the embryos of other animals appear to be quite similar. For example, the embryos of fish, amphibians, reptiles, birds and mammals all initially have gill slits. As the embryos develop further, the gill slits disappear in all but fish. It is thought that gill slits were a characteristic that all these animals once shared with a common ancestor.

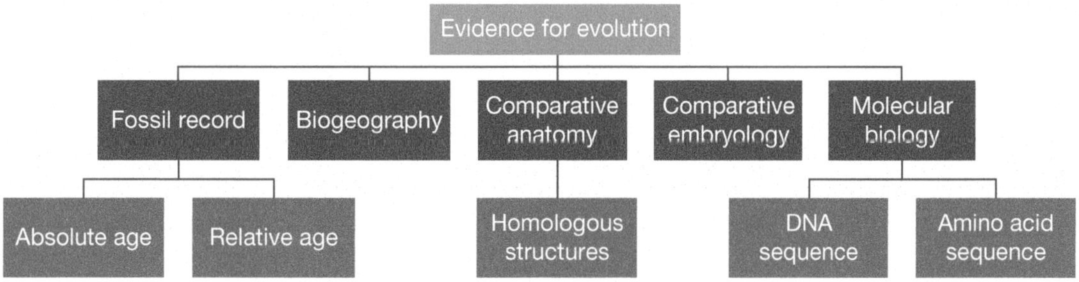

3.10.4 Molecular biology

How amazing is it that all living things share the same overall genetic coding system or language? Although the sequences may vary, the possible letters or nucleotides and the rules of reading them are basically the same. This is one of the reasons that we can cut DNA out of one organism and paste it into another so that it will make a protein it did not previously have the genetic instructions for.

We can use this concept of a universal genetic code to determine the evolutionary relationships between species. The similarities and differences between their DNA sequences and amino acid sequences in proteins can be used to determine how closely they are related and to estimate the period since they shared a common ancestor.

3.10.5 Linking proteins, amino acids, DNA and evolution

Proteins are universally important chemicals that are essential to the survival of organisms. In chapter 2 we looked at the coding and synthesis of proteins (see subtopic 2.4).

The genetic message to make proteins is stored in DNA. A section of the DNA (gene) is transcribed into messenger RNA (mRNA), which is then translated into proteins. Each of the DNA triplets and mRNA codons code for a specific amino acid, and the sequence of the nucleic acids determines the sequence of the amino acids that will make up a specific protein.

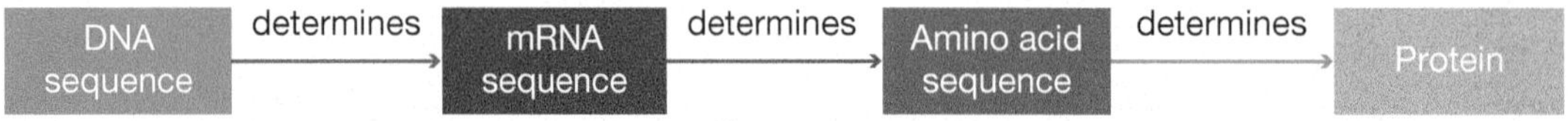

3.10.6 DNA sequences

DNA hybridisation is a technique that can be used to compare DNA sequences in different species to determine how closely related they are. The tree diagram below left shows the evolutionary relationships within a group of primates. Which primate is most closely related to humans and which is least closely related?

Tree showing inferred evolutionary relationships between primates based on DNA hybridisation evidence

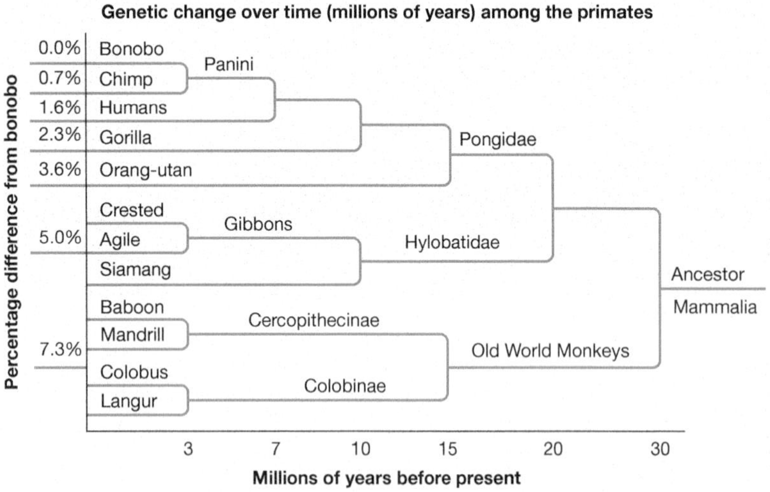

DNA hybridisation. The more closely related organisms are, the more similar their DNA sequences.

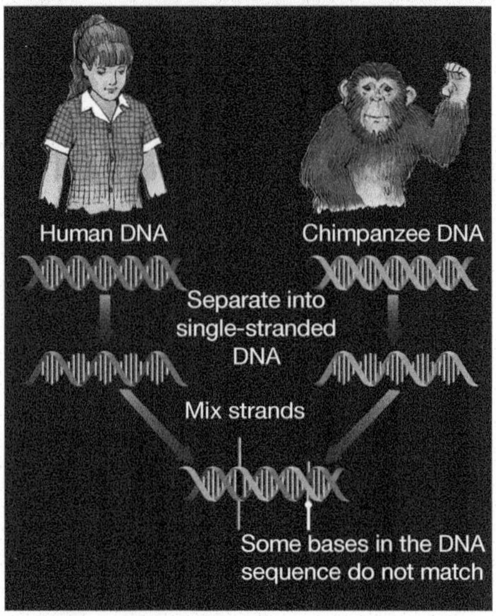

3.10.7 Amino acid sequences

A change in the DNA sequence may lead to a change in the type and sequence of amino acids in the protein produced. This idea is used in comparing the amino acid sequences of specific proteins in organisms. The graph on the next page provides an example of the number of differences in the amino acid sequence of a protein called cytochrome C. This protein is found in many species, but different species have slightly or significantly different versions of the protein.

3.10.8 The molecular clock

In 1966, biochemists Vincent M. Sarich and Allan Wilson noticed that changes in the amino acid sequences of particular proteins in related species appeared to occur at a steady rate. They found more amino acid sequence differences the longer that two species had existed separately. From these observations the concept of the molecular clock arose. This concept used differences in two species' amino acid sequences to estimate the time since the species had diverged. Based on the analysis of immunological evidence, Sarich and Wilson concluded that humans and African apes shared a common ancestor a lot later (no more than five million years ago) than was suggested by palaeontologists (15–25 million years ago).

This figure illustrates the molecular clock concept. It suggests, on the basis of amino acid sequence differences, species A and B are more closely related than A and C or B and C. It also suggests that species A and B diverged from a common ancestor just over 4 million years ago and that species A, B and C shared a common ancestor about 10 million years ago.

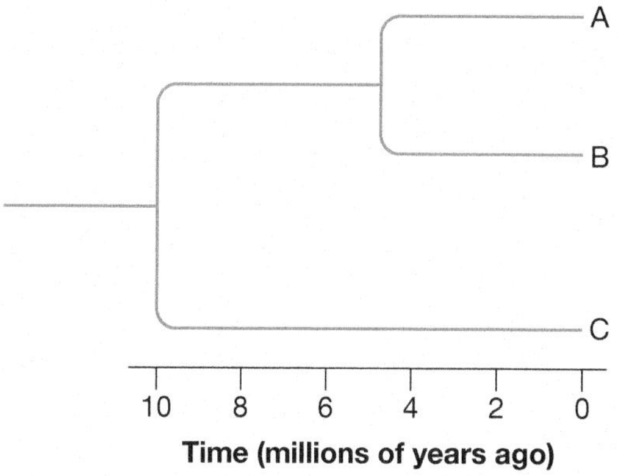

Cytochrome C is an important protein involved in the conversion of energy into a form that the cell can use. Although a part of the cytochrome molecule maintains a specific shape, over time other parts of the molecule have mutated.

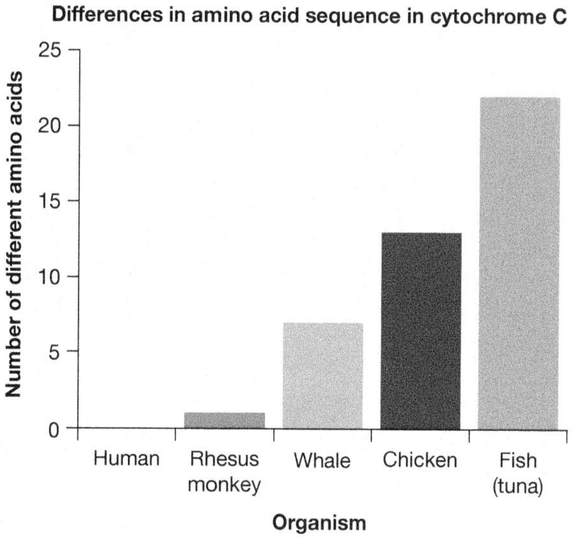

3.10 Exercises: Understanding and inquiring

To answer questions online and to receive **immediate feedback** and **sample responses** for every question, go to your learnON title at www.jacplus.com.au. *Note:* Question numbers may vary slightly.

Remember

1. List five types of evidence that support Darwin's theory of evolution.
2. Define the following terms.
   (a) Homology
   (b) Homologous structures
3. Provide examples of homologous structures.
4. Describe how the following can be used as a source of evidence for evolution.
   (a) Homologous structures
   (b) Comparative embryology
   (c) DNA sequences
   (d) Amino acid sequences
5. What is DNA hybridisation and why is it useful?
6. Suggest a link between proteins, amino acids, DNA and evolution.
7. Suggest the purpose of the molecular clock.

Investigate and create

8. Find out the names of several species of Australia's indigenous flora and fauna and the reasons they are unique.
9. Trace the pedigree of a horse, cat or dog. Explain reasons for some of the matings in the pedigree.
10. How is artificial selection different from natural selection?
11. Explain the significance of adaptations of organisms in relation to their survival.
12. Find out what the environment was like at some particular time in the past. Which traits or features would have given an organism increased chances of survival? Would these features still be an advantage in modern times? Explain.
13. What will Earth be like in the year 3000? Brainstorm and then record a summary of the ideas. Design a human that would be best suited to this futuristic environment. Present your suggestion as a poster or web page with descriptive labels to explain the functions or advantages of your futuristic human's body.
14. Cytochrome C is known primarily for its function in mitochondria and its involvement in ATP synthesis. Scientists are also researching its involvement in the process of programmed cell death (apoptosis) and to determine evolutionary relationships.
    (a) Find out more about the structure and functions of cytochrome C.
    (b) Construct a model of a cytochrome C protein.
    (c) Report your findings in a creative multimedia format.

Think and discuss

15. Examine the table below showing the DNA sequence from part of a haemoglobin gene from four different mammalian species.
    (a) T, G, C and A represent nitrogenous bases. Suggest what they are abbreviations for.
    (b) (i) In terms of the first 11 nitrogenous bases, which mammalian species is most similar to humans?
        (ii) How is this species different from humans?
    (c) (i) In terms of the first 11 nitrogenous bases, which mammalian species is least similar to humans?
        (ii) How is this species different from humans?
    (d) On the basis of the data in the table, rank these species in terms of how long ago they may have shared a common ancestor with humans.
    (e) Suggest how these differences in the sequence of nitrogenous bases in DNA may have arisen.

DNA sequence from part of a haemoglobin gene from four mammalian species

Species	DNA sequence				
Human	TGACAAGAACA	-	GTTAGAG	-	TGTCCGAGGACCAACAGATGGGTACCTGGGTCCCAAGAAACTG
Orang-utan	TCACGAGAACA	-	GTTAGAG	-	TGTCCGAGGACCAACAGATGGGTACCTGGGTCTCCAAGAAACTG
Rhesus monkey	TGACGAGAACA	A	GTTAGAG	-	TGTCCGAGGACCAACAGATGGGTACCTGGGTCTCCAAGAAACTG
Rabbit	TGGTGATAACA	A	GACAGAG	A	TATCCGAGGACCAGCAGATAGGAACCTGGGTCTCTAAGAAGCTA

Differences between the human DNA sequence and those of other species are shown by coloured letters. The dash (-) is used to keep the sequences aligned. Note that there are two sequence differences between the human and some other primates (orang-utan and monkey), but there are more between the human and the rabbit DNA. Why?

16. Examine the data in the table at right and then complete the following:
    (a) Reorder the information so that the species are arranged from the most to the least related to humans.
    (b) Which species is most closely related to humans?
    (c) Which species is less likely to be closely related to humans?
    (d) Use the information in the table to construct a flowchart, graph or diagram to show the likely relationship between the seven listed species.

DNA differences among closely related species

Species tested against human DNA	Percentage differences
Human	–
Gorilla	1.8
Green monkey	9.5
Orang-utan	3.6
Chimpanzee	1.4
Capuchin monkey	15.8
Gibbon	5.3

3.11 Origin of whose species?

3.11.1 Changing views

There are many alternative cultural and religious views as to the origin of life and where humans fit into it. The current scientific view is based on Darwin's and Wallace's theory of evolution of species by natural selection. This theory changed the way many viewed the origin of life and its diversity on our planet.

A modern reconstruction of a Neanderthal. The excerpt at left from the famous book *Clan of the Cave Bear* refers to the Clan, who are Neanderthals, and Ayla, a Cro-Magnon or early modern human.

Ayla examined her son again, trying to remember the reflection of herself. My forehead bulges out like that, she thought, reaching up to touch her face. And that bone under his mouth, I've got one too. But, he's got brow ridges, and I don't. Clan people have brow ridges. If I'm different, why shouldn't my baby be different? He should look like me, shouldn't he? He does a little, but he looks a little like Clan babies, too. He looks like both. I wasn't born to the Clan, but my baby was, only he looks like me and them, like both mixed together.

Source: The Clan of the Cave Bear by Jean M. Auel

3.11.2 Where do humans fit in?

Until recently, in western cultures, life on Earth was considered as unchanging and due to the unrelated products of special creation. It is no surprise that the theory of evolution caused such outrage.

In 1858, Alfred Wallace sent a letter to Charles Darwin that summarised his own independently constructed theory of evolution. This finally prompted Darwin to publish his own controversial theory of evolution in *The Origin of the Species* later that year. This publication resulted in the beginning of many heated debates about where humans fit into the pattern of life on Earth. The development of this theory and its impact is further explored in subtopic 3.4.

Alfred Russel Wallace (1823–1913)

Charles Darwin (1809–1882) aged 45 in 1854

Darwin published *The Descent of Man* in 1871. In this book, he suggested that humans and other species on Earth were related. At the time, this idea was met with outrage and disbelief. Just over a hundred years later, we have biochemical evidence to support Darwin. DNA sequencing has shown that the DNA in humans and chimpanzees differs by only about 1 per cent! There are also shared patterns of relatedness with other primates and organisms from other levels of classification.

3.11.3 Humans are primates

Humans, orang-utans, gorillas and chimpanzees all belong to the primate order of classification. Primates are placental mammals, and many of their features relate to their ancestors having an arboreal (tree-dwelling) lifestyle. Primate hands (and sometimes feet) are able to grasp and manipulate objects using their five digits; they have a prehensile thumb or toe, and nails instead of claws. Most primates also rely more on sight than smell. This is why their faces are flatter than many other types of mammals; their forward-facing eyes enable stereoscopic vision. Unlike many other groups of mammals, they have colour vision so they are able to detect when particular foods may be ready to eat, and their teeth allow for a varied diet.

Humans did not evolve from apes or monkeys. Evidence suggests, however, that we do share a common ancestor.

Where humans fit into the classification system

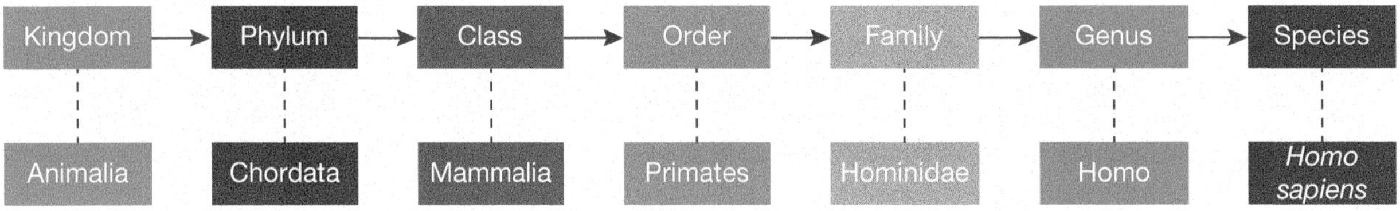

Who else is in our family?

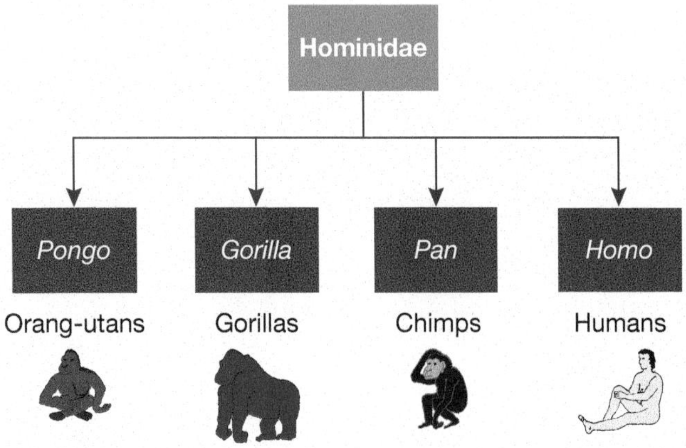

Primates possess particular features that gave their tree-dwelling ancestors a selective advantage.

3.11.4 The missing link?

Once Darwin's theory of evolution became more widely accepted, one of the next questions was where the missing link is between humans and apes. Research and discoveries suggest that, rather than one missing link, there are many different ancestral species in our history. The diagrams on the following page provide examples of some important fossil finds. While these have helped us to discover parts of the jigsaw that make up our evolutionary history, there is considerable debate about how these pieces fit together.

The variety of interpretations of fossilised skulls and other parts has led to the development of different theories and timelines to describe the evolutionary relationships between humans and our erect walking ancestors. (mya = million years ago)

Australopithecus afarensis
Dated: 7–3 mya
Located: Africa
Known for being: First group classifield as hominid and believed to be a common ancestor of both humans and living apes.
Alias: 'Lucy'

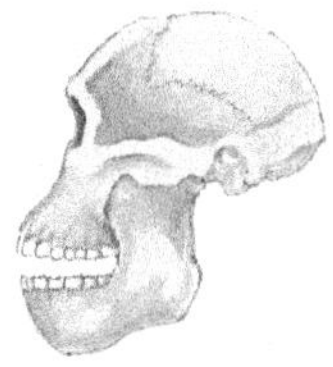

Homo habilis
Dated: 2.2–1.6 mya
Located: South and East Africa
Known for being: First member of genus *Homo*

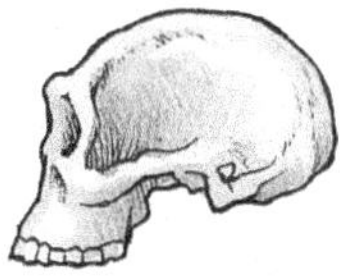

Homo ergaster
Dated: 1.8–1.2 mya
Located: Africa
Known for: Migrating out of Africa to Asia at least 1.8 mya. Could be a different population of *Homo erectus*?

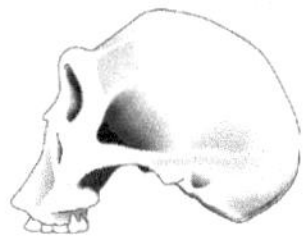

Homo erectus
Dated: 1 800 000–20 000 years ago
Located: Africa, Asia (and Europe?)
Known for: Walking upright
Alias: 'Upright man', Java Man (Indonesia), Peking Man (China)

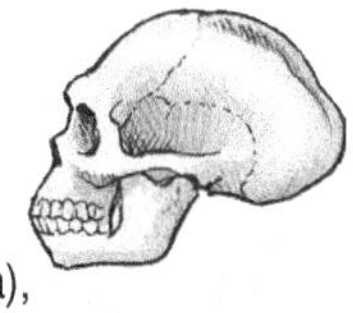

Homo neanderthalensis
Dated: 230 000–29 000 years ago
Located: Europe and western Asia
Known for: Being well adapted to very cold climates and used spears and sharp tools — and possibly basic words.
Alias: Neanderthal man

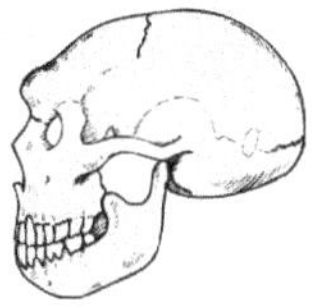

Early Homo sapiens
Dated: 100 000 years ago
Located: ?
Known for: Possibly being our direct ancestor and producing complex tools, jewellery and paintings on cave walls.
Alias: Cro-Magnon man

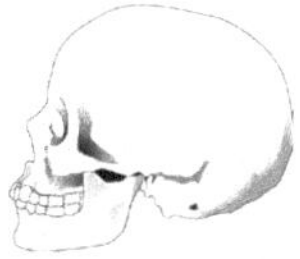

HOW ABOUT THAT!

Scientists are investigating genetic patterns in mitochondrial DNA (mtDNA) in various populations. There is a controversial theory that suggests that Earth's current human population evolved from one female (Mitochondrial or African Eve) who lived in Africa 200 000 years ago.

Ancient DNA is telling us new stories about Neanderthals and also about how closely we may be related to them. In 2007, DNA studies suggested that some Neanderthals had red hair and pale skin with similar pigmentation to some modern-day humans. In 2010, another DNA study suggested that most humans have at least 1–4 per cent Neanderthal DNA within their genome. This suggests that Homo sapiens interbred with Neanderthals. Some studies estimate that Neanderthal DNA is 99.7 per cent identical to modern human DNA, compared to 98.8 per cent similarity between humans and chimpanzees. What will the next fossils discovered and new technologies dig up?

A Scottish doctor, Robert Broom (1866–1951), made important fossil discoveries. One of these was of this famous *Australopithecus africanus* fossil known as Mrs Ples.

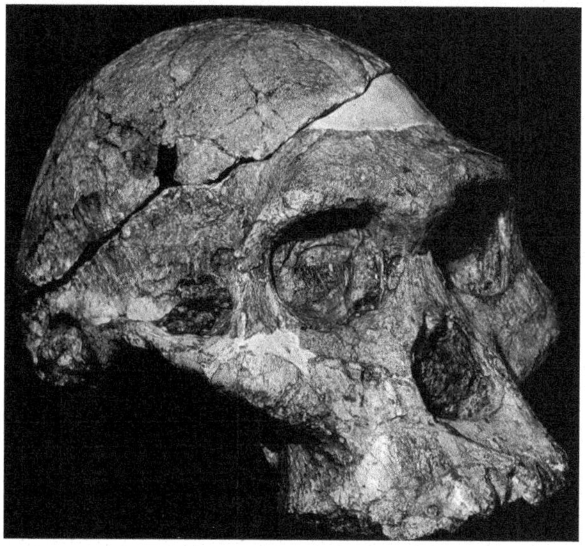

3.11 Exercises: Understanding and inquiring

To answer questions online and to receive **immediate feedback** and **sample responses** for every question, go to your learnON title at www.jacplus.com.au. *Note:* Question numbers may vary slightly.

Remember

1. Suggest why the theory of evolution caused such outrage.
2. Suggest what prompted Charles Darwin to finally publish his theory of evolution.

3. State the following classifications for humans.
   (a) Class
   (b) Order
   (c) Family
   (d) Genus
   (e) Species
4. (a) Describe features that are common to all primates.
   (b) Suggest why primates share these features.
5. List three other members of the Hominidae family.
6. Did humans evolve from apes? Justify your response.

Investigate, think and create

7. (a) Select one of the skulls described in this subtopic and research its owner.
   (b) Construct a skull model out of plasticine, dough, clay or another suitable material.
   (c) Create a story about a week in the owner's life.

Investigate and discuss

8. Find out more about Neanderthals and the research that is linking them even closer to our species. Compare these findings with previous theories about where they fitted into our family tree.
9. Find out more about the debates regarding interpretation of fossils to develop an evolutionary tree for *Homo sapiens*. What's the problem?
10. A very important characteristic that palaeontologists look for in fossils is whether the individual was bipedal (capable of walking on two legs). Find out more about features that they look for in a fossil to determine whether this is the case.

Analyse, think and discuss

11. Examine the skulls illustrated on the right of a chimpanzee, *Australopithecus africanus* and *Homo sapiens*.
    (a) Describe how they are similar.
    (b) Describe how they are different.
    (c) Suggest reasons for the differences.

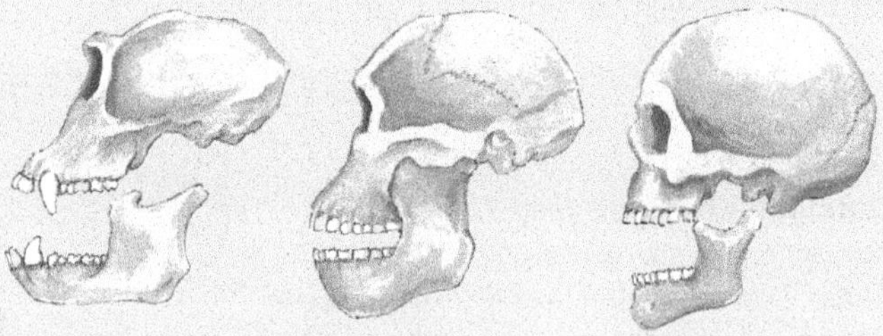

(a) Chimp (b) *Australopithecus africanus* (c) Human

3.12 See you later, alligator

3.12.1 Extinction

It has been estimated that about four species become extinct every hour. Over the next few decades, as many as one million species could be lost forever. How will the loss of biodiversity affect life on Earth?

Extinction, or the disappearance of a species and the resulting loss of genetic information from the gene pool, is a natural occurrence. The fossil record shows several times when huge numbers of species became extinct. Species that cannot successfully reproduce under changed environmental conditions will cease to exist. It is estimated that 99 per cent of all the species that have ever lived are now extinct.

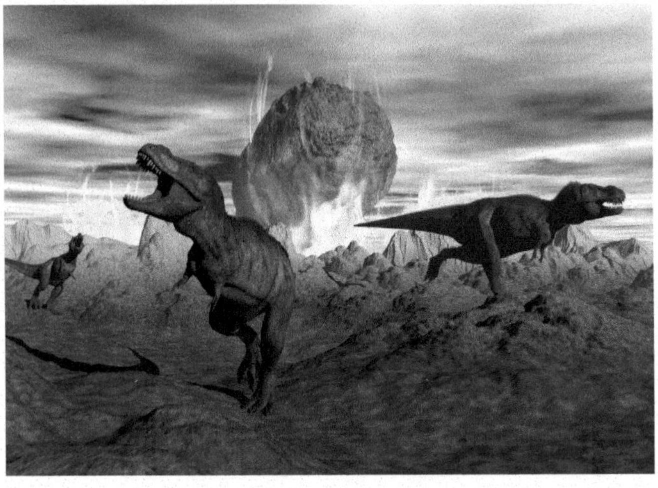

Mass extinction is the term used to describe the dying out of thousands of species all over the world at the same time. Suggested reasons include gradual changes in the climate or meteor strikes. The most devastating mass extinction of all occurred about 225 million years ago, in which 96 per cent of all species died out. This included about 90 per cent of the marine life present at the time. Around 65 million years ago, more than 75 per cent of all species died out. This included more than half the marine species, many terrestrial plants and animals such as the dinosaurs. About 10 000 years ago another mass extinction occurred, this time involving many species of giant mammals. During this extinction, many large marsupials in Australia and placental species in other countries became extinct.

How do you think these mass extinctions affected the species that survived? (mya = millions years ago)

Main event	Epoch	Mass extinction
Modern humankind	Quaternary 2 mya – present	Human activities threaten mass extinction in modern times
Early flowering plants; origin of birds	Cretaceous 146 – 65 mya	76% of species lost 47% of genera lost
Spread of dinosaurs, ammonites and cycads	Triassic 245 – 208 mya	Extinction of many marine species
Primitive reptiles and paramammals; appearance of beetles and conifers	Permian 286 – 245 mya	96% of species lost 84% of genera lost
Rise of mosses and ferns, ammonites and spiders	Devonian 410 – 360 mya	82% of species lost 55% of genera lost
Appearance of primitive fishes, trilobites, molluscs and crustacea	Ordovician 505 – 440 mya	85% of species lost 60% of genera lost

3.12.2 Why are species disappearing?

Over the last 100 years, the number of extinctions has dramatically increased. Some of the reasons for this include the following:

- Rainforests provide a home for a large number of the world's species. Over 60000 species of spiders and insects live in the Amazonian rainforest. **Deforestation** over the past 50 years has removed half of the world's forests. This has been done to provide land for farming, woodchips, timber, fuel and sites for urban development as well as mines and dams. In the past 200 years, Australia has cleared two-thirds of its rainforests.
- **Introduced species**, such as rabbits, may be better able to hunt for food and living space than native species. In this event, there is an increase in the population of introduced species and a decrease in the number of native species.
- Some species are hunted by people for meat, hide, horns, feathers or eggs, or because they are a threat to domesticated species.

3.12.3 Why worry? Be happy

The majority of Australians live in cities, far removed from forests. Would the reduction in the number of species and the resulting loss of biodiversity in forests ever affect city dwellers?

Some of the ways in which wild species affect our lives are described below.

- Wild plant and animal species provide a source of wonder and beauty for large numbers of people.
- Rainforests provide a huge store of untapped genetic material, much of which may be useful to humans.
- Each organism in the food web holds a very important place. Removal of one species has a major effect on the rest of the organisms in that food web.
- Most of our modern crop plants were domesticated from wild plants. With the increase in the world's population, finding suitable food crops from wild species may be important in the future.
- Wild species help to recycle nutrients in the soil, providing us with fertile soil for crop growth.
- Many wild species help to filter and remove poisonous substances from the air, water and soil.

- The greater the genetic diversity within a species, the better its chances of surviving in changing environmental conditions. Research has found that if a species loses genetic diversity, it will eventually dwindle in numbers and perish.
- All species have a right to exist without their survival being threatened by human interference.

3.12.4 What is being done about the loss of species?

We have at last realised the importance of genetic biodiversity on our planet. The following approaches in Australia and overseas are attempting to save some of our endangered species from extinction.

- National parks are being established so that species numbers can be monitored. For example, the number of Chinese giant pandas is increasing as more reserves are set up where they can breed free from human interference.
- Botanical gardens and zoos are being maintained so that plants and animals can be bred in captivity.
- Existing areas are being protected by fencing them off. For example, sand dunes at beaches are sometimes closed off to allow the growth of plants and the reintroduction of animal species.

3.12.5 Dead as a dodo

The dodo was a bird with a large, hooked beak and a plume of white feathers on its rear. It was first sighted on Mauritius, an island in the Indian Ocean, in 1598. In 1681, only 83 years later, all dodos were extinct!

Dodos are now alive only in storybooks.

Although the dodo could not fly, it was believed to have evolved from an ancestor (similar to a pigeon) that could fly. At the time that this ancestor landed on Mauritius, over four million years ago, its new habitat had a plentiful food supply and contained no predators. Can you suggest a reason for the dodo's loss of the ability to fly? How could this be advantageous to its survival? Scientists believe that with no natural predators the dodo did not need to fly and so evolved into a flightless bird.

The dodo's nests were on the ground since it was flightless. This and a lack of natural predators made the dodo easy prey for the Dutch sailors who discovered them, and subsequent Portuguese invaders would club the birds to death for sport or food. However, one of the main causes of extinction was the destruction of the forest in which dodos lived. This led to a reduced food supply. The other major cause was the introduction by humans of new species, such as cats, rats and pigs, which destroyed their nests.

3.12.6 Losing more than your stripes

Will quaggas make a reappearance?

Quaggas formerly lived in South Africa. They were a variety of zebra with a distinctive patch of stripes around the head, neck and front portion of their body. The unusual stripes, like those on zebras, may have rendered the animal less visible and given some protection against predators. Can you suggest a reason for the variation of stripes in the quagga compared to other types of zebra?

Like other grazing mammals, quaggas were hunted by settlers who saw them as competitors for the grazing of their sheep, goats and other livestock.

It has been recorded that their flesh was eaten and their skin was used as grainbags and leather. Quaggas became extinct when a quagga mare at Amsterdam Zoo died on 12 August 1883. This realisation did not register until many years later.

A project in South Africa involves an attempt to bring back the quagga from extinction and reintroduce it into reserves in its former habitat. This project is possible because DNA analysis has shown that the quagga was actually a subspecies of a type of zebra that is still alive. It is hoped that some quagga genes still exist in the populations of these zebras. By selectively breeding zebra individuals to concentrate the quagga genes, they may be able to bring back some of the features (e.g. colouration) of the extinct quagga.

HOW ABOUT THAT!

Melbourne Zoo and Healesville Sanctuary are actively involved in the breeding in captivity of endangered species. This program enables species to be returned to the wild, and some members of the species to be preserved in case wild populations die out. The long-footed potoroo, Leadbeater's possum, helmeted honeyeater and orange-bellied parrot have all been saved from extinction in this manner, and large breeding populations are now established in the wild.

The Leadbeater's possum is an endangered animal.

3.12.7 Is extinction permanent?

The Tasmanian tiger or thylacine (*Thylacinus cynocephalus*) looked like a large, long dog with stripes, a heavy stiff tail and a large head. An adult thylacine was about 58 centimetres tall, about 180 centimetres long and could weigh up to 30 kilograms. It was carnivorous, had large powerful jaws, and fed on kangaroos and rodents. Although it looked like a dog, it was a marsupial with a pouch and was more closely related to kangaroos and koalas.

The introduction of sheep as livestock in 1824 resulted in conflict between the settlers and the thylacines. In 1830, thylacine bounties were introduced by Van Diemens Land Company and, by 1888, the Tasmanian parliament had placed a price of one pound per thylacine's head. It was not until 1909, after 2184 bounties were paid, that the government bounty scheme was terminated. By 1910 thylacines were rare, and in 1933 the last thylacine to be captured was sold to Hobart Zoo. This thylacine died on 7 September 1936 and it was in this year that thylacines were added to the list of protected wildlife. In 1986, thylacines were declared extinct by international standards.

On 4 May 2000, the first piece of DNA was extracted from a thylacine fetus that had been preserved in alcohol since 1866. These DNA fragments were then inserted into bacteria and multiplied to create a library of DNA for research.

A preserved thylacine fetus at the Australian Museum. What are the implications of cloning extinct animals?

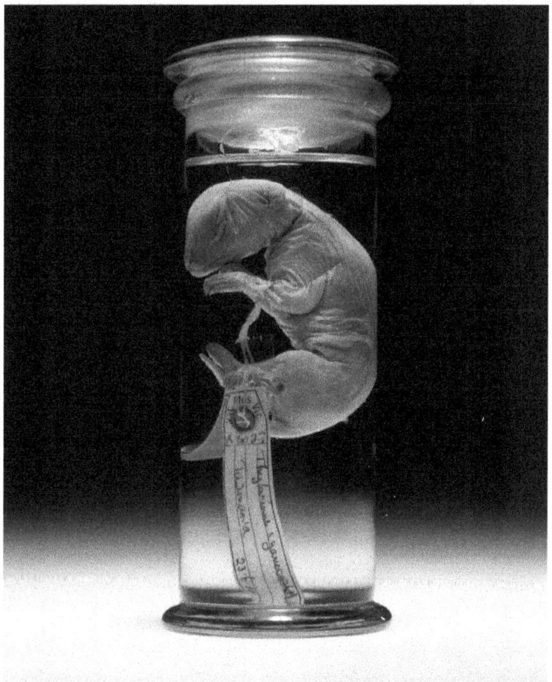

Australian Museum scientists tried to determine the entire thylacine genome and clone the animal, but the DNA was highly degraded and the project was abandoned. Some scientists still hope that one day we will bring the thylacine back to life.

3.12 Exercises: Understanding and inquiring

To answer questions online and to receive **immediate feedback** and **sample responses** for every question, go to your learnON title at www.jacplus.com.au. *Note:* Question numbers may vary slightly.

Remember

1. What are the three major factors leading to the extinction of species?
2. Why is it important to humans that genetic biodiversity be as large as possible?
3. List three ways in which humans are attempting to preserve endangered species.
4. Why is breeding in captivity being undertaken?
5. What is a mass extinction? Give three examples of mass extinctions.
6. What is a dodo and how is it thought to have become extinct?
7. Why did the dodo build its nests on the ground and lose its ability to fly?
8. Draw a timeline to show the history of events that contributed to the extinction of thylacines.
9. Summarise ways in which humans are attempting to bring back extinct animals.

Think and discuss

10. Consider the reasons that deforestation occurs. What alternatives to deforestation are available that could help to preserve our tropical forests?
11. Which one of the three approaches to preserve genetic diversity do you prefer? Explain your choice.
12. Apart from the factors already mentioned, what other factors may be contributing to the extinction of our native animals and plants?
13. As a class or in groups, brainstorm other outcomes or effects on humans if genetic biodiversity is reduced.

Investigate

14. Evaluate theories about the causes of the extinction of the dodo, the quagga and the thylacine.
15. Investigate the rate of extinction of Australian animals or plants since European settlement and suggest causes and implications of this.
16. Choose one of the following (the approximate year of extinction is given in brackets) and describe the animal, its lifestyle and the theory of the cause of its extinction: flightless ibis (1000), giant lemur (500), giant moa (1500), aurochs (1627) or an animal of your choice.

Create

17. Role-play a situation in which developers want to remove trees in a forest to provide land for housing. They have provided the home-buyers with free seeds to revegetate the area. Conservationists disapprove of the plan.

3.13 Tapestries within our biosphere

Science as a human endeavour

3.13.1 Interwoven threads

Throughout history, life on Earth has been connected to global climate change. The evolution of species has been linked not only to their environments but also to each other. Like threads in a three-dimensional tapestry, the components that make up our Earth's biosphere are interwoven on many different levels.

3.13.2 Prokaryotic cells change our planet

When the first signs of life appeared on Earth, its atmosphere was not as it is today. There was no oxygen for cellular respiration and no ozone layer to protect organisms from the sun's harmful ultraviolet radiation.

The first cellular organisms to appear were prokaryotes, such as bacteria. Fossils of prokaryotes have been found in 3.5-billion-year-old rocks, and fossil records suggest that mounds of these bacteria once covered the Earth.

Mutations, biodiversity and oxygen

Various types of mutations occurred in these prokaryotes. This resulted in an increasingly diverse range of new life forms. The selection of a sequence of these mutations enabled some bacteria to harvest energy from the sun and use carbon dioxide in the atmosphere to make their own food. Using this process of photosynthesis, they released oxygen back into the atmosphere. Over time, this changed the composition of Earth's atmosphere, with some of the oxygen also being converted into ozone, which later formed the ozone layer.

3.13.3 Eukaryotic cells finally appear

Eukaryotic cells finally made their entrance on Earth around 1.8 billion years ago. These cells differed from **prokaryotic cells** in that they contained a variety of membrane-bound **organelles**. These included a nucleus, endoplasmic reticulum and Golgi bodies. Some scientists have suggested that these may have originated from deep folding of the plasma membrane.

The fossil record suggests that organisms made up of many eukaryotic cells appeared about one billion years ago. Cells within these multicellular organisms became specialised for particular functions.

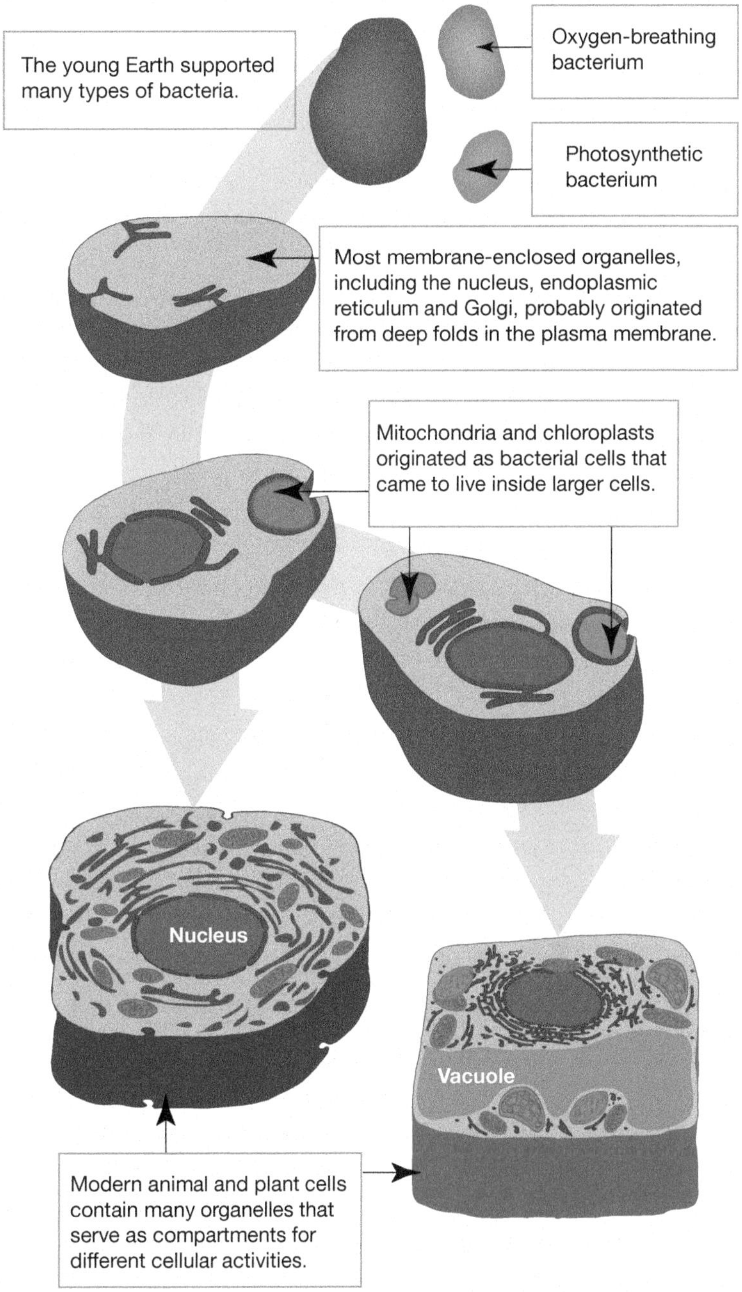

The young Earth supported many types of bacteria.
Oxygen-breathing bacterium
Photosynthetic bacterium
Most membrane-enclosed organelles, including the nucleus, endoplasmic reticulum and Golgi, probably originated from deep folds in the plasma membrane.
Mitochondria and chloroplasts originated as bacterial cells that came to live inside larger cells.
Nucleus
Vacuole
Modern animal and plant cells contain many organelles that serve as compartments for different cellular activities.

3.13.4 Endosymbiotic theory of evolution

Symbiosis describes a relationship between two different species in which they both benefit from living and working together. When one of these organisms lives within the other it is called **endosymbiosis**. The **endosymbiotic theory** describes how an ingested bacteria and its larger host cell could become so dependent on each other that after many years of evolution they could not live without each other.

3.13.5 Chloroplasts and mitochondria contain their own DNA

Chloroplasts and **mitochondria** are membrane-bound organelles that are found in eukaryotic cells. They possess striking similarities to prokaryotic cells. They are surrounded by a double membrane and contain their own DNA, which is different and separate from the

DNA located within the nucleus. Their DNA is used to produce many of the proteins (such as enzymes) that are essential for their function. These organelles also reproduce like bacteria and coordinate their own DNA replication and division.

It is thought that animals and plants shared a common ancestor that had acquired mitochondria by the process of endosymbiosis. Later, plants also acquired chloroplasts, and their evolutionary path diverged from that of animals.

With increased numbers of photosynthetic organisms, there was an increase in the amount of oxygen in the atmosphere. This provided an environment with conditions suitable for the evolution of a variety of organisms that could use this oxygen in the process of cellular respiration. This process removed oxygen from the atmosphere and released carbon dioxide back into it.

3.13.6 Parasitic superpowers

In various science fiction and superhero stories there are numerous characters (usually the nemesis or enemy of a superhero) that can steal superpowers from others. Did you know that there are organisms that can actually do this?

Amid the diversity of life currently on our planet, some animals have evolved mechanisms to be able to incorporate organelles (such as chloroplasts) or specialised cells from other species into their own bodies and then use the functions of what they have stolen.

One recent discovery is a type of sea slug, *Elysia chlorotica*. These sea slugs steal chloroplasts from algae and then start photosynthesising themselves. Some scientists view this as a type of symbiosis at a genetic level, with partners sharing genes.

The nudibranch *Phidiana crassicornis* can extract stinging cells from jellyfish and coral to use as a weapon.

Sea slugs of the species *Elysia chlorotica* (top) steal chloroplasts from algae (bottom) and gain the ability to photosynthesise.

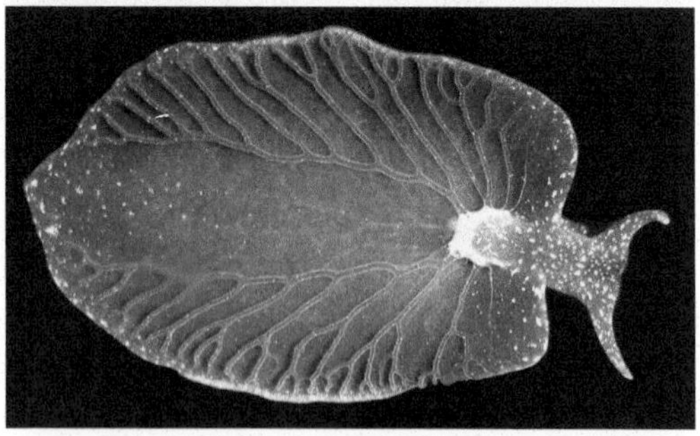

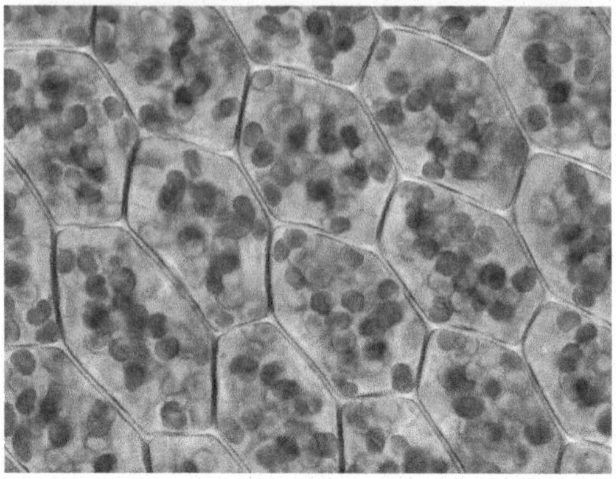

Stealing stinging powers

A type of nudibranch, *Phidiana crassicornis*, can extract the stinging organelles (cnidocytes) from jellyfish and coral and insert them into their own tissue, giving them a weapon that they didn't have before! Once they use these cells, however, they need to find a fresh supply. This is not the case with *Elysia chlorotica.*

Solar-powered animals

After just one single meal of algae (*Vaucheria litorea*), *Elysia chlorotica* possesses the ability to photosynthesise for life. Scientists have found that the slug actually cuts open the algal filaments and sucks out the contents. It then transfers the living chloroplasts into cells lining its gut. This phenomenon is sometimes referred to as **kleptoplasty** and the captured plastids as **kleptoplasts**.

As these chloroplasts are in the somatic cells and not the gametes of the sea slugs, they are not inherited when the slugs reproduce. Their offspring need to have their own meal of algae to gain the ability to photosynthesise.

We can't survive without you!

Scientists have also discovered that not only is the association between *Elysia chlorotica* and *Vaucheria litorea* specific, it is also obligate. This means that the algae are essential to the life cycle of the sea slug. *Elysia chlorotica* will not complete metamorphosis and develop into an adult in the absence of its algal prey.

3.13.7 A mystery yet to solve

Scientists are currently exploring questions about how these chloroplasts can continue to function without the algal nucleus on which they were previously dependent. Some scientists suggest that the slug's genome may contain genes transferred from the alga without which the chloroplasts could not function, making it a **holobiont** (combined genome) of slug genes and algal genes.

Other scientists consider that the survival of the chloroplasts is a result of multiple endosymbiotic events, gene transfer and the evolution of modern-day chloroplasts and mitochondria. Scientists are still asking questions and will hopefully be able to use genomic, biochemical, molecular and cellular approaches to unravel the mystery. Will you be the one to solve it?

3.13.8 Looking at the past

There are many scientific careers involved in exploring Earth's history, unlocking mysteries of life from the past and relating it to the present and future. These include palaeogeology, palaeobiology, geology and archaeology.

3.13 Exercises: Understanding and inquiring

To answer questions online and to receive **immediate feedback** and **sample responses** for every question, go to your learnON title at www.jacplus.com.au. *Note:* Question numbers may vary slightly.

Remember

1. Identify the first type of cellular organisms to appear on Earth.
2. How long ago does the fossil record suggest that cellular organisms on Earth appeared?
3. Suggest an environmental effect caused by photosynthesis.
4. Approximately how long ago did eukaryotic cells first appear on Earth?
5. Identify a way in which prokaryotic and eukaryotic cells differ from each other.
6. Outline the difference between symbiosis and endosymbiosis.
7. Describe features that mitochondria and chloroplasts have in common.
8. Explain how *Elysia chlorotica* adults can contain chloroplasts when their sex cells (eggs) do not.
9. The relationship between *Elysia chlorotica* and *Vaucheria litorea* is specific and obligate. What does this mean?
10. State the types of scientific approaches that scientists may need to use to solve the mysteries of the evolutionary relationship between *Elysia chlorotica* and *Vaucheria litorea*.

Investigate, think and discuss

11. Find out more about the evolution of:
    (a) chloroplasts and mitochondria
    (b) *Elysia chlorotica* and *Vaucheria litorea*
    (c) prokaryotic cells and eukaryotic cells.
12. A 268-million-year-old fossil found on the southern coast of NSW by Professor Guang Shi of Deakin University is considered to contain two species of fossilised bacteria that lived in symbiosis with a burrowing animal. Shi suspects that the animal acted like a gardener, cultivating the bacterial species best suited to changing climatic conditions. Find out more about this research and other research related to palaeobiology, palaeogeology and global change.

13. Examine the figure on the right and read the relevant information in this subtopic.
    (a) At what age do these sea slugs reproduce?
    (b) Do the eggs of *Elysia chlorotica* contain chloroplasts?
    (c) Describe what happens five days after the larva feed on algae.
    (d) Suggest an evolutionary advantage of the relationship being selected for.
    (e) Suggest a future application of this knowledge.
    (f) Suggest a hypothesis about this process that could be investigated.

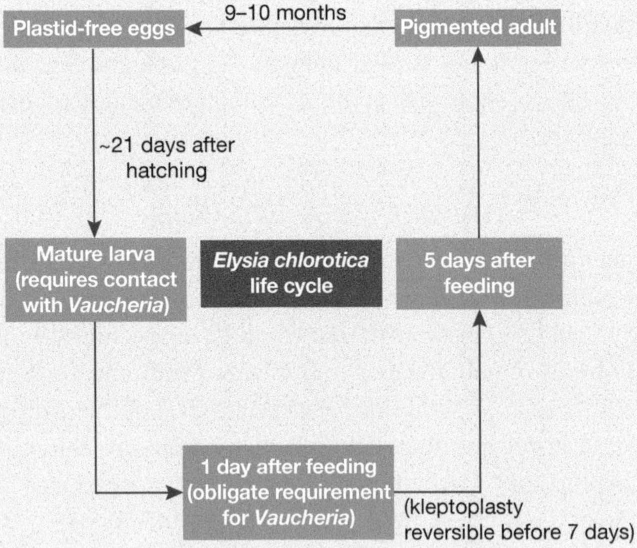

14. Research and report on scientific contributions or research from one of the following eminent Australian palaeontologists: Geoffrey Archer, Malcolm Walter, Neil Archbold, Neil Marshall, Alan Partridge, David Haig, John Laurie, Dana Milder, John Mortimer.
15. In 2008, palaeontologist Kate Trinajstic from Curtin University in Perth made one of Australia's most significant fossil discoveries. The famous 'mother fish' fossil, *Materpiscis attenboroughi* (named after the well-known science presenter David Attenborough) pushed back the earliest known live birth an astonishing 200 million years. The analysis of her discovery brought together scientists from different science disciplines and showed the need to develop new techniques for further investigations.
    (a) Find out more about Dr Trinajstic's 'mother fish' fossil and how it changed our evolutionary theories.
    (b) Find out examples of science disciplines that are involved in analysing fossils.
    (c) Dr Trinajstic is now working with Kliti Grice, professor of chemistry at Curtin University and Director of the Western Australia Organic and Isotope Geochemistry Centre, to identify biomarkers in fossils that will reveal what the fish were made of and what they ate. Research and report on:
        (i) the use of biomarkers to find out more about fossils
        (ii) organic and isotope geochemistry.

The 'mother fish' fossil

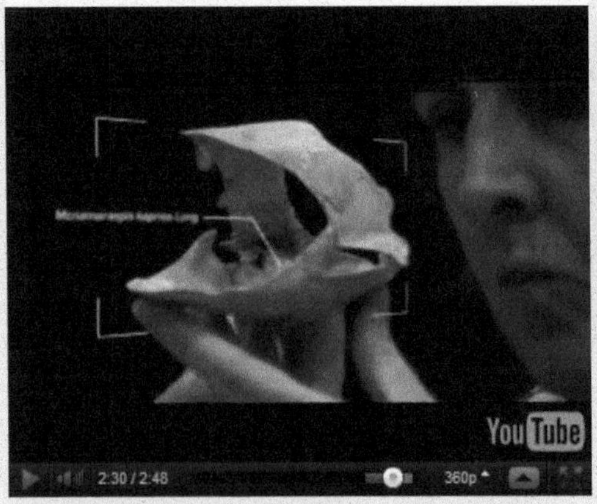

Dr Kate Trinajstic

Investigate, imagine and create

16. Construct models to link the endosymbiotic evolution theory to either mitochondria or chloroplasts.
17. Find out more about organisms that steal functions from other species. Create a science fiction story that links what you have found to your imagination.
18. Imagine that you are a palaeobiologist or palaeogeologist. Go online to do some research.
    (a) Write a story describing an exciting week in your life.
    (b) Suggest three research questions that you may be involved in investigating.

3.14 Storyboards and Gantt charts

3.14.1 Storyboards and Gantt charts

1. Decide how many scenes you need in your story. Often, 6–8 is a good number. Divide your page into this number of equal sections.
2. Consider which will be the three main events in your story and draw them roughly in the first, middle and last sections of your page.
3. Brainstorm the scenes that fit between these. Select the most appropriate and add them as intermediate
4. Mentally stand back and examine your story outline; make any desired changes to enhance its dramatic impact.

how to ...?

What are the main scenes in a story or event?

question

Helps you to use both your imagination and organisational skills to capture and share thoughts

why use?

Storyboard

A Outline of scene 1	**B** Outline of scene 2	**C** Outline of scene 3
D Outline of scene 4	**E** Outline of scene 5	**F** Outline of scene 6

also called

Comic strip

comparison

Similarity

Both show the sequence of events.

Difference

Storyboards use sketches or diagrams while Gantt charts use tables.

Gantt chart

example

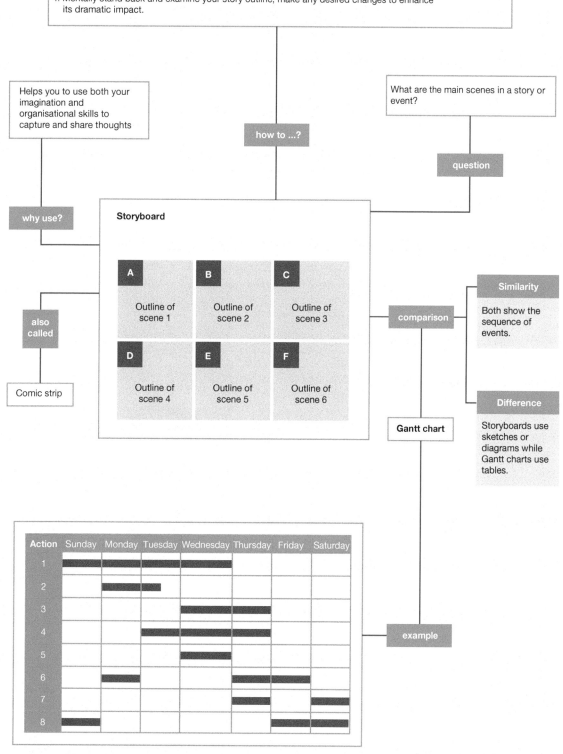

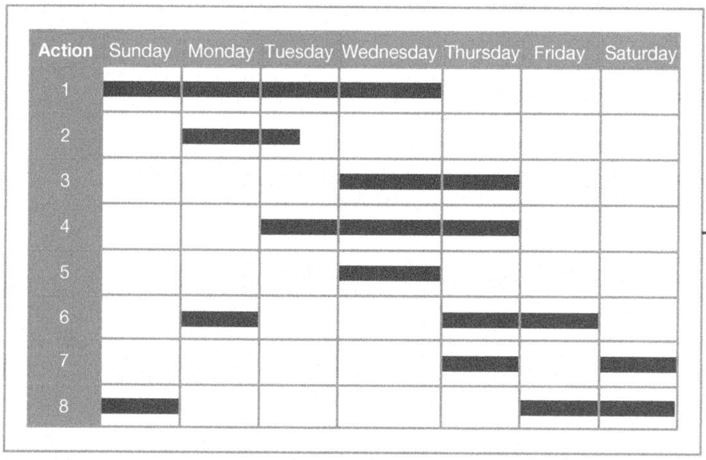

learn on

3.15 Project: Natural selection — the board game!

3.15.1 Scenario

There are few people in Australia today who haven't played a board game such as *Monopoly*, *Scrabble* or *Civilisation* sometime in their life. Even today, when computer games such as *Halo* or online games such as *World of Warcraft* are regularly played by tens of thousands of Australians, sales of old-school board games such as *Snakes and Ladders*, *Kingmaker* and, yes, *Monopoly* are still a healthy component of the income for a toy and game store. Apart from the fact that they are a great choice when there's no electricity and can be played and enjoyed by people from completely different generations, psychologists suggest that their continued popularity can also be attributed to the fact that there is just as much luck as skill in determining the winner. In this way, board games are much like real life!

The effectiveness of using game play as a way of teaching concepts is the stock-in-trade of the educational game company BrainGames, who produce computer games that teach science, maths, history and geography concepts. Games such as *The Revenge of Pavlov's Dogs* and *Where in the World is Amerigo Vespucci?* have made them the leader in the educational games market. However, keen to exploit the non-computer-equipped market sector, BrainGames now want to branch out into board games and the first board game they want to produce will be based on one of the key ideas of biology.

Will the organism you create be more successful than the dodo?

3.15.2 Your task

As part of the Games Development Division at BrainGames, you and your team are to develop a prototype board game based on the idea of natural selection and evolution. In this game, players will be able to select a variety of characteristics to give an organism and then, over the course of the game, see whether these organisms survive intact as their environment is changed. Your prototype must include:

- a game board
- game pieces
- a rule book.

You may also choose to include game mechanics such as cards, spinners or dice.

Individual pathways

ACTIVITY 3.1	ACTIVITY 3.2	ACTIVITY 3.3
Evolution doc-8467	Investigating evolution doc-8468	Investigating evolution further doc-8469

learnON ONLINE ONLY

3.16 Review

3.16.1 Study checklist

Biodiversity

- outline some sources or causes of genetic diversity
- suggest the relationship of biodiversity to evolution
- describe the potential impact of reproductive technologies and genetic engineering on genetic diversity
- describe a possible consequence of artificial selection on biodiversity

The history of life on Earth

- extract information from diagrams and tables relating to the history of life on Earth
- construct a timeline for the history of life on Earth
- sequence the major events in the evolution of life on Earth

Fossils

- define the term 'fossil'
- outline conditions necessary for fossilisation
- distinguish between the following types of fossils: moulds, casts, imprints and petrified fossils
- distinguish between relative and absolute dating of fossils

The theory of evolution by natural selection

- define the following terms: evolution, selection pressure, natural selection
- suggest how genetic characteristics may have an impact on survival and reproduction
- describe the process of natural selection using examples
- explain the importance of variations in the process of evolution

Evidence supporting the theory of evolution

- describe how the fossil record provides evidence for evolution
- outline how comparative anatomy has been used to support the theory of evolution and to determine evolutionary relationships between species
- describe examples of molecular biology techniques and how they are used to work out evolutionary relationships
- evaluate and interpret a variety of different types of evidence used to support the theory of evolution

3.16 Review 1: Looking back

To answer questions online and to receive **immediate feedback** and **sample responses** for every question, go to your learnON title at www.jacplus.com.au. *Note:* Question numbers may vary slightly.

1. Identify the term from the list at right that matches the meaning below.
   (a) The process by which the individuals with the most advantageous variation survive and reproduce more successfully than others
   (b) A special characteristic that improves an organism's chance of surviving in an environment
   (c) The range of structural and behavioural differences in a species
   (d) Evidence of past life
   (e) Structures that may look similar due to comparable selection pressures rather than shared ancestry
   (f) A technique used to compare the similarity of DNA
   (g) The struggle for resources between members of the same species
   (h) The geographical distribution of species
   (i) Comparing the structure of organisms
   (j) A process by which unrelated organisms living in similar environments develop similar features
   (k) A technique that uses measurements of isotopes to determine the age of rocks and fossils
   (l) Variation among organisms at the ecosystem, species and gene level
   (m) The process in which organisms with particular features are selected and bred together
   (n) A genetically identical organism
   (o) Made up of amino acids

Variation Competition Adaptation
Fossil Biodiversity Biogeography
DNA hybridisation Radiometric dating
Analogous structures Comparative anatomy
Natural selection Artificial selection
Convergent evolution Clone Protein

2. Examine the figures at right of the *Archaeopteryx* and a modern flying bird.
   (a) How are they similar?
   (b) How are they different?
   (c) Suggest reasons for the similarities and differences.
3. Refer to the animal kingdom evolutionary tree below right to answer the following questions.
   (a) Which animal group (or phylum) is most closely related to the Chordata?
   (b) Which is more closely related to the Mollusca phylum: Arthropoda or Nematoda?
   (c) Which animals have radial symmetry?
   (d) Do you think that Platyhelminthes would have more or less in common with Annelida than with Cnidaria? Explain.
   (e) Suggest the significance of adaptations of organisms in relation to their survival.

Skeleton of (a) *Archaeopteryx* and (b) a modern flying bird. The black regions on the skeletons show distinctive reptilian features (at left) and bird features (at right).

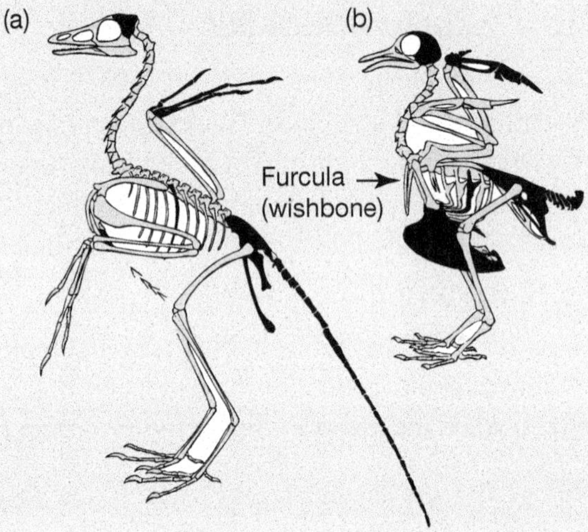

The animal kingdom evolutionary tree based on genetic and structural information

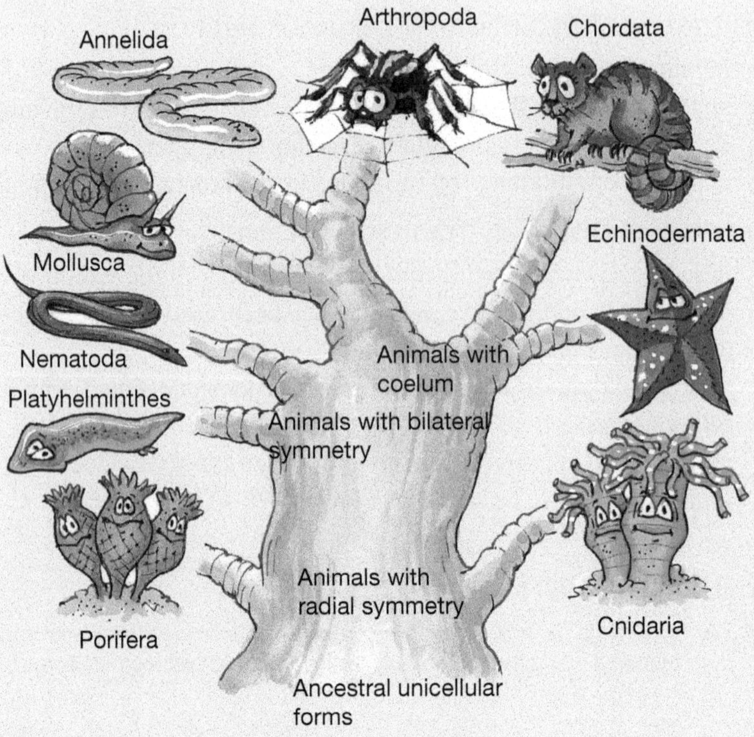

4. Not all people accept the theory of evolution. Find out why some people do not support this theory. What are some other theories? What do you believe? Why?
5. Identify the following.
   (a) The scientific name for humans
   (b) The abbreviation for deoxyribonucleic acid
   (c) The southern supercontinent
   (d) My name is included in a system of nomenclature.
   (e) Around 30 of these make up the surface of the Earth.
   (f) A permanent change in DNA
   (g) The type of dating that assumes that lower layers of rock are older than the ones above
   (h) The rank between phylum and order
   (i) Members of the same species living together in the same place at the same time
6. Examine the figure below and identify the epoch in which:
   (a) Aborigines arrived in Australia
   (b) the first marsupials appeared in Australia
   (c) swimming and flying mammals appeared
   (d) the dinosaurs became extinct.

A timeline of some marsupial fossil finds and major mammal events

Some marsupial fossil finds and events		Epoch (millions of years ago)		Major mammal events
Present	Cenozoic era	**HOLOCENE** **0.01–present**	Quaternary period	Humans investigate Earth's history
Most of the large Pleistocene marsupials became extinct about 15 000–30 000 years ago.		**PLEISTOCENE** **1.64–0.01 mya**		Aborigines arrived in Australia about 55 000 years ago.
Many giant browsing marsupials became extinct; there were grazing kangaroos and lots of diprotodons.		**PLIOCENE** **5.2–1.64 mya**	Tertiary period	*Homo habilis*, the earliest known human, appeared in East Africa.
Primitive marsupial 'mice' and 'tapirs' were found at Lake Eyre, South Australia, and diprotodons at Bullock Creek, Northern Territory.		**MIOCENE** **23.5–5.2 mya**		Lots of marsupial mammals were living in Australia and South America.
First Australian marsupials occurred about 23 million years ago. Fossils of diprotodons and a relative of the pygmy possum were found in Tasmania.		**OLIGOCENE** **35.5–23.5 mya**		First marsupials appeared in Australia. First primates appeared.
Lots of marsupial fossils of this age were found in South and North America.		**EOCENE** **56.5–35.5 mya**		Swimming and flying mammals appeared.
Dinosaurs became extinct about 65 million years ago.		**PALAEOCENE** **65–56.5 mya**		More mammals appeared after dinosaurs became extinct.

7. Copy and complete the Venn diagram at right using the terms below.

function similar structure not
common different divergent convergent
structure function

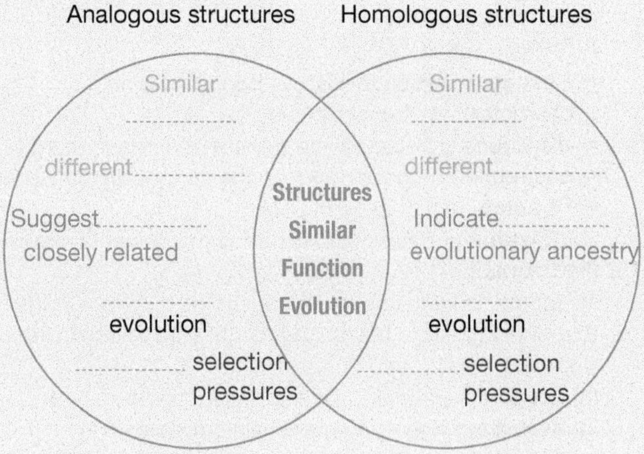

8. Examine the diagram at right and deduce the answers to the following questions.
   (a) Write down the names of the fossils in order from youngest to oldest.
   (b) Which layer is the same age as layer 3, layer 4 and layer 5 respectively? How can you tell?
   (c) Out of all the fossil layers numbered 1–11, which layer is oldest?
9. A 2006 study showed that, because Australia had banned the use of one particular group of new antibiotics (the fluoroquinolones) in livestock, the level of resistance to the human antibiotic cyprofloxacin was only 2 per cent. In countries where these antibiotics are used in livestock, the human resistance level is around 15 per cent. Explain these findings.
10. It is 100 000 years ago, and you are a Cro-Magnon individual whose tribe has captured some wild dogs and are interested in breeding them. Write a story describing which dogs you choose to breed from and why. How do the pups turn out — is there much variation? How do you decide which dogs to keep? Does everyone agree?
11. Biologists make inferences about evolutionary relationships based on comparative anatomy. Molecular biology techniques such as DNA hybridisation and protein sequencing may support or contradict these inferences. Explain how molecular biology techniques can be used to work out evolutionary relationships.
12. When Darwin visited the Galapagos Islands he found that longer-beaked species of finches were found in areas where insects were plentiful and shorter-beaked finches were found where seeds were the main food source. Account for this observation using the theory of evolution by natural selection.

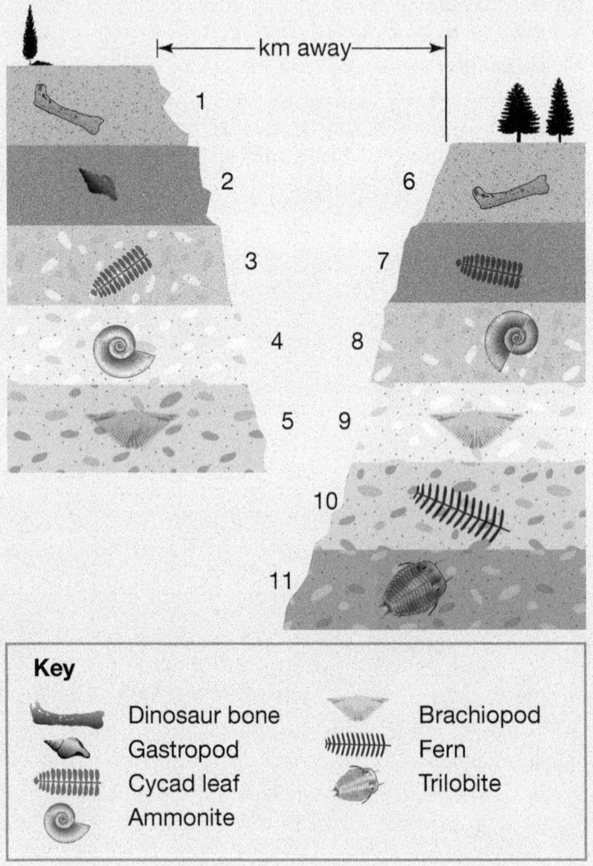

13. Sequence the following into the correct evolutionary order.
   - Flowering plants evolve
   - Early dinosaurs evolve
   - Mammals, flowering plants, insects, fish and birds dominate
   - Bacteria evolve
   - All living things are in the ocean; massive increase in multicellular organisms
   - Most dinosaurs become extinct
   - Greatest mass extinction of all time
   - Dinosaurs dominate the planet

14. Insert the following labels in the diagram on the right to produce a model of how natural selection brings about genetic change in a population.
    - Selective agent acts
    - Next generation contains more of the favourable characteristic
    - Individuals best suited to the environment (fittest) survive and reproduce more successfully
    - Population contains genetically different individuals

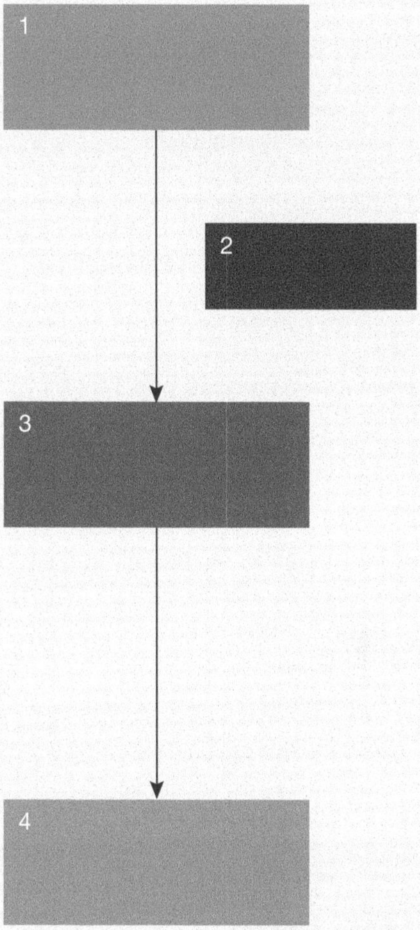

15. Find out about difficulties that one of the following scientists had in being able to express their scientific opinions because of society during their lifetime.
    (a) Gregor Mendel
    (b) Jean Baptiste de Lamarck
    (c) Charles Darwin
16. Are humans still evolving? Organise a class debate to discuss this issue.
17. Read through the boxed text *A tangled family tree* and then answer the following questions.
    (a) Outline a possible implication of the findings of the genome comparisons on humans and chimps.
    (b) If speciation is the creation of new species from existing ones, suggest a definition for *reverse speciation*.
    (c) Carefully study the image of the Toumaï skull. List features that are chimp-like and features that are human-like. Do you think it more closely resembles a human or a chimp? Give reasons for your suggestion.
    (d) How do we currently define a species? Suggest implications of these findings on our current definition. Suggest a definition that you think should be used in the future.
18. Find out more about the Toumaï fossil and the different interpretations that are suggested about it. Construct a storyboard to outline two of these interpretations.
19. Research current theories on human evolution. Use a Gantt chart to summarise some of your findings.
20. Find out more about an aspect of human evolution that interests you and construct a storyboard to share what you have found out.
21. (a) Find out more about the 'hobbit-like' human ancestor (*Homo floresiensis*) and the various scientific discussions about its lineage.
    (b) Use a Gantt chart to summarise your findings.
    (c) Create a storyboard for a possible day in the life of *Homo floresiensis*.
22. Find out more about one of the following topics and report your findings using a thinking tool of your choice.
    - Reverse speciation
    - How the X chromosome can be used in evolution studies
    - Research on the human genome
    - The Piltdown Man hoax
    - The Laetoli footprints
    - An *Australopithecus afarensis* called Lucy
    - Mitochondrial Eve
    - Neanderthals
    - Human cultural evolution

A TANGLED FAMILY TREE

A recent genetic study has suggested that human and chimp ancestors may have continued to interbreed after evolution had split them apart.

The genomes of humans, chimps and gorillas were compared using a molecular clock to estimate how long ago the three groups diverged.

It had been thought that the last common ancestor of chimps and humans lived over 7 million years ago. However, patterns in genes on the X chromosome suggest that the two lineages may have split between 5.4 and 6.3 million years ago.

The research suggested that after humans and chimps initially split into separate species, they continued to interbreed and in fact rehybridised into a single species in a reverse speciation event.

Other research has contested these results, suggesting that the genetic changes can be explained by differences in mutation rates between species and that the reverse speciation did not occur.

The Toumai fossil shown here is thought to be the earliest fossil from the human family tree, but may also be a common ancestor to chimps and humans, or even fit elsewhere on the evolutionary tree.

What effect may this research have on our current interpretations of fossils and theories of human evolution? What will future research tell us about our family tree?

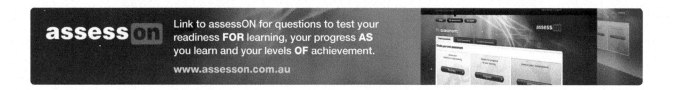

learnON RESOURCES — ONLINE ONLY

Complete this digital doc: Worksheet 3.6: Evolution: Puzzles (doc-19428)

Complete this digital doc: Worksheet 3.7: Evolution: Summary (doc-19429)

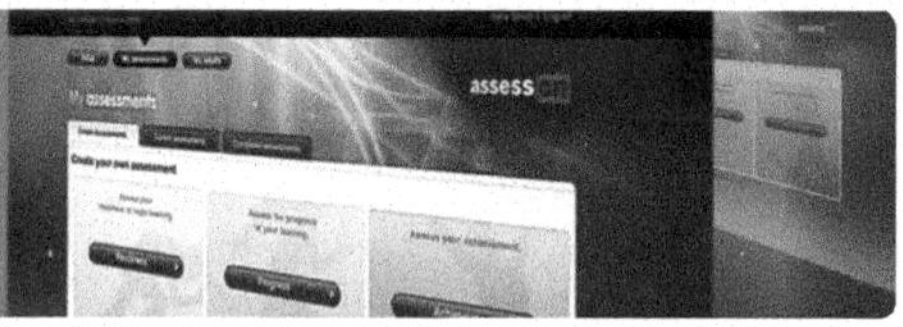

TOPIC 4
Chemical patterns

4.1 Overview

4.1.1 Why learn this?

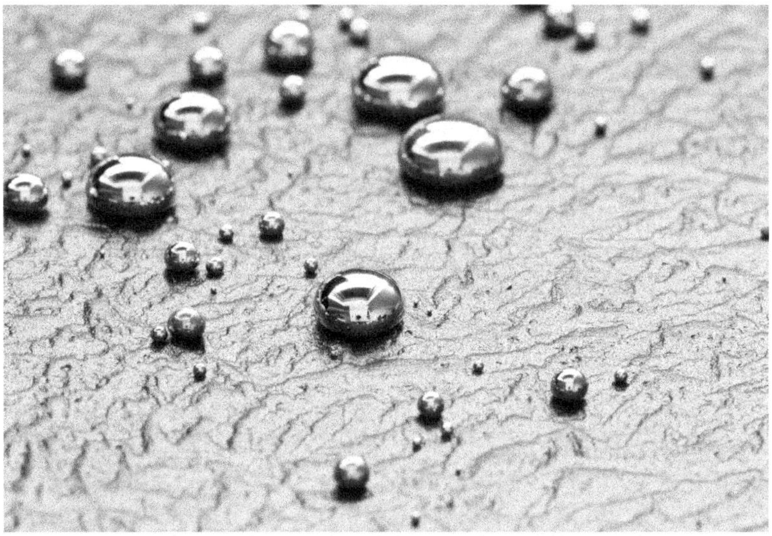

To understand the way that chemicals react with each other, you need to take a look inside the atoms of chemical elements. When you do, you can find patterns that help explain the properties of the elements and the way in which elements and compounds behave when they react with each other. One property of elements is their physical state. The mercury shown in this photo is a metal, but it has such a low melting temperature that it is a liquid at room temperature.

4.1.2 Think about chemical patterns

assessON

- Who was Dmitri Mendeleev and how was he able to predict the future?
- What are metalloids?
- Why is the petrol used in motor vehicles unleaded?
- Why do we talk about shells when describing electrons?
- Why are you more likely to find pure gold on or near the Earth's surface than pure copper or iron?
- What is the connection between the reactivity of metals and the ancient Roman Empire?

LEARNING SEQUENCE

Numerous **videos** and **interactivities** are embedded just where you need them, at the point of learning, in your learnON title at www.jacplus.com.au. They will help you to learn the content and concepts covered in this topic.

Your quest

Inside the elements

Atoms are the building blocks of the chemical elements. They are, therefore, also the building blocks of compounds and mixtures. For thousands of years, alchemists and scientists have searched for patterns in the substances that make up the universe. Many of them succeeded to some extent. But the discovery by Lord Rutherford in 1911 that most of the atom was empty space, and subsequent discoveries about the particles inside that atom by Niels Bohr and other scientists, provided the missing links in the patterns. Answer the questions below to find out what you already know about the atom and the chemical elements.

Think and remember

1. Identify the subatomic particle or particles that:
   (a) orbit the nucleus
   (b) can be found inside the nucleus
   (c) has/have no electric charge
   (d) has/have a positive electric charge
   (e) has/have a negative electric charge
   (f) is/are lightest.
2. The atom shown in the diagram below belongs to a single chemical element.
   (a) What is the atomic number of the element?
   (b) Which particles are counted to determine the atomic number of the element?
   (c) Identify the element in the diagram.

A simplified model of an atom

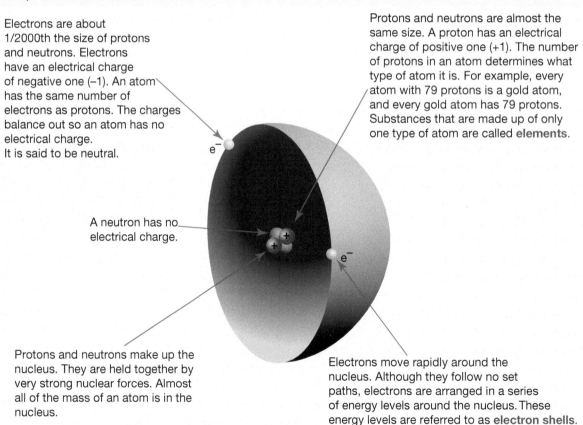

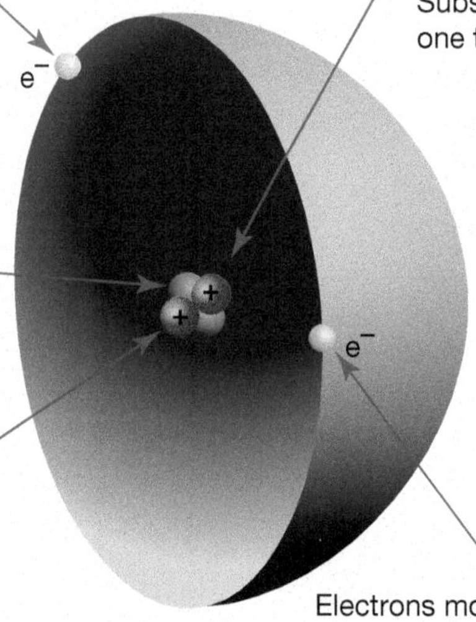

Electrons are about 1/2000th the size of protons and neutrons. Electrons have an electrical charge of negative one (–1). An atom has the same number of electrons as protons. The charges balance out so an atom has no electrical charge. It is said to be neutral.
Protons and neutrons are almost the same size. A proton has an electrical charge of positive one (+1). The number of protons in an atom determines what type of atom it is. For example, every atom with 79 protons is a gold atom, and every gold atom has 79 protons. Substances that are made up of only one type of atom are called **elements**.
e⁻
A neutron has no electrical charge.
+
+
e⁻
Protons and neutrons make up the nucleus. They are held together by very strong nuclear forces. Almost all of the mass of an atom is in the nucleus.
Electrons move rapidly around the nucleus. Although they follow no set paths, electrons are arranged in a series of energy levels around the nucleus. These energy levels are referred to as **electron shells**.

3. What is the electric charge of the nucleus of every atom?

4. Identify the chemical element or elements that match each of the following descriptions.
   (a) They combine chemically to produce water.
   (b) It is neither a metal nor a non-metal and is used in electric circuits inside devices such as computers and mobile phones.
   (c) It has the symbol Na.
   (d) They combine chemically to produce the compound that we know as table salt.
   (e) It is the only metal that exists as a liquid at normal room temperatures.

Investigate

5. Research and report on the contributions of Lord Rutherford, Niels Bohr and Sir James Chadwick to our knowledge of the atom.

4.2 The periodic table

4.2.1 The periodic table

Russian chemist Dmitri Mendeleev confidently predicted the properties of the element germanium 15 years before it was discovered. He was able to do this because he and other scientists had arranged all known elements into a set of rows and columns called the **periodic table**.

The periodic table on the following pages shows 112 elements. At the time of publication, scientists have reported the discovery of elements with atomic numbers up to 118. However, some of the discoveries have not been confirmed by the International Union of Pure and Applied Chemistry (IUPAC). Until they are, their existence is 'unofficial'. Those yet to be confirmed are elements 113, 115, 117 and 118. The discoveries of elements 114 and 116 were confirmed in June 2011. The properties of new elements are predicted before their discovery, just as they were in Mendeleev's time.

learnon RESOURCES — ONLINE ONLY

Try out this interactivity: Periodic table (int-0758)

4.2.2 The patterns emerge

Two thousand years ago, only 10 elements had been identified. They were carbon, sulfur, iron, copper, zinc, silver, tin, gold, mercury and lead. By the early nineteenth century, scientists had identified over 50 elements. Chemists had already begun to search for patterns among the elements in the hope of finding a way to classify them. It was difficult at that time to find patterns because many elements were still undiscovered.

In 1864, British chemist John Newlands arranged the elements in order of increasing atomic weight and found that every eighth element shared similar properties. In 1869, Mendeleev, building on the work of Newlands and other scientists, discovered a way of organising the elements into rows and columns. This arrangement formed the basis of what we now know as the periodic table. The elements were arranged in rows in order of increasing mass or atomic weight. Mendeleev called the rows of elements **periods** and the columns, which each contained a family of elements, **groups**. This arrangement is called the periodic table because elements with similar properties occur at regular intervals or periods. In a strange twist of fate, German chemist Lothar Meyer, who worked independently of Mendeleev, came up with a similar arrangement of the elements at about the same time.

The observation that the physical and chemical properties of the elements recur at regular intervals when elements are listed in order of atomic weight is known as the **periodic law**.

4.2.3 An educated guess

Mendeleev was so confident about the periodic law that he deliberately left gaps in his periodic table and was able to predict the properties of the unknown elements that would fill the gaps. Mendeleev predicted the existence of germanium, which he called eka-silicon. This element was discovered in 1886, 15 years later. The table on the following page shows the accuracy of Mendeleev's predictions.

Alkali metals → Group 1

Alkaline earth metals → Group 2

Period 1		
	1 Hydrogen **H** **1.008**	2 Helium **He** **4.003**

Key
- Atomic number
- Name
- Symbol
- Relative atomic mass

Transition metals: Groups 3–9

	Group 1	Group 2	Group 3	Group 4	Group 5	Group 6	Group 7	Group 8	Group 9
Period 2	3 Lithium **Li** **6.94**	4 Beryllium **Be** **9.02**							
Period 3	11 Sodium **Na** **22.99**	12 Magnesium **Mg** **24.31**							
Period 4	19 Potassium **K** **39.10**	20 Calcium **Ca** **40.08**	21 Scandium **Sc** **44.96**	22 Titanium **Ti** **47.87**	23 Vanadium **V** **50.94**	24 Chromium **Cr** **52.00**	25 Manganese **Mn** **54.94**	26 Iron **Fe** **55.85**	27 Cobalt **Co** **58.93**
Period 5	37 Rubidium **Rb** **85.47**	38 Strontium **Sr** **87.62**	39 Yttrium **Y** **88.91**	40 Zirconium **Zr** **91.22**	41 Niobium **Nb** **92.91**	42 Molybdenum **Mo** **95.96**	43 Technetium **Tc** **98.91**	44 Ruthenium **Ru** **101.1**	45 Rhodium **Rh** **102.91**
Period 6	55 Caesium **Cs** **132.9**	56 Barium **Ba** **137.3**	57–71 Lanthanides	72 Hafnium **Hf** **178.5**	73 Tantalum **Ta** **180.9**	74 Tungsten **W** **183.8**	75 Rhenium **Re** **186.2**	76 Osmium **Os** **190.2**	77 Iridium **Ir** **192.22**
Period 7	87 Francium **Fr**	88 Radium **Ra**	89–103 Actinides	104 Rutherfordium **Rf**	105 Dubnium **Db**	106 Seaborgium **Sg**	107 Bohrium **Bh**	108 Hassium **Hs**	109 Meitnerium **Mt**

Lanthanides

57 Lanthanum **La** **138.91**	58 Cerium **Ce** **140.122**	59 Praseodymium **Pr** **140.91**	60 Neodymium **Nd** **144.24**	61 Promethium **Pm** **(145)**	62 Samarium **Sm** **150.4**	63 Europium **Eu** **151.96**

Actinides

89 Actinium **Ac** **(227)**	90 Thorium **Th** **232.04**	91 Protactinium **Pa** **231.04**	92 Uranium **U** **238.03**	93 Neptunium **Np** **237.05**	94 Plutonium **Pu** **(244)**	95 Americium **Am** **(243)**

The period number refers to the number of the outermost shell containing electrons.

Properties of eka-silicon and germanium

Properties of eka-silicon as predicted by Mendeleev	Properties of germanium, discovered in 1886
A grey metal	A grey-white metal
Melting point of about 800 °C	Melting point of 958 °C
Relative atomic mass of 73.4	Relative atomic mass of 72.6
Density of 5.5 g/cm^3	Density of 5.47 g/cm^3
Reacts with chlorine to form compounds with four chlorine atoms bonded to each eka-silicon atom	Reacts with chlorine and forms compounds in a ratio of four chlorine atoms to every germanium atom

Mendeleev's work led many scientists to search for new elements. By 1925, scientists had identified all 92 naturally existing elements.

The periodic table shown at the beginning of this section includes the names, **symbols** and **atomic numbers** of the first 112 elements. The symbols are a form of shorthand for writing the names of the

The elements in the purple cells adjacent to the bold black border are neither metals nor non-metals. They are called metalloids.

Non-metals → | Halogens ↓ (Group 17) | Noble gases ↓ (Group 18)

Group 10	Group 11	Group 12	Group 13	Group 14	Group 15	Group 16	Group 17	Group 18
			5 Boron **B** **10.81**	6 Carbon **C** **12.01**	7 Nitrogen **N** **14.01**	8 Oxygen **O** **16.00**	9 Fluorine **F** **19.00**	10 Neon **Ne** **20.18**
			13 Aluminium **Al** **26.98**	14 Silicon **Si** **28.09**	15 Phosphorus **P** **30.97**	16 Sulfur **S** **32.06**	17 Chlorine **Cl** **35.45**	18 Argon **Ar** **39.95**
28 Nickel **Ni** **58.69**	29 Copper **Cu** **63.55**	30 Zinc **Zn** **65.38**	31 Gallium **Ga** **69.72**	32 Germanium **Ge** **72.63**	33 Arsenic **As** **74.92**	34 Selenium **Se** **78.96**	35 Bromine **Br** **79.90**	36 Krypton **Kr** **83.80**
46 Palladium **Pd** **106.4**	47 Silver **Ag** **107.9**	48 Cadmium **Cd** **112.4**	49 Indium **In** **114.8**	50 Tin **Sn** **118.7**	51 Antimony **Sb** **121.8**	52 Tellurium **Te** **127.8**	53 Iodine **I** **126.9**	54 Xenon **Xe** **131.3**
78 Platinum **Pt** **195.1**	79 Gold **Au** **197.0**	80 Mercury **Hg** **200.6**	81 Thallium **Tl** **204.4**	82 Lead **Pb** **207.2**	83 Bismuth **Bi** **209.0**	84 Polonium **Po** **(209)**	85 Astatine **At** **(210)**	86 Radon **Rn** **(222)**
110 Darmstadtium **Ds**	111 Roentgenium **Rg**	112 Copernicium **Cn**						

Metals ←

64 Gadolinium **Gd** **157.25**	65 Terbium **Tb** **158.93**	66 Dysprosium **Dy** **162.50**	67 Holmium **Ho** **164.93**	68 Erbium **Er** **167.26**	69 Thulium **Tm** **168.93**	70 Ytterbium **Yb** **173.04**	71 Lutetium **Lu** **174.97**

96 Curium **Cm** **(247)**	97 Berkelium **Bk** **(247)**	98 Californium **Cf** **(251)**	99 Einsteinium **Es** **(254)**	100 Fermium **Fm** **(257)**	101 Mendelevium **Md** **(258)**	102 Nobelium **No** **(255)**	103 Lawrencium **Lr** **(256)**

elements and are recognised worldwide. Some periodic tables describe the properties of each element, including its physical state at room temperature, melting point, boiling point and **relative atomic mass**. Most elements exist as solids under normal conditions and a few exist as gases. Only two elements exist as liquids at normal room temperature — bromine and mercury.

4.2.4 Counting sub-atomic particles

The periodic table is organised on the basis of atomic numbers. The atomic number of an element is the number of protons present in each atom. Atoms with the same atomic number have identical chemical properties. Because atoms are electrically neutral, the number of protons in an atom is the same as the number of electrons. The **mass number** of an atom is the sum of the number of protons and neutrons in the atom. The number of neutrons in an atom can therefore be calculated by subtracting the atomic number from the mass number. This information is usually shown in the following way:

$$^{A}_{Z}E$$

where A = the mass number (number of protons and neutrons), Z = the atomic number (number of protons) and E = the symbol of the element.

For example, the element iron has a mass number of 56 and an atomic number of 26. It can be represented as follows:

$$^{56}_{26}\text{Fe}$$

Once you know the atomic number and mass number of an element, you can work out how many electrons and neutrons it has.

The atomic number of iron is 26 because all iron atoms have 26 protons. Iron's mass number of 56 indicates that most iron atoms have a total of 56 protons and neutrons. To calculate the number of neutrons, the atomic number is subtracted from the mass number to give 30 neutrons. Since atoms are electrically neutral and protons have a positive charge, each iron atom has 26 electrons.

4.2.5 How heavy are atoms?

Measuring and comparing the masses of atoms is difficult because of their extremely small size. Chemists solve this problem by comparing equal numbers of atoms, rather than trying to measure the mass of a single atom.

A further problem arises because not all atoms of an element are identical. Although all atoms of a particular element have the same atomic number, they can have different numbers of neutrons. Hence, some elements contain atoms with slightly different masses.

These different masses are used to calculate an average or **weighted mean**, which is based on the relative amounts of each type of atom. This number is referred to as the relative atomic mass and is usually not a whole number. The mass number (A) of an element can usually be found by rounding the relative atomic mass.

4.2.6 Families of elements

The periodic table contains eight groups (or families) of elements, some of which have been given special names. (Remember that these groups form columns in the periodic table.)

- Group 1 elements are known as **alkali metals**. The alkali metals all react strongly with water to form basic solutions.
- Group 2 elements are referred to as **alkaline earth metals**.
- Group 17 elements are known as **halogens**. The halogens are brightly coloured elements. Chlorine is green, bromine is red-brown and iodine is silvery-purple.
- Group 18 elements are known as **noble gases**. The noble gases are inert and do not readily react with other substances.
- The block of elements in the middle of the table is known as the **transition metal block**.

The line that zigzags through the periodic table separates the **metals** from the **non-metals**. About three-quarters of all elements are classified as metals, which are found on the left-hand side of the table. The non-metals are found on the upper right-hand side of the table. Eight elements that fall along the line between metals and non-metals have properties belonging to both. They are called **metalloids**.

Illuminated signs use tubes filled with the noble gas neon.

4.2.7 Metals

Metals have several features in common.

- They are solid at room temperature, except for mercury which is a liquid.
- They can be polished to produce a high shine or **lustre**.
- They are good **conductors** of electricity and heat.
- They can all be beaten or bent into a variety of shapes. We say they are **malleable**.
- They can be made into a wire. We say they are **ductile**.
- They usually melt at high temperatures. Mercury, which melts at −40° C, is one exception.

WHAT DOES IT MEAN?

The word *malleable* comes from the Latin word *malleus*, meaning 'hammer'.

4.2.8 Non-metals

Only 22 of the elements are non-metals. At room temperature, eleven of them are gases, ten are solid and one is liquid. The solid non-metals have most of the following features in common.

- They cannot be polished to give a shine like metals; they are usually dull or glassy.
- They are **brittle**, which means they shatter when they are hit.
- They cannot be bent into shape.
- They are usually poor conductors of electricity and heat.
- They usually melt at low temperatures.
- Many of the non-metals are gases at room temperature.

Common examples of non-metals are sulfur, carbon and oxygen.

INVESTIGATION 4.1

Chemical properties of metals and non-metals

AIM: To investigate and compare the chemical properties of metals and non-metals

Materials:

safety glasses, gloves and laboratory coat
1M hydrochloric acid
water
magnesium
iron filings
copper filings
sulfur powder
universal indicator
4 test tubes
4 gas jars filled with oxygen gas
4 deflagrating spoons
dropping pipette
spatula
Bunsen burner, heatproof mat and matches

Method and results

- Place a small quantity of magnesium in a test tube. Add about 2 mL of hydrochloric acid.

1. Record your observations in a suitable table.

CAUTION

The heating part of this experiment should be done in a fume cupboard. Safety glasses, gloves and laboratory coats must be worn at all times.

2. Repeat using the iron filings, copper filings and sulfur powder. Record your observations.
   - Place a small amount of magnesium in a deflagrating spoon and heat it. When hot, place it into the gas jar full of oxygen gas. **Do not look directly at the flame**.
3. Record your observations.
4. Repeat using the iron and copper filings. Record your observations.
5. Repeat using a small amount of sulfur powder. Record your observations. **This part of the experiment must be performed in a fume cupboard**.
   - Add about 10 mL of water to each jar and shake. Add 3 drops of universal indicator.
6. Record the colour and determine the pH of the solution.

Burning sulfur in oxygen in a gas jar

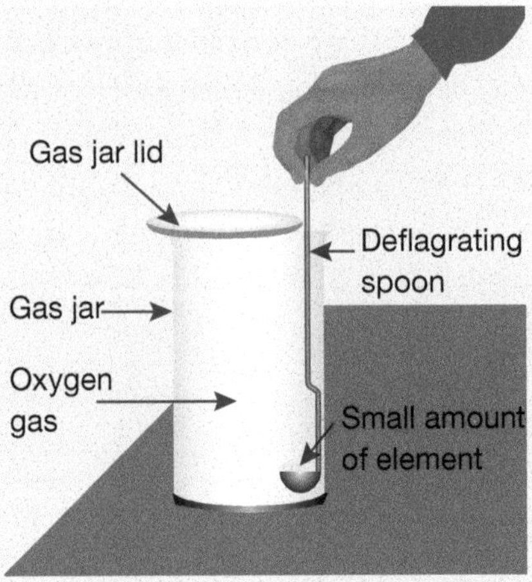

Discuss and explain

7. Use the periodic table to determine which of the elements tested were metals and which were non-metals.
8. Describe any differences between the effect of acids on metals and non-metals.
9. Describe what happened when the metals and non-metals reacted with oxygen.
10. The metal or non-metal oxides formed in the gas jars dissolved in water to form acidic and basic solutions. What type of solution did the metals form? What type of solution did the non-metals form?

INVESTIGATION 4.2

Comparing the properties of two metal families

Calcium and magnesium are from group 2 of the periodic table (alkali earth metals). Copper and iron are transition metals.

AIM: To investigate and compare the chemical properties of metals from two different groups of the periodic table

Materials:

small samples of magnesium, iron and copper
'rice grain' equivalent amounts of calcium chloride, magnesium chloride, iron chloride and copper chloride
spatula
5 test tubes and a test-tube rack
electric circuit to measure conductivity (2-volt power supply, 3 connecting leads, 2 alligator clips, and a light globe and holder)
2M hydrochloric acid
water
matches
stirring rod
safety glasses and laboratory coat

Method and results

1. Record the results of each of the following experiments in an appropriate table.
   - Describe the physical state (solid, liquid or gas) of each element.
   - Describe the physical appearance of each element.
   - Set up the circuit as shown in the diagram at right and determine whether each of the elements conducts electricity.

Circuit used to measure electrical conductivity

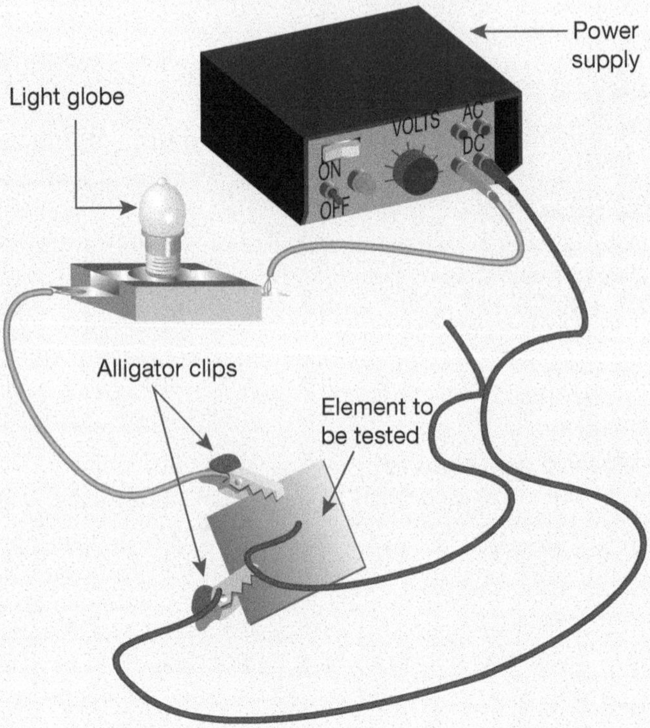

- Determine whether any of the elements react with water by placing a small sample in 2 mL of water in a test tube. Record any changes that occur in your table.
- Determine whether the metals react with acid by placing a small sample of each metal in 1 mL of 2M hydrochloric acid in a test tube. If a gas is produced, test it by holding a lit match at the mouth of the test tube. Make sure the test tube is pointed away from you. If hydrogen is present, you will hear a 'pop'. If oxygen is present, the match should burn more brightly. If carbon dioxide is present, the match should go out.
- Your teacher may show or describe to you how the metal calcium responds to some of the tests described.
- Add a small amount of each of the metal compounds (magnesium chloride, calcium chloride, iron chloride and copper chloride) to 5 mL of water. Comment on their solubility and the colour of any solution made.

Discuss and explain

2. What are the properties of copper and iron? Are there any similarities?
3. What are the properties of calcium and magnesium? Are there any similarities?
4. List the metals in order of reactivity with water and acids. List them from most reactive to least reactive.
5. Were there any differences between solubilities of the metal compounds or the colours of the solutions they formed? Describe these differences.
6. What could you infer about the properties of elements in the same group? Give reasons for your answer.

4.2.9 Metalloids

Some of the elements in the non-metal group look like metals. One example is silicon. While it can be polished like a metal, silicon is a poor conductor of heat and electricity and cannot be bent or made into wire. Elements that have features of both metals and non-metals are called metalloids. There are eight metalloids altogether: boron, silicon, arsenic, germanium, antimony, polonium, astatine and tellurium.

Metalloids are important materials often used in electronic components of computer circuits.

4.2.10 Following a trend

There are a number of repeating patterns in the periodic table. The most obvious is the change from metals on the left of each period to non-metals on the right. Other patterns exist in the physical and chemical properties of elements in the same group or period. Some of these trends are shown in the table below.

Patterns in the periodic table

Characteristic	Pattern down a group	Pattern across a period
Atomic number and mass number	Increases	Increases
Atomic radius	Increases	Decreases
Melting points	Decreases for groups 1 to 5 and increases for groups 15 to 18	Generally increases then decreases
Reactivity	Metals become more reactive and non-metals become less reactive	Is high, then decreases and then increases. Group 18 elements are inert and do not react.
Metallic character	Increases	Decreases

Atomic size or radius increases down a group and decreases across a period. Can you suggest why?

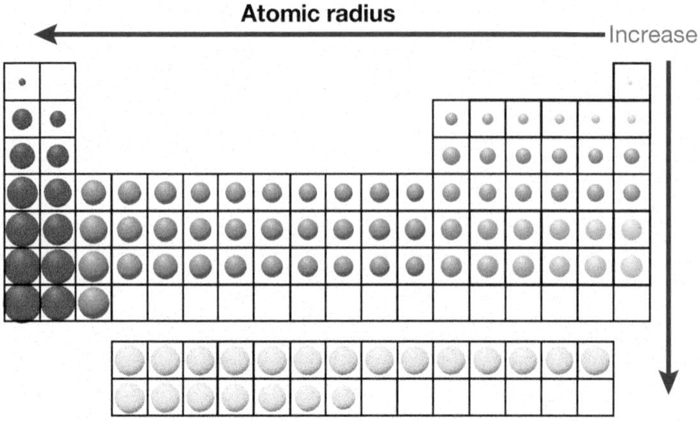

HOW ABOUT THAT!

Lead poisoning was a common occurrence in ancient Rome because the lead the Romans used to make water pipes and cooking utensils slowly dissolved into the water. Acute lead poisoning causes mental impairment and personality changes. The effects are not immediately noticeable, but occur gradually as the amount of lead in the body accumulates. Some historians attribute the strange behaviour of several Roman emperors to lead poisoning.

In the Middle Ages, plates, cups and other drinking vessels were often made from pewter, an alloy of lead and tin. The acids in food and drinks caused lead to leach out and cause poisoning.

Until 1986, lead was added to petrol to stop 'knocking' in car engines. Unleaded fuel allows a catalytic converter to prevent pollutants such as nitrous oxides and carbon monoxide from being emitted. With lead in the petrol, these devices couldn't work. Lead emissions from cars were possibly causing a build-up of lead in humans in built-up areas.

The word *plumber* is derived from the Latin word *plumbum*, meaning 'lead'. Look up the symbol for lead in the periodic table. Where do you think this symbol came from?

Unleaded petrol was introduced to Australia in 1986.

4.2 Exercises: Understanding and inquiring

To answer questions online and to receive **immediate feedback** and **sample responses** for every question, go to your learnON title at www.jacplus.com.au. *Note:* Question numbers may vary slightly.

Remember

1. State whether the following statements are true or false.
   (a) The noble gases are found in group 18.
   (b) The non-metals are found in the upper right-hand side of the periodic table.
   (c) There are more metals than non-metals.
   (d) Few elements are found naturally as liquids.

2. What is the name of the element in:
   (a) group 2, period 3
   (b) group 17, period 2
   (c) group 1, period 4
   (d) group 18, period 3?
3. Draw an outline of the periodic table showing where you would find the following elements: the noble gases, the alkali metals, the alkaline earth metals, the halogens and the transition metals.
4. In the outline of the periodic table shown at right, some of the elements have been replaced by letters. Using the correct chemical symbols, write down which of these elements fit the following categories.
   (a) Two elements that are gases at room temperature
   (b) Two elements that are metals
   (c) Two elements that are transition elements
   (d) An element that is a noble gas
   (e) Two elements that are in the same group
   (f) Two elements that are in the same period
   (g) The elements that are alkali metals
   (h) The element that is a halogen

Use this outline of the periodic table to answer question 4.

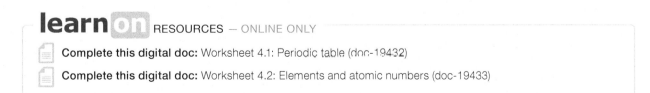

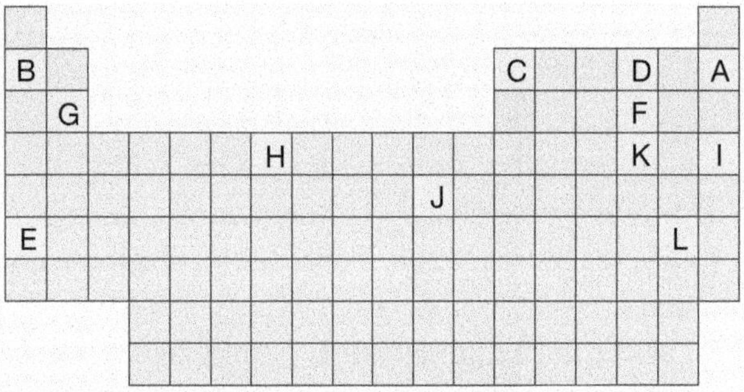

5. What is the difference between the mass number and the relative atomic mass of an element?
6. Describe what happens to the metallic character of the elements as you go across the periodic table.
7. Construct a table showing the name; mass number; atomic number; and number of protons, neutrons and electrons of the following elements.
   (a) ${}^{12}_{6}C$ (b) ${}^{65}_{13}Zn$ (c) ${}^{40}_{18}Ar$ (d) ${}^{197}_{79}Au$ (e) ${}^{238}_{92}U$

Think

8. Explain how Mendeleev was able to predict the properties of elements even before they were discovered.
9. At room temperature, which group of the periodic table consists exclusively of gases?
10. Compose a rhyme, poem or song that can help you learn the names of the first 20 elements of the periodic table in order.

Design and create

11. Design and create a poster, multimedia presentation or web page about one of the groups of the periodic table of elements. Include images of each of the elements in the group and a list of the properties that they have in common.

Investigate

12. Some elements with atomic numbers greater than 92 have been artificially created in laboratories. Find out how they are made and named, and describe some of their common properties.
13. It is said that the stars are the 'element factories of the universe'; that is, stars make the elements. Do some research and find out how the stars make elements.
14. Find out which single element makes up about three-quarters of the mass of the universe.
15. Choose an element and research the following information about it:
    - when it was discovered
    - who discovered it
    - how it is found in nature
    - its properties and uses.

learnON RESOURCES — ONLINE ONLY

Complete this digital doc: Worksheet 4.1: Periodic table (doc-19432)

Complete this digital doc: Worksheet 4.2: Elements and atomic numbers (doc-19433)

4.3 Small but important

4.3.1 The influence of electrons

When atoms come into contact with one another, they often join together to form **molecules**. Other atoms join together to form giant **crystals** that contain billions of atoms. The electrons in each atom account for the chemical behaviour of all matter, because they form the outermost part of the atom.

Shells of electrons

Drawing an accurate picture of an atom using a diagram is difficult because electrons cannot be observed like most particles. Their exact location within the atom is never known — they tend to behave like a 'cloud' of negative charge. Furthermore, an atom is many times larger than its nucleus so it is not practical to draw a diagram to scale. Nonetheless, diagrams are useful because they help us to understand how atoms combine.

An **electron shell diagram** is a simplified model of an atom. In these diagrams, the nucleus of the atom, containing protons and neutrons, is drawn in the middle. Electrons are arranged in a series of energy levels around the nucleus. These energy levels are called **shells** and are drawn as concentric rings around the nucleus. The electrons in the inner shells are more strongly attracted to the nucleus than those in the outer shells.

Each shell contains a limited number of electrons. The first (or K) shell holds a maximum of two electrons. The second (or L) shell holds up to eight electrons. The third (M) shell holds up to 18 electrons. The fourth (N) shell holds up to 32 electrons. The maximum number of electrons in each shell can be calculated using the rule:

the nth shell holds a maximum of $\mathbf{2n^2}$ electrons.

For example, the fourth shell holds a maximum of 2×4^2 electrons, which is 32 electrons.

Electron configuration

The **electron configuration** of an element is an ordered list of the number of electrons in each shell. The electron configuration is determined from the atomic number of the element, which is the same as the number of protons in the nucleus of each atom. In a **neutral** atom, the total number of electrons is the same as the number of protons.

To work out the electron configuration of a particular atom, you need to remember that electrons occupy the innermost shells first. Once the first two shells are filled, the remaining electrons begin to fill the third shell. For example, the element sodium has an atomic number of 11. Each atom has 11 protons and 11 electrons. The electrons will fill the two innermost shells first — two in the first shell and eight in the second shell. That accounts for ten. The remaining electron must be in the third shell because the first two have already been filled.

An electron shell diagram of a sodium atom

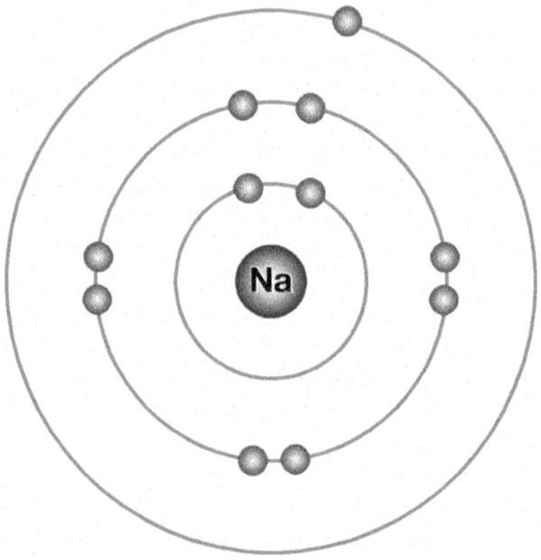

The electron configuration of an atom is written by showing the number of electrons in each shell separated by commas. For example: sodium 2, 8, 1.

4.3.2 The periodic table explained

When Mendeleev and Meyer grouped elements on the basis of their similar chemical properties, they were not aware of the existence of electrons. We can now explain many of their observations using our understanding of electron shells.

Atoms in the same group of the periodic table have similar properties because they have the same number of electrons in their outer shells. (The outer shell is the last shell to be filled by electrons.) The number of electrons in the outer shell relates to the group number in the periodic table. Hence, all elements in group 1 have one electron in their outer shell and all elements in group 18 (with the exception of helium) have eight electrons in their outer shell.

Filling up in turn

The largest atoms contain up to seven shells of electrons. Thus, there are seven periods (rows) in the periodic table. (Look at the periodic table in subtopic 4.2 to confirm this.) The period number tells you the number of shells containing electrons. The first shell can hold up to two electrons, so there are two elements in the first period, with hydrogen containing one electron and helium containing two. The second shell holds up to eight electrons, so there are eight elements in the second period.

Even though the third shell can hold up to 18 electrons, there are only eight elements in the third period. This is because the outer shell of an atom can never hold more than eight electrons as the atom would then become unstable. Therefore, while the third shell is yet to be filled completely, electrons begin to fill the fourth shell in both potassium and calcium atoms. This stabilises the atoms because the third shell is no longer the outer shell. The filling of the third shell resumes in the block of elements from scandium to zinc (the transition metals). Once the third shell is full, the fourth shell continues to fill from gallium to xenon.

Element	Symbol	Atomic number	Electron configuration
Oxygen	O	8	2, 6
Fluorine	F	9	2, 7
Neon	Ne	10	2, 8
Sodium	Na	11	2, 8, 1
Magnesium	Mg	12	2, 8, 2
Sulfur	S	16	2, 8, 6
Chlorine	Cl	17	2, 8, 7
Argon	Ar	18	2, 8, 8
Potassium	K	19	2, 8, 8, 1

Note that the fourth shell of the potassium atom begins filling before the third shell is full.

learnon RESOURCES — ONLINE ONLY

Try out this interactivity: Shell-shocked (int-0676)

4.3.3 Upwardly mobile electrons

If enough energy is supplied to an atom, electrons can move from one shell (or energy level) to another (higher) energy level. This may occur when atoms are heated by a flame. When electrons move between energy levels, they either absorb or emit an amount of energy related to the difference in energy between the energy levels. Electrons returning to a lower energy level emit this energy in the form of light. The size of the difference in energy levels determines the colour of the light. Thus, flame colours can be used to identify elements.

Various metal ions produce characteristic colours when they are volatilised in a flame.

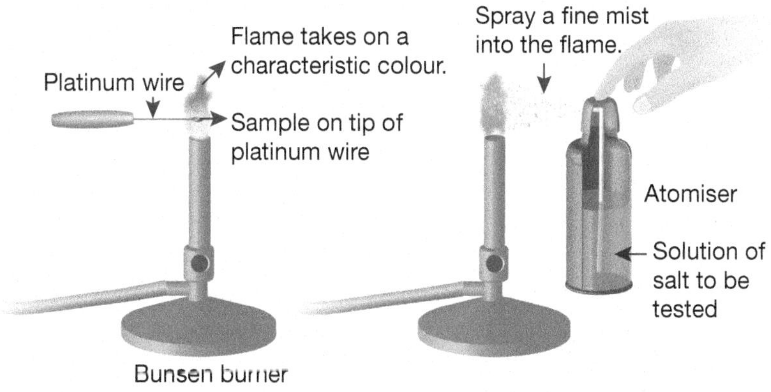

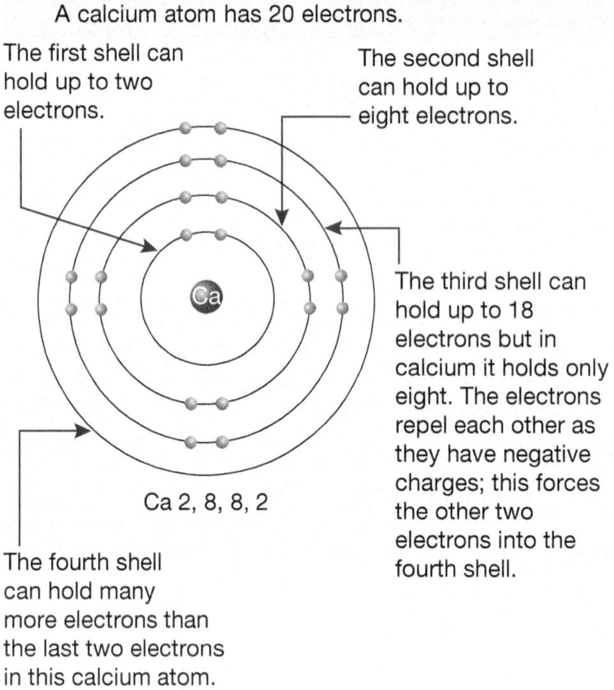
A calcium atom has 20 electrons.
The first shell can hold up to two electrons.
The second shell can hold up to eight electrons.
The third shell can hold up to 18 electrons but in calcium it holds only eight. The electrons repel each other as they have negative charges; this forces the other two electrons into the fourth shell.
Ca
Ca 2, 8, 8, 2
The fourth shell can hold many more electrons than the last two electrons in this calcium atom.

INVESTIGATION 4.3

Flame tests

AIM: To observe evidence of electrons dropping from one energy level to another

Materials:

safety glasses and laboratory coat
2M hydrochloric acid
Bunsen burner, heatproof mat and matches
5 evaporating dishes
barium carbonate
sodium carbonate
copper carbonate
potassium carbonate
strontium carbonate
10 mL measuring cylinder
spatula

CAUTION

Laboratory coats and safety glasses must be worn at all times.

Method and results

- Place 10 mL of 2M hydrochloric acid in an evaporating dish and place the dish on a heatproof mat.
- Add a spatula full of the barium carbonate to the evaporating dish.
- Carefully hold the lit Bunsen burner at an angle over the spray produced by the reacting acid and carbonate as shown in the diagram at right. Observe the change in the colour of the flame.
- Repeat using the other carbonates. Use a different evaporating dish each time.

1. Record the colours produced by the different carbonates in a suitable table.

Heatproof mat
Spray from reacting acid and carbonate
Evaporating dish
Bunsen burner

Discuss and explain

2. Flame tests provide evidence that electrons do actually occupy different energy levels. Why do different elements produce different colours?
3. Is it the metal part of the compound or the carbonate part (carbon and oxygen) that produces the colour? How do you know?

4.3 Exercises: Understanding and inquiring

To answer questions online and to receive **immediate feedback** and **sample responses** for every question, go to your learnON title at www.jacplus.com.au. *Note:* Question numbers may vary slightly.

Remember

1. What is the name given to the different energy levels in which electrons can be found?
2. How many electrons are needed to fill:
   (a) the first shell
   (b) the second shell
   (c) the third shell
   (d) the fourth shell?
3. What is meant by the term *outer shell*?
4. What information about the electron arrangement is given by the group number of an element?
5. What information about the electron arrangement is given by the period number of an element?

Think

6. Name the elements that have the following electron arrangements.
   (a) 2, 4
   (b) 2, 8, 5
   (c) 2
   (d) 2, 8, 8, 2
7. Write the electron arrangement for each of the following elements.
   (a) Boron
   (b) Neon
   (c) Potassium
   (d) Fluorine
   (e) Silicon
8. (a) If an element has one electron in its outer shell, is it a metal or a non-metal? Explain your answer.
   (b) If an element has seven electrons in its outer shell, is it a metal or a non-metal? Explain your answer.
   (c) What is special about elements that have eight electrons in their outer shell?
9. What experimental evidence is there to show that electron shells actually exist?

Investigate

10. The electron arrangement of elements is more complex than the explanation in this subtopic. Find out about subshells and orbitals and how they are involved in determining how electrons are arranged in atoms.
11. A lithium atom has three protons, two neutrons and three electrons. Make a 3-dimensional model of this atom.

learnon RESOURCES — ONLINE ONLY

Complete this digital doc: Worksheet 4.3: Electron shells (doc-19435)

Complete this digital doc: Worksheet 4.4: The structure of the atom (doc-19436)

4.4 When atoms meet

4.4.1 It's the shell structure that counts

Knowledge of the electron shell structures of atoms helps us to understand how compounds such as sodium chloride (table salt) form. When atoms react with each other to form compounds, it is the electrons in the outer shell that are important in determining the type of reaction that occurs.

It's great to be noble

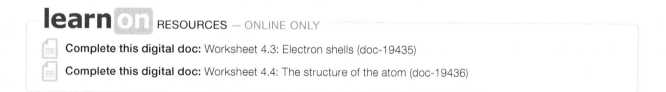

In 1919, Irving Langmuir suggested that the noble gases do not react to form compounds because they have a stable electron configuration of eight electrons in their outer shell. Most other atoms react because their electron arrangements are less stable than those of the noble gases. Atoms become more stable when they attain an electron arrangement that is the same as that of the noble gases. Chemical reactions can allow atoms to obtain this arrangement. The table of electron arrangements in subtopic 4.3 shows that the two noble gases neon and argon have eight electrons in their outer shells. The atoms of other elements must gain or lose electrons to attain full outer shells. In this way they become more stable, ending up with the electron arrangement of the nearest noble gas in the periodic table.

4.4.2 Some gain, some lose

Atoms that have lost or gained electrons and therefore carry an electric charge are called **ions**. Metal atoms, such as sodium, magnesium and potassium, have a small number of outer shell electrons. They form ions by losing the few electrons in their outer shell. This means that metal ions have more protons than electrons and so are positively charged. For example, the magnesium atom loses its two outer shell electrons to become a positively charged magnesium ion. The symbol for the magnesium ion is Mg^{2+}. The '2+' means that two electrons have been lost to form the ion. Positively charged ions are called **cations**.

Non-metal atoms form ions by gaining electrons to fill their outer shell. These ions contain more electrons than protons, so they are negatively charged. For example, the chlorine atom gains one electron to fill its outer shell, becoming a negatively charged chloride ion. Its symbol is Cl^-. The '–' means that one electron has been gained to form the ion. Negatively charged ions are called **anions**.

The diagram below shows how sodium and chlorine atoms lose and gain electrons respectively to form ions. Note that the sodium atom becomes a sodium ion and that the chlorine atom becomes a chlor*ide* ion. (When non-metals form ions, the suffix '-ide' is used.)

How sodium and chlorine atoms form ions

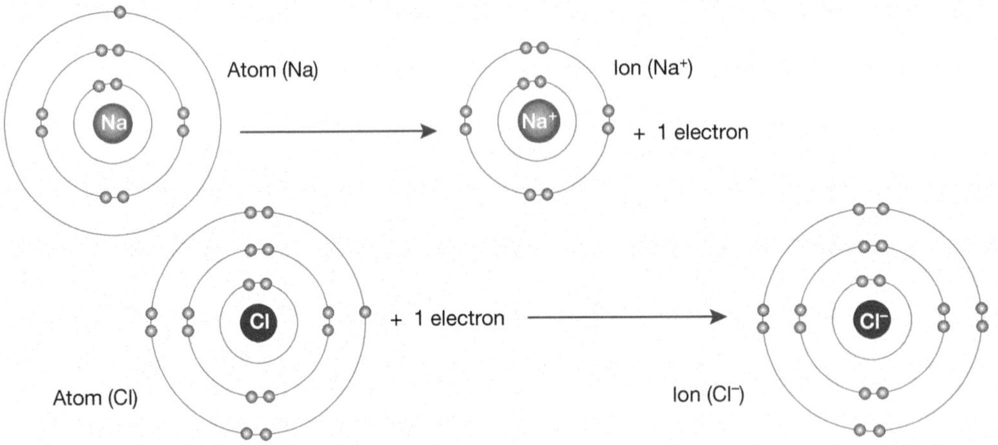

4.4.3 It's a game of give and take

Compounds such as sodium chloride, copper sulfate, calcium carbonate and sodium hydrogen carbonate all form when atoms come in contact with each other and lose or gain electrons. Compounds formed in this way are called **ionic compounds**.

Ionic compounds form when metal and non-metal atoms combine. A sodium atom loses an electron to form an ion and a chlorine atom gains an electron to form an ion. The electrons are transferred from one atom to the other, and the oppositely charged ions produced attract each other and form a compound. This electrical force of attraction between the ions is called an **ionic bond**.

The diagram on the following page shows some examples of the transfer of electrons that occurs when ionic compounds are formed. Note that more than two atoms may be involved to ensure that all the elements achieve eight electrons in their outer shell. For example, when magnesium reacts with chlorine to form magnesium chloride, each magnesium atom loses two electrons. Since each chlorine atom needs to gain only one electron, a magnesium atom gives one electron to each of two chlorine atoms. The resulting Mg^{2+} and Cl^{-} ions are attracted to each other to form the compound $MgCl_2$.

learnon RESOURCES — ONLINE ONLY

Try out this interactivity: Pass the salt (int-0675)

4.4.4 What do ionic compounds have in common?

Ionic compounds have the following properties.

- They are made up of positive and negative ions.
- They are usually solids at room temperature.
- They normally have very high melting points because the electrostatic force of attraction between the ions is very strong.
- They usually dissolve in water to form **aqueous solutions**.
- Their aqueous solutions normally conduct electricity.

WHAT DOES IT MEAN?

The word *aqueous* comes from the Latin word *aqua*, meaning 'water'. Other words beginning with the prefixes aque- or aqua- relate to water (for example, aqueduct, aquatic, aqualung).

The give and take of electrons that occurs in the formation of the ionic compounds sodium chloride, magnesium chloride and sodium oxide

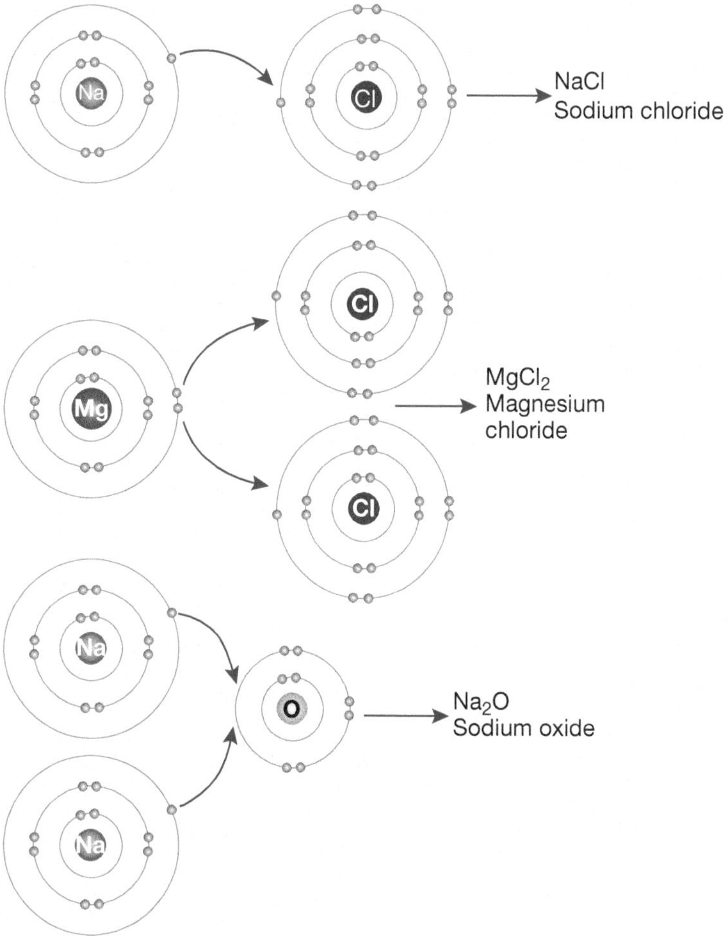

Positive ions			Negative ions		
Atom name	**Ion name**	**Chemical symbol**	**Atom name**	**Ion name**	**Chemical symbol**
lithium	lithium	Li^{+}	iodine	iodide	I^{-}
sodium	sodium	Na^{+}	fluorine	fluoride	F^{-}
potassium	potassium	K^{+}	chlorine	chloride	Cl^{-}
calcium	calcium	Ca^{2+}	oxygen	oxide	O^{2-}
aluminium	aluminium	Al^{3+}	nitrogen	nitride	N^{3-}

(a) A stick and ball representation of the lattice structure of sodium chloride; the sticks represent the bonds between the atoms.

(b) The ions in the lattice are effectively held in a tight arrangement.

(c) Individual salt crystals form regular square blocks because of the ionic lattice.

(a)

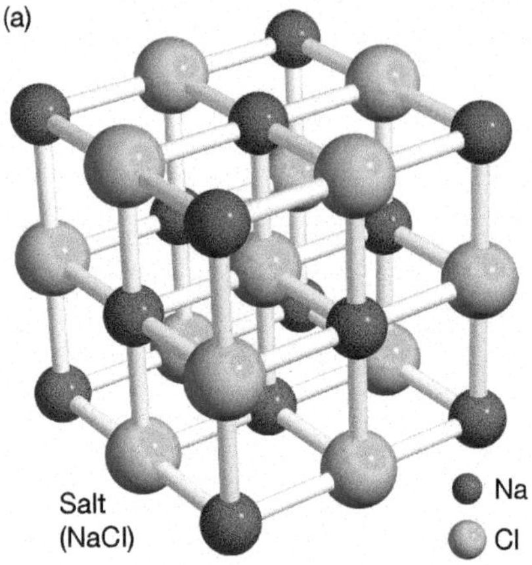

(b)

(c)

4.4 Exercises: Understanding and inquiring

To answer questions online and to receive **immediate feedback** and **sample responses** for every question, go to your learnON title at www.jacplus.com.au. *Note:* Question numbers may vary slightly.

Remember

1. Why do ions form?
2. What is a positively charged ion called?
3. What is a negatively charged ion called?
4. What properties do most ionic compounds have in common?
5. What types of elements combine to form ionic compounds?

Think

6. Write the symbol for the ion formed by each of the following elements.
   (a) Sodium
   (b) Potassium
7. Copy and complete the following table using the information here as well as the periodic table on pages 152 and 153. The first entry has been done for you.

Ion symbol	Ion name	Gained electron or lost electrons?	Number of electrons lost/gained	Total number of electrons in ion
F^-	fluoride	gained	1	10
Be^{2+}				
N^{3-}				
Cl^-				
Sn^{2+}				
Ag^+				

8. How many electrons have been gained or lost by the following ions?
   (a) Pb^{4+}
   (b) Br^-
   (c) Cr^{3+}
   (d) Se^{2-}
9. Draw diagrams like those on page 165 to show how each of the following ionic compounds form.
   (a) Magnesium fluoride
   (b) Lithium chloride
   (c) Aluminium sulfide
   (d) Calcium oxide

Imagine

10. Imagine that you are the outer shell electron of a sodium atom and you are going to form the ionic compound sodium chloride. Describe your experiences in a piece of creative writing. Discuss details such as the physical states and properties of the elements and compound involved, their atomic structure, reasons for forming ions and, finally, the reasons why the ions form the ionic compound.

learnon RESOURCES — ONLINE ONLY

Complete this digital doc: Worksheet 4.5: Ionic bonding (doc-19437)

Complete this digital doc: Worksheet 4.6: Writing formulae for ionic compounds (doc-19438)

Complete this digital doc: Worksheet 4.7: Electron configurations (doc-19439)

4.5 When sharing works best

4.5.1 Covalent bonding

Ionic compounds form when atoms lose or gain electrons. Atoms can also achieve stable electroni configurations by sharing electrons with other atoms to gain a full outer shell. When two or more atoms share electrons, a molecule is formed. A chemical bond formed by sharing electrons is called a **covalent bond**. The compounds formed are called **covalent** or **molecular compounds**. Non-metal atoms share electrons to form covalent bonds.

Molecules can be made of more than one type of atom or made of atoms of the same element. For example, oxygen gas consists of molecules formed when two oxygen atoms share electrons. Individual atoms of oxygen are not stable and become more stable by sharing electrons with each other.

learnon RESOURCES — ONLINE ONLY

Watch this eLesson: Perkin's mauve (eles-1774)

4.5.2 Electron dots: what's the point?

It is possible to draw diagrams to show how elements share electrons to form covalent compounds. These diagrams are called **electron dot diagrams**. They show the symbol for the atom and dots for the electrons in the outer shell of atoms. The table at right shows electron dot diagrams for some elements. Note that the electrons in the diagrams are arranged in four regions around the atom. Wherever possible, they are grouped in pairs.

When elements combine to form covalent compounds, they share electrons to achieve a full outer shell with eight electrons. Hydrogen has a full outer shell when it has two electrons, but all the other elements in the table need eight electrons to fill the outer shell.

Electron dot diagrams for some elements

Symbol	Electron configuration	Electron dot diagram
H	1	
C	2, 4	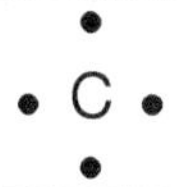
O	2, 6	
S	2, 8, 6	
Cl	2, 8, 7	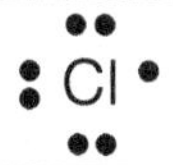
N	2, 5	
F	2, 7	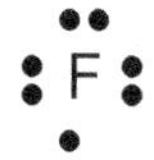

The formation of covalent molecules

Name and formula	Atoms	Compound	Structural formula	Explanation
Chlorine Cl_2	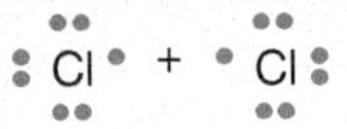		Cl — Cl *Note:* The line represents a sharing of two electrons and is called a single covalent bond.	Each chlorine atom needs to share one electron to gain a full outer shell.
Hydrogen chloride HCl	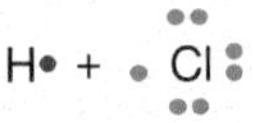		H — Cl	Both the hydrogen and chlorine atom need to share one electron to gain a full outer shell.
Oxygen O_2	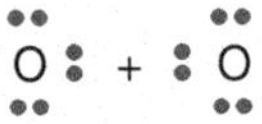	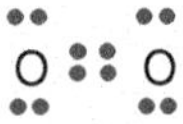	O = O *Note:* The double line represents a double covalent bond.	Each oxygen atom needs to share two electrons to gain a full outer shell.
Nitrogen N_2			N≡N *Note:* The triple line represents a triple covalent bond.	Each nitrogen atom shares three electrons to gain a full outer shell.
Water H_2O	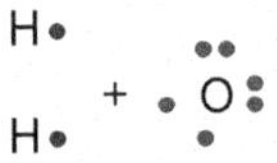	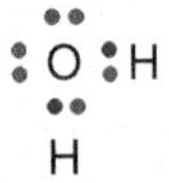	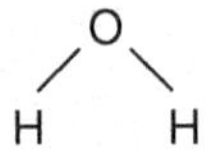	Each hydrogen atom needs one electron and the oxygen atom needs two to gain a full outer shell.
Carbon dioxide CO_2	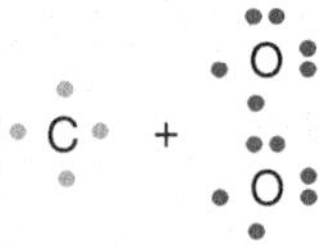		O=C=O	Each oxygen atom needs two electrons and the carbon atom needs four electrons to gain a full outer shell.

The table above shows how some covalent compounds form. The shared electrons are called **bonding electrons**. It is also possible to draw a **structural formula**, where a dash is used to represent these shared electrons. The dash represents the covalent bond and the other electrons are not drawn. It is also possible for double or triple covalent bonds to form. The way electrons are shared determines the ratio in which elements combine to form a covalent compound. It also determines the **chemical formula** of the compound.

4.5.3 Covalent compounds

Most covalent compounds have the following properties.

- They exist as gases, liquids or solids with low melting points because the forces of attraction between the molecules are weak.
- They generally do not conduct electricity because they are not made up of ions.
- They are usually insoluble in water.

4.5 Exercises: Understanding and inquiring

To answer questions online and to receive **immediate feedback** and **sample responses** for every question, go to your learnON title at www.jacplus.com.au. *Note:* Question numbers may vary slightly.

Remember

1. What kinds of elements combine to form covalent compounds?
2. What is a covalent bond?
3. What does an element's electron dot diagram represent?
4. What properties do most covalent compounds have in common?

Think

5. What is the difference between a single covalent bond and a triple covalent bond in terms of the number of electrons involved?
6. For the covalent compounds shown below, state whether their bonds are single, double or triple covalent bonds.

(a)

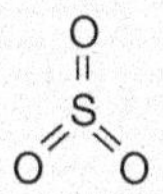

Sulfur trioxide — a gas used to make sulfuric acid

(b)

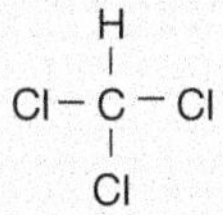

Chloroform — a liquid once used as an anaesthetic

(c)

Acetylene — a colourless gas used in welding

7. (a) Draw electron dot diagrams to show how the following covalent compounds form.
   (i) Hydrogen fluoride (HF)
   (ii) Methane (CH_4)
   (iii) Phosphorus chloride (PCl_3)
   (iv) Hydrogen sulfide (H_2S)
   (v) Tetrachloromethane (CCl_4)
   (vi) Ammonia (NH_3)
   (vii) Carbon disulfide (CS_2)
   (b) What pattern emerges between the structural formula of the compound and the number of electrons involved in bonding?
   (c) State whether the covalent bonds in the compounds are single, double or triple bonds.
8. Why don't the noble gases form covalent compounds?
9. Explain why CO_2 (a compound) and O_2 (an element) are both molecules.

Investigate

10. Silicon dioxide, commonly known as silica or sand, is a hard, solid, covalent compound with a very high melting point. Find out about its structure.
11. Although carbon and graphite are both made up of carbon atoms, they have very different properties. Investigate their properties and explain why they are so different in terms of their covalent structure.

learnON RESOURCES — ONLINE ONLY

Try out this interactivity: Making molecules (int-0228)

4.6 How reactive?

Science as a human endeavour

4.6.1 Reactivity of metals

Few metals are found as elements like gold; most are found as compounds or ores.

Have you ever wondered why gold can be found lying near the surface of the Earth and yet we need to mine iron ore and smelt it in large furnaces before we can obtain iron? The answer lies in the reactivity of the metals. Gold is a very unreactive element. It does not combine readily with other elements to form compounds. Most metals are much more reactive than gold.

When the Earth formed, the more reactive metals — including aluminium, copper, zinc and iron — reacted with other elements to form ionic compounds. These compounds are the **mineral ores** from which the metal elements are obtained. Iron ores include haematite (Fe_2O_3), magnetite (Fe_3O_4), siderite ($FeCO_3$), pyrite (FeS_2) and chalcopyrite ($CuFeS_2$).

The reactivity of metals is dependent on how easily they are able to give up their outer shell electrons. For example, it is easier for an atom to give up a single electron from an outer shell than to give up two electrons.

The reactivity of metals can be investigated by observing their reactions with acids. A metal reacts with hydrochloric acid according to the equation:

$$\text{metal} + \text{hydrolic acid} \longrightarrow \text{salt} + \text{hydrogen gas}$$

In these reactions electrons are transferred away from the metal atoms to the hydrogen in the acid, forming positive metal ions and hydrogen gas. The metal is said to have displaced the hydrogen from the acid. For this reason, these reactions are also **displacement reactions**.

4.6.2 Metals in ancient times

The most powerful ancient civilisations succeeded and prospered because they developed better weapons than their enemies by using metals such as copper, tin and iron. The Mesopotamians, who occupied a large region of the Middle East, learned almost 5000 years ago how to separate copper and tin from their ores using a process called **smelting**. Smelting is a chemical process in which carbon reacts with molten ore to separate the relatively pure metal. In ancient times, charcoal was used in furnaces to provide the carbon. They combined molten copper and tin to produce an **alloy** known as **bronze**, which is resistant to corrosion and harder than both copper and tin. The ancient Egyptians, Persians and Chinese also used bronze in weapons, ornaments, statues and tools.

INVESTIGATION 4.4

Investigating reactivity

AIM: To investigate the reactivity of a range of metals

Materials:

5 test tubes and a test-tube rack
safety glasses
1 cm × 4 cm piece of magnesium ribbon (or equivalent amount)
1 cm × 4 cm piece of zinc, copper, aluminium and iron
1M hydrochloric acid
measuring cylinder, small funnel, thermometer and steel wool

CAUTION

Wear safety glasses.

Method and results

- Polish each of the metal pieces with steel wool.
- Pour 10 mL of acid into each test tube.

1. Measure and record the temperature.
   - Add one metal to each test tube. Look for the presence of bubbles on the surface of the metals.
2. Arrange the test tubes in order of increasing bubble production and record your observations.

Discuss and explain

3. List the metals in order of increasing reactivity.
4. Discuss the limitations of this experiment.

HOW ABOUT THAT!

The ancient Romans used the smelting process to separate iron from iron ore. They strengthened it by pounding it with a hammer and used it to produce weapons, shields and armour that was harder and stronger than bronze. The use of iron weapons allowed the Roman legions to rule the Mediterranean world and beyond for over 400 years.

The gladius (a short iron sword), together with a long iron shield, gave the Roman army a huge advantage over its enemies. The shields were often used by groups of soldiers to form a protective wall and roof known as a *testudo* (tortoise) around themselves.

4.6.3 The activity series

The **activity series** places the elements in decreasing order of reactivity:

$$Li \rightarrow K \rightarrow Na \rightarrow Ca \rightarrow Mg \rightarrow Al \rightarrow Mn \rightarrow Cr \rightarrow Zn \rightarrow Fe \rightarrow$$
$$Ni \rightarrow Sn \rightarrow Pb \rightarrow H \rightarrow Cu \rightarrow Hg \rightarrow Ag \rightarrow Au \rightarrow Pt.$$

In order to react with acid and release hydrogen gas, the metal must be before hydrogen in the activity series.

Lithium, potassium, sodium and calcium are the most reactive metals. They will react with water to produce hydrogen gas. Magnesium through to lead will react with acid to form hydrogen gas, but copper, mercury and silver will not. Gold and platinum are even less reactive than copper and silver. Most of the elements at the top of the activity series were discovered much later than those at the bottom. Gold, silver, mercury and copper were all discovered over 2000 years ago. Potassium, sodium and calcium were not discovered until 1808. Why do you think this is so?

INVESTIGATION 4.5

Quantified reactivity

AIM: To quantify and measure the reactivity of metals

Materials:

safety glasses, heatproof mat, steel wool and gas syringe
1 cm × 4 cm piece of zinc, copper, aluminium and iron
1 cm × 4 cm piece of magnesium ribbon (or equivalent amount)
1M hydrochloric acid
retort stand, bosshead and gas syringe clamp
1 cm × 6 cm length of plastic tubing
250 mL side-arm conical flask
rubber stopper to fit conical flask
stopwatch or clock with second hand
50 mL measuring cylinder
distilled water

CAUTION

Wear safety glasses.

Method and results

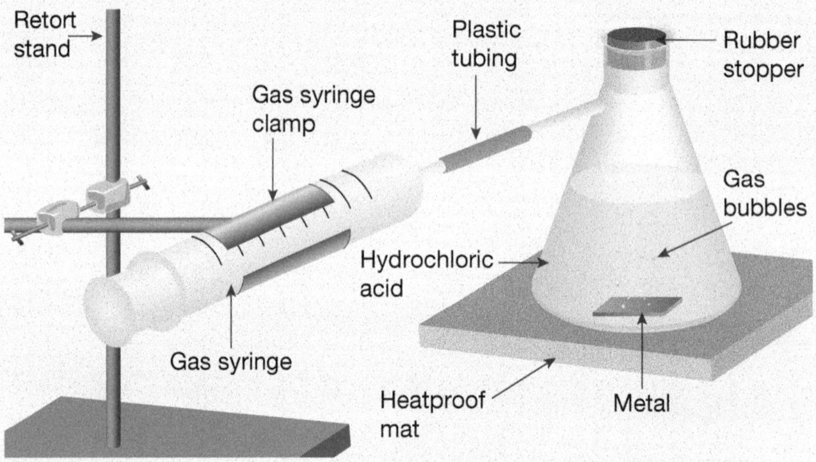

- Polish each of the metal pieces with steel wool.
- Mount the gas syringe in the clamp as shown in the diagram at right. Your teacher will tell you if the syringe needs to be lubricated. Push the plunger fully in and attach the plastic tubing to the nozzle.
- Pour 50 mL of acid into the flask.
- Connect the other end of the plastic tubing to the conical flask.
- Place one of the pieces of metal in the conical flask and quickly seal with the rubber stopper.
- Have one student act as a timer and another as a recorder.
- As soon as the metal is dropped in, start timing.

1. Using a suitable table, record the volume of gas in the syringe every 30 seconds until gas is no longer produced, the syringe is full or 10 minutes has passed, whichever occurs first.
   - Repeat the procedure with the other metals, taking care to rinse out the flask carefully each time with distilled water.
2. In your workbook, plot the results for all of the metals on one set of axes. Put the volume of gas on the vertical axis and time on the horizontal axis.

Discuss and explain

3. Use your graph to list the five metals in increasing order of reactivity and explain your reasoning.
4. Write a word equation for the reaction of each of the metals with the acid. If no reaction occurred, write 'no reaction'.
5. Write an equation using formulae for the reaction of each of the metals with the acid. If no reaction occurred, write 'no reaction'.
6. To which general reaction type or types do reactions between metals and acids belong?
7. Some of the variables in this investigation were not carefully controlled. List them and explain how this may have affected your results and conclusions.

learnon RESOURCES — ONLINE ONLY

Watch this eLesson: Davey and potassium (eles-1773)

4.6 Exercises: Understanding and inquiring

To answer questions online and to receive **immediate feedback** and **sample responses** for every question, go to your learnON title at www.jacplus.com.au. *Note:* Question numbers may vary slightly.

Remember

1. Name the gas produced in the reaction of a metal with hydrochloric acid.
2. Why is iron usually found in the form of a compound in the Earth's crust?

Think

3. Explain why the reactivity of metals decreases from left to right across the periods of the periodic table.

Investigate

4. Design and carry out an experiment that investigates the reactivity of alloys, such as stainless steel and brass. Compare these results with those obtained for the metal elements.
5. Research and report on the science of metallurgy and the role of metallurgists in the mining industry.

Create

6. When scientists attend conferences, they often present the results of their investigations as a poster. A poster can describe their work with photographs, drawings and concise written summaries. Present the findings of your investigation into the reactivity of metals as a poster to display in your classroom.

4.7 Finding the right formula

4.7.1 Chemical ID

The chemicals used in your school science laboratory are usually identified by both a name and a formula. Most people are able to recognise the formula of common compounds such as water (H_2O) and carbon dioxide (CO_2). A chemical formula (plural *formulae*) is a shorthand way to write the name of an element or compound. It tells us the number and type of atoms that make up an element or compound. Writing the correct formula is of paramount importance in chemistry. Most chemical problems cannot be solved without the knowledge of chemical formulae.

A hydrogen molecule and an oxygen molecule

4.7.2 It's elementary

Often the formula of a substance is simply the symbol for the element. Metals such as iron and copper, which contain only one type of atom, are identified simply by the symbol for that element (for example, Fe and Cu). Noble gases such as neon (Ne) have a similar formula.

Some non-metal elements such as hydrogen, oxygen and nitrogen exist as simple molecules. These molecules form when atoms of the same non-metal are joined together by covalent bonds. For example, the formula for the element hydrogen is H_2, indicating that two hydrogen atoms are joined together to make each molecule of hydrogen. A **molecular formula** is a way to describe the number and type of atoms that join to form a molecule.

Some common non-metal molecules and their molecular formulae

Name	Formula
Hydrogen	H_2
Nitrogen	N_2
Chlorine	Cl_2
Bromine	Br_2
Oxygen	O_2
Sulfur	S_2
Phosphorus	P_2

4.7.3 Formulae of compounds

The formula of a compound shows the symbols of the elements that have combined to make the compound and the ratio in which the atoms have joined together. For example, the chemical formula for the covalent compound methane, a constituent of natural gas, is CH_4 — one carbon atom for every four hydrogen atoms. The formula for the ionic compound calcium chloride, which is used as a drying agent, is $CaCl_2$ — two chlorine ions for every calcium ion.

4.7.4 Valency: formulae made easy

Knowledge of the **valency** of an element is essential if we wish to write formulae correctly.

The valency of an element is equal to the number of electrons that each atom needs to gain, lose or share to fill its outer shell. For example, the chlorine atom has only seven electrons in its outer shell, which can hold eight electrons. By gaining one electron, its outer shell becomes full. Chlorine therefore has a valency of one. The magnesium atom has two electrons in its outer shell. By losing two electrons, it is left with a full outer shell. Magnesium therefore has a valency of two.

A simple guide to remembering the valency of many elements is to remember which group in the periodic table they belong to. The number of outer shell electrons allows you to work out how many electrons are required to fill the outer shell. The table at right provides a simple guide to the valency of many elements.

Valency of groups in the periodic table

Group	Valency
Group 1	1
Group 2	2
Group 13	3
Group 14	4
Group 15	3
Group 16	2
Group 17	1

4.7.5 Writing formulae for covalent compounds

To write the formula of a non-metal compound made up of only two elements, use the valency of each element and follow the steps shown below.

Example 1

Write the formula for carbon dioxide.

Step 1	*Determine the valency of the elements involved.* Carbon has a valency of four; oxygen a valency of two. (That is, carbon needs to share four electrons, while oxygen needs to share two electrons.)
Step 2	*Determine the ratio of atoms that need to combine so that each element can share the same number of electrons.* A ratio of one carbon atom to two oxygen atoms would result in both sharing four electrons.
Step 3	*Write the formula using the symbols of the elements and writing the ratios as subscripts next to the element. (The number 1 can be left out as writing the symbol for the element assumes that one atom is present.)* The formula for carbon dioxide is CO_2.

Example 2

Write the formula for phosphorus chloride.

Step 1	*Determine the valency of the elements involved.* Phosphorus has a valency of three; chlorine has a valency of one.
Step 2	*Determine the ratio of atoms that need to combine so that each element can share the same number of electrons.* A ratio of one phosphorus atom to three chlorine atoms would result in both sharing three electrons.
Step 3	*Write the formula using the symbols of the elements and writing the ratios as subscripts next to the element.* The formula for phosphorus chloride is PCl_3.

Example 3

Write the formula for hydrogen oxide (water).

Step 1	*Determine the valency of the elements involved.* Hydrogen has a valency of one; oxygen has a valency of two.
Step 2	*Determine the ratio of atoms that need to combine so that each element can share the same number of electrons.* A ratio of two hydrogen atoms to one oxygen atom would result in both sharing two electrons.
Step 3	*Write the formula using the symbols of the elements and writing the ratios as subscripts next to the element.* The formula for hydrogen oxide is H_2O.

4.7.6 Writing formulae for ionic compounds

The formulae for ionic compounds can be written from knowledge of the ions involved in making up the compound. In ionic compounds, metal ions combine with non-metal ions. The tables below list common positive and negative ions and their names.

Metal atoms usually form positive ions. The number of positive charges on the ion is called its **electrovalency**. For example, a sodium ion has one positive charge (Na^+), the calcium ion has two positive charges (Ca^{2+}) and the aluminium ion has three (Al^{3+}). Note that some of the transition metals (e.g. iron) have more than one valency, as shown in the table above right. The Roman numerals in brackets after iron and copper identify the valency.

Electrovalencies of some common positive ions

Number of positive charges in each ion		
+1	+2	+3
Hydrogen (H^+)	Calcium (Ca^{2+})	Aluminium (Al^{3+})
Potassium (K^+)	Copper(II) (Cu^{2+})	Iron(III) (Fe^{3+})
Silver (Ag^+)	Iron(II) (Fe^{2+})	
Sodium (Na^+)	Lead (Pb^{2+})	
Ammonium (NH_4^+)	Magnesium (Mg^{2+})	
	Zinc (Zn^{2+})	

Electrovalencies of some common negative ions

Number of negative charges in each ion		
−1	−2	−3
Bromide (Br^-)	Carbonate (CO_3^{2-})	Phosphate (PO_4^{3-})
Chloride (Cl^-)	Oxide (O^{2-})	Nitride (N^{3-})
Hydrogen carbonate (HCO_3^-)	Sulfate (SO_4^{2-})	
Hydroxide (OH^-)	Sulfide (S^{2-})	
Iodide (I^-)		
Nitrate (NO_3^-)		

Non-metals usually form negative ions. The number of negative charges is the electrovalency of the ion. For example, chloride has one negative charge (Cl^-), oxide has two negative charges (O^{2-}) and phosphide has three (P^{3-}). There are also some more complex negative ions called **molecular ions**, such as hydroxide ions (OH^-) and sulfate ions (SO_4^{2-}). These groups of atoms have an overall negative charge and are treated as a single entity. Note that the hydrogen ion, although a non-metal ion, exists as a positive ion.

The following examples show how to determine the formulae for ionic compounds.

Example 1

Write the formula for sodium chloride.

Step 1	*Determine the electrovalency of the ions that comprise the compound and write down their symbols.* The symbol for the sodium ion is Na^+ and the symbol for the chloride ion is Cl^-.
Step 2	*Determine the ratio of ions required in order to achieve electrical neutrality. (Remember compounds have no overall charge.)* The ratio of negative to positive charges for sodium and chloride ions is 1 : 1. That is, it takes one negatively charged chloride ion to balance the charge of one positively charged sodium ion.
Step 3	*Write the formula for the compound using the numbers in the ratios as subscripts. (Remember the number 1 does not need to be included.)* The formula for the compound is NaCl.

Example 2

Write the formula for aluminium oxide.

Step 1	*Determine the electrovalency of the ions that comprise the compound and write down their symbols.* The symbol for the aluminium ion is Al^{3+} and the symbol for the oxide ion is O^{2-}.
Step 2	*Determine the ratio of ions required in order to achieve electrical neutrality. (Remember compounds have no overall charge.)* The ratio of negative to positive charges for aluminium and oxide ions is 2 : 3. That is, it takes three negatively charged oxide ions to balance the charge of the two positively charged aluminium ions.
Step 3	*Write the formula for the compound using the numbers in the ratios as subscripts.* The formula for aluminium oxide is Al_2O_3.

Example 3

Write the formula for calcium phosphate.

Step 1	*Determine the electrovalency of the ions that comprise the compound and write down their symbols.* The symbol for the calcium ion is Ca^{2+} and the symbol for the phosphate ion is PO_4^{3-}.
Step 2	*Determine the ratio of ions required in order to achieve electrical neutrality. (Remember compounds have no overall charge.)* The ratio of negative to positive charges for calcium and phosphate ions is 3 : 2. That is, it takes two negatively charged phosphate ions to balance the charge of the three positively charged calcium ions.
Step 3	*Write the formula for the compound using the numbers in the ratios as subscripts.* The formula for calcium phosphate is $Ca_3(PO_4)_2$.

Note the use of brackets in the formula to show that more than one molecular ion is needed to balance the electric charge.

INVESTIGATION 4.6

The ionic compound formula game

AIM: To practise deriving the chemical formulae for ionic compounds

Materials:

a set of playing cards with the name and valency of each of the positive and negative ions listed in the electrovalency tables in this section. You will need four identical cards for each ion.

Method and results

- Organise a group of four students to play the card game. The aim of this game is to collect as many cards as possible by producing compounds with correct chemical formulae.
- Shuffle the cards and distribute them to the players.
- The dealer puts down one card.
- The rest of the players try to produce a chemical formula using their cards. The first person to come up with a correct chemical formula wins the hand and keeps the cards. They are put aside until the end of the game. The dealer will decide the winner of the hand.
- The person to the left of the dealer then puts down one of their cards.
- The other players in the game now try to produce a chemical formula using the cards they have in their hands. Again, the person to come up with a correct chemical formula wins that hand and the cards are put aside until the end of the game.
- The game continues moving to the next person until no one is able to produce a chemical formula. The game stops at this point.
- Each player then counts the number of cards they have produced formulae with. The winner is the person with the most cards.

1. Write a list of the formulae and the name of the compounds formed.

Discuss and explain

2. What is the best strategy to win the game?
3. Did you find the game useful in learning the formulae of compounds? Explain.

4.7 Exercises: Understanding and inquiring

To answer questions online and to receive **immediate feedback** and **sample responses** for every question, go to your learnON title at www.jacplus.com.au. *Note:* Question numbers may vary slightly.

Remember

1. What is a chemical formula?
2. What is a molecular formula?
3. What does the formula of a compound tell you about the compound?
4. Write the symbols for the following elements: sodium, hydrogen, potassium, lead, chlorine, iodine and sulfur.
5. Which elements are present in each of the following compounds?
   (a) HNO_3 (b) $NaHCO_3$ (c) FeS
6. How is the valency of an element determined?
7. How many chloride (Cl^-) ions would be required to combine with each of the following ions to form an ionic compound?
   (a) calcium (Ca^{2+}) (b) aluminium (Al^{3+}) (c) silver (Ag^+)
8. Write down the valencies for the following elements: sodium, hydrogen, lead, chlorine, iodine, magnesium and sulfur.

Think

9. The ions listed below can combine in many different ways to form 25 different compounds. Write the formulae and names of these compounds.
   Na^+ Fe^{3+} Li^+ Cu^{2+} Al^{3+}
   Cl^- OH^- N^{3-} O^{2-} SO_4^{2-}
10. The chloride ion has the same valency as the sodium ion. However, it has a different electrovalency. Why?
11. Write a formula for each of the following.
   (a) Oxygen gas
   (b) Chlorine gas
   (c) Lead
   (d) Nitrogen oxide
   (e) Zinc oxide
   (f) Potassium sulfate
   (g) Calcium hydroxide
12. Name the following compounds.
   (a) NH_4Cl
   (b) KI
   (c) $Al(NO_3)_3$
   (d) $Fe(OH)_3$
   (e) $KHCO_3$
   (f) $MgCO_3$
   (g) HNO_3
13. Explain why group 18 is not listed in the table in this subtopic showing valency of groups in the periodic table.

Imagine

14. Imagine that there was no recognised system for naming elements and compounds. Describe some of the problems this would lead to.

Create

15. Create your own ionic compound formula game. It could be an improved version of Investigation 4.6; however, it does not have to be a card game.

learnON RESOURCES — ONLINE ONLY

Complete this digital doc: Worksheet 4.8: Covalent bonding (eles-19443)

Complete this digital doc: Worksheet 4.9: Chemical formulae (eles-19444)

4.8 Concept and mind maps

4.8.1 Concept and mind maps

1. On small pieces of paper, write down all the ideas you can think of about a particular topic. Concept maps and mind maps can also be drawn with software or apps.
2. Select the most important ideas and arrange them under your topic. Link these main ideas to your topic and write the relationship along the link.
3. Choose ideas related to your main ideas and arrange them in order of importance under your main ideas, adding links and relationships.
4. When you have placed all of your ideas, try to find links between the branches and write in the relationships.

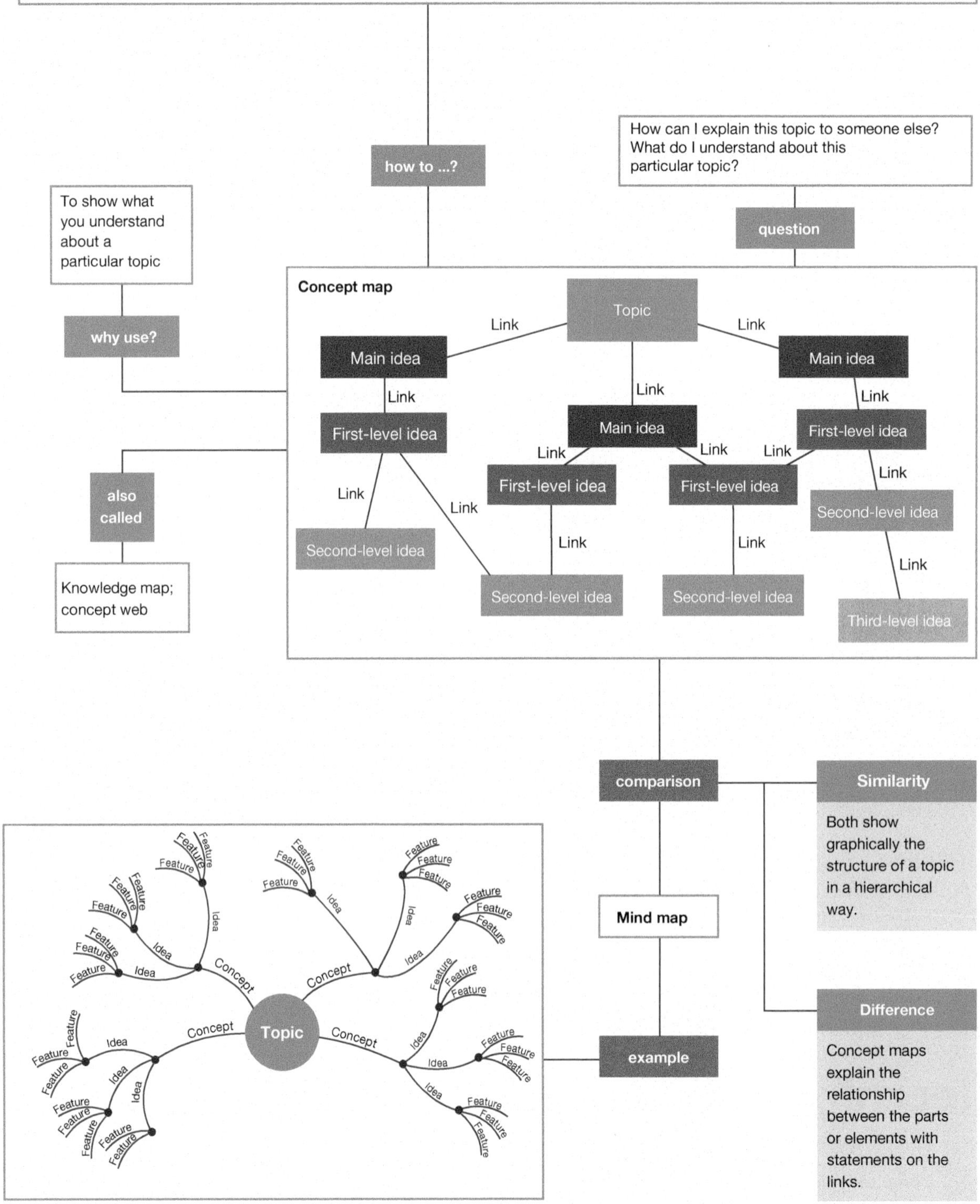

4.8 Exercises: Understanding and inquiring

To answer questions online and to receive **immediate feedback** and **sample responses** for every question, go to your learnON title at www.jacplus.com.au. *Note:* Question numbers may vary slightly.

Think and create

1. A concept map can be used to illustrate some of the important ideas associated with the atom and the links between them.
   a. Copy the concept map at right into your workbook and complete it by adding links between the ideas.
   b. Construct your own concept map to show how ideas about what is inside substances are linked. Begin by working in a group to brainstorm the main ideas of the topic.

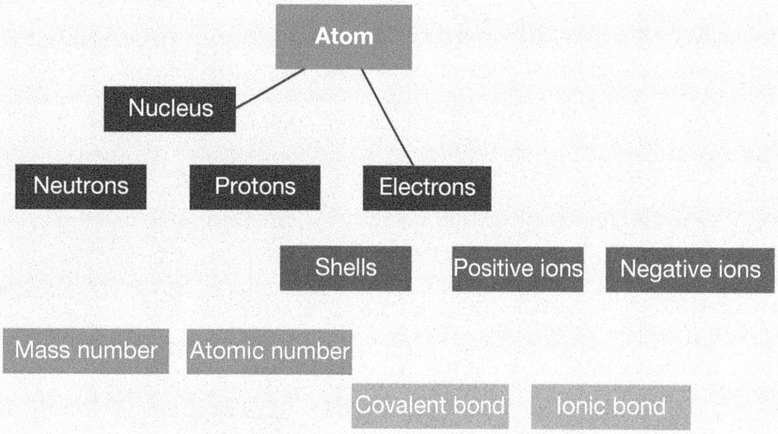

2. Create a concept map to illustrate ideas and links related to:
   a. the structure of the atom
   b. the periodic table.

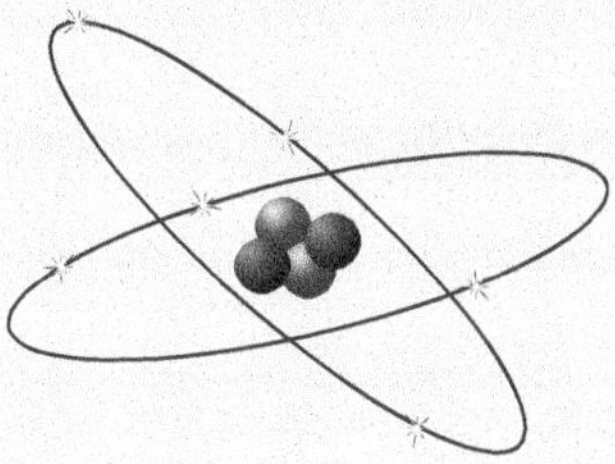

3. A mind map is similar to a concept map, but doesn't explain the links between the major concepts and ideas. Complete the mind map at right to represent your knowledge of metals, non-metals and metalloids.

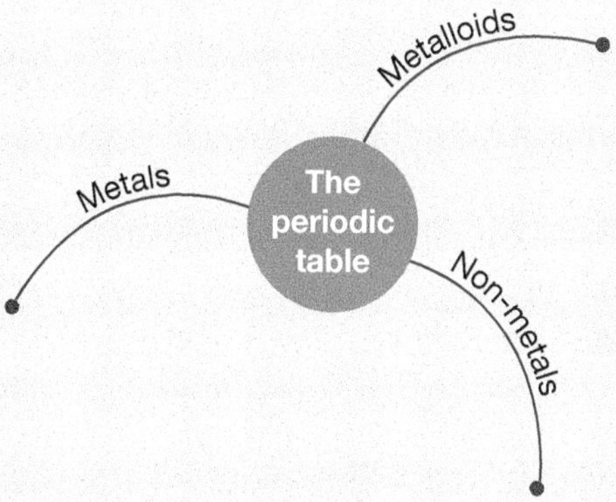

4. In a small group, brainstorm a list of important words, concepts and ideas associated with covalent bonding and ionic bonding. Use the list to create either a concept map or a mind map beginning with the term *chemical bonding*.

Concept map

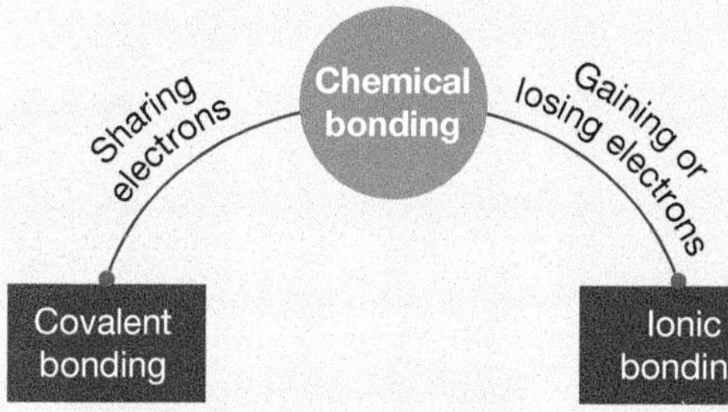

Mind map

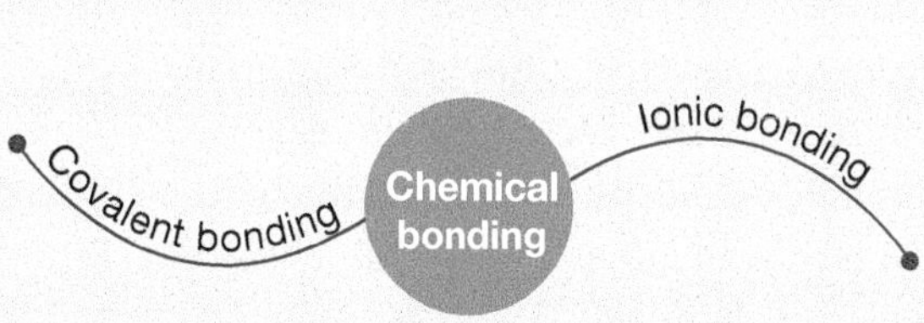

HOW ABOUT THAT!

Oxygen gas consists of molecules in which two oxygen atoms share electrons. The formula for oxygen gas is therefore O_2. Ozone gas, which exists naturally in the upper atmosphere, consists of 'triplets' of oxygen atoms sharing electrons. The formula for ozone is therefore O_3.

learn**on**

4.9 Project: The mystery metal

Scenario

Your eccentric aunt loves combing through junk shops in search of overlooked treasures, and every time you spend a day with her she'll make you go into one grubby store smelling of mangy mink coats after another. One day during the school holidays, you are wandering idly in one of these old junk shops while your aunt haggles for an old vase with the owner. You find a lump of metal in a drawer of an old dresser. The shopkeeper says that you can keep it and you put it in your pocket. Occasionally over the next few days you wonder what the metal is. Is it something valuable like platinum, or useful like aluminium? Or is it just an old lump of lead? By the end of the holidays, you've forgotten all about the lump of mystery metal.

When you get back to school, your science teacher announces that everyone in your class is to enter a competition that the Australian Chemistry Teachers' Association is running. The competition requires you to write an online 'Choose your own adventure' story that has a chemistry theme. You and your friends are scratching your heads trying to come up with an idea when, suddenly, you remember that lump of mystery metal you found in the junk shop. Maybe you could use that as the theme for your competition entry …

Your task

Either on your own or as part of a group, develop a 'Choose your own adventure' story exploring the identification of the mystery metal. Create a series of interconnected PowerPoint slides that can be uploaded. A player starting at the first screen will advance through a storyline according to the choices they make at each slide. The choices will relate to various chemical and physical characteristics of the metal. The right sequence of choices will eventually lead to the correct identification of the mystery metal.

4.10 Review

4.10.1 Study checklist

Atoms and the periodic table

- recall the characteristics and location in the atom of protons, neutrons and electrons
- explain how the electron structure of an atom determines its position in the periodic table and its properties
- recognise that elements in the same group of the periodic table have similar properties
- recognise that the atomic numbers of elements in the periodic table increase from left to right across each period
- distinguish between the atomic number, mass number and relative atomic mass of an atom
- describe common properties of elements in each of the alkali metals, halogen and noble gas groups of the periodic table
- distinguish between the properties of metals, non-metals and metalloids

Electron shells and bonding

- describe the structure of atoms in terms of electron shells
- relate the energy of electrons to shells
- explain the movement of electrons to higher energy levels and the emission of light when they return to a lower level
- describe covalent bonding in terms of the sharing of electrons in the outer shells of atoms
- describe ionic bonding in terms of the formation of ions and relate it to the number of electrons in the outer electron shells of atoms
- relate the reactivity of metals to the electron shell structure of their atoms and their location in the periodic table

Valency and chemical formulae

- define the valency of an element as the number of electrons an atom needs to gain, lose or share to fill its outer shell
- relate the valency of an atom to its group in the periodic table
- deduce the formula of a variety of simple covalent and ionic compounds from the valency of their constituent elements

Science as a human endeavour

- investigate how the periodic table was developed and how this depended on experimental evidence at the time
- describe the hazards associated with the use of lead and describe recent attempts to reduce its use
- relate the reactivity of metals to the mining industry
- describe the extraction of copper, tin and iron by ancient civilisations
- describe the uses of bronze and iron by ancient civilisations

Individual pathways

ACTIVITY 4.1
Revising chemical patterns
doc-8470

ACTIVITY 4.2
Investigating chemical patterns
doc-8471

ACTIVITY 4.3
Investigating chemical patterns further
doc-8472

learnon ONLINE ONLY

4.10 Review 1: Looking back

To answer questions online and to receive **immediate feedback** and **sample responses** for every question, go to your learnON title at www.jacplus.com.au. *Note:* Question numbers may vary slightly.

1. Explain why it is more useful to display the elements as a periodic table than as a list.
2. The periodic table is an arrangement of all the known elements. What information is given by the group and period numbers on the periodic table?
3. Explain how the periodic table has helped chemists of both the past and present when they are searching for new elements.
4. Explain why water does not appear in the periodic table.
5. Write the atomic number and mass number of the following atoms and then calculate the number of protons, neutrons and electrons they have.
   (a) $^{28}_{14}Si$
   (b) $^{52}_{24}Cr$
   (c) $^{197}_{79}Au$
   (d) $^{206}_{82}Pb$
   (e) $^{242}_{94}Pu$
6. To which group of elements in the periodic table does the neon used in lighting belong?
7. List five properties that all (or almost all) metals have in common.
8. List five properties that most solid non-metals have in common.
9. As you move down the groups in the periodic table, how does the reactivity change for:
   (a) metals
   (b) non-metals?
10. As you move across the periodic table, what changes occur in:
    (a) atomic number
    (b) mass number
    (c) melting points
    (d) metallic character?
11. Although they look very different from each other and have very different uses, arsenic, germanium and silicon belong to the group of elements known as metalloids. How are metalloids different from all of the other elements in the periodic table?
12. Copy and complete the following table.

Name	Symbol	Atomic number	Electron configuration
Lithium	Li	3	2, 1
	C	6	
			2, 6
Neon			
	Na		
		13	
			2, 8, 5
Chlorine			
	K		2, 8, 8, 1
	Ca	20	

13. All atoms of the element magnesium have 12 protons. Of those atoms, 80 per cent have 12 neutrons.
    (a) State the atomic number of magnesium.
    (b) What is the mass number of most magnesium atoms?
    (c) How many electrons orbit a neutral magnesium atom?
    (d) Explain why all magnesium atoms don't have the same mass number.

14. Copy and complete the following table.

Ion	Ion symbol	Atomic number	Electron configuration
		3	2
	Na^+		
		12	2, 8, 8
	N^{3-}		
		9	2, 8
Sulfide			

15. The electron shell diagram at right has its first two shells filled. It could represent a neutral atom, a positive ion or a negative ion. Identify the names and symbols of the atom or ion if it represents:
    (a) a neutral atom (identify one)
    (b) a positive ion (identify two)
    (c) a negative ion (identify two).

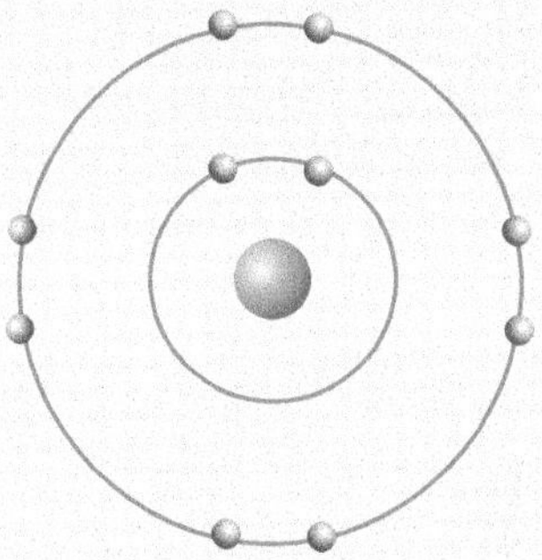

16. Show how the following ionic compounds form.
    (a) Lithium fluoride (LiF)
    (b) Sodium oxide (Na_2O)
17. Show how the following covalent compounds form.
    (a) Hydrogen chloride (HCl)
    (b) Ammonia (NH_3)
18. What are the differences between the properties of ionic and covalent compounds?
19. Explain why you are more likely to find pure gold than pure copper in the ground.
20. Explain why metals such as gold, silver and copper were discovered about 2000 years ago while the metals potassium, sodium and calcium were not discovered until about 200 years ago.
21. Write formulae for the following substances.
    (a) Oxygen gas
    (b) Carbon dioxide gas
    (c) Aluminium oxide
    (d) Sodium fluoride
    (e) Calcium carbonate
    (f) Zinc chloride
    (g) Iron(III) sulfide
    (h) Sulfur dioxide
    (i) Carbon
    (j) Lead

learnON RESOURCES — ONLINE ONLY

Complete this digital doc: Worksheet 4.10: Chemical patterns: Summary (eles-19445)

TOPIC 5 Chemical reactions

5.1 Overview

5.1.1 Why learn this?

Engineers developing polymers for spacesuits or the next generation of passenger aircraft need knowledge and understanding of chemical reactions to produce new materials that are strong, light and capable of resisting high temperatures. Other useful substances and materials such as fuels, metals and pharmaceuticals are also products of chemical reactions.

5.1.2 Think about chemical reactions

assesson

- How does zinc slow down the rusting of iron?
- Spectacles that turn into sunglasses — how do they do it?
- Bottled gas, candle wax and petrol all come from one substance. What is it?
- Cow manure to replace petrol — how is that possible?
- In what part of your body do you find catalytic catalase?
- How do light sticks produce light?
- What do plastic fruit juice containers, a toilet seat and a $100 note have in common?
- Why are some plastics harder than others?

LEARNING SEQUENCE

Numerous **videos** and **interactivities** are embedded just where you need them, at the point of learning, in your learnON title at www.jacplus.com.au. They will help you to learn the content and concepts covered in this topic.

5.1.3 Your quest

Chemical reactions

You should already know quite a lot about chemical reactions. Answer the questions below to review your knowledge.

Remember and explain

1. Each of the photos below depicts an energy transformation that occurs as a result of a chemical reaction.
   (a) What one-word name is given to all three chemical reactions?
   (b) Which of the chemical reactions are exothermic?
   (c) Name the one reactant that participates in all three chemical reactions.
   (d) Identify the fuel in each of the three chemical reactions.
   (e) Identify one chemical product of the reactions depicted in photos A and C.
   (f) Name one chemical product that results from the reaction in photo B.

2. Chemical reactions take place in all living things to keep them alive.
   (a) Which chemical reaction takes place in all green plants in the presence of sunlight?
   (b) Identify the only solid product of the chemical reaction referred to in part (a).
   (c) Name the chemical reaction that takes place in every cell of all animals to transform stored energy to other forms of energy.

Inside chemical reactions

Chemical reactions take place when the bonds between atoms are broken and new bonds are formed. This creates a new arrangement of atoms and therefore at least one new substance.

3. Explain what happens to the chemical bonds during the chemical reaction between oxygen and hydrogen as illustrated in the diagram on the right.

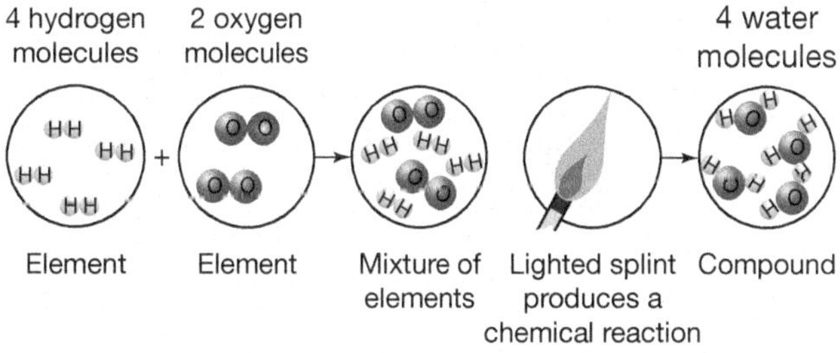

5.2 The language of chemical reactions

5.2.1 Chemical equations

In order to communicate with each other easily about chemical reactions, scientists all over the world need to use the same language. That language involves chemical symbols, formulae and **equations**.

Word equations provide a simple way to describe chemical reactions by stating the reactants and products. Chemical equations that use formulae provide more information. They show how the atoms in the reactants combine to form the products.

Writing chemical equations involves some simple mathematics and a knowledge of chemical formulae. Chemical equations are set out in the same way as word equations, with the reactants to the left of the arrow and products to the right. However, they are different from word equations in three ways:

- Formulae are used to represent the chemicals involved.
- The physical states of the chemicals are often included.
- Numbers are written in front of the formulae in order to balance the numbers of atoms on each side of the equation.

The rules of a 'game' of balancing equations are described below. Read through the rules very carefully before you play the game.

GAME RULES

GAME RULE 1. Know your products

The products of a reaction must be known from either **observation** or reliable sources (such as chemists). For example, it is well known that the product of the reaction between hydrogen gas and oxygen gas is water vapour (gas).

GAME RULE 2. Know your formulae

You need to know the formulae of all the reactants and products. For example:

- formula of hydrogen gas H_2
- formula of oxygen gas O_2
- formula of water vapour H_2O.

Remember! Because each substance has only one correct chemical formula, it **cannot** be changed by altering the subscript numbers.

GAME RULE 3. Write down the formulae

The formulae must be written according to the word equation, with reactants on the left-hand side of the arrow and products on the right-hand side.

$$H_2 + O_2 \longrightarrow H_2O$$

GAME RULE 4. Balance the numbers of atoms

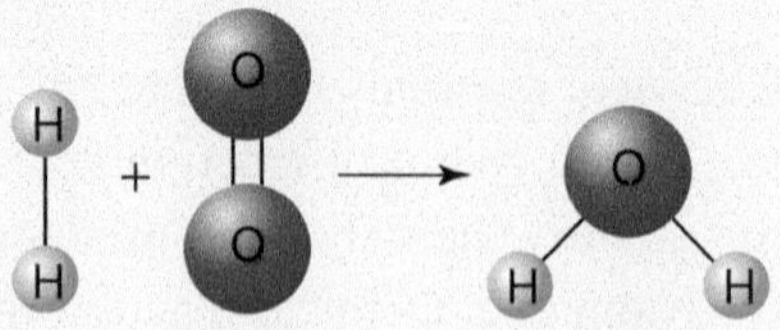

First, make a list of the elements present in the formulae under the heading 'Element', as shown below. Then count up how many atoms of each element are represented by the formulae under the headings 'Reactants' and 'Products'.

Element	Reactants	Products
H	2	2
O	2	1

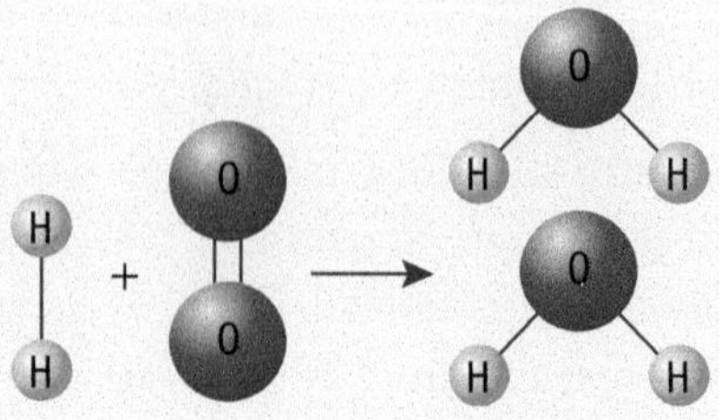

You can see that there are not enough oxygen atoms on the product side. The only way this can be adjusted is by writing numbers in front of the chemical formulae.

When we write a number **in front** of a formula, it **multiplies all the atoms** in that formula. Let's increase the number of oxygen atoms on the product side by placing a 2 in front of the formula for water.

$$H_2 + O_2 \longrightarrow \mathbf{2}H_2O$$

Recounting the atoms we find:

Element	Reactants	Products
H	2	4
O	2	2

The oxygen atoms are now balanced, but the hydrogen atoms are not. Let's try writing a 2 in front of hydrogen's formula on the reactant side to increase the number of hydrogen atoms.

$$\mathbf{2}H_2 + O_2 \longrightarrow \mathbf{2}H_2O$$

Counting the atoms again we find:

Element	Reactants	Products
H	4	4
O	2	2

The numbers of each of the elements are the same on both sides of the equation. The equation is balanced!

GAME RULE 5. Include the states

To indicate the physical state of each chemical involved in the reaction, the following symbols are used:

- solid (s)
- liquid (l)
- gas (g).

The symbol (aq) is used to represent an **aqueous solution** of a substance. An aqueous solution is obtained when a substance is dissolved in water.

Write the correct symbol representing the physical state of each reactant and product.

$$2H_2(g) + O_2(g) \longrightarrow 2H_2O(l)$$

Formulae correct!
Number of atoms balanced!
States correct!
Formula equation complete!
Game over!

The reaction between hydrogen and oxygen

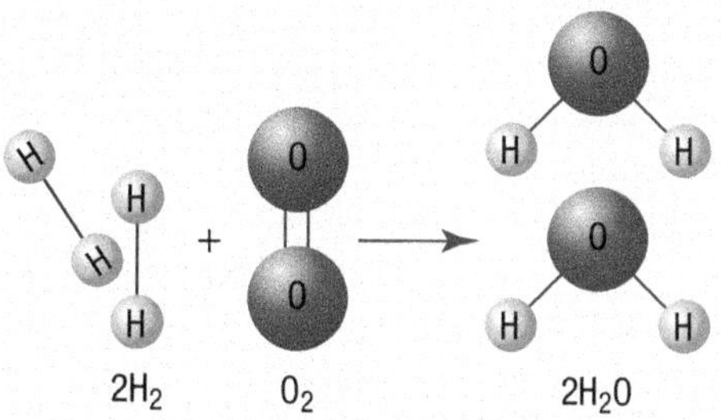

PLAY THE GAME

- Write a word equation and an equation using formulae for each of the six reactions listed. An example is provided on the next page. See the tables over the page for the correct formulae.
  1. Carbon monoxide gas and oxygen gas react to form carbon dioxide gas.
  2. Sodium hydroxide solution and hydrochloric acid solution react to form sodium chloride solution and water.
  3. Mercury metal and oxygen gas react to form solid mercury(II) oxide.
  4. Magnesium metal and hydrochloric acid solution react to form hydrogen gas and magnesium chloride solution.
  5. Sodium metal and water react to form hydrogen gas and sodium hydroxide solution.
  6. Copper sulfate solution and sodium hydroxide solution react to form solid copper hydroxide and sodium sulfate solution.

The formulae of some common ionic compounds

Compound	Formula
Sodium hydroxide	NaOH
Sodium chloride	NaCl
Magnesium chloride	$MgCl_2$
Copper hydroxide	$Cu(OH)_2$
Sodium sulfate	Na_2SO_4
Copper sulfate	$CuSO_4$
Sodium hydrogen carbonate	$NaHCO_3$
Mercury(II) oxide	HgO
Sodium citrate	$C_6H_5O_7Na_3$

The formulae of some common covalent substances

Compound	Formula
Water	H_2O
Citric acid	$C_6H_8O_7$
Carbon dioxide	CO_2
Oxygen	O_2
Hydrochloric acid	HCl
Carbon monoxide	CO
Hydrogen	H_2

<table>
<tr><th>Balancing a chemical equation</th><th>Example (Methane gas will burn in air. This is an example of a combustion reaction. This type of reaction produces CO_2 and H_2O.)</th></tr>
<tr><td>Step 1: Start with the word equation and name all of the reactants and products.</td><td>Methane gas + oxygen gas ⟶ carbon dioxide + water</td></tr>
<tr><td>Step 2: Replace the words in the word equation with formulae and rewrite the equation.</td><td>Methane gas = CH_4 Oxygen gas = O_2 (reactants)
Carbon dioxide = CO_2 Water vapour = H_2O (products)
$CH_4 + O_2 \longrightarrow CO_2 + H_2O$</td></tr>
<tr><td>Step 3: Count the number of atoms of each element (represented by the formulae of the reactants and products).</td><td><table><tr><th>Element</th><th>Reactants</th><th>Products</th></tr><tr><td>C</td><td>1</td><td>1</td></tr><tr><td>H</td><td>4</td><td>2</td></tr><tr><td>O</td><td>2</td><td>3</td></tr></table></td></tr>
<tr><td>Step 4: If the number of atoms of each element is the same on both sides of the equation, the equation is already balanced. If not, numbers need to be placed in front of one or more of the formulae to balance the equation. These numbers are called coefficients and they multiply all of the atoms in the formula.</td><td>To balance the hydrogen atoms, put a 2 in front of H_2O.
$CH_4 + O_2 \longrightarrow CO_2 + 2H_2O$
The oxygen atoms can be balanced by putting a 2 in front of the O_2 on the left.
$CH_4 + O_2 \longrightarrow CO_2 + 2H_2O$
The equation is now balanced. It can be checked by counting the number of atoms of each element on both sides of the new equation.<table><tr><th>Element</th><th>Reactants</th><th>Products</th></tr><tr><td>C</td><td>1</td><td>1</td></tr><tr><td>H</td><td>4</td><td>4</td></tr><tr><td>O</td><td>4</td><td>4</td></tr></table></td></tr>
<tr><td>Step 5: Add physical state symbols.</td><td>$CH_4(g) + 2O_2(g) \rightarrow CO_2(g) + 2H_2O(g)$
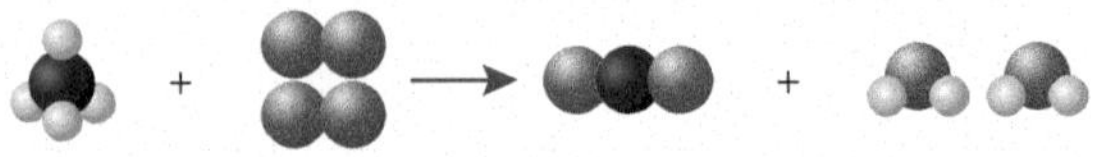
Methane Oxygen Carbon dioxide Water</td></tr>
</table>

5.2 Exercises: Understanding and inquiring

To answer questions online and to receive **immediate feedback** and **sample responses** for every question, go to your learnON title at www.jacplus.com.au. *Note:* Question numbers may vary slightly.

Remember

1. Describe three differences between word equations and equations in which formulae are used.
2. How are the states (solid, liquid and gas) indicated in a chemical equation?
3. What is an aqueous solution and how is it represented in a chemical equation?

Think

4. Which symbols would you use in a chemical equation to represent the metals iron, mercury, zinc and aluminium?
5. Try writing a balanced equation using formulae for the reaction that occurs when you eat a sherbet lolly. These sweets commonly contain citric acid and sodium hydrogen carbonate. In the mouth, these chemicals dissolve in your saliva and react together to form sodium citrate solution, carbon dioxide gas and water. Use the table in section 5.2.1 to help you.
6. Explain why it is necessary to balance chemical equations.
7. Test your ability to balance chemical equations by completing the **Checking for balance** interactivity.

learnon RESOURCES — ONLINE ONLY

Try out this interactivity: Balancing chemical equations (int-0677)

Complete this digital doc: Worksheet 5.1 Chemical equations (doc-19447)

Complete this digital doc: Worksheet 5.2 Balancing chemical equations (doc-19448)

Complete this digital doc: Worksheet 5.3 A world of reactions (doc-19449)

5.3 Precipitation reactions

5.3.1 Aqueous solution

When table salt (sodium chloride) is dissolved in water to form an **aqueous solution**, it seems to disappear. The ions in the salt no longer bond together as a large array of positive and negative ions like they do as a solid. The sodium ions and the chloride ions separate when they dissolve.

Sodium chloride dissolving in water can be represented by the equation:

$$NaCl(s) \xrightarrow{(H_2O)} Na^+(aq) + Cl^-(aq)$$

Ions in aqueous solutions are therefore separate entities and are able to react independently.

Ionic compounds dissolve in water to varying degrees. Some are soluble, others slightly soluble and others insoluble. The box on the next page outlines some handy rules for predicting whether or not a compound is soluble.

The formation of the brilliant yellow precipitate, lead iodide, from the colourless solutions lead nitrate and potassium iodide

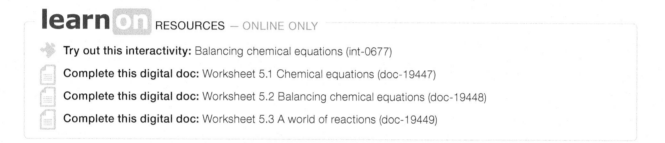

SOLUBLE OR NOT?

1. All compounds containing either the Na^+, NH_4^+, K^+, or NO_3^- ion will dissolve in water. Compounds containing these ions never form precipitates.
   Example: This rule tells us that NaCl, NH_4Cl, K_2SO_4 and $AgNO_3$ are all soluble in water and therefore do not form precipitates.
2. Compounds containing the Cl^-, Br^- and I^- ions are soluble, except when they contain the Ag^+, Pb^{2+} or Hg^{2+} ions.
   Example: This rule tells us that $FeCl_3$, $ZnBr_2$ and AlI_3 are soluble, but that AgCl, $HgBr_2$ and PbI_2 are not soluble.
3. Compounds containing the SO_4^{2-} ion are soluble, except for $BaSO_4$, $PbSO_4$ and $CaSO_4$.
   Example: This rule tells us that $ZnSO_4$ will dissolve, but $BaSO_4$ will form a precipitate.
4. Compounds containing OH^-, CO_3^{2-} and PO_4^{3-} are insoluble except when they contain the ions Na^+, NH_4^+ or K^+.
   Example: This rule tells us that $BaCO_3$ and $Zn(OH)_2$ will form insoluble precipitates, but Na_2CO_3 and KOH will not.
5. Some compounds are slightly soluble. These include $Ca(OH)_2$, $PbCl_2$, $PbBr_2$, $CaSO_4$ and Ag_2SO_4.

5.3.2 Suddenly it appeared!

When two solutions containing dissolved ions are mixed together, the ions are able to come into contact with each other. Oppositely charged ions attract. In some cases, the attraction is strong enough to form ionic bonds and hence a new ionic compound. Some of these compounds are insoluble (unable to dissolve in water) and so a solid called a **precipitate** forms. Chemical reactions in which precipitates form are called **precipitation reactions**. When colourless lead nitrate solution and colourless potassium iodide solution are added together, a brilliant yellow precipitate is formed.

Changing partners

Another example of a precipitation reaction occurs between silver nitrate solution and sodium chloride solution. When these two colourless solutions are added together in a test tube the contents become cloudy, indicating that a precipitate has formed. If the tube is allowed to stand for a while, the solid settles to the bottom and we can see that a clear solution is also present. The products of the reaction are insoluble solid silver chloride (the precipitate) and sodium nitrate (not visible because it is soluble in water). This reaction can be represented by the equation:

$$\text{silver nitrate} + \text{sodium chloride} \longrightarrow \text{silver chloride} + \text{sodium nitrate}$$
$$AgNO_3(aq) + NaCl(aq) \longrightarrow AgCl(s) + NaNO_3(aq)$$

Silver nitrate, sodium chloride and sodium nitrate all dissolve in water. Therefore, they have the symbol (aq). Silver chloride does not dissolve in water, so it has the symbol (s) to indicate that it is solid.

The equation shows that the ions in the reactants have changed partners. The silver ion is paired with the chloride ion on the product side of the reaction and the sodium ion is paired with the nitrate ion. The opposite is the case on the reactant side. A positive ion can pair up only with a negative ion because oppositely charged ions are attracted to each other. When writing the formula of any new compound, the positive ion is always written first.

Ions sometimes change partners when a chemical reaction takes place.

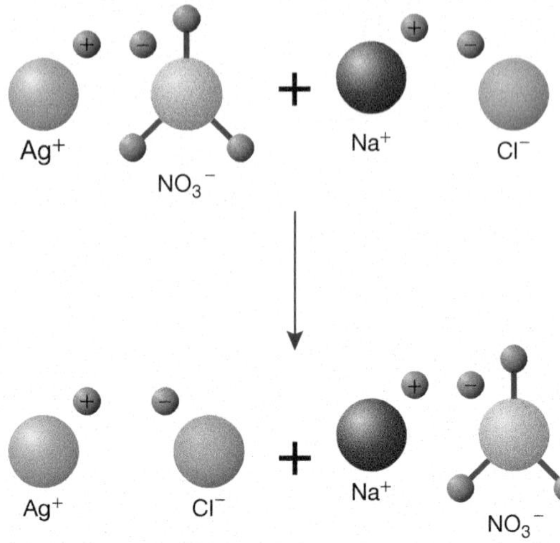

INVESTIGATION 5.1

Will it precipitate?

AIM: To predict and test for precipitation when a variety of solutions are added to each other

Materials:

5 semi-micro test tubes and a test-tube rack
a white tile and a black tile
safety glasses
dropping bottles of the following solutions: copper sulfate, sodium chloride, silver nitrate, cobalt chloride, sodium hydroxide, potassium iodide

CAUTION

Wear safety glasses.

Method and results

- Place 10 drops of copper sulfate solution in each test tube.
- Add 10 drops of sodium chloride to the first test tube, 10 drops of silver nitrate to the second, and so on until each tube contains copper sulfate solution and one other solution. Hold a black or white tile behind the test tube if necessary to detect the presence of a precipitate.

1. If there is a visible reaction, record your observations in a table.
   - Tip the residues into a waste bottle. Wash out the test tubes thoroughly and this time place 10 drops of sodium chloride in each of the test tubes. Again add one of the other solutions to each test tube (but not copper sulfate, which has already been tested).
2. Record your observations in your table.
   - Repeat until all possible pairs of solutions have been tested.

Discuss and explain

3. Write word equations for each of the pairs that reacted to form a precipitate.
4. Use formulae to write chemical equations for each of the pairs that reacted to form a precipitate.
5. You could have predicted which pairs of solutions would form a precipitate using the box of solubility rules under the heading *Soluble or not?* in this subtopic. Check to see if the rules match your results.

5.3 Exercises: Understanding and inquiring

To answer questions online and to receive **immediate feedback** and **sample responses** for every question, go to your learnON title at www.jacplus.com.au. *Note:* Question numbers may vary slightly.

Remember

1. What is a precipitate?

Think

2. Write an equation for the reaction that occurs when the salt copper sulfate dissolves in water.
3. Which two of the following compounds will be soluble in water?
   (a) $NaNO_3$
   (b) KI
   (c) PbI_2
   (d) $Zn(OH)_2$
4. Which of the following compounds will be insoluble in water?
   (a) $CuCO_3$
   (b) AgI
   (c) NaCl
   (d) $Mg(OH)_2$
5. Write down the possible combinations of ions formed when the following solutions are mixed together.
   (a) Sodium chloride and copper sulfate
   (b) Sodium hydroxide and copper sulfate
   (c) Lead nitrate and sodium hydroxide
   (d) Potassium iodide and sodium carbonate
6. For each of the reactions listed in question 5, name the precipitate that would form. If you believe that no precipitate would form, write 'no precipitate'.
7. Create a table with three columns headed 'Soluble', 'Insoluble' and 'Slightly soluble'. Use the information in the box headed *Soluble or not?* to fill the table.

learnon RESOURCES — ONLINE ONLY

Complete this digital doc: Worksheet 5.4: Precipitation (doc-19451)

5.4 Chemicals can be a health hazard

Science as a human endeavour

5.4.1 Dangerous goods and hazardous substances

Many of the chemicals used in industry, medicine, schools, universities and homes can be hazardous to your health. The hazards come about because these chemicals can react with parts of your body — inside or out. Apart from the dangers to your own health, chemicals can react with common substances such as water and air or have properties that cause great damage to property and the environment.

Laws exist, at both national and state level, to ensure that people who use harmful chemicals are informed about how to handle them safely. For this purpose, harmful chemicals are placed within one or both of the **dangerous goods** or **hazardous substances** groups.

learnon RESOURCES — ONLINE ONLY

Watch this eLesson: Saving acid wetlands (eles-1072)

5.4.2 Dangerous goods

Chemicals in the dangerous goods group are those that could be dangerous to people, property or the environment. Most dangerous goods are grouped into one of nine classes according to the greatest immediate risk they present. Some of the classes are divided into subclasses. Dangerous goods must be identified with the appropriate dangerous goods sign on their labels. The table on the following page lists the classes and subclasses, along with their respective label signs.

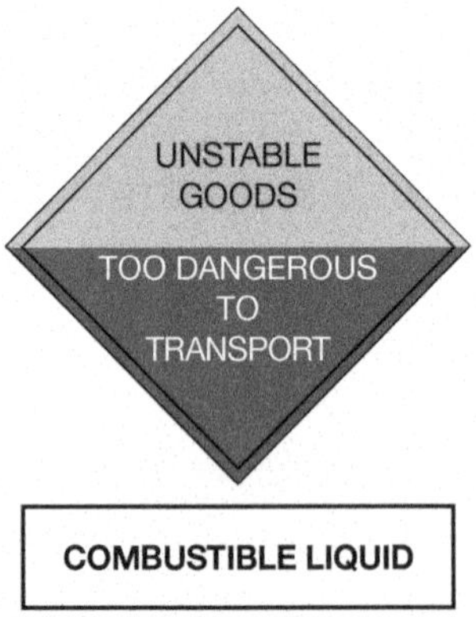

Outside these nine classes, there are two other groups of dangerous goods:

1. goods too dangerous to be transported
2. combustible liquids (C1), which includes liquids that are not as easily ignited as flammable liquids, but which will ignite at temperatures below their boiling point.

5.4.3 Hazardous substances

Chemicals in the hazardous substances group are those that have an effect on human health. The effect may be immediate, such as poisoning and burning, or long term, such as causing liver disease or cancer. Hazardous substances can enter the body in a number of ways. They can be inhaled, absorbed through the skin, ingested (swallowed) or injected.

Hazardous substances are identified on their labels by a signal word providing a warning about the substance, or the word 'Hazardous' printed in red. Signal words include 'dangerous poison', 'poison', 'warning' and 'caution'. Labels of hazardous substances also include:

- information about the risks of the substance
- directions for use
- safety information
- first aid instructions and emergency procedures.

If the substance is also in the dangerous goods group, the label will include the appropriate diamond sign showing its class.

A hazardous substance label

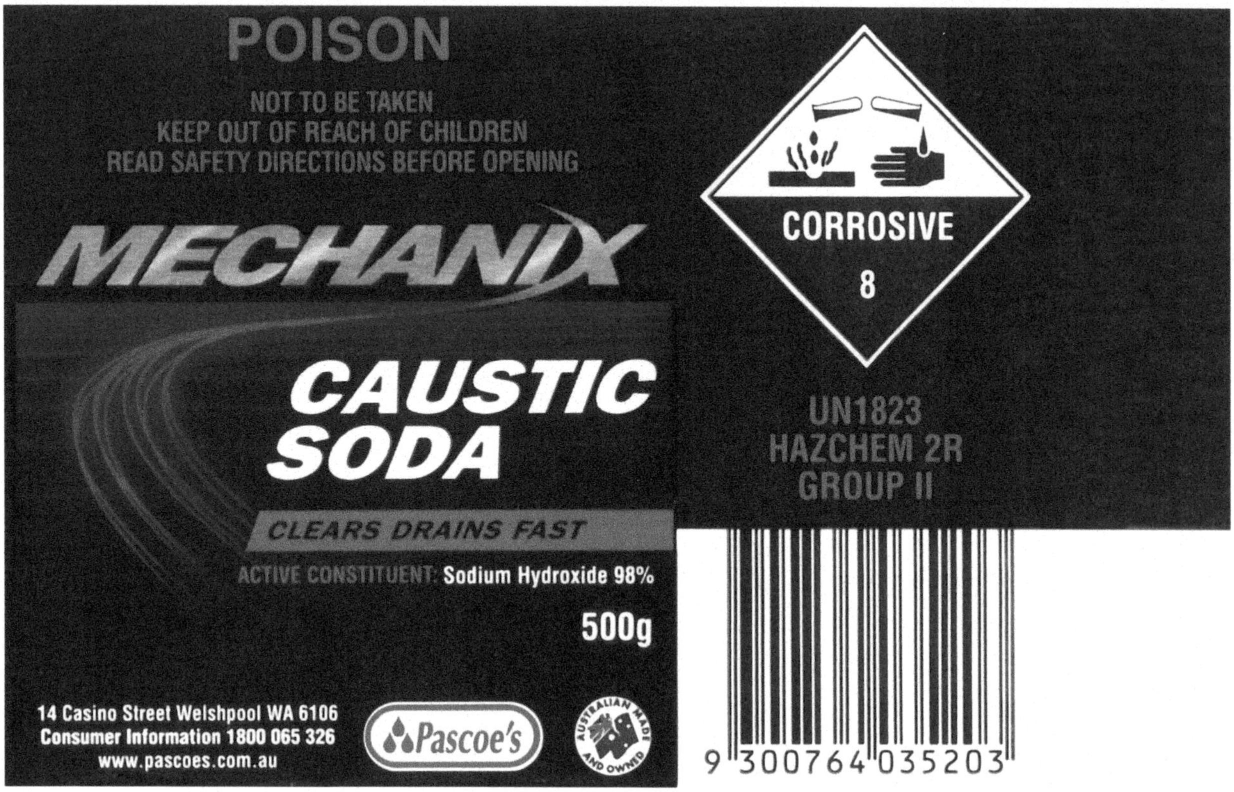

Classes and subclasses of dangerous goods

Class	Description	Sign
Class 1	Explosive substances or articles used to produce explosions	
Class 2.1	Flammable gases: gases that ignite in air if in contact with a source of ignition such as a spark or flame	
Class 2.2	Non-flammable, non-toxic gases: these gases may cause suffocation	
Class 2.3	Toxic gases: gases likely to cause death, serious illness or injury if inhaled	
Class 3	Flammable liquids: liquids with vapours that can ignite on contact with air at temperatures below 60.5 °C	
Class 4.1	Flammable solids: solids that are easily ignited by a source of ignition such as a spark or flame	

Class	Description	Sign
Class 4.2	Substances liable to spontaneous combustion: solids that can ignite without an external source of ignition	
Class 4.3	Substances that emit flammable or toxic gases on contact with water	
Class 5.1	Oxidising agents: substances that may contribute to the combustion of other substances, increasing the risk of fire	
Class 5.2	Organic peroxides: substances that undergo exothermic decomposition reactions	
Class 6.1	Toxic substances: chemicals likely to cause death, serious illness or injury if swallowed, inhaled or brought into contact with skin	
Class 6.2	Infectious substances: substances containing micro-organisms likely to cause diseases in humans or animals	
Class 7	Radioactive material	
Class 8	Corrosive substances: substances that corrode metals or cause injury by reacting on contact with living tissue	
Class 9	Miscellaneous dangerous goods and articles: dangerous substances and objects that do not belong to the other classes	

5.4.4 Keeping you informed

All employers are required by law to make sure that their employees are fully informed about the chemicals in the workplace that are classified as dangerous goods and/or hazardous substances. A list of such chemicals stored or used in the workplace must be kept, along with a copy of each chemical's **material safety data sheet (MSDS)**. Chemical suppliers are required to provide an MSDS for each of the hazardous substances or dangerous goods they supply. In turn, employers are required to make the MSDS accessible to employees who are exposed to the chemicals.

An MSDS is likely to consist of several A4 pages. Many of them can be downloaded directly from the internet. The information on an MSDS should include:

- the ingredients of the product
- the date of issue — an up-to-date MSDS should be no more than five years old
- information about health hazards and first aid instructions
- precautions that need to be taken when using the product
- information about safe storage and handling of the product.

5.4.5 Assessing risk

A **risk assessment** identifies the potential hazards of an experiment and gives protective measures to minimise the risk. Before any experiment involving chemicals is conducted in your school laboratory, a risk assessment is carried out. The form of a risk assessment varies from school to school, but will always contain:

- a summary of the experiment
- a list of the risks and safety precautions for each chemical
- information about whether the chemical is classified as a hazardous substance or dangerous good
- a list of protective measures to be taken. These might include the use of a fume hood and/or the wearing of safety glasses or other protective items.
- first aid information.

Most of the information used in a risk assessment is obtained from the MSDS for each chemical used. The date on the MSDS used for each chemical must be stated to ensure that the risk assessment is up to date.

Part of a risk assessment sheet is shown on the following page. Risk assessment sheets in schools are usually completed and signed by a qualified science teacher or laboratory technician. Your science teacher is required to carefully read the risk assessment sheet before allowing an experiment involving chemicals to commence.

5.4 Exercises: Understanding and inquiring

To answer questions online and to receive **immediate feedback** and **sample responses** for every question, go to your learnON title at www.jacplus.com.au. *Note:* Question numbers may vary slightly.

Remember

1. What do chemicals listed as dangerous goods have in common?
2. If a chemical in the dangerous goods group is explosive, toxic and corrosive, how is the decision about which class it is placed in made?
3. What do chemicals listed as hazardous substances have in common?
4. List four signal words used on the labels of hazardous substances.
5. What is an MSDS and what should it include?
6. Where do employers obtain MSDSs for hazardous substances and dangerous goods?
7. Whose responsibility is it to make sure that people have access to an MSDS for each of the hazardous chemicals and dangerous goods that they store or use?

Think

8. What characteristics do chemicals listed as both dangerous goods and hazardous substances have in common?
9. Explain the difference between flammable liquids (Dangerous goods, Class 3) and combustible liquids (Dangerous goods, C1).
10. Explain the difference between the purposes of an MSDS and a risk assessment.
11. Why should every chemical used in a laboratory (including water) be considered to be a health hazard?

Investigate

12. Many chemical suppliers provide access to MSDSs online. Use the internet to search for an MSDS on hydrochloric acid and use it to answer the following questions.
    (a) What are some alternative names for hydrochloric acid?
    (b) What are the health hazards of hydrochloric acid?
    (c) What first aid treatment is recommended if hydrochloric acid:
        (i) is ingested (swallowed)
        (ii) is inhaled
        (iii) makes contact with an eye
        (iv) makes contact with the skin?
    (d) What recommendations are made for the storage of hydrochloric acid?

Part of a risk assessment sheet for Investigation 4.4

RISK ASSESSMENT SHEET						
ACTIVITY	Investigating reactivity					
REFERENCE	Science Quest 10					
CHAPTER 4	Chemical patterns			INVESTIGATION 4.4		
SUMMARY OF EXPERIMENT						
REACTIVITY OF METALS						
1. Place pieces of magnesium, copper, zinc, aluminium and iron in test tubes.						
1. Add 1M hydrochloric acid to the test tubes and observe the reaction.						
PROTECTIVE MEASURES						
Glasses	Gloves	Dust mask		Lab coat	Fume hood	
x				x	x	
SAFETY INFORMATION						
Reactant	HS	DG	CLASS	MSDS	UN	HAZCHEM
Hydrochloric acid 1M	N	Y	8			
		• Do not breathe gas/fumes/vapour/spray. • Wear suitable protective clothing. • Avoid contact with skin.				
FIRST AID						
SWALLOWED	Contact doctor or poisons centre. Give glass of water.					
EYE	Wash with running water for 15 minutes. Seek medical attention.					
SKIN	Remove contaminated clothing. Wash with soap and water.					
INHALED	Fresh air. Rest. Keep warm.					
REACTANT	HS	DG	CLASS	MSDS	UN	HAZCHEM
Magnesium	N	Y	4.1		1869	4(Y)
• Flammable		• Wear suitable clothing and eye protection. • Do not breathe dust. • Never add water to this product. • Keep locked up. • Avoid contact with skin.				
FIRST AID						
SWALLOWED	Rinse mouth with water.					
EYE	Wash with running water.					
SKIN	Wash with soap and water. For burns: Immerse in cold running water. Bandage lightly. Seek medical attention.					
INHALED	Blow nose. Rinse mouth with water.					
REACTANT	HS	DG	CLASS	MSDS	UN	HAZCHEM

5.5 A world of reactions

5.5.1 Classifying chemical reactions

In a world where countless chemical reactions take place, it is helpful to classify the reactions. They can be classified according to whether they release or absorb energy. They can also be grouped according to the nature of the reactants, the nature of the products, the way in which charged particles in atoms rearrange themselves, or even the number of reactants. Any one reaction can fall into several different groups.

5.5.2 Corroding away

Corrosion is a chemical reaction in which a metal is 'eaten away' by substances in the air or water. The tarnishing of silver jewellery and cutlery, rust, and the green coating that appears on copper are all examples of corrosion.

The rusting of these old car wrecks is an example of corrosion.

Rust protection

If you look at a sheet of **galvanised** iron, you will notice that it does not have a shiny metallic surface. Galvanised iron has been coated with a layer of zinc metal. The zinc prevents the iron underneath from reacting with oxygen and water in the air and rusting. Instead, the zinc corrodes first. It reacts with oxygen and a dull layer of zinc oxide forms on the surface. The equation for this reaction is:

$$2Zn(s) + O_2(g) \longrightarrow 2ZnO(s)$$

5.5.3 Displacement reactions

In the classroom laboratory, waste solutions containing silver ions are never poured down the sink. They are collected and sent to commercial laboratories where the valuable silver is recovered from the solutions. Silver metal can be recovered from silver nitrate solution simply by adding a piece of copper wire. This happens according to the equation:

$$\underset{\text{atoms}}{Cu(s)} + \underset{\text{ions}}{2AgNO_3(aq)} \longrightarrow \underset{\text{atoms}}{2Ag(s)} + \underset{\text{ions}}{Cu(NO_3)_2(aq)}$$

Reactions of this type, where an element displaces another element from a compound, are called **displacement reactions**. In this example, copper has displaced the silver from the silver nitrate solution. The reactions of metals with acids are examples of displacement reactions.

5.5.4 Combustion – a burning question

Combustion reactions are those in which a substance reacts with oxygen, and heat is released. Examples of combustion reactions include the burning of petrol in a motorcycle engine, wax vapour in a candle flame and natural gas in a kitchen stove. In each of these cases **hydrocarbons** (compounds containing only the elements carbon and hydrogen) combine with oxygen in the air to form carbon dioxide gas and water vapour. This is shown in the following equation for the burning of methane (natural gas) in a gas jet.

$$\underset{\text{methane molecule}}{CH_4(g)} + \underset{\text{oxygen molecule}}{2O_2(g)} \longrightarrow \underset{\text{carcon dioxide molecule}}{CO_2(g)} + \underset{\text{water molecule}}{2H_2O(g)}$$

5.5.5 Breaking down

In **decomposition reactions** one single compound breaks down into two or more simpler chemicals. An example of this is the decomposition of zinc carbonate. This is represented by the equation:

$$ZnCO_3(s) \longrightarrow Zn(s) + CO_2(g)$$

5.5.6 Getting together

Often two elements combine in chemical reactions to form a compound. Such reactions are called **combination reactions**. The reaction of magnesium with oxygen is a spectacular example. Magnesium burns in air, producing a brilliant flash of white light. The equation for this combination reaction is:

$$\underset{\text{atoms}}{2Mg(s)} + \underset{\text{molecules}}{O_2(g)} \longrightarrow \underset{\text{molecules}}{2MgO(s)}$$

Notice that this combination reaction is also a combustion reaction. It is also an exothermic reaction because it transfers energy to the surroundings. (Endothermic reactions are chemical reactions that absorb energy from the surroundings.)

5.5.7 Redox reactions

In many chemical reactions, electrons are either completely or partially moved from one atom, ion or molecule to another. This process is known as **electron transfer**. Chemical reactions that involve electron transfer are called **redox reactions**. Redox reactions are extremely important in industry and in our everyday lives.

A redox reaction is really two reactions occurring simultaneously. In the electron transfer process, one reactant loses electrons and another gains electrons. Loss of electrons is known as **oxidation**. Gain of electrons is called **reduction**. Oxidation and reduction always occur together, thus the two words are combined to form the word *redox*, which is used to describe reactions where electrons are transferred.

The mnemonic **OIL RIG** may help you to remember these processes: oxidation is loss, reduction is gain.

Each of the corrosion, displacement, combustion and combination reactions described earlier are examples of redox reactions. Oxidation and reduction can be clearly seen in the reaction when zinc corrodes.

INVESTIGATION 5.2

Decomposing powder

AIM: To observe a decomposition reaction

Materials:

laboratory coat and safety glasses
zinc carbonate powder
spatula
Bunsen burner, heatproof mat and matches
large Pyrex test tube and test-tube rack
test-tube holder
electronic balance
marking pen
stereo microscope
Petri dish

CAUTION

Wear safety glasses and a laboratory coat.

Method and results

- Place two spatulas of zinc carbonate powder in the test tube. Weigh the test tube and record the mass.
- Mark the level of the powder in the test tube with the marking pen.
- Heat the test tube gently in a blue Bunsen burner flame for 5 to 10 minutes.
- While heating the test tube, hold a lit match at the mouth of the tube. Record your observations.
- Allow the test tube to cool down. Note any change in the level of powder and then reweigh the test tube. Record the mass.
- Place small amounts of zinc carbonate and the powder from the test tube in the Petri dish. Examine them using a stereo microscope. Record your observations.

CAUTION

Make sure the test tube is not pointing at anyone.

Discuss and explain

1. Which gas was given off during the reaction?
2. Explain any change that occurred in the mass.
3. Write word and formula equations for the reaction.

HOW ABOUT THAT!

In the early days of chemistry, oxidation was defined as the combination of a chemical with oxygen or as the removal of hydrogen from a compound. Reduction was defined as the opposite of oxidation; that is, the removal of oxygen or the combination of hydrogen with a chemical. Today we know that oxygen and hydrogen are not necessarily involved at all in a redox reaction. An example is when sodium metal and chlorine gas are produced by passing an electric current through molten sodium chloride.

$$\underset{\text{ions}}{\text{Sodium}} + \underset{\text{ions}}{\text{chloride}} \longrightarrow \underset{\text{metal}}{\text{sodium}} + \underset{\text{gas}}{\text{chloride}}$$

$$2Na^+(l) + 2Cl^-(l) \longrightarrow 2Na(s) + Cl_2(g)$$

In this redox reaction, sodium is reduced and chlorine is oxidised. Oxidation is now defined as the transfer of electrons from a reactant. In the above redox reaction, chlorine is oxidised and sodium is reduced. However, when sodium metal is produced, it is extremely reactive and will explode on contact with air or water as shown here.

Sodium metal exploding on water

Corrosion of zinc

In the corrosion of the layer of zinc used to protect iron from rusting, a transfer of electrons from the zinc atoms to the oxygen molecules occurs. This causes the formation of positive zinc ions and negative oxide ions. These oppositely charged ions attract and bond together to form the ionic compound zinc oxide.

$$2Zn^{2+} + 2O^{2-} \longrightarrow 2ZnO$$

$$2\,\text{zinc atoms} + 1\,\text{oxygen molecule} \longrightarrow 2Zn^{2+}\,\text{ions} + 2O^{2-}\,\text{ions}$$

In this reaction, zinc atoms lose electrons; thus zinc is oxidised. Oxygen molecules gain electrons; thus oxygen is reduced. Remember that oxidation and reduction always occur together.

Displacement of silver

The chemical equation for the displacement of the silver ion from silver nitrate by copper is:

$$\text{copper} + \text{silver nitrate} \longrightarrow \text{silver} + \text{copper nitrate}$$

$$\underset{\text{atom}}{Cu(s)} + \underset{\text{ions}}{2AgNO_3(aq)} \longrightarrow \underset{\text{atoms}}{2Ag(s)} + \underset{\text{ion}}{Cu(NO_3)_2(aq)}$$

In this reaction, electrons are transferred from the copper atoms to the silver ions. Silver ions (Ag^+) in the solution gain electrons to form atoms of solid silver. Thus, silver ions are reduced. Copper atoms ($Cu(s)$) lose electrons, forming copper ions ($Cu^{2+}(aq)$), which dissolve into a solution. The formation of copper ions changes the colour of the solution from colourless to blue. The copper atoms are oxidised. The nitrate ion is not involved in the electron transfer.

Combustion of methane

The chemical equation for the burning of methane in a gas jet is:

$$CH_4(g) + 2O_2(g) \longrightarrow CO_2(g) + 2H_2O(g)$$

In this redox reaction, electron transfer is not complete. The reactants are molecules and the products are also molecules. In each molecule, electrons are shared by the atoms. However, the oxygen atoms in the products attract the electrons more strongly than the carbon and hydrogen atoms. Therefore, the shared electrons spend more time close to the oxygen atoms. The electrons have been partially transferred to the oxygen atoms. Thus, oxygen is reduced and the carbon in methane is oxidised.

Combination of magnesium and oxygen

The chemical equation describing the combination of magnesium and oxygen as a result of burning is:

$$2Mg(s) + O_2(g) \longrightarrow 2MgO(s)$$

In this reaction, electrons are transferred from the atoms in the magnesium metal to the oxygen atoms in the oxygen molecule. This forms positive metal ions and negative oxide ions. These ions are attracted to each other because of their opposite charges and form the white ionic solid magnesium oxide. Magnesium, which loses electrons, is oxidised, and oxygen is reduced.

5.5.8 Light and shade

People who wear glasses often don't want to swap over to sunglasses when they go outside. **Photochromic** lenses solve the problem by darkening as the wearer moves from indoors into bright sunshine. They lighten again when the wearer moves back into an area of low light. Plastic photochromic glasses use organic material that darkens the lenses when exposed to ultraviolet light.

Glass photochromic glasses work due to the presence of silver chloride (AgCl) crystals in the glass. When in sunshine, ultraviolet light is absorbed by the silver chloride crystals and a redox reaction occurs. Electrons are transferred from the chloride ions to the silver ions according to the equation:

$$\underset{\text{ion}}{Ag^+} + \underset{\text{ion}}{Cl^-} \longrightarrow \underset{\text{atom}}{Ag} + \underset{\text{atom}}{Cl}$$

Silver particles then form in the glass, darkening the lens so that visible light is absorbed and reflected.

The fading of the dark glass is more complicated. The chlorine atoms are very reactive. To stop them reacting with the silver atoms and reversing the process too quickly, singly charged copper ions are dissolved in the molten glass during the manufacturing process. These ions react with the chlorine atoms to form chloride ions and doubly charged copper ions in the reaction:

$$Cu^+ + Cl \longrightarrow Cu^{2+} + Cl^-$$

When the glasses are no longer in the sunlight, the doubly charged copper ions accept an electron from the silver atom. The silver ion re-forms and the dark lens becomes light again:

$$Cu^{2+} + Ag \longrightarrow Cu^+ + Ag^+$$

5.5.9 Reactions with a zap!

The chemical reactions that produce electrical energy in electric cells (more commonly known as batteries) are redox reactions. In electric cells, electrons are transferred from one reactant to another through the wires that make up the electric circuit. This is very useful because the moving electrons can provide the energy to operate our appliances. Thus, chemical energy from the redox reaction is converted to electrical energy. The reactants in the cells are not in direct contact with each other. In an ordinary carbon battery or dry cell, the reactants are separated by a paste that allows the movement of electric charge. The electrons flow from one reactant at the negative electrode, through the electric circuit to the other reactant at the positive electrode. Chemical products are formed at both electrodes.

WHAT DOES IT MEAN?

The word photochromic comes from the Greek words *photo*, meaning 'light', and *khroma*, meaning 'colour'.

5.5 Exercises: Understanding and inquiring

To answer questions online and to receive **immediate feedback** and **sample responses** for every question, go to your learnON title at www.jacplus.com.au. *Note:* Question numbers may vary slightly.

Remember

1. Construct a table like the one below and use it to summarise each of the groups of reactions discussed in this subtopic. List one example of a reaction for each group.

Reaction type	Description	Example

2. What do all redox reactions have in common?
3. What is oxidation?
4. What is reduction?
5. Where does the word *redox* come from?
6. Consider the reaction:

$$2Zn(s) + O_2(g) \longrightarrow 2ZnO(s)$$

(a) From which reactant are the electrons being transferred?
(b) Which reactant are they transferred to?

7. Give an example of a redox reaction where electron transfer is not complete.

Think

8. (a) Refer to the tables in subtopic 5.2. Write a balanced equation using formulae for the following reactions:
   (i) copper metal + zinc sulfate solution ⟶ zinc metal + copper sulfate solution
   (ii) sodium metal + oxygen gas ⟶ solid sodium oxide
   (iii) carbon monoxide gas + oxygen gas ⟶ corbon dioxide gas
   (iv) hydrogen peroxide(H_2O_2) solution decomposes to form hydrogen gas and oxygen gas

   (b) State the type of each of the reactions in part (a).
9. Explain how it can be said that the reaction between magnesium and oxygen is four reactions in one: a combustion reaction, a combination reaction, a redox reaction and an exothermic reaction.

learnon RESOURCES — ONLINE ONLY

Complete this digital doc: Worksheet 5.5: Metal displacement (doc-19453)

Complete this digital doc: Worksheet 5.6: Corrosion and combustion (doc-19454)

5.6 Producing salts

5.6.1 Neutralisation reactions

Neutralisation is the name given to the chemical reaction in which an acid and a base react with each other to produce water. The other substance produced in a neutralisation reaction is called a **salt**. Many neutralisation reactions occur in water. These reactions are said to occur 'in solution'.

Your stomach contains hydrochloric acid, which helps to break up food for digestion. Too much acid, however, can be a problem. If your stomach produces too much acid, you may need to take an antacid such as milk of magnesia. This medicine has the solid base magnesium oxide (MgO) suspended in it. This base reacts with the hydrochloric acid in your stomach according to the equation:

$$\underset{\text{base}}{MgO(s)} + \underset{\text{acid}}{2HCl(aq)} \longrightarrow \underset{\text{salt}}{MgCl_2(aq)} + \underset{\text{water}}{H_2O(l)}$$

The products are the salt magnesium chloride and water. The salt contains the positive metal ion from the base and the negative non-metal ion from the acid.

The base sodium hydrogen carbonate, commonly known as bicarb, is a component of baking powder. It has the formula $NaHCO_3$ and contains the hydrogen carbonate ion HCO_3^-. When bases containing this ion react with acids, carbon dioxide gas is produced as well as salt and water. When hydrochloric acid and bicarb are mixed together, the following reaction takes place:

$$\underset{\text{base}}{NaHCO_2(s)} + \underset{\text{acid}}{HCl(aq)} \longrightarrow \underset{\text{salt}}{NaCl(aq)} + \underset{\text{carbon dioxide}}{CO_2(g)} + \underset{\text{water}}{H_2O(l)}$$

In both of the reactions mentioned, the salts formed were metal chlorides, because they contained the chloride ion (Cl^-) from the hydrochloric acid. Neutralisation reactions between many different acids and bases are possible; therefore, it is possible to produce many different salts. Some of these reactions are summarised in the table below.

Base	Acid	Negative ion present in salt	Salt
Sodium hydroxide	**Sulfuric** acid	Sulfate SO_4^{2-}	Sodium **sulfate**
Magnesium oxide	Hydro**chloric** acid	Chloride Cl^-	Magnesium **chloride**
Sodium oxide	**Acetic** acid	Acetate CH_3COO^-	Sodium **acetate**
Copper(II) oxide	**Nitric** acid	Nitrate NO_3^-	Copper(II) **nitrate**

HOW ABOUT THAT!

Many salts are brightly coloured and can be highly poisonous — not at all suitable for sprinkling on your fish and chips! Salts containing copper ions are usually blue, those containing nickel are pale green, those containing iron can be green or orange, and cobalt salts are pink.

INVESTIGATION 5.3

Pass the salt!

AIM: To identify the products of a reaction between an acid and a base

Materials:

safety glasses and laboratory coat
50 mL burette
retort stand, bosshead and clamp
tripod and gauze mat
Bunsen burner, heatproof mat and matches
20 mL pipette
100 mL conical flask
pipette bulb
white tile
dropping bottle of phenolphthalein indicator
wire shaped into a loop with a handle
small funnel
1M hydrochloric acid solution
1M sodium hydroxide solution
evaporating dish
silver nitrate solution in a dropping bottle
sample of sodium chloride
test tube

Method and results

- Rinse the burette with the hydrochloric acid solution and then, using the funnel, fill the burette with the hydrochloric acid solution.
- Rinse the pipette with sodium hydroxide solution using the pipette bulb.
- Set up the equipment as shown in the diagram bottom right. Use the pipette and bulb to transfer 20 mL of the sodium hydroxide solution into the conical flask.
- Add a few drops of phenolphthalein indicator to the sodium hydroxide.
- Add the acid from the burette carefully until the pink colour of the indicator disappears. The colour change indicates that the neutralisation reaction is complete.
- Pour the contents of the flask into an evaporating dish. Heat the dish with the Bunsen burner and gently evaporate the water. Be careful — spattering may occur.
- When the water has nearly evaporated, turn off the Bunsen burner and allow the dish to cool and the remaining water to evaporate without further heating.
- Test the white crystals for the presence of sodium ions by placing a few crystals on a wire loop and heating in a Bunsen burner flame. Compare this flame colour with a known sample of sodium chloride.

1. Record your observations.

- Test for the presence of chloride ions by dissolving a few crystals in half a test tube of water and adding a few drops of silver nitrate. A white cloudiness indicates that chloride ions are present.

2. Record your observations.

CAUTION

Wear safety glasses and a laboratory coat.

CAUTION

Never pipette using your mouth.

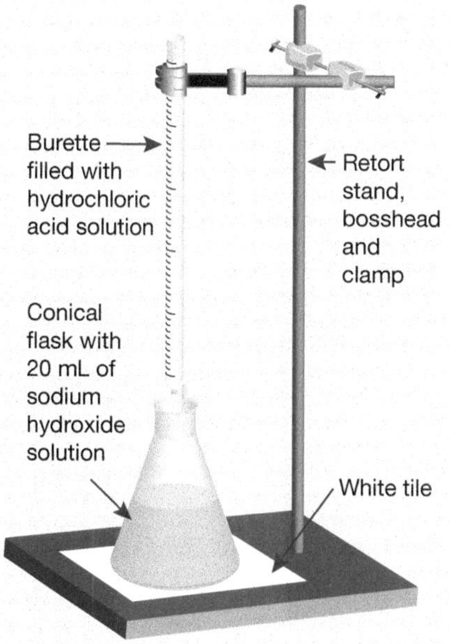

Discuss and explain

3. Comment on the information that the flame and silver nitrate tests provided. What conclusion can you draw?
4. Write a word equation for the neutralisation reaction.
5. Write a balanced equation, using formulae, for the neutralisation reaction.
6. Design a test to show that water was the other product of the reaction.

5.6 Exercises: Understanding and inquiring

To answer questions online and to receive **immediate feedback** and **sample responses** for every question, go to your learnON title at www.jacplus.com.au. *Note:* Question numbers may vary slightly.

Remember

1. What is a salt?
2. What are the products of the reaction between an acid and a base that contains the hydrogen carbonate ion?

Think

Use the tables in subtopic 5.2 and the table at right to answer the following questions.

3. Using formulae, write equations for the following reactions.
   (a) Solid sodium hydrogen carbonate and sulfuric acid react to form a sodium sulfate solution, carbon dioxide and water.
   (b) Solid potassium hydroxide and hydrochloric acid react to form a solution of potassium chloride and water.
   (c) Solid copper oxide reacts with sulfuric acid to form a solution of copper sulfate and water.
4. Name the salts that form from the reaction between:
   (a) magnesium hydroxide and hydrochloric acid
   (b) potassium hydroxide and acetic acid
   (c) sodium carbonate and sulfuric acid.

Some common laboratory bases

Base	Formula
Sodium hydroxide	$NaOH$
Copper hydroxide	$Cu(OH)_2$
Potassium hydroxide	KOH
Magnesium hydroxide	$Mg(OH)_2$
Sodium carbonate	Na_2CO_3
Sodium bicarbonate	$NaHCO_3$

5.7 Fuelling our lifestyle

Science as a human endeavour

5.7.1 It begins with the sun

No cars, no streetlights, no heating in cold weather, no television … We would not have these 'necessities' if we didn't have access to the chemical energy stored in **fossil fuels**.

The term *fossil fuels* refers to coal, natural gas and oil, all of which were formed from decaying plants or animals over tens or hundreds of millions of years.

The formation of fossil fuels begins with a chemical reaction called **photosynthesis**. This reaction requires energy from the sun and a pigment in plants called **chlorophyll**. Photosynthesis is described by the chemical equation below:

The flame burning on a gas stove marks the end of a long journey for a chemical called methane.

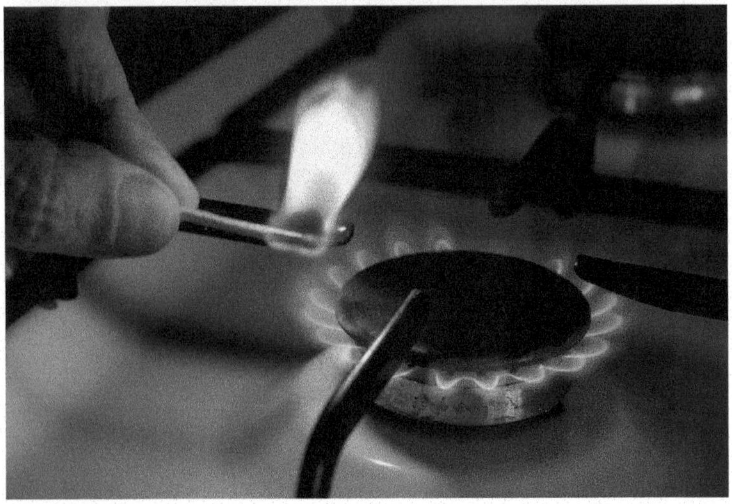

$$6CO_2(g) + 6H_2O(l) \xrightarrow[\text{chlorophyll}]{\text{light energy}} C_6H_{12}O_6(s) + 6O_2(g)$$

Animals eat plants (or other animals that eat plants), so all energy in food consumed by animals begins with photosynthesis.

During the formation of fossil fuels, dead plants and animals decay, but the chemical energy stored within them remains in the fossil fuels. When fossil fuels are burnt, that chemical energy is converted to other forms of energy in combustion reactions (see subtopic 5.5).

5.7.2 Gas fuel

The flame you see on a gas stove results from the burning of natural gas. This gas was formed millions of years ago when the remains of tiny marine and freshwater plants and animals were transformed into natural gas that became trapped in rock. When natural gas is burned to heat water, your home or to cook food, methane reacts with oxygen to produce carbon dioxide and water vapour. During the reaction, the chemical energy stored in the methane molecules is transformed, heating the surrounding air, water or food.

$$\text{methane} + \text{oxygen} \longrightarrow \text{carbon dioxide} + \text{water vapour}$$
$$CH_4(g) + 2O_2(g) \longrightarrow CO_2(g) + 2\,H_2O(g)$$

5.7.3 Solid fuel

Another fuel that is formed over a period of millions of years is **coal**. Coal is formed from the remains of plants that were buried in sediments between 20 million and 300 million years ago. As coal (which consists mainly of carbon) is burned, it reacts with oxygen to produce carbon dioxide gas.

$$\text{carbon} + \text{oxygen} \longrightarrow \text{carbon dioxide}$$
$$C(s) + O_2(g) \longrightarrow CO_2(g)$$

In Australia, coal is used mainly in the generation of electricity. In power stations, the energy released in the chemical reaction is used to change water into steam. The rapidly moving steam turns turbines that generate electricity.

Coal is also used to make other fuels such as coal gas and to make methanol and coke. Coke is used in the refining of steel.

5.7.4 Liquid fuel

Crude oil is a sticky, dark, smelly liquid. Most of the chemicals in it are hydrocarbons — compounds of hydrogen and carbon atoms. Crude oil was formed from the remains of marine plants and animals that died over 200 million years ago.

Hydrocarbons

Hydrocarbon	No. of carbon atoms	Structure	Boiling point (°C)
Methane	1	CH_4	−164
Ethane	2	CH_3–CH_3	−89
Propane	3	CH_3–CH_2–CH_3	−42
Butane	4	CH_3–CH_2–CH_2–CH_3	0
Pentane	5	CH_3–CH_2–CH_2–CH_2–CH_3	36
Hexane	6	CH_3–CH_2–CH_2–CH_2–CH_2–CH_3	69
Heptane	7	CH_3–CH_2–CH_2–CH_2–CH_2–CH_2–CH_3	98
Octane	8	CH_3–CH_2–CH_2–CH_2–CH_2–CH_2–CH_2–CH_3	126
Nonane	9	CH_3–CH_2–CH_2–CH_2–CH_2–CH_2–CH_2–CH_2–CH_3	151
Decane	10	CH_3–CH_2–CH_2–CH_2–CH_2–CH_2–CH_2–CH_2–CH_2–CH_3	174

Crude oil is a mixture of chemicals that includes diesel fuel, petrol, aviation fuel, tar, kerosene and many more hydrocarbons. So many chemicals make up crude oil that it has to be separated into the different hydrocarbons before it can be useful.

The table of hydrocarbons on the previous page shows that the longer the molecule becomes, the higher its boiling point. Chemists use this property to separate the components of crude oil in a process called **fractional distillation**.

When crude oil is heated to 370 °C, most of its components are changed into a gaseous state. In fractional distillation, this gaseous crude oil is passed into the bottom of a fractionating tower, which becomes gradually cooler further up the tower. The hot vapours cool as they rise up the fractionating tower. The heaviest hydrocarbons condense back to a liquid near the base of the tower. The other hydrocarbons, still gases, rise through the tower until they cool off enough to condense back to a liquid (at a temperature just below their boiling point). The different hydrocarbons are separated at different places up the tower according to their different boiling points. Each **fraction** is piped away for processing.

The fractional distillation of crude oil

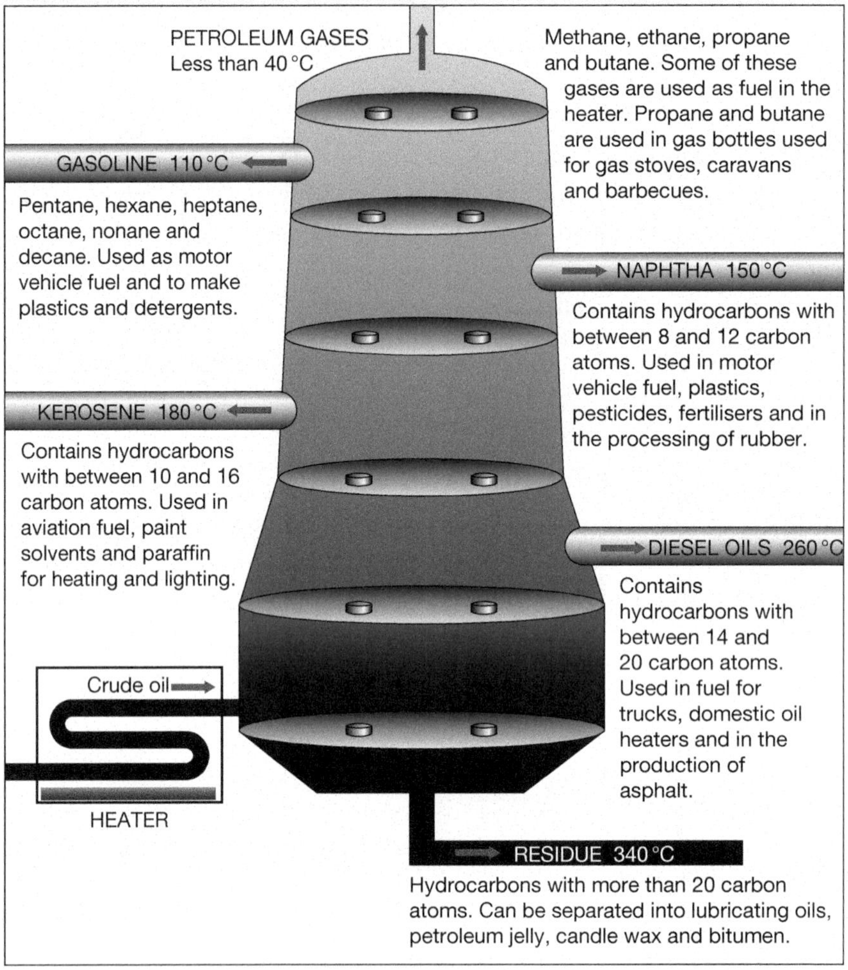

PETROLEUM GASES
Less than 40 °C
Methane, ethane, propane and butane. Some of these gases are used as fuel in the heater. Propane and butane are used in gas bottles used for gas stoves, caravans and barbecues.
GASOLINE 110 °C
Pentane, hexane, heptane, octane, nonane and decane. Used as motor vehicle fuel and to make plastics and detergents.
NAPHTHA 150 °C
Contains hydrocarbons with between 8 and 12 carbon atoms. Used in motor vehicle fuel, plastics, pesticides, fertilisers and in the processing of rubber.
KEROSENE 180 °C
Contains hydrocarbons with between 10 and 16 carbon atoms. Used in aviation fuel, paint solvents and paraffin for heating and lighting.
DIESEL OILS 260 °C
Contains hydrocarbons with between 14 and 20 carbon atoms. Used in fuel for trucks, domestic oil heaters and in the production of asphalt.
Crude oil
HEATER
RESIDUE 340 °C
Hydrocarbons with more than 20 carbon atoms. Can be separated into lubricating oils, petroleum jelly, candle wax and bitumen.

5.7.5 Alternatives to fossil fuels

The world's reserves of fossil fuels are limited, and eventually they will run out. At present, in Australia, we obtain 95 per cent of our energy needs from fossil fuels. It is important that as well as conserving energy, we search for alternatives to fossil fuels.

Biofuels

Biofuels are fuels made from **biomass**. Biomass is the name given to plant and animal tissue. Living things, including bacteria, animals and plants, are all energy converters. When they convert energy, some waste is produced. The chemical energy in waste and other chemicals made by living things can be converted into more useful forms of energy.

This biogas facility at Melbourne Water's Western Treatment Plant at Werribee is used to generate electricity.

These bubbles of methane are trapped in a frozen pond. The methane gas is produced by rotting organic matter that has sunk to the bottom of the pond.

It's not really waste

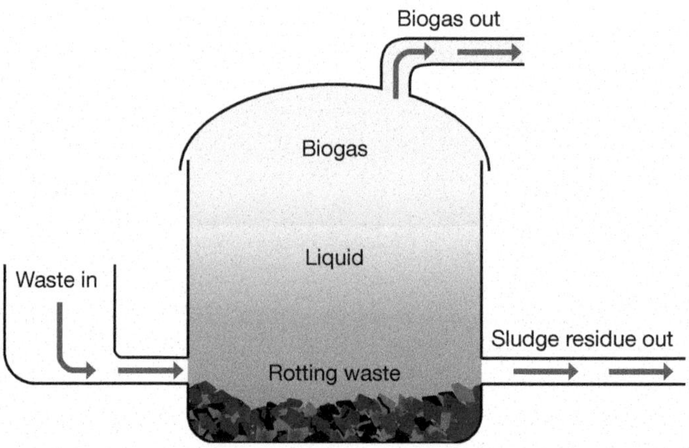

A simple household biogas digester

Imagine producing fuel from human sewage! It would not only help solve the problem of disposing of human waste in big cities, but also reduce the demand for fossil fuels. You might be surprised to know that sewage is already being used to make fuel.

The fuel produced from biomass can be a solid (like the wood burned in an open fireplace or barbecue), a liquid (like ethanol) or a gas. The gas produced from biomass is known as **biogas**.

Plant and animal wastes can be converted to biogas in a biogas digester. Biogas is mainly methane and carbon dioxide. The methane produced can then be used for heating and to power homes and farms.

Biogas is the product of the chemical reaction that takes place when wastes rot in the absence of oxygen. The solids left over from the production of biogas can be used as fertiliser or combined with compost and sand to make organic soils.

Sewage treatment plants, and farms on which animals graze, are ideal locations for biogas digesters because of the availability of animal waste. Farms also provide a use for the leftover solids from biogas digesters.

China leads the way

In China, where about 70 per cent of the population lives on farms or in villages, more than 12 million households use biogas digesters to provide energy for lighting and cooking. Human and animal waste is fed into the digester. Many digesters are directly connected to toilets and pigsties. The waste is allowed to rot, producing a gas that is about 60 per cent methane, which bubbles to the top of the digester. The dark and surprisingly odourless sludge residue is drained from the digester and used as fertiliser.

China also has more than 1500 larger biogas plants that produce gas for heating and the generation of electricity.

Alcohol — a petrol alternative

Alcohol produced from fermented sugar or corn is a biofuel that can be used to fuel motor vehicles. The alcohol produced from the fermentation of sugar is ethyl alcohol, commonly called **ethanol**.

$$\text{sugar} \longrightarrow \text{ethanol} + \text{carbon dioxide}$$
$$C_6H_{12}O_6 \longrightarrow 2C_2H_5OH + 2CO_2$$

Ethanol as a fuel source has been most successfully adopted in Brazil, where there is a large source of sugar cane and conditions are suitable for fermenting and distilling the sugar cane. In Brazil, all passenger and light commercial vehicles are powered by ethanol or a blend of petrol and ethanol. New cars manufactured in Brazil are now designed to run on fuels that are made up of anything between 20 per cent and 100 per cent ethanol. They are known as 'flexible-fuel' or 'flex' vehicles.

Motor vehicles account for about two-thirds of the petroleum used in Australia. As reserves of petroleum become more scarce and expensive, ethanol is becoming a more desirable alternative. The use of biofuels like ethanol also helps to improve air quality and may reduce the production of greenhouse gases.

A blend of ethanol and petrol is now available in Australia as E10, which is 10 per cent ethanol and 90 per cent unleaded petrol. Most new cars manufactured in Australia are able to run on E10.

HOW ABOUT THAT!

When cows burp or pass wind, they release methane gas. In fact, cows are responsible for up to about 20 per cent of the methane in the atmosphere. Imagine if the methane could be used as fuel!

5.7 Exercises: Understanding and inquiring

To answer questions online and to receive **immediate feedback** and **sample responses** for every question, go to your learnON title at www.jacplus.com.au. *Note:* Question numbers may vary slightly.

Analyse and evaluate

1. Use the information in the table of hydrocarbons to draw a graph showing the relationship between the number of carbon atoms in a hydrocarbon and the boiling point of the hydrocarbon. Explain how this relationship is useful in the refining of crude oil.

Remember

2. Explain how all fossil fuels are formed.
3. Identify the type of chemical reaction in which fossil fuels are converted to other forms of energy.
4. Write a word equation for the burning of methane.
5. What is a hydrocarbon?
6. Which property of the different hydrocarbons in crude oil allowed chemists to develop the process of fractional distillation?
7. List six components of crude oil and state one use for each component.
8. Distinguish between biomass and biogas.
9. Describe the process of producing biogas.
10. Which two gases are the main components of biogas?

Think

11. *Fossil fuels can be referred to as stores of solar energy.* Explain the meaning of this statement.
12. Which has the higher boiling point: diesel oil or kerosene?
13. Which type of chemical reaction is used to produce ethanol by fermentation?

Create

14. Write an article for *Farmers Weekly* to convince farmers, who usually put cow manure from their milking sheds onto their farms as fertiliser, to use cow manure to make biogas.

Discuss and report

15. In a group, discuss and produce a report on each of the following questions.
    (a) Why should the use of biogas as a fuel for heating be encouraged?
    (b) Wood is an example of a biofuel. What are the disadvantages in the use of wood as a fuel?
    (c) Suggest a range of alternatives to fossil fuels that could be used to supply our energy needs in Australia. List the advantages and disadvantages of each alternative.

Investigate

16. Carbon capture is the process of separating carbon dioxide from the emission of fossil-fuel-fired plants and biogas digesters so that it can be stored in a suitable location instead of adding to the carbon dioxide already in the atmosphere. Research and report on current methods of:
    (a) capturing carbon dioxide
    (b) storing carbon dioxide.

5.8 Rates of reaction

5.8.1 Reaction rate

The rate at which chemical reactions occur varies. Some reactions occur within a fraction of a second, while others may take days or even years. Sometimes it is necessary or convenient to speed up a chemical reaction.

The speed, or **rate**, of a reaction can be very important. We need to know how long food will take to cook or how long it will take for a medicine such as an antacid to make us feel better. Controlling the rate at which reactions occur is therefore of great interest to scientists. Increasing the temperature, the surface area of solid reactants, the concentration of the reactants or, in some cases, the exposure to light can increase the rate of chemical reactions.

Change the amount

The burning of wood is a combustion reaction. The chemical word equation for this reaction is:

$$\text{wood} + \text{oxygen} \xrightarrow{\text{heat}} \text{water} + \text{carbon dioxide}$$

When you're out camping you might want to boil your billy quickly or maybe just get warm by the fire. You could speed up the rate of burning by fanning the flames of the fire. This increases the amount of oxygen reaching the wood. This is an example of changing the amount, or concentration, of a reactant.

Change the temperature

Many types of organism are found in food. Chemical reactions that spoil food take place in microbes. Refrigeration cools the food and the microbes. This makes the chemical reactions inside the microbes slow down and the food keeps for longer.

Heating things up can make a reaction happen more quickly. Think of frying an egg. If you do it on a low flame, it takes longer than if you use a high flame. Heating makes the particles in the reactants move faster and collide more often. This helps to speed up the reaction.

Change the surface area

Bath bombs are sold as solid balls. When they are added to water, the chemicals inside them begin to dissolve. The ball slowly disappears. But what if the same bath bomb was crushed into smaller pieces? A much larger surfce area comes in contact with the water, and the bath bomb dissolves much more quickly.

learnon RESOURCES — ONLINE ONLY

Try out this interactivity: Reaction rates (int-0230)

5.8.2 Catalysed reactions

Another way to increase the rate of a reaction is to use a **catalyst**. Catalysts are not changed by the reaction. There is always as much catalyst present at the end of a reaction as there was at the start. Catalysts work by helping bonds to break more easily; therefore, the reactants need less energy to react and the reaction is faster. A catalyst can be recovered and used again and again. We all make use of catalysts every day. Cars have catalytic converters; contact lenses are cleaned using a catalysed chemical reaction; and there are catalysts in the food you eat every day. There are also thousands of catalysts in your body without which you could not live. These biological catalysts are called **enzymes**.

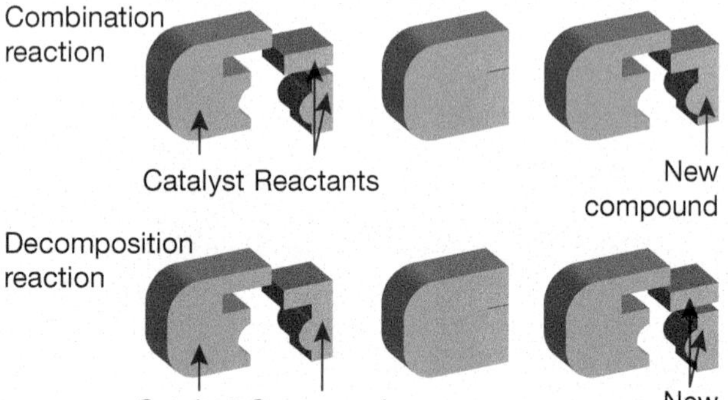

Catalysts in industry

Industry makes use of many catalysts. For example:

- iron and iron oxide are used to catalyse the production of ammonia gas. Ammonia is used to make fertilisers and explosives.
- vanadium oxide (V_2O_5) is used in the production of sulfuric acid. One important reaction in this process, between sulfur dioxide gas and oxygen, has a very slow rate at room temperature. However, it proceeds rapidly at 450 °C in the presence of a vanadium oxide catalyst according to the equation:

$$2SO_2(g) + O_2(g) \xrightarrow[450^\circ C]{V_2O_5} 2SO_3(g)$$

Note that the catalyst is written above the arrow and not on the side of the reactants. It is not changed as the reaction takes place.

- crystalline substances made of aluminium, silicon and oxygen called **zeolites** are used to 'crack' (break up) the large molecules in crude oil to form the smaller molecules, such as octane, found in petrol.

Everyday catalysts

In the confined space of the internal combustion engine, the fuel does not completely react with oxygen. As a result, carbon monoxide (CO), a highly poisonous gas, is produced. Nitrogen oxides are other harmful gases produced by car engines. In order to reduce the amount of pollution from these gases, cars are fitted with catalytic converters as part of the exhaust system. These converters have a honeycombed surface that is coated with the metals platinum and rhodium, and with aluminium oxide. At the catalyst surface, the nitrogen oxides are converted to less harmful gases and the carbon monoxide is reacted with more oxygen to form carbon dioxide according to the equation:

$$2CO_2(g) + O_2(g) \longrightarrow 2CO_2(g)$$

Catalysts can also help clean your contact lenses. One cleaning product makes use of a platinum catalyst. A solution of hydrogen peroxide (H_2O_2) is poured into a small container that contains a platinum-coated disc. The platinum causes the peroxide to decompose according to the reaction:

$$2H_2O_2(aq) \xrightarrow{Pt} 2H_2O(l) + O_2(g)$$

Any microbes not tolerant to oxygen on the contact lenses are killed by the oxygen released.

Catalytic converters have a large surface area and are coated with a metal catalyst.

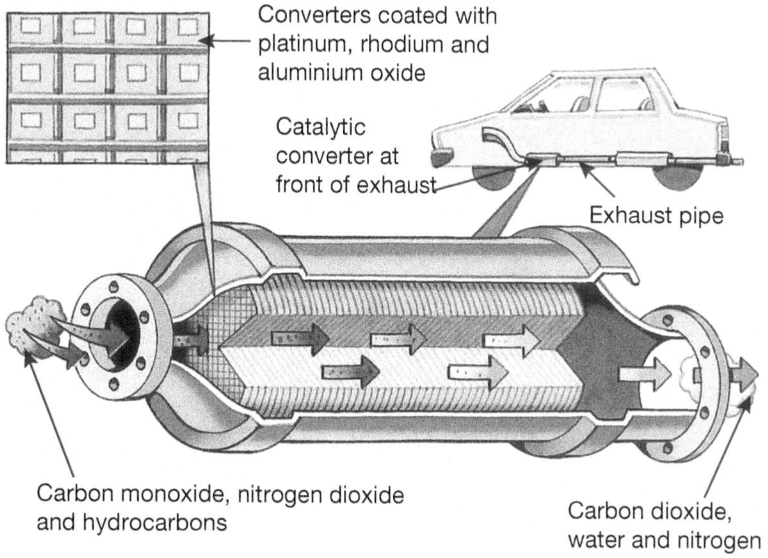

The honeycombed surface of catalytic converters helps to maximise their surface area and help convert harmful gases into harmless ones.

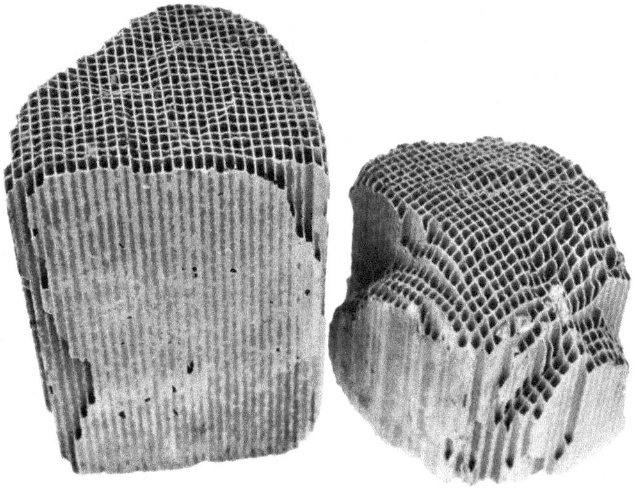

Catalysts in living things

Almost every one of the chemical reactions that take place in your body is controlled by an enzyme. Enzymes are large protein molecules essential for digesting food, breaking down toxic waste products, and numerous other chemical processes that keep you alive and healthy. The enzyme **amylase**, which is present in your saliva, is involved in the breakdown of starch into sugar.

Your liver contains an enzyme called **catalase**. Catalase speeds up the breakdown of hydrogen peroxide, a toxic waste product produced in your cells.

Enzymes are also used to make bread, cheese, vinegar and many other food products.

Enzymes are used to make many food products, including bread, cheese and vinegar.

INVESTIGATION 5.4

A liver catalyst

AIM: To observe the effect of a catalyst on a decomposition reaction

Materials:

heatproof mat
2 test tubes and test-tube rack
20% hydrogen peroxide solution
spatula
fresh liver
safety glasses
mortar and pestle

Method and results

- Pour hydrogen peroxide to a depth of 3 cm into the test tubes. Label the test tubes 1 and 2.
- Grind a small piece of liver in the mortar and pestle. Add liver to test tube 1 only.

1. Record your observations.

Discuss and explain

2. What effect did the liver have on the breakdown of hydrogen peroxide?
3. What evidence is there to suggest that a chemical reaction has taken place?
4. Suggest a reason why the liver was ground up before it was placed in the hydrogen peroxide solution.
5. What is the function of test tube 2?

5.8 Exercises: Understanding and inquiring

To answer questions online and to receive **immediate feedback** and **sample responses** for every question, go to your learnON title at www.jacplus.com.au. *Note:* Question numbers may vary slightly.

Remember

1. Explain what a catalyst does and give three examples.
2. Why are catalysts important to industry?
3. What are enzymes?

Think

4. Why do catalytic converters have a honeycombed surface?
5. Give one reason why the enzyme amylase is added to bread.

Investigate

6. Find out why leaded petrol must not be used in cars fitted with catalytic converters.
7. Research the production of cheese. Find out the enzymes used to produce a range of cheeses. Present your findings as a poster or wall chart.
8. Several diseases are caused by the body's failure to produce a particular enzyme. Phenylketonuria (PKU) and galactosemia are two such diseases. Research one of these diseases and present a report about their causes and treatment.

learnon RESOURCES — ONLINE ONLY

Complete this digital doc: Worksheet 5.7: Speeding up reactions (doc-19457)

5.9 Form and function: Polymers

5.9.1 Plastics

Which material is strong, light in weight and cheap to make, comes in a huge range of colours and can be moulded into any shape?

It could only be one of the **synthetic** (manufactured) materials we know as **plastics**. All plastics are products of chemical reactions. They are used to manufacture food containers and packaging, ballpoint pens, plumbing materials, car parts, rubbish bins, cling films such as GLAD®Wrap and a multitude of other items.

WHAT DOES IT MEAN?

The word *plastic* comes from the Greek word *plastikos*, meaning 'able to be moulded'.

5.9.2 Monomers and polymers

All of the synthetic materials we call plastics are **polymers**. However, not all polymers are synthetic. Cotton, wool, leather and rubber are examples of natural polymers.

Polymers are very large molecules that consist of many repeating units called **monomers**. Monomers are small molecules and most contain the element carbon. Polymer molecules may, therefore, contain thousands of carbon atoms. The other elements commonly found in monomers and polymers include hydrogen, oxygen, chlorine, fluorine and nitrogen.

A polymer is made up of smaller units called monomers bonded together.

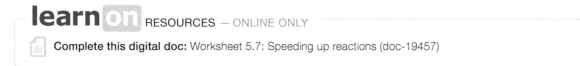

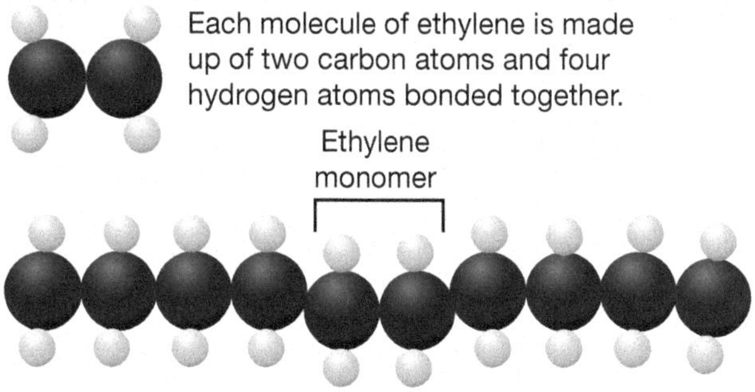
Each molecule of ethylene is made up of two carbon atoms and four hydrogen atoms bonded together.
Ethylene monomer

The polymer polyethylene is made of thousands of ethylene monomers all bonded together in long chains.

The chemical reactions that occur when polymers form can be modelled using plastic beads or blocks that click together to form a long chain. Each plastic bead or block represents a single monomer molecule. The long chain, which may contain thousands of monomers, represents a polymer molecule.

The prefix *poly* (meaning 'many') is often used when naming polymers. For example, polyvinyl acetate (PVA) is a polymer made from the monomer known as vinyl acetate.

Synthetic polymers

Polyvinyl chloride, also known as PVC, is formed from the monomer vinyl chloride. PVC is light, rigid and doesn't corrode, making it ideal for use in drainage and sewerage pipes.

Polyethylene is formed from the monomer ethylene. It is light, tough and resistant to most acids and bases. Polyethylene is used to make plastic bags, soft-drink bottles, buckets, cling wrap and many other household products.

Nylon is one of a group of polymers known as polyamides, which are formed from monomers joined by amide bonds. It is used to make fabrics for clothing, ropes, guitar strings, machine parts and much more.

PVC pipes, polyethylene bottle and nylon rope

5.9.3 Co-polymers

Some polymers form when identical monomer units link together. Teflon, which is used as a non-stick coating on pans and baking trays, is a polymer of this type. Other polymers, such as nylon and synthetic rubber, are formed when two different monomers alternate in the chain, creating a **co-polymer**. The chemical reactions that join monomers together are known as **polymerisation** (polymer-forming) reactions.

5.9.4 Versatile polymers

A plastic fruit juice container and a toilet seat are both made from polymers, or plastics. Some plastics are flexible and soften when they are heated. They can be moulded easily into a variety of useful products such as milk and fruit juice containers, rubbish bins, spectacle lenses, electrical insulation and laundry baskets. Other plastics are quite rigid and do not soften when heated. These plastics are used to make items such as toilet seats, electrical switches, bench tops, outdoor furniture, lampholders and other products that require strength and rigidity.

A tetrafluoroethene molecule. Tetrafluoroethene is the monomer from which teflon is formed.

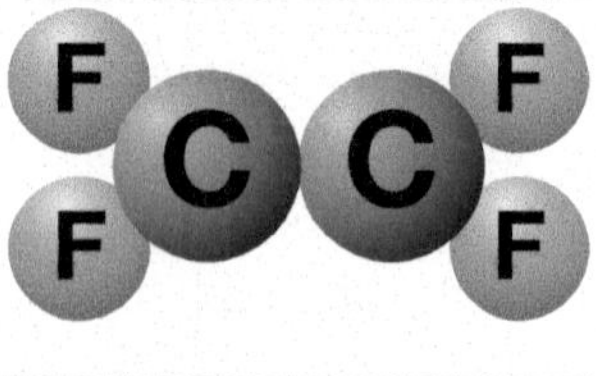

HOW ABOUT THAT!

Australia was the first country in the world to use only plastic notes. These notes are more difficult to counterfeit and last much longer than the old paper notes. They can also be recycled into other products.

5.9.5 Two of a kind

Thermoplastic polymers soften when they are heated. They melt easily and can be moulded into useful products when hot. Polythene and PVC are examples of thermoplastic or thermosoftening polymers. Polythene (also called polyethylene or, more correctly, polyethene) is a soft plastic used to make cling film, squeeze bottles, milk crates and many other useful items. PVC has many different uses, for example as shoe soles, mouthguards, drainpipes, floor tiles and packaging. The chains of monomer molecules in a thermoplastic polymer are able to slide past each other when the polymer is heated, allowing the plastic to soften and melt.

Thermosetting polymers do not soften when they are heated, but **char** (blacken) instead. They are hard, rigid and sometimes brittle. Bakelite, which is used to make electrical switches, doorhandles and lampholders, is a thermosetting polymer. Other thermosetting polymers such as melamine are used to make laminates for benchtops. In thermosetting polymers, the chains of monomer molecules are locked together firmly by chemical bonds between the chains, known as **crosslinks**. Strong heating can break down their structure, leaving the black element carbon.

Structure of a thermosetting polymer. The chains of monomer molecules are locked firmly together. When the polymer is heated it does not melt, but eventually breaks down (decomposes).

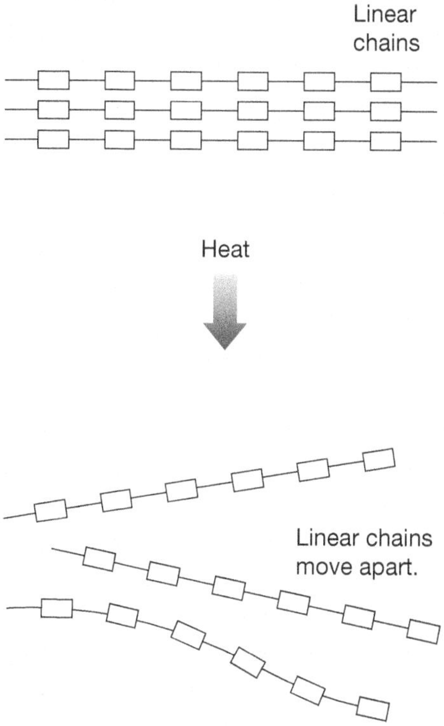

Structure of a thermoplastic polymer. The chains of monomer molecules are able to slide past one another when the polymer is heated.

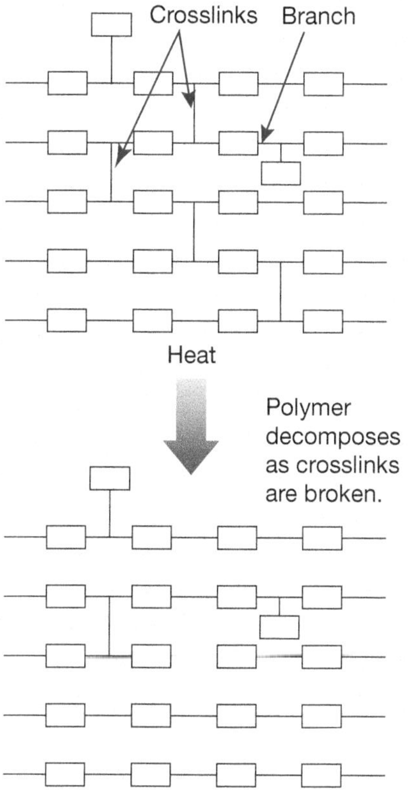

Lego blocks are produced using a thermoplastic polymer that is kept molten at 235 °C and then moulded under pressure. It cools to form a solid in about 15 minutes.

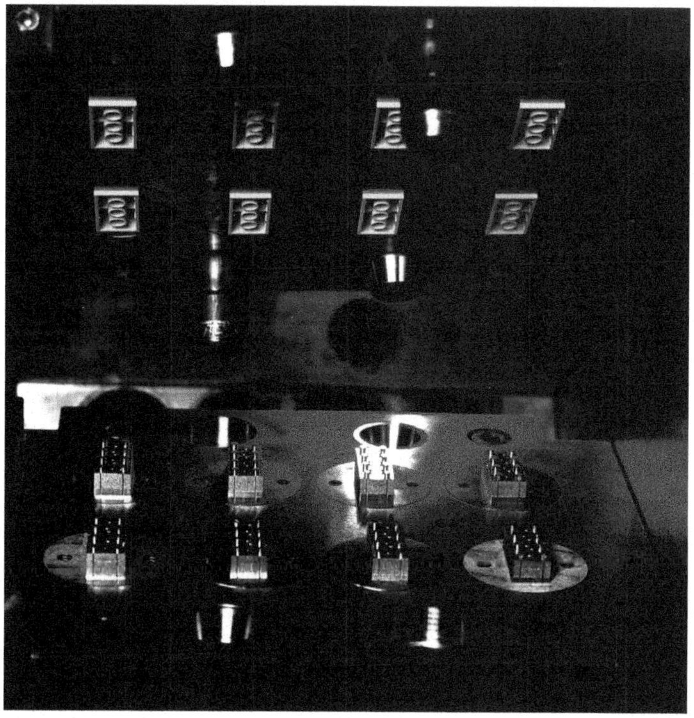

5.9 Exercises: Understanding and inquiring

To answer questions online and to receive **immediate feedback** and **sample responses** for every question, go to your learnON title at www.jacplus.com.au. *Note:* Question numbers may vary slightly.

Remember

1. What is the meaning of the word *plastic*?
2. What is a polymer?
3. Which element is contained in most monomers and polymers?
4. What do thermosetting and thermoplastic polymers have in common? In what ways are they different?
5. Classify the following polymers as thermosetting or thermoplastic: bakelite, PVC, polyethene, melamine, polyethylene terephthalate (PET).
6. What is a crosslink?
7. What is another word that means the same as thermoplastic?

Think

8. Name the polymers formed from the following monomers.
   (a) Ethene
   (b) Styrene
   (c) Propene
9. Explain why thermosetting polymers do not melt as easily as thermoplastic polymers.
10. Would you use a thermosetting or thermoplastic polymer to make the following? Give a reason for your choice in each case.
    (a) Bucket
    (b) Doorhandle
    (c) Saucepan handle
    (d) Toothbrush

Imagine

11. Imagine if all the objects made from plastics suddenly disappeared. How would your life be changed? Write an imaginative piece describing this. You may wish to talk to someone over the age of 70 about life before plastics.

Investigate

12. Design an experiment to compare the strength of plastic supermarket bags with the strength of other types of bags. Ensure your tests are fair.
13. Thermosoftening plastics are usually sold to manufacturers in the form of powders or small granules. Find out about the following ways of moulding the granules into useful products.
    (a) Vacuum forming
    (b) Calendering
    (c) Blow moulding
14. Most plastics undergo photodegradation. What does this mean?

5.10 A cool light

5.10.1 Light from a chemical reaction

Night golf can be played using golf balls that glow in the dark. Glow necklaces and light sticks are glowing plastic tubes that are popular in the evening at outdoor events and amusement parks. Light sticks can also be included in survival kits as a light source. Where does the light come from in these glowing devices? The answer lies in **chemiluminescence** — the production of light from a chemical reaction.

A light stick consists of two distinct parts: an outer plastic tube and an inner glass vial. The outer plastic tube is sealed and contains a solution of a chemical called an ester and a **fluorescent** dye. The inner glass vial is thin and breakable. It contains a solution of hydrogen peroxide. When the light stick is bent, the inner glass vial breaks, causing the two solutions to mix. The chemical reaction (a redox reaction) between the two solutions produces the light.

The chemical reaction between the hydrogen peroxide solution and the ester solution releases energy that is transferred to the fluorescent dye molecules. The excited dye molecules give off their excess energy as light without any noticeable heat. That is why the light is referred to as cool light.

How a glow stick works

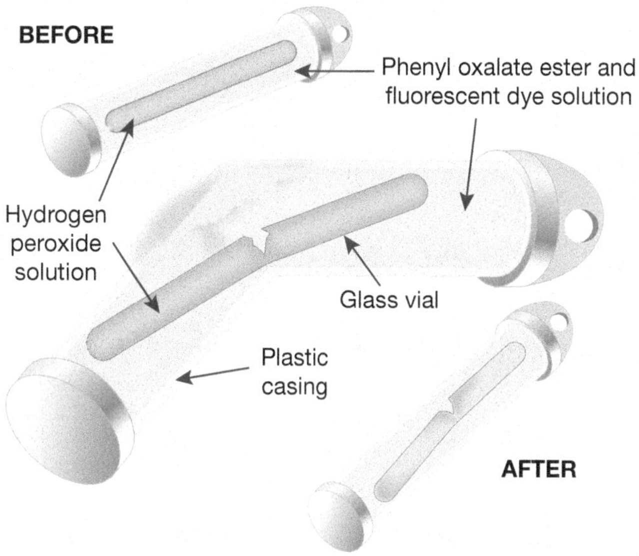

5.10.2 Light from living things

Some living things can produce light in a process called **bioluminescence**. One of the most common examples of bioluminescence is seen in **fireflies**, insects of the family *Lampyridae*. The abdomens of fireflies glow during the mating season to attract potential mates. This type of light is also a form of cool light, produced by chemical reactions in the cells of the firefly. During these chemical reactions, energy is transferred to luminescent molecules, which, like the molecules of dye in the light stick, become excited and emit energy as light. The chemical reactions in the firefly are controlled by special enzymes called **luciferases**. The luciferase enzymes are produced in the cells of the firefly's abdomen and allow the light-producing chemical reactions to occur.

Light sticks can be used for safety and for fun.

Many living things use bioluminescence to light their dark surroundings, attract their prey and camouflage themselves. Organisms such as bacteria, protozoa, fungi, sponges, crustaceans, insects, fish, squid, jellyfish and simple plants have been found to be bioluminescent.

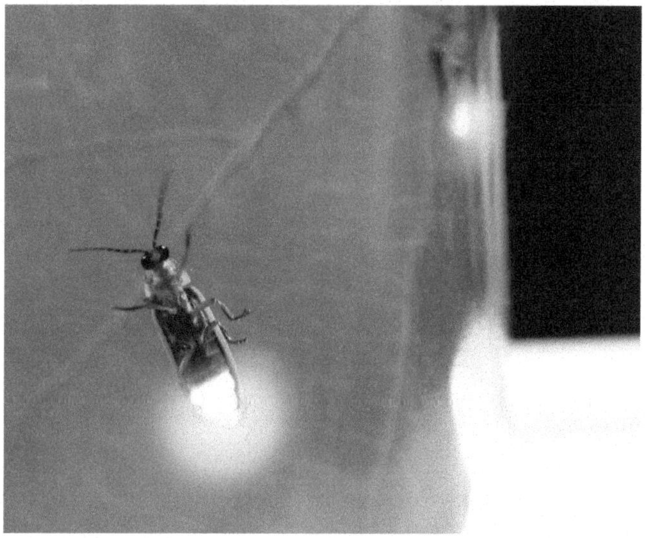

5.10.3 Mimicking bioluminescence

The production of cool light by fireflies has been used as a model for the development of chemiluminescent materials. Although the production of light by chemiluminescence has been possible for some time, commercial applications

Bioluminescent fungi at night. Some scientists believe that bioluminescent fungi use their light to attract insects.

Some jellyfish use bioluminescence to startle predators or to attract a mate.

were often not developed because the reactions were relatively inefficient. The firefly is able to produce light very efficiently through the chemical reactions in the cells of its abdomen. However, in recent years chemical research has uncovered new, more efficient chemiluminescent reactions. This has enabled the commercial production of chemiluminescent items and the use of chemiluminescence techniques in scientific research.

5.10.4 Using chemiluminescence and bioluminescence

The reactions that occur in chemiluminescence and bioluminescence have been adapted for use in scientific research, medicine, ecology, hygiene and food quality control. Bioluminescence is used when testing for tuberculosis to determine the most suitable antibiotic to give to the patient. Scientists have used gene transfer technology to insert the firefly's gene for making luciferase enzymes into bacteria from tuberculosis patients. These bioluminescent bacteria are then tested for their resistance to different antibiotics. The effectiveness of the antibiotics can be easily determined by the amount of bioluminescence remaining. Bioluminescent bacteria have also been used to test for mercury pollution in water. No doubt future scientists will find many more uses for chemiluminescence and bioluminescence.

5.10 Exercises: Understanding and inquiring

To answer questions online and to receive **immediate feedback** and **sample responses** for every question, go to your learnON title at www.jacplus.com.au. *Note:* Question numbers may vary slightly.

Remember

1. In what ways are chemiluminescence and bioluminescence similar?
2. Draw a diagram to explain how light is produced in a chemiluminescent light stick.

Investigate

3. Use the library and internet to find out how scientists use chemiluminescence and bioluminescence. Produce a web page of your findings, including links to some of the websites that you used for your investigation.
4. Organisms can use bioluminescence to light their dark surroundings, to attract a mate, to attract their prey and to camouflage themselves. Find examples of plants or animals that use bioluminescence for each of these reasons.

5.11 Target maps and single bubble maps

5.11.1 Target maps and single bubble maps

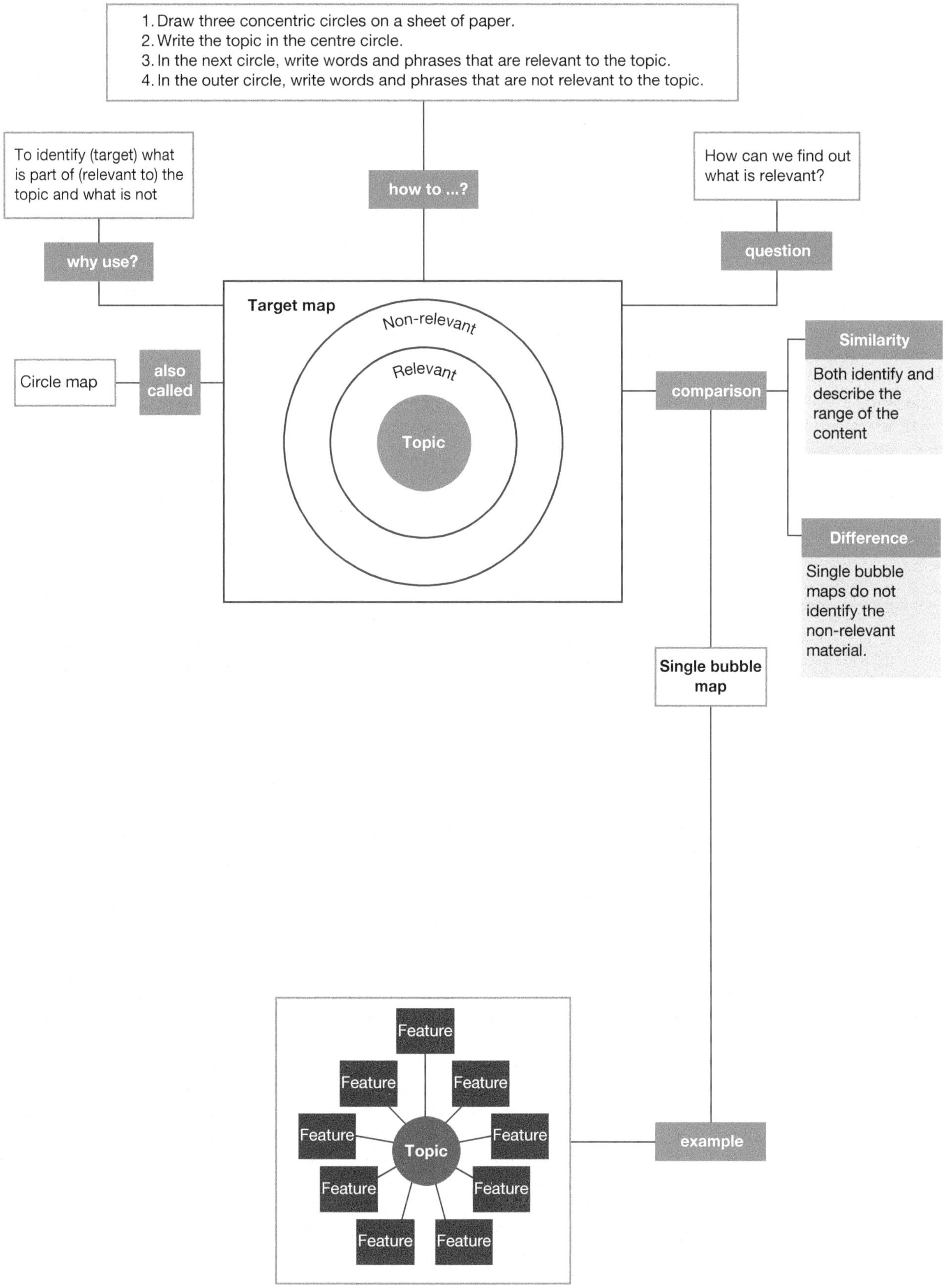

5.11 Exercises: Understanding and inquiring

To answer questions online and to receive **immediate feedback** and **sample responses** for every question, go to your learnON title at www.jacplus.com.au. *Note:* Question numbers may vary slightly.

Think and create

1. The target map at right separates the content that is relevant to a particular group of chemical reactions. Which group of chemical reactions is represented?

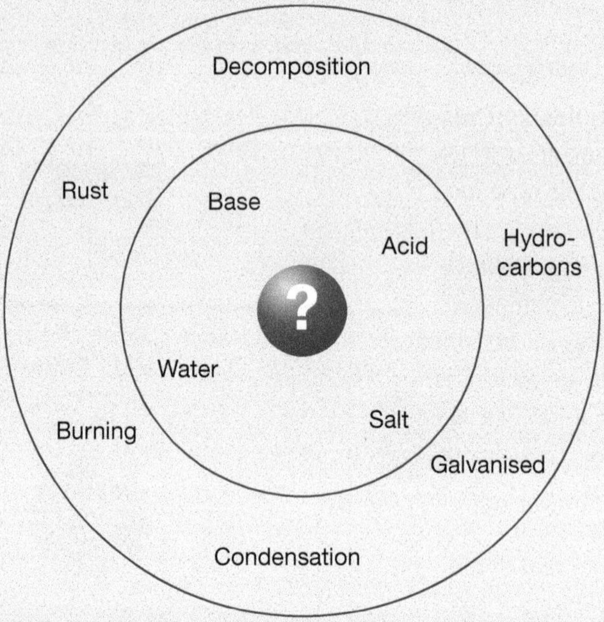

2. Use the terms in the target map in question 1 and the box below to construct target maps that are relevant to:
   (a) all chemical reactions
   (b) combustion reactions
   (c) corrosion reactions.

precipitate
oxygen
melting
energy absorbed
reactants
products
evaporation
broken bonds

3. (a) Work in a small group to write down as many ideas associated with chemical reactions as you can.
   (b) Once your group has completed a list, work on your own to create a single bubble map that represents what you believe to be the ten most important ideas related to chemical reactions.
4. Create single bubble maps to represent the important ideas associated with:
   (a) precipitation reactions
   (b) redox reactions
   (c) dangerous goods
   (d) hazardous substances.
5. Create a single bubble map on the topic 'Plastics'.
6. Sort the following substances into three groups to create single bubble maps relevant to the topics:
   (a) salts
   (b) fuels
   (c) catalysts.

biogas	methane
sodium chloride	platinum
enzymes	propane
magnesium chloride	sodium acetate
coal	ethanol
zeolites	copper nitrate
hydrocarbons	vanadium oxide
amylase	sodium sulfate

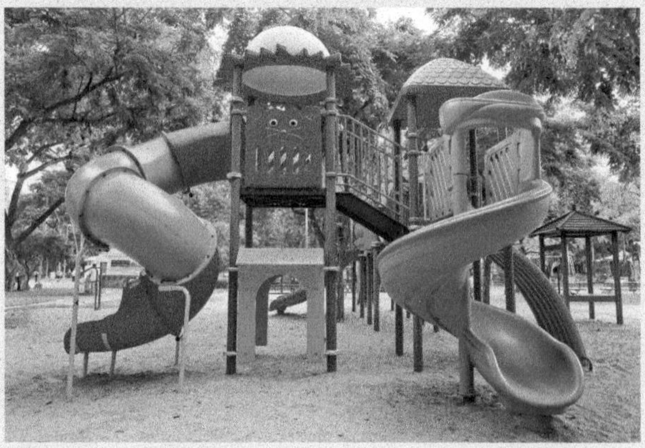

learnon

5.12 Project: Flavour fountain

Scenario

The Sparky Cola Corporation has a series of advertisements for which they are famous that always involve a Mentos lolly being dropped by various means into a bottle of Sparky Cola, causing a huge foaming jet to burst out of the bottle. As a good science student, you know that the jet is the result of the carbon dioxide dissolved in the cola being able to form sizable gas bubbles very quickly on the rough surface of the lolly. You may well have even done this trick yourself.

Sparky Cola have decided that they want bigger jets than ever before and they pride themselves on using real video footage rather than CGI. They are providing a special prize at the next Science Fair for the project that determines how the biggest jet can be produced from a single Mentos lolly and 600 mL of cola. You are to provide not only a scientific report, but also video footage of your highest fountain that they can use in their next ad. You and your friends are determined to win the cash prize at the Science Fair and the TV fame for your Flavour Fountain footage!

Your task

You will design and carry out an investigation that will test a number of different factors (for example, regular cola or diet cola) to determine which will produce the highest cola fountain from a 600 mL bottle of cola and a Mentos lolly. Your findings will be presented in the form of a scientific report. You will also produce video footage of the highest fountain that you can make using what you have discovered.

5.13 Review

5.13.1 Study checklist

Making and using plastics

- describe the molecular structure of polymers
- explain how polymers are derived from monomers
- relate the uses of polymers to their properties

Writing chemical equations

- describe chemical reactions using chemical formulae
- write correctly balanced chemical equations

Types of chemical reactions

- describe and compare the characteristics of precipitation, corrosion, displacement, combustion, decomposition, combination and neutralisation reactions
- describe the transfer of electrons to and from different atoms in redox reactions
- distinguish between oxidation and reduction in redox reactions
- describe examples of a range of redox reactions
- describe the use of fractional distillation in the refining of crude oil for use in combustion reactions
- outline some examples of the use of biofuels as alternatives to fossil fuels

Chemical reaction rates

- describe the effect of temperature, surface area of solid reactants and concentration of reactants on the rate of chemical reactions
- describe the role of catalysts in chemical reactions
- investigate examples of the use of naturally occurring catalysts in the human body

Science as a human endeavour

- appreciate the role of plastics in everyday life
- recognise the need to use laboratory chemicals safely and responsibly
- describe the dangers associated with each of the classes and subclasses of substances classified as dangerous goods
- recognise and interpret an MSDS and a risk assessment
- investigate the technologies associated with carbon capture and storage
- discuss and debate the advantages and disadvantages of the use of fossil fuels and their alternatives for the supply of energy
- investigate the use of catalysts in motor vehicles and food production
- investigate the causes and treatment of diseases caused by enzyme deficiencies

Individual pathways

ACTIVITY 5.1	ACTIVITY 5.2	ACTIVITY 5.3
Revising chemical reactions	Investigating chemical reactions	Investigating chemical reactions further
doc-8473	doc-8474	doc-8475

learnON ONLINE ONLY

5.13 Review 1: Looking back

To answer questions online and to receive **immediate feedback** and **sample responses** for every question, go to your learnON title at www.jacplus.com.au. *Note:* Question numbers may vary slightly.

1. What is the only reliable evidence indicating that a chemical reaction has taken place?
   (a) A change in temperature
   (b) A change in state
   (c) Formation of a new substance
   (d) Disappearance of one or more reactants
2. Describe one characteristic that is common to all materials that we call plastics.
3. Explain how polymers are made, using the terms *monomers* and *chemical bonds* in your explanation.
4. Two types of plastics are thermoplastic polymers and thermosetting polymers.
   (a) Describe the differences in the properties of these two types of plastics.
   (b) Explain the differences in their properties in terms of bonding between the chains of monomers of which they are made.
   (c) List two examples of each type of plastic.

5. In an experiment to test the effect of the amount of liver on the breakdown of hydrogen peroxide, the results shown in the table at right were obtained.
   (a) Write a word equation for the reaction occurring in this experiment.
   (b) Use formulae to write an equation for this chemical reaction.
   (c) Graph these results on graph paper.
   (d) What does the graph show about the effect of the liver on the rate of this reaction?
   (e) Why does the liver affect this reaction?

Mass of liver (g)	Volume of oxygen released (cm^3)
0.5	2.5
1.0	5.1
2.0	9.8
2.5	11.5

6. Complete the following table, then write the final balanced equation and show physical state symbols.

Balancing a chemical equation	**Example: Ethene gas will burn in air in a combustion reaction. This type of reaction produces CO_2 and H_2O.**
Step 1: Start with the word equation and name all of the reactants and products.	________ gas + __________ gas carbon dioxide + water
Step 2: Replace the words in the word equation with formulae and rewrite the equation.	Ethene gas = C_2H_4 Oxygen gas = ____ (reactants)________ = CO_2 Water vapour = ________ (products)____________ $\longrightarrow$ __________
Step 3: Count the number of atoms of each element (represented by the formulae of the reactants and products).	(see table below)

Element	Reactants	Products
C		
H		
O		

7. When an aqueous solution of barium hydroxide reacts with an aqueous solution of ammonium hydroxide, the temperature of the products becomes low enough to freeze water.
   (a) What is an aqueous solution?
   (b) Is this an example of an exothermic or endothermic chemical reaction? Explain your answer.
   (c) Where does the energy transferred to or from the reactants go?

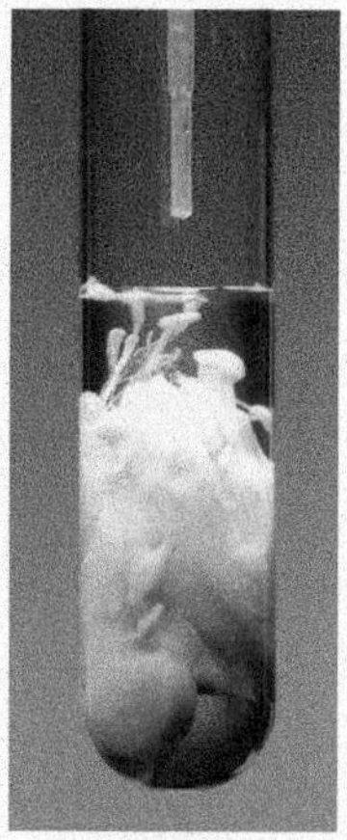

8. The two reactants in the chemical reaction taking place in the test tube shown at right are aqueous solutions. There is enough evidence in the photograph to identify the type of chemical reaction taking place.
   (a) What type of chemical reaction is it?
   (b) What evidence in the photograph identifies the type of reaction?
9. Which of the following is a balanced equation?
   (a) Na + 2Cl $\longrightarrow$ 2NaCl
   (b) MgO + 2HCl $\longrightarrow$ MgCl + H_2
   (c) MgO + 2HCl $\longrightarrow$ $MgCl_2$ + H_2O
   (d) 2Na + Cl $\longrightarrow$ 2NaCl
10. Which of the following are products of the reaction between silver nitrate and sodium chloride?
    (a) Silver nitrate and sodium chloride
    (b) Nitrogen chloride and silver sodium
    (c) Do not react so there will be no products
    (d) Silver chloride and sodium nitrate
11. Many chemicals are classified as dangerous goods and/or hazardous substances.
    (a) Describe the differences between these two categories of chemicals.
    (b) What do these two categories of chemicals have in common?

Fractional distillation columns in an oil refinery

12. What is an MSDS and what is it used for?
13. Write balanced equations using formulae for the following reactions.
    (a) Aluminium metal + oxygen gas ⟶ solid aluminium oxide
    (b) Potassium metal + oxygen gas ⟶ solid potassium oxide
    (c) Solid carbon + oxygen gas ⟶ carbon dioxide gas
    (d) Solid copper carbonate ⟶ solid copper oxide + carbon dioxide gas
    (e) Iron metal + sulfur powder (S_8) ⟶ solid iron sulfide (FeS_2)
    (f) Copper sulfate solution + zinc metal ⟶ copper metal + zinc sulfate solution
    (g) Copper(II) sulfate solution + sodium hydroxide solution ⟶ solid copper(II) hydroxide + sodium sulfate solution
    (h) Solid magnesium hydroxide + hydrochloric acid ⟶ magnesium chloride + water
14. State the reaction type (displacement, combination, decomposition, precipitation, combustion or neutralisation) for each of the reactions in question 13.
15. Which of the reactions in question 13 are redox reactions?
16. The chemical reaction between acids and metals is a displacement reaction. An experiment was carried out to measure how long it took for equivalent amounts of different metals to dissolve in 100 mL of hydrochloric acid. The results shown in the table below were obtained.
    Unfortunately, the person recording the data on the computer accidentally changed the order of the metals recorded in the table.

Metal	Time taken for the metal to dissolve (min)
Iron	1.0
Magnesium	3.0
Tin	2.5
Aluminium	3.5
Nickel	2.0
Zinc	1.5

    (a) Using the known activity series in subtopic 4.6, redraw the table so that the correct metal is matched with the correct time.
    (b) Explain why all of these reactions are called displacement reactions.
    (c) Which element is displaced in the reactions?
17. Predict the salts that would result from the neutralisation reaction between:
    (a) magnesium oxide and hydrochloric acid
    (b) copper(II) oxide and sulfuric acid
    (c) sodium hydroxide and acetic acid
    (d) sodium oxide and nitric acid.
18. During fractional distillation, what differing observable property of hydrocarbons is used to separate them from crude oil?
19. How is the molecular structure of methane different from that of octane?
20. Which hydrocarbons used in fuel production are separated from crude oil at the highest temperature?
21. Identify four gases that are separated from crude oil at the very top of the distillation tower.
22. Describe (in words) the chemical transformation that takes place in a biogas digester.
23. Write the chemical word equation that describes the fermentation process used to produce ethanol.
24. One of the chemical reactions used during the production of sulfuric acid makes use of the catalyst vanadium oxide (V_2O_5). The chemical equation for this reaction is:

$$2SO_2(g) + O_2(g) \xrightarrow{V_2O_5} 2SO_3(g)$$

    (a) What is a catalyst?
    (b) Why doesn't V2O5 appear as one of the reactants?
    (c) What name is given to catalysts in living things?
    (d) Describe at least one chemical reaction that is caused by light and how the reaction is useful.
    (e) What is chemiluminescence?
    (f) How is bioluminescence different from other types of chemiluminescence?

learnON RESOURCES — ONLINE ONLY

Complete this digital doc: Worksheet 5.8: Chemical reactions: Summary (eles-19458)

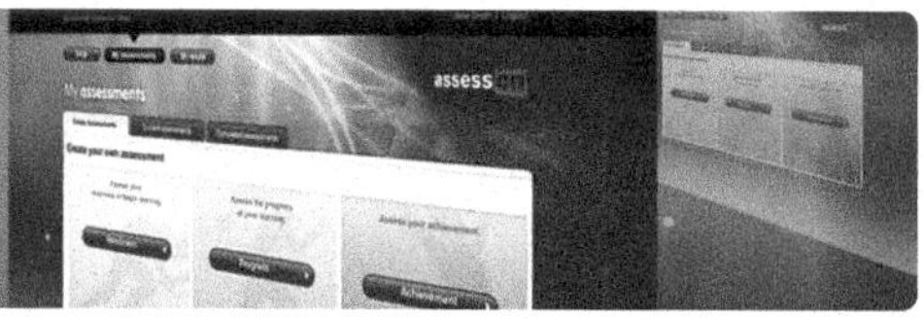

TOPIC 6
The mysterious universe

6.1 Overview

6.1.1 Why learn this?

On any cloudless night, a pattern of stars, galaxies and clouds of gas appears to spin above our heads. Yet against this backdrop, changes are taking place — often hard to see and sometimes spectacular, but always raising questions in our minds about the past and the future. How and when did it all begin?

6.1.2 Think about the universe

assess on

- Where are stars formed?
- Why do stars appear to show different colours?
- How old is the universe?
- How does a red giant become a white dwarf?
- What can we actually see from space?
- The universe may have started with a 'big bang', but what is the 'big crunch'?

LEARNING SEQUENCE

Numerous **videos** and **interactivities** are embedded just where you need them, at the point of learning, in your learnON title at www.jacplus.com.au. They will help you to learn the content and concepts covered in this topic.

6.1.3 Your quest

Awesome stars

When they gazed at the night sky, the earliest humans would have been in awe of the stars. What questions would they have asked? What are the stars made up of? Where did they come from? Are they alive? Humans were driven to find a way to explain the stars' existence.

Think

1. Why do we know so much more about stars and the universe than the earliest humans?
2. What is a star? Write your own description of what a star is.
3. What is the name of the nearest star to the planet Earth?
4. How are stars formed?
5. Does a star ever die?
6. List all of the objects other than stars that you can see in the night sky.

6.1.4 Looking back in time

The object in photograph (a), below, is not a star. It is a quasar called PG 0052+251. It emits much more light than any star could. Quasars are found only at very large distances from the solar system. Observations of distant objects like quasars provide clues about how the universe began.

Think

7. Astronomers believe that quasars are formed when black holes at the centre of galaxies begin to pull in gas and stars from the galaxy.
   (a) What is a black hole?
   (b) What is a galaxy?
   (c) To which galaxy does the solar system belong?
8. The photograph of PG 0052+251 was taken by the Hubble Space Telescope.
   (a) Where is the Hubble Space Telescope?
   (b) Why are the photographs taken by the Hubble Space Telescope clearer than those taken by larger telescopes on the Earth's surface?

(a) The quasar PG 0052+251 is 1.4 billion light-years away. That is, when you look at its image, you are seeing it as it was 1.4 billion years ago.

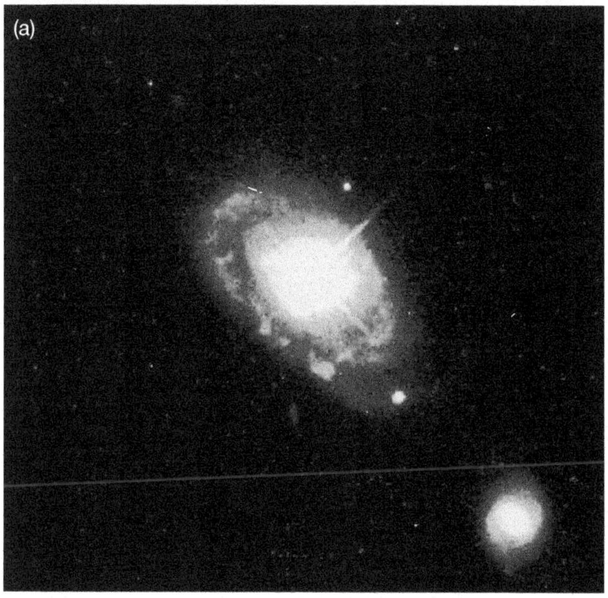

(b) The Hubble Space Telescope. Even though it is much smaller than many telescopes on the ground, it can see much further into the universe because it is above the Earth's atmosphere.

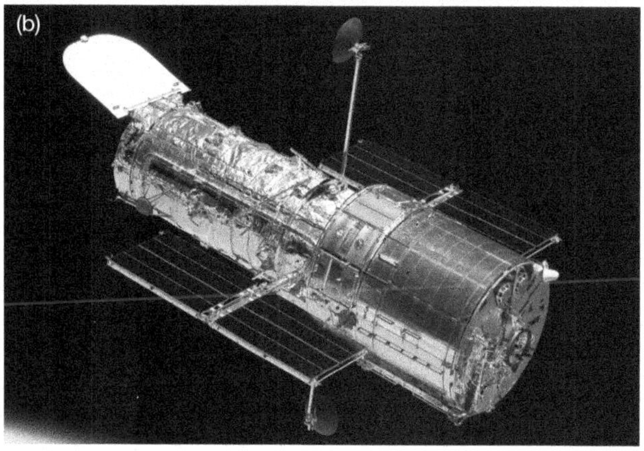

6.1.5 Where Earth fits into the universe

Until almost 400 years ago, most astronomers believed that the Earth was at the centre of the universe. It was surrounded by a 'celestial sphere' on which the stars were attached. The Moon orbited the Earth. The sun and planets were also believed to orbit the Earth. Then, quite quickly, the idea that the sun was the centre of the universe became accepted. We now know that the Earth is just a tiny part of the solar system, which is a tiny speck in a galaxy known as the Milky Way.

The sun is one of up to 400 billion stars in the Milky Way, and the Milky Way galaxy is one of more than 100 billion galaxies in the universe.

The solar system is just a tiny part of the rotating Milky Way galaxy.

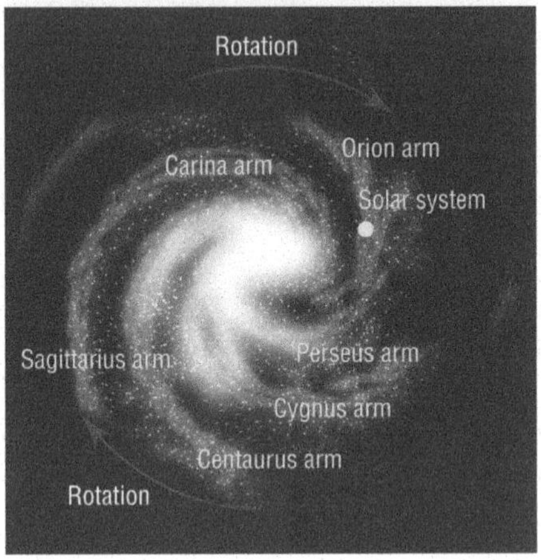

Think

9. Which people and events caused the change in thinking about the place of the Earth in the universe about 400 years ago?
10. How do we know so much more about the distant parts of the universe now, in the twenty-first century, than we did 400 years ago when people were arguing about whether the Earth or the sun was the centre of the universe?
11. Given that the Earth is such a tiny speck, would you expect to find other, similar planets in the universe? If so, where would you expect to find them?

6.2 Observing the night sky

6.2.1 Seeing stars

When you look up into the sky on a clear night, you will see countless specks of light stretching from horizon to horizon.

Looking again later the same night, you should clearly see many of the same recognisable patterns as before, but they will have moved to a different position in the sky. From these simple observations, it is easy to conclude that the sky is a crystal-clear sphere dotted with the tiny lights we call stars. This 'celestial sphere' seems to rotate above our heads, carrying with it the fixed patterns or **constellations** of gleaming stars.

6.2.2 Constellations

Ancient astronomers grouped stars according to the shapes they seemed to form. The shapes were usually of gods, animals or familiar objects. The most well-known constellations are the 12 groups we know as the signs of the zodiac. These constellations follow the **ecliptic** and their names include Taurus (the bull), Leo (the lion) and Sagittarius (the archer). You probably know the rest. If not, a discussion with your friends will help.

A time-lapse photograph of the sky clearly shows the apparent movement of the stars.

Today, astronomers recognise 88 constellations. When observed from Earth, the stars in each constellation appear to be very close to each other. But the stars that make up constellations can be located at very different distances from Earth. For example, the star Betelgeuse in the Orion constellation is approximately 650 light-years from Earth, whereas the star Bellatrix in the same constellation is about 240 light-years from us.

6.2.3 Constellations 'on the move'

We know now that the celestial sphere proposed by the Greek astronomer Ptolemy in 150 CE was wrong. The apparent motion of the fixed pattern of stars at night, shown in the time-lapse photograph on the previous page, is due to the rotation of the Earth.

The apparent change in position of the constellations is due to the Earth's orbit around the sun. Sky charts, sometimes called sky maps, star maps or star charts, show the position of constellations, stars and the **planets** from different locations for each month of the year.

In 150 AD, the Greek astronomer Ptolemy suggested that the stars were attached to a 'celestial sphere' that rotated above our heads. According to Ptolemy, the sun, the planets and the Moon also orbited the Earth.

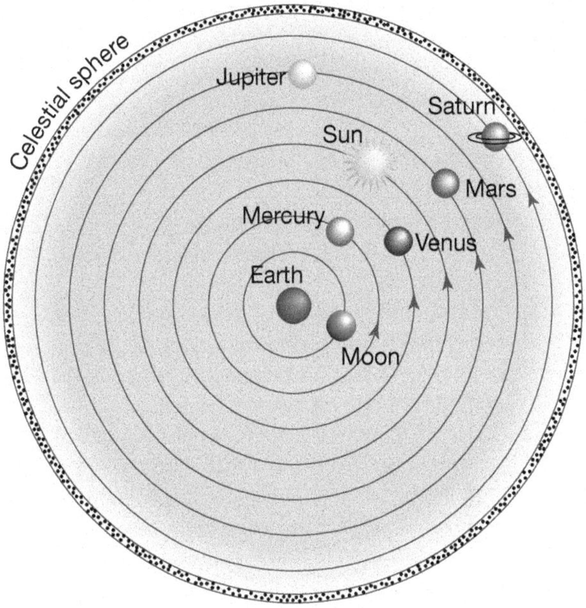

The constellations Crux, Orion and Taurus as seen from Australia

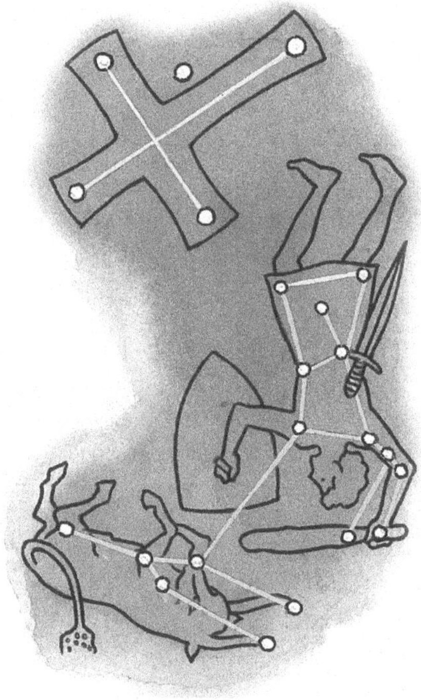

6.2.4 A closer view

The development of the telescope in the sixteenth century allowed Earth-bound astronomers to see objects in the sky with much greater precision than ever before. Observations using telescopes showed that many different types of objects in the sky could be identified. These included single or double stars, groups of stars called **galaxies**, clusters of galaxies, and clouds of gas and dust called **nebulae**.

In 1718, English astronomer Edmond Halley, who is well-known for identifying the comet named after him, used his telescope to check three particularly bright stars: Sirius, Procyon and Arcturus. He found that the position of each one relative to surrounding stars was noticeably different from the positions recorded by ancient Greek astronomers centuries before. There were even slight differences between Halley's observations and those of Danish astronomer Tycho Brahe about 150 years earlier. Never again could the stars be described as 'fixed in the heavens'.

learnon RESOURCES — ONLINE ONLY

Explore more with this weblink: Star maps

INVESTIGATION 6.1

Using a sky chart

AIM: To use a sky chart to locate some constellations, stars, planets and other celestial features

Materials:

torch covered with red cellophane
compass
star chart for the current month
highlighter pen

Method and results

- Use the **Star maps** weblink in your Resources section to download and print a star chart for the current month. The stars on the star chart are represented by dots. The size of the dot for each star is a rough indication of its brightness.
- On a clear night, find a location away from bright lights, buildings and other obstacles that will block your view of the night sky. To get the best view you will need to wait for 10–15 minutes to adjust to the dark. The red cellophane over the torch will also assist with your night vision.
- North and south are labelled at the bottom of each star chart. Use the compass to ensure that you are facing either north or south while observing the sky.
- Initially facing south, locate the Southern Cross (Crux) constellation and the two pointers Alpha Centauri and Beta Centauri.

1. Highlight the stars and constellations on your star chart as you find them.
   - Find the constellation Taurus and identify its brightest star Aldebaran.
   - Find Orion and locate the stars Rigel and Betelgeuse.
   - Return to the Southern Cross and use it, along with the pointers, to locate and highlight the South Celestial Pole on your chart. It can be located by following the line of the long arm of the Southern Cross and finding where it intersects with a perpendicular line to the line joining the two pointers.
   - Try to locate at least one of the planets. Unlike the stars, they do not 'twinkle'. Daily newspapers usually list the rising and setting times of the planets or they can be found on the internet. Like the sun and the Moon, the planets rise in the east and set in the west. They follow the same path as the sun across the sky.
2. Locate and highlight as many other stars and constellations on the star chart as you can.

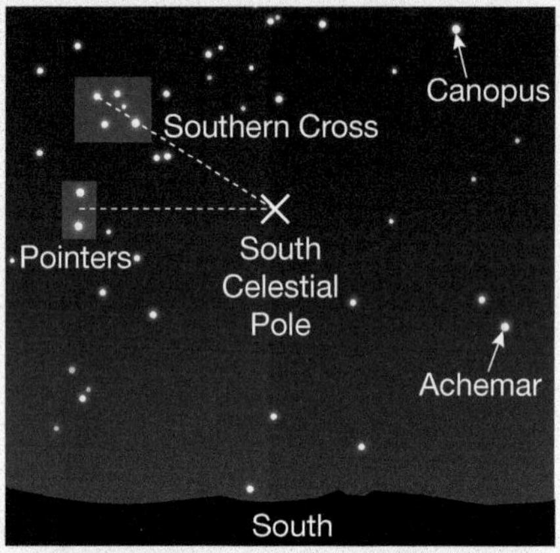

6.2.5 Questions about stars

Halley's observations raised some new questions about stars. Why should only a few stars move quickly enough for their motion to be noticed? Why do they happen to be among the very brightest stars? Perhaps some stars are closer to Earth than others. Being closer, they would appear brighter than other stars and their motion would be detectable against the backdrop of more distant, and therefore dimmer, stars.

The Horsehead Nebula in the constellation of Orion. A nebula is a cloud of dust and gas, visible as a glowing or dark shape in the sky against a background of stars.

It's all relative

The apparent movement of objects at different distances is due to the actual movement of the observer. It is an effect called **parallax**. In 1837, German astronomer Friedrich Bessel became the first person to provide proof of a parallax effect when observing stars. As the Earth orbits the sun, the positions of stars change very slightly relative to each other. If all the stars were the same distance from the Earth, this would not happen.

Observations of a stellar parallax effect indicate that some stars are relatively close to us while others are much further away. The transparent celestial sphere idea of the past must be banished, replaced by an even more awe-inspiring image — that of star-studded space stretching before us with no known boundary or end.

HOW ABOUT THAT!

A light-year is not a measure of time! I'a measure of distance. In one year, light travels a distance of 9 500 000 000 000 or 9.5×10^{12} kilometres. This distance is called a light-year.

USING LARGE NUMBERS

Very large numbers are often written using **scientific notation**. This allows us to avoid writing lots of zeros and also makes the numbers easier to read, because the reader does not have to count the zeros. For example, the distance between the Earth and the sun averages 150 million kilometres. This could be written 150 000 000 km or, in scientific notation, as 1.5×10^8 km.

Some other examples are:

- $45\,000\,000\,000 = 4.5 \times 10^{10}$
- $700\,000\,000\,000\,000\,000 = 7.0 \times 10^{17}$.

INVESTIGATION 6.2

The effect of parallax

AIM: To observe the effect of parallax

Materials:

a number of traffic cones (witch's hats)
pencil and paper

Method and results

- Mark a circle on the school oval to represent Earth's orbit around the sun.
- Place a series of traffic cones at different distances from the circle to represent stars nearby and far away.
- Take a walk around Earth's 'orbit' and, at several different points, sketch the appearance of the 'stars' relative to one another and to even more distant objects such as trees and fence posts.

Discuss and explain

1. Looking at your sketches, did the positions of the stars relative to one another appear to change as you moved around the orbit?
2. Can you see any difference between the relative movements of the nearby stars compared with those of the more distant stars?

learnON RESOURCES — ONLINE ONLY

Watch this eLesson: Twinkle, twinkle (eles-0071)

6.2 Exercises: Understanding and inquiring

To answer questions online and to receive **immediate feedback** and **sample responses** for every question, go to your learnON title at www.jacplus.com.au. *Note:* Question numbers may vary slightly.

Remember

1. Explain why stars appear to rotate during the night.
2. Explain why stars gradually change their position in the night sky throughout the year.
3. What do the constellations of the zodiac have in common?
4. How did the invention of the telescope change our view of the night sky from Earth?
5. Explain why the planets were given a name that means *wanderer*.
6. What do we mean by the term *parallax*?
7. How did observations of a stellar parallax effect change our ideas about the universe?

Evaluate

8. The estimated distances from Earth to some stars and galaxies are listed below. How long would it take to reach each of them, travelling at the speed of light (about 300 000 km/s)?

Sun	Our own star	1.5×10^8 km
Proxima Centauri	The closest star after the sun	4.0×10^{13} km
Centre of Milky Way	Our own galaxy	2.5×10^{17} km
Magellanic Clouds	One of the closest galaxies	1.5×10^{18} km
Andromeda galaxy	One of the closest galaxies	1.4×10^{19} km
Quasars	Very distant objects	1.4×10^{23} km

Think

9. Explain why it is easier to observe the night sky in rural areas than in the city.
10. All of the stars in the constellation Orion appear to be the same distance from Earth. Explain how observers on Earth know that they are not.
11. Explain why the planets that are visible to the naked eye appear to change position against the fixed patterns of other stars.
12. Radio waves travel through space at the same speed as light, which is about 300 000 km/s. How long would it take a radio message from Earth to reach the solar system's nearest neighbouring star?

Imagine

13. Is it likely that a spacecraft from Earth will ever venture out to planets orbiting the closest stars? Present some calculations to support your answer.

learnon RESOURCES — ONLINE ONLY

Complete this digital doc: Worksheet 6.1: Observing stars (doc-19462)

6.3 Stars — a life story

6.3.1 Stars come and go

Movie stars come and go. Some have brief careers while others seem to go on forever. It's much the same with the stars in the sky. Stars come and go — they don't last forever. However, their 'careers' are usually much longer than those of the movie variety.

A star is born

Dust and gas are not evenly distributed in interstellar space. There appear to be currents of denser material swirling throughout the universe. Within these currents, the density sometimes reaches the critical figure of 100 atoms per cubic centimetre. At this point, gravity takes hold and the gas and dust begin to collapse, forming a cloud. Such clouds of interstellar matter are called nebulae and are really like star nurseries. The Great Nebula in the constellation of Orion is a nebula large enough to be seen with the naked eye. The collapse continues under the influence of gravity, forming visible globules in the nebula cloud. As the globules collapse further, the formation of any original gas cloud is accelerated. The now dense cloud is known as a **protostar**. At the same time, the increasing pressure causes the temperature to rise. This effect is modelled in Investigation 6.3.

INVESTIGATION 6.3

Heat produced by compressing a gas

AIM: To model the generation of heat during the formation of a star

Materials:

a bicycle pump
a tyre with inner tube

Method and results

- Using an energetic pumping action, inflate a tyre with the bicycle pump. Alternatively, just pump the bicycle pump with your finger partially covering the open end so the air does not escape.
- Now feel the body of the pump.

Discuss and explain

1. What change has been observed?
2. How does an increase in air pressure affect the temperature of the surroundings?
   (The opposite effect can be observed when carbon dioxide gas is released from a soda bulb.)

6.3.2 Nuclear fusion

The temperature and pressure of the gases in a protostar eventually rise high enough for atomic nuclei to become joined together by a process called **nuclear fusion**. As a result of fusion, two isotopes of hydrogen, deuterium (hydrogen-2) and tritium (hydrogen-3) combine to form helium nuclei, neutrons and vast amounts of energy.

This nuclear fusion reaction in stars releases vast amounts of energy.

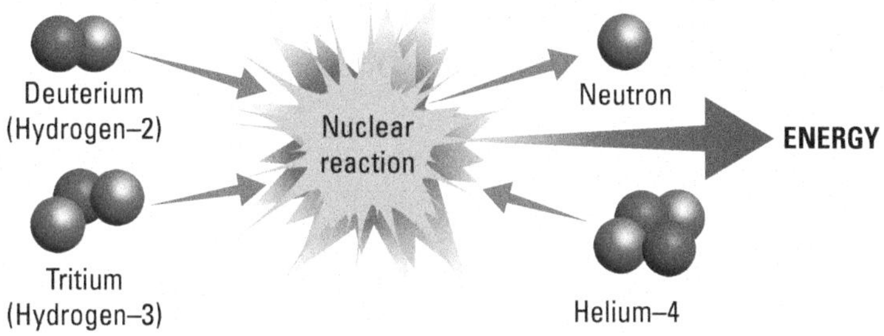

6.3.3 The young, the old and the dead

A quick glance around the night sky shows us that stars differ quite noticeably from one another, both in how bright they appear to us and in their colour (see Investigation 6.4). Some of them are relatively close to the Earth, while others are much further away. There are young stars, middle-aged stars like the sun, old and dying stars, and exploded stars. By collecting details of a wide range of stars, we can trace the various stages of development of typical and unusual stars.

6.3.4 Star light, star bright

How bright a star appears to us (its **apparent magnitude**) depends on its actual brightness (its **absolute magnitude**) and its distance from Earth. A dim star close to us may appear brighter than a really bright star a long way away. To calculate the absolute magnitude of a star, astronomers must know how far away it is. The colour of a star depends on its surface temperature: red stars are cool, and white and blue stars are hot.

The Great Nebula of Orion

A question of magnitude

The Hertzsprung–Russell diagram sorts stars according to their absolute magnitude (or luminosity) and surface temperature.

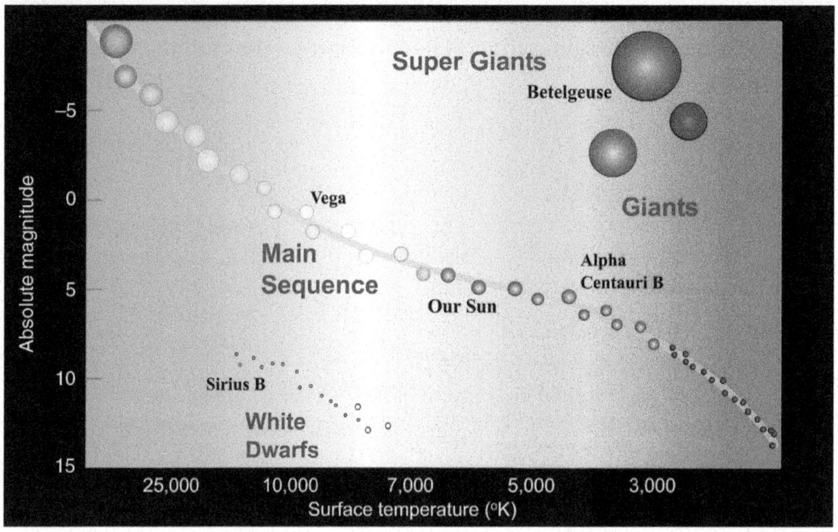

How bright a star or planet appears as viewed from Earth is measured on a scale of apparent magnitude. On this scale, brighter objects have the lowest apparent magnitudes. For example, the sun has an apparent magnitude of –27. A full moon has apparent magnitudes of approximately –13. The brightest stars have apparent magnitudes between –1 and 1. The weakest objects visible with the naked eye have an apparent magnitude of approximately 6.

The absolute magnitude is a measure of how much light an object emits. The sun is much smaller than Rigel in Orion and it emits a lot less light. However, it appears brighter to us because it is much closer than Rigel. The Moon emits no light of its own.

The table below shows some typical values of apparent and absolute magnitudes.

Star and constellation	Apparent magnitude	Absolute magnitude
Sun	–27	+4.7
Sirius (Canis Major)	–1.5	+8.7
Canopus (Carina)	–0.73	–4.6
Alpha Centauri (Centaurus)	–0.33	+4.7
Rigel (Orion)	+0.11	–7.5
Beta Centauri (Centaurus)	+0.60	–5.0
Betelgeuse (Orion)	+0.80	–5.0
Aldebaran (Taurus)	+0.85	–0.3
Alpha Crucis (Southern Cross)	+0.90	-3.9

Colour and brightness

An interesting way of displaying the data collected about stars was developed independently by two astronomers, Ejnar Hertzsprung from Denmark and Henry Norris Russell from America. This method has now been named after both of them. In the Hertzsprung–Russell diagram, the absolute brightness of a star is plotted against its surface temperature, which is deduced from its colour. When data for many stars are plotted, most of them, including our sun, fall into what is known as the **main sequence**. Exactly where a star is found along the main sequence is determined by its mass. Low-mass stars tend to be cooler and less bright than high-mass stars.

Other types of stars show up very clearly on the Hertzsprung–Russell diagram but in much smaller numbers than in the main sequence. The names of these stars — white dwarfs, red giants, blue giants and super giants — clearly describe their characteristics. Astronomers suggest that all stars begin their existence in the main sequence and spend the largest part of their life there. This explains why most of the stars observed at a particular time are found in the main sequence. The rarer types are stars passing relatively quickly through later stages of development on the way to extinction as their nuclear fuel runs out.

Large stars follow a different evolutionary sequence to smaller stars like the sun.

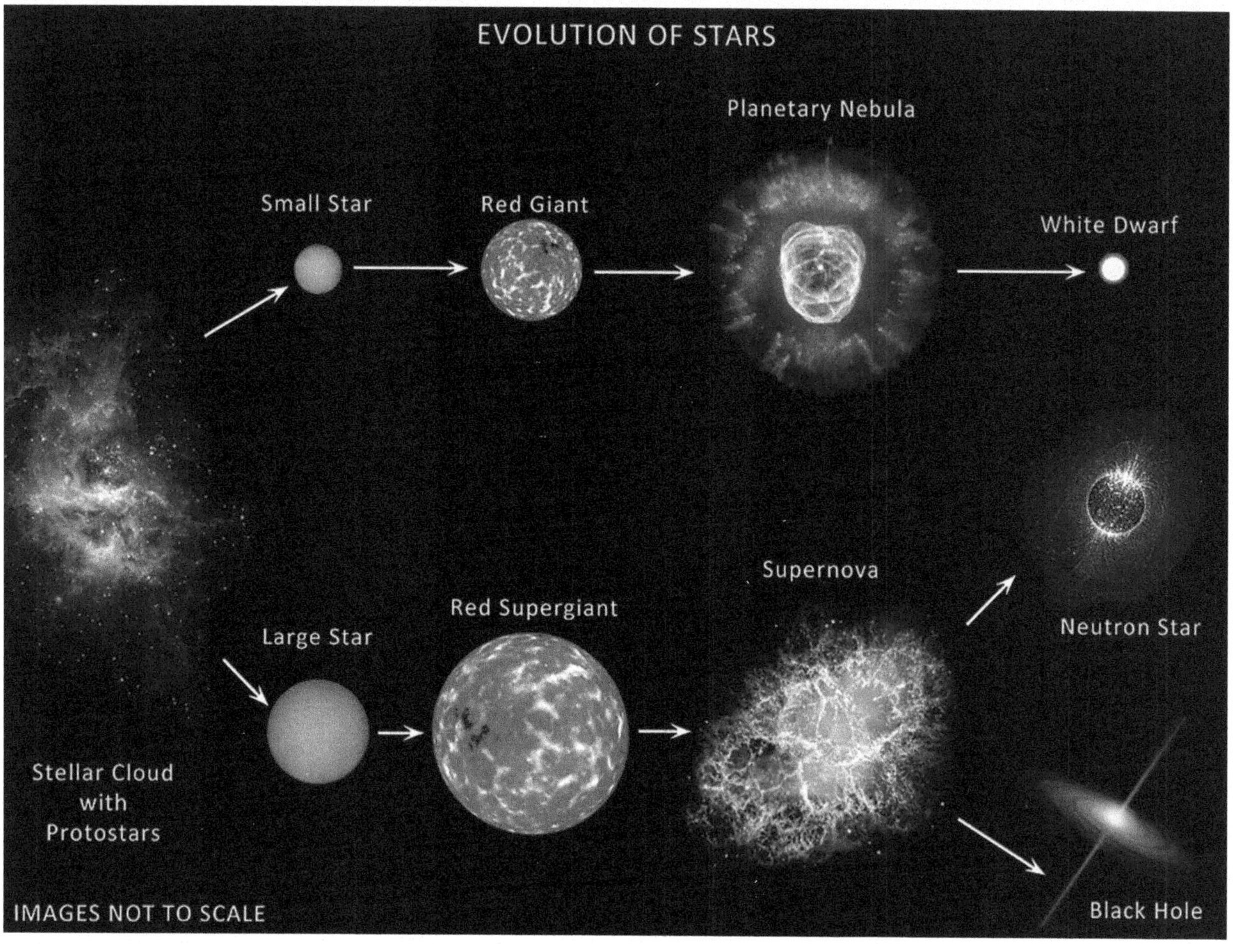

Red giants

In a stable main sequence star, hydrogen is steadily turned into helium by the process of fusion. As helium builds up in the core of the star, the region where energy is produced by the fusion of hydrogen becomes a shell around the core. The shell gradually expands and the star swells to 200 or 300 times its original size, cooling as it does so, to become a **red giant**. This will eventually happen to our sun, which will grow large enough to swallow up the inner planets, including Earth.

The brightness of many red giants varies greatly because they have become unstable after many millions of years of stability. The red giant Mira in the constellation Cetus (the Whale) was the first variable red giant discovered. The brightness of Mira increases and decreases over a huge range in a cycle that averages 320 days. Not surprisingly, it is known as a **pulsating star**. The shorter cycles of some pulsating stars are so predictable that they can be used as markers to measure vast interstellar distances in the universe.

In the core of a red giant, new fusion processes take place, turning helium into heavier elements such as beryllium, neon and oxygen. This increases the rate of energy production and raises the star's temperature. A sun-like star might shine 100 times more brightly than it did in its stable period as part of the main sequence.

The Crab Nebula, the remains of a supernova observed almost a thousand years ago. At the centre is a rapidly spinning neutron star called a pulsar.

INVESTIGATION 6.4

Seeing the colours of stars

AIM: To investigate the relationship between star colour and brightness

Materials:

star atlas (optional)
pair of binoculars (optional)

Method and results

- Use the information below, a star atlas or an astronomy computer program to help you to find the constellation Orion (the Hunter). Alternatively, find a colour photograph of the constellation Orion.

The star α-Orionis (also known as Betelgeuse) is a red giant that has a diameter bigger than Earth's orbit. It appears quite visibly red to the naked eye and this distinctive colour shows up even more clearly through binoculars. The star β-Orionis (also known as Rigel) is 60 000 times as bright as the sun.

- Compare the brightness and colours of Betelgeuse and Rigel.
- Try to locate the Orion Nebula using the following information.

The constellation Orion is visible from every inhabited place on Earth. It is most easily recognised from the line of three stars that represent the hunter's belt. Remember, the constellations were named by observers in the Northern Hemisphere, so to Southern Hemisphere observers the constellations appear upside down. This is why Orion's sword points upwards from the belt when viewed from Australia. This group of stars, making up the sword and the belt, is often known as the Saucepan.

Orion's sword, pointing upwards from the belt, contains a misty patch visible to the naked eye. This is the Orion Nebula, labelled M42 by the astronomer Messier, who prepared a catalogue of such objects in an attempt to prevent observers being distracted by them. Through binoculars, stars can be seen embedded in the gas and dust of the Orion Nebula, and new stars have been seen as they begin to emit light. The Orion Nebula and other similar formations are the birthplaces of the stars.

Discuss and explain

1. How do the colours of Betelgeuse and Rigel compare?
2. Which of Betelgeuse and Rigel appears to be brighter? Relate your observation to the table in this section under the heading *A question of magnitude*.
3. Which of the two stars is cooler?

6.3.5 The death of a star

Eventually, the rapid pulsations lead to the destruction of the star. The nature of its death depends on the size of the star.

White dwarfs

For stars less than about eight times the mass of our sun, the destruction begins when the outer layers are thrown off into space and the core flares brightly, forming a ring of expanding gas called a **planetary nebula**. The name *planetary nebula* is misleading because it is not related to planets. But it does have the cloud-like nature of other nebulae. The name came about because astronomers using very early telescopes thought that the clouds resembled the planets Uranus and Neptune.

The remaining star fades to become a **white dwarf**, typically about the size of the Earth but with a very high density and a surface temperature of about 12 000 °C. It then slowly cools, becomes a cold black dwarf and disappears from view.

Coming to a violent end

Stars that are more than about eight times the mass of our sun come to a much more violent end. They swell into much larger red giants called **supergiants** and blow up in a huge explosion called a **supernova**. The matter making up the star is hurled into space along with huge amounts of energy. A supernova can emit as much energy in a month as the sun radiates in a million years. Supernova events are very rare, being seen only every 200 to 300 years on average and fading within a few years. They are extremely important in the universe because it is within these violent explosions that the heavy elements such as iron and lead are produced.

What remains of a supernova is extremely dense; the pull of gravity becomes so great that even the protons and electrons in atoms are forced together. They combine to form neutrons, and the resulting solid core is known as a **neutron star**. If the remaining core has a mass more than about three times that of our sun, the force of gravity is great enough to suck in everything — even light. Such a core becomes a **black hole**.

learnON RESOURCES — ONLINE ONLY

Watch this eLesson: Biggest bang (eles-1074)

Try out this interactivity: Star cycle (int-0679)

6.3.6 Galaxies

Stars group together in groups to form galaxies, attracted towards each other by gravitational forces. Our own sun is one of an estimated 200–400 billion stars in the **Milky Way** galaxy. We think there are more than 100 billion other galaxies of different sizes and shapes throughout the universe. Each of these galaxies is home to stars at all stages of their life cycles.

The Milky Way galaxy, shown at right, is a spiral galaxy. Our solar system is found on the Orion arm of the spiral. Due to the rotation of the galaxy, our solar system orbits the centre of the galaxy at a speed of about 200 kilometres per second.

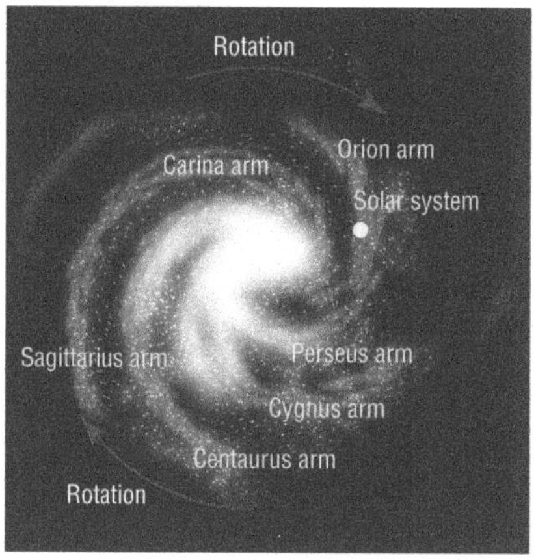

6.3 Exercises: Understanding and inquiring

To answer questions online and to receive **immediate feedback** and **sample responses** for every question, go to your learnON title at www.jacplus.com.au. *Note:* Question numbers may vary slightly.

Remember

1. Explain why most stars are found in the main sequence of the Hertzsprung–Russell diagram.
2. To which group of stars shown on the Hertzsprung–Russell diagram does the sun belong?
3. How does a red giant become a white dwarf?
4. Why is the term *planetary nebula* a misleading way to describe the ring of expanding gas thrown out by a red giant during its transformation into a white dwarf?
5. Explain how galaxies are formed.
6. In which galaxy can the Earth and the rest of our solar system be found?

Evaluate

7. The table at right lists information about three bright stars.
   (a) Which star has the greatest actual brightness?
   (b) Which star is the faintest as seen from Earth?

Star	Apparent magnitude	Absolute magnitude
Rigel	0.11	–7.5
Aldebaran	0.86	–0.3
Canopus	–0.73	–4.6

Think

8. Why are nebulae often referred to as star nurseries?
9. Deuterium and tritium, both isotopes of hydrogen, combine in nuclear fusion reactions in stars. How are the atoms of these two isotopes different from each other?
10. Is it likely that our own star, the sun, will become a supernova? Explain your answer.
11. What would the night sky look like if you had eyes that could see like the Hubble Space Telescope?

Investigate

12. Find out more about the formation and destruction of a supernova. For example, when was the last supernova seen? Can we predict when the next one will be seen?

learnon RESOURCES — ONLINE ONLY

Complete this digital doc: Worksheet 6.2: The brightness of stars (doc-19465)

Complete this digital doc: Worksheet 6.3: Star life cycles (doc-19466)

6.4 The changing universe

6.4.1 Stars on the move

Will the sky you see tonight ever be the same again? Within a person's lifetime, the patterns of stars and galaxies in the night sky do not seem to change, but each night's sky is different. Photographic techniques show us the movements of the stars and tell us that what we see as permanent is actually a universe in a state of continuous and often violent movement and change.

The movement of stars towards or away from the Earth can be measured using the **Doppler effect**. Christian Johann Doppler was an Austrian physicist who noted the change in pitch that results from a source of sound approaching or moving away. We hear the same effect when a high-speed train or aeroplane passes us or when we hear the pitch of a fire-engine's siren drop as it goes by.

Doppler suggested that the same effect we notice in sound waves might be seen in light as well. The Doppler effect would produce a change in the **frequency** of light waves emitted from a moving source. In 1851, the French physicist Armand Fizeau suggested that this change in frequency might be seen by comparing the spectrum of light from a moving source with that from a stationary one.

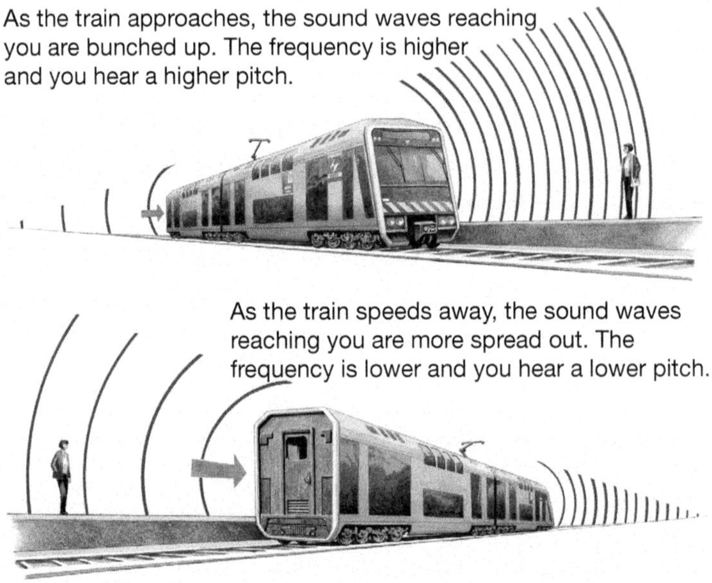

As the train approaches, the sound waves reaching you are bunched up. The frequency is higher and you hear a higher pitch.
As the train speeds away, the sound waves reaching you are more spread out. The frequency is lower and you hear a lower pitch.

INVESTIGATION 6.5

Doppler effect using rotating sound source

AIM: To observe the Doppler effect on sound

Materials:

a source of sound that can easily be spun in a circle, e.g. a battery-powered electronic buzzer that produces a single note or a whistle fastened securely in the end of a length of rubber tubing

a length of strong string

a partner

Method and results

- Ask your partner to spin the sound source around in a circle on the end of the piece of string. If you are using a whistle, your partner should blow through the attached rubber tubing to produce a sound. Listen carefully to the note produced.

CAUTION

Take care while spinning the source. Ensure that the string is strong enough and that no-one is in the path of the rotating source of sound.

1. What can you hear happening to the pitch of the buzzer?
2. When is the pitch highest? When is it lowest?

A whistle can be used as a rotating sound source.

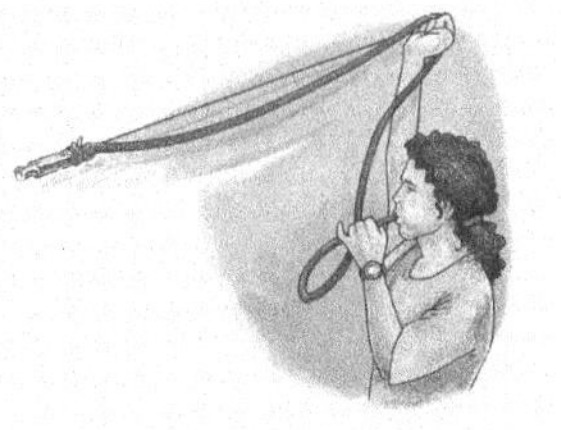

6.4.2 The spectra of stars

When the **spectrum** of the light from a star is analysed, some dark lines are observed. These correspond to colours of light that have been absorbed by substances in the star. Different substances absorb different colours of light. By identifying the **wavelengths** of the colours missing from the spectrum, astronomers can find out which elements are present in the star.

The spectrum of white light from a nearby star. The black lines show which colours have been absorbed by elements in the star. The numbers indicate the wavelength of the light in nanometres.

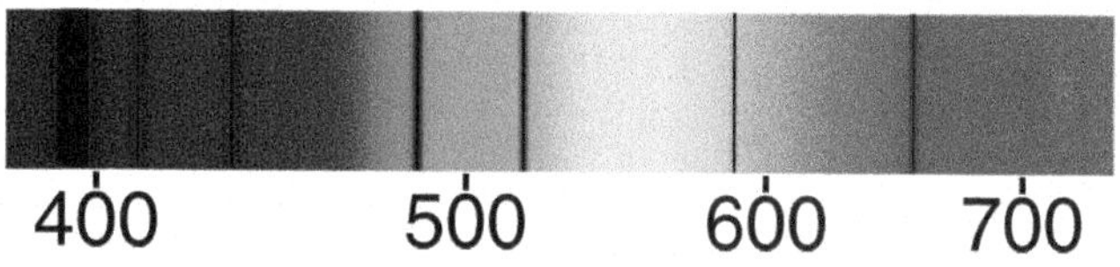

In many cases, missing colours in the spectra of stars are shifted from their expected positions. A shift to lower or 'redder' frequencies is called a **red shift** and results from a star's movement away from the Earth. A shift to higher or 'bluer' frequencies is called a **blue shift** and is caused by a star's movement towards the Earth.

Nearby objects show a range of Doppler shifts. Some stars, like Sirius (the Dog star), are moving away from us and others are moving slowly towards us. Some even show alternate red and blue shifts in step with changes in brightness, suggesting that these stars have an invisible dark companion orbiting them. The brightness is reduced as the circling star passes between us and the main star, while the Doppler shift is caused by the main star moving in response to the gravitational pull of its dark companion.

6.4.3 Retreating galaxies

On a much larger scale, the study of the Doppler shifts of galaxies provides us with an amazing picture of the universe. A relatively small number of galaxies, including the nearby Andromeda galaxy, are moving towards the Earth, but the majority of galaxies are moving away from us at a considerable speed. Even more extraordinary is the relationship between the size of the red shift and the distance from Earth. This was first investigated by the astronomer Edwin Hubble and is now referred to as Hubble's law. Hubble proposed this law in 1929 while working at what was then the largest telescope in the world in California. The law states that the further away a galaxy is, the greater its red shift and so the faster it is moving away from us.

While this finding appears to put the Earth in a very special position at the centre of a rapidly expanding universe, it is in fact an illusion. Observers anywhere in the universe will see the surrounding galaxies moving away from them at a speed that is consistent with Hubble's law.

6.4.4 Estimating the size of the known universe

We can only say that the universe is as big as we can see, so the size of the known universe has steadily increased over the centuries as observation techniques have improved. The Hubble Space Telescope is finding more and more distant objects and so the known universe is still getting bigger. Objects have been seen at distances estimated to be about 14 billion light-years from Earth, which suggests that the universe is at least 14 billion years old.

6.4 Exercises: Understanding and inquiring

To answer questions online and to receive **immediate feedback** and **sample responses** for every question, go to your learnON title at www.jacplus.com.au. *Note:* Question numbers may vary slightly.

Remember

1. Why are there black lines in the spectra of the light emitted by stars?
2. Which colour of light has the higher frequency — red or blue?
3. What is a red shift? What does it tell us about how a star is moving relative to the Earth?
4. What is Hubble's law?

Think

5. The light from a star is often analysed by its wavelength instead of its frequency. Long wavelengths correspond to low frequencies and short wavelengths correspond to high frequencies. The spectrum of colours emitted by excited atoms of hydrogen on Earth contains the wavelength 656.285 nm. This same wavelength is observed in the spectrum of light from the bright star Vega at 656.255 nm. Is Vega moving towards or away from Earth?
6. The most distant objects in the universe are estimated to be about 14 billion light-years from Earth. Explain why the age of the universe is thought to be at least 14 billion years.

Imagine

7. Collect and summarise a media report about a new discovery about space outside the solar system.

learnON RESOURCES — ONLINE ONLY

Try out this interactivity: Shifting spectral lines (int-0678)

Complete this digital doc: Worksheet 6.4: The expanding universe (doc-19468)

6.5 How it all began

6.5.1 How it all begin

When and how did the universe begin? Was there a beginning? Perhaps it was always there. If there was a beginning, will there be an end? The study of the answers to these questions is called **cosmology**.

Following Edwin Hubble's discoveries about the expanding universe, two major theories about the beginning of the universe became popular — the **big bang theory** and the **steady state theory**.

learnON RESOURCES — ONLINE ONLY

Watch this eLesson: The expanding universe (eles-0038)

6.5.2 The big bang

According to the most commonly accepted theory among cosmologists, the universe began about 15 billion years ago with a 'big bang'.

1. **The big bang ($t = 0$)**

   It's hard to imagine, but at this moment there was no space and no time. All that existed was energy, which was concentrated into a single point called **singularity**.

2. **One ten million trillion trillion trillionths of a second later ($t = +\frac{1}{10^{43}}$ s)**

   Time and space had begun. Space was expanding quickly and the temperature was about 100 million trillion trillion degrees Celsius. (The current core temperature of the sun is 15 million degrees Celsius.)

3. **One ten billion trillion trillionths of a second after the big bang ($t = +\frac{1}{10^{34}}$ s)**
The universe had expanded to about the size of a pea. Matter in the form of tiny particles such as electrons and **positrons** *(positively charged electrons) had formed. Particles collided with each other, releasing huge amounts of energy in the form of light. Until this moment there was no light.*
4. **One ten thousandth of a second after the big bang ($t = +\frac{1}{10^{4}}$ s)**
Protons and neutrons had formed as a result of collisions between smaller particles. The universe was very bright because light was trapped as it was continually being reflected by particles.
5. **One hundredth of a second after the big bang ($t = +\frac{1}{100}$ s)**
The universe was still expanding and cooling rapidly. It had grown to the same size as our solar system, but there was still no such thing as an atom.
6. **One second after the big bang (t = +1 s)**
The universe was probably more than a trillion trillion kilometres across. It had cooled to about ten billion degrees Celsius.
7. **Five minutes after the big bang (t = +5 min)**
The nuclei of hydrogen, helium and lithium had formed among a sea of electrons.
8. **Three hundred thousand years after the big bang (t = +300 000 years)**
The universe was about one thousandth of its current size. It had cooled to about 3000 °C. Electrons had slowed down enough to be captured by the nuclei of hydrogen, helium and lithium, forming the first atoms. There was now enough empty space in the universe to allow light to escape to the outer edges. For the first time, the universe was dark.
9. **Two hundred million years after the big bang (t = +200 000 000 years)**
The first stars had appeared as gravity pulled atoms of hydrogen, helium and lithium together. **Nuclear reactions** *took place inside the stars, causing the nuclei of the atoms to fuse together to form*

heavier nuclei. Around some of the newly forming stars, swirling clouds of matter cooled and formed clumps. This is how planets began to form.

10. **One billion years after the big bang (t = +1 000 000 000 years)**

    The universe was beginning to become a little 'lumpy'. The force of gravity pulled matter towards the 'lumpier' regions, causing the first galaxies to form.

The Einstein connection

The big bang theory would not make any sense at all if it were not for Albert Einstein's famous equation. How could matter be created from nothing? Well, the singularity before the big bang was not 'nothing'. It was a huge amount of energy (with no mass) concentrated into a tiny, tiny point.

Einstein proposed that energy could be changed into matter. His equation $E = mc^2$ describes the change.

E represents the amount of energy in joules.

m represents the mass in kilograms.

c is the speed of light in metres per second (300 000 000 m/s).

Einstein's equation also describes how matter can be changed into energy. That is what happens in nuclear power stations, nuclear weapons and stars.

WORKING WITH BILLIONS AND TRILLIONS

One billion is equal to one thousand million; that is, 1 000 000 000 or 10^9.

One trillion is equal to one thousand billion; that is, 1 000 000 000 000 or 10^{12}.

So one billion trillion is 1 000 000 000 000 000 000 000 or 10^{21}.

When numbers get that large, there are too many zeros to count. It is much easier to use powers of ten notation, also known as scientific notation.

6.5.3 The steady state theory

According to the steady state theory, which was proposed in 1948, there was no beginning of the universe. It was always there. The galaxies are continually moving away from each other. In the extra space left between the galaxies, new stars and galaxies are created. These new stars and galaxies replace those that move away, so that the universe always looks the same.

6.5.4 The great debate

A huge debate between those who supported the steady state theory and those who supported the big bang theory raged from 1948 until 1965. During that period, the evidence supporting the big bang theory grew.

The red shift

The red shift provides evidence for an expanding universe. This evidence supports the big bang theory and causes problems for those supporting the steady state theory. A steady state universe could expand only if new stars and galaxies replaced those that moved away. There is no way to explain how these new stars and galaxies could be created from nothing. Apart from that, these young stars and galaxies have not been found by astronomers.

The elements

The amounts of hydrogen and helium in the universe support the big bang theory. According to the steady state theory, the only way that helium can be produced is by the nuclear reactions taking place in stars. About 8.7 per cent of the atoms in the universe are helium. This is far more than could be produced by the stars alone. The percentage of helium atoms can, however, be explained by their creation as a result of the big bang.

The afterglow

When George Gamow and Ralph Alpher proposed their version of the big bang theory in 1948, they calculated that the universe would now, about 15 billion years after creation, have a temperature of 2.7 °C above **absolute zero**. That's −270 °C. Anything with a temperature above absolute zero emits radiation. The nature of the radiation depends on the temperature. Gamow predicted that, because of its temperature, the universe would be emitting an 'afterglow' of radiation. This afterglow became known as cosmic microwave background radiation.

HOW ABOUT THAT!

The big bang theory was first proposed in 1927 by Georges Lemaitre, a Catholic priest from Belgium. But it wasn't called the 'big bang theory' then. Ironically, the name 'big bang' was invented by Fred Hoyle, one of the developers of the steady state theory. He used the name to try to ridicule the cosmologists who proposed the big bang theory.

In 1929, Edwin Hubble's observations and law provided convincing evidence to support Lemaitre's proposal. In 1933, Lemaitre presented the details of his theory to an audience of scientists in California. Albert Einstein, by then recognised as one of the greatest scientists of all time, was in the audience. At the end of Lemaitre's presentation Einstein stood, applauded and announced, 'That was the most beautiful and satisfactory explanation of creation that I have ever heard'.

This radiation was discovered by accident in 1965. Engineers trying to track communications satellites picked up a consistent radio noise that they couldn't get rid of. The noise wasn't coming from anywhere on Earth, because it was coming from all directions. It was the cosmic microwave background radiation predicted by Gamow. Its discovery put an end to the steady state theory, leaving the big bang theory as the only theory supported by evidence currently available. Even Fred Hoyle, who had ridiculed the idea of a 'big bang', admitted that the evidence seemed to favour the big bang theory.

6.5.5 Mapping the universe

In 1989, a satellite named COBE (COsmic Background Explorer) was put into orbit around Earth to accurately measure the background radiation and temperature of the universe. COBE could detect variations as small as 0.00003 °C. As predicted by Gamow, it detected an average temperature of −270 °C.

In 2001, a probe called WMAP (Wilkinson Microwave Anisotropy Probe) was sent into orbit around Earth at a much greater distance to gather even more accurate data, detecting temperatures within a millionth of a degree. WMAP's first images were released by NASA in February 2003.

The computer-enhanced image of cosmic microwave background radiation shown at right was produced by the WMAP mission. The background radiation detected was released only 380 000 years after the big bang — the first radiation to escape. The image shows how the temperature varied across the universe as it was 380 000 years after the big bang. The blue parts of the map are the cooler regions. These regions were cool enough for atoms, and eventually

WMAP image of cosmic microwave background radiation

galaxies, to form. The red parts are warmer regions. The map shows that galaxies are not evenly spread throughout the universe. They support the theory of an expanding universe that began with a big bang.

The Wilkinson Microwave Anisotropy Probe (WMAP). Its main mission was to gather evidence to help cosmologists find out how the universe began and predict what will happen in the future.

6.5.6 Will it ever end?

Will the expansion of the universe continue forever? If the universe does stop expanding, what will happen to it? There are several competing theories about the answers to these questions. One theory suggests that there is not enough mass in the universe for gravity to be able to pull it all back, so it will continue to expand forever. Other theories suggest that the universe will eventually end. According to these theories, the end will come when:

- the universe snaps back onto itself in a 'big crunch' (the **big crunch theory**). If this happens, the end result will be a single point — singularity. Some cosmologists believe that the big crunch will be followed by another big bang.
- the expansion of the universe continues and stars use up their fuel and burn out, causing planets to freeze (the **big chill theory**). The universe would then consist of scattered particles that never meet again.
- the universe rips itself apart violently as a result of expanding at an increasing speed (the **big rip theory**). According to this theory, the end of the universe will also be the end of time itself.

learnon RESOURCES — ONLINE ONLY

Watch this eLesson: Entropy (eles-1073)

6.5 Exercises: Understanding and inquiring

To answer questions online and to receive **immediate feedback** and **sample responses** for every question, go to your learnON title at www.jacplus.com.au. *Note:* Question numbers may vary slightly.

Remember

1. What is the science of cosmology?
2. How old is the universe believed to be?
3. According to the big bang theory, what was there at the time of the universe began?
4. Why could there not have been anything before the big bang?
5. Approximately how long after the big bang did:
   (a) time and space begin to exist
   (b) matter appear
   (c) protons and neutrons form
   (d) neutral atoms first exist
   (e) galaxies begin to form?
6. How did galaxies begin to form?
7. What does Einstein's famous equation have to do with the big bang theory?
8. Which of the two theories about the 'beginning' of the universe proposed that there was no beginning?
9. How did the steady state theory explain that the universe was expanding, yet remained the same?
10. What evidence put an end to the steady state theory?
11. List three major pieces of evidence that supported the big bang theory.
12. Name and describe three theories about how the universe might end.

Think

13. What would have happened to the universe if, one million years after the big bang, the matter in it was perfectly evenly distributed and not moving?
14. Explain why neutral atoms were not likely to form during the first five minutes after the big bang.
15. WMAP is able to provide a picture of the universe as it was 380 000 years after the big bang. Why is it unable to provide a map of the universe as it was before that time?
16. Why go to the expense of measuring background radiation with a satellite or space probe when it could be done from Earth?

Create

17. Draw flowcharts to describe:
    (a) the big bang theory
    (b) how the big bang and big crunch cycle might work together.

learnon RESOURCES — ONLINE ONLY

Complete this digital doc: Worksheet 6.5: The big bang (doc-19469)

6.6 Eyes on the universe

Science as a human endeavour

6.6.1 Detecting radio waves

For hundreds of years, light telescopes have been used to observe what lies beyond the solar system. To find out what's in deep space, in the most distant parts of the universe, observing visible light is not enough. We rely on other parts of the electromagnetic spectrum.

Until the accidental discovery in 1931 that stars emitted radio waves as well as light, the only way to observe distant stars and galaxies was with light telescopes. Like light and other forms of **electromagnetic radiation**, radio waves travel through space at a speed of 300 000 kilometres per second. Radio waves from deep in space are collected by huge dishes and reflected towards a central antenna. The waves are then analysed by a computer, which produces an image that we can see.

The Arecibo dish in Puerto Rico is the largest single radio telescope in the world. It is 305 metres across.

Radio telescopes can detect tiny amounts of energy. In fact, the total amount of energy detected in ten years by even the largest radio telescopes would light a torch globe for only a fraction of a second. They can detect signals from much further away than light telescopes.

Unlike light waves, radio waves can travel through clouds in the Earth's atmosphere, and can be viewed in daylight as well as at night. Radio waves also pass through clouds of dust and gas in deep space.

Sharpen up!

Images produced by single radio telescopes are not very sharp. To solve this problem, signals from groups of telescopes pointed at the same object are combined to produce sharper images.

6.6.2 Learning from radio waves

As well as telling us about the size, shape and movement of every type of star (from our own sun to stars at the outer edges of the universe), radio telescopes reveal information about a star's temperature and the substances from which it is made. Radio telescopes can work out what a star is made up of by using the fact that different elements emit different frequencies of radio waves.

Radio waves have, among other things, allowed us to:

- analyse the distribution of stars in the sky
- discover **quasars**, which, before 1960, were believed to be normal stars. They are like stars, but emit a lot more radiation and are travelling away from us at huge speeds. Quasars are believed to be the most distant objects in the universe.
- discover **pulsars**, which are huge stars that have collapsed, emitting radio waves. Because pulsars spin rapidly — a bit like a lighthouse — the radio waves reach the Earth as radio pulses.

6.6.3 Eyes in orbit

There are more than 2500 satellites currently orbiting the Earth, many of them constantly watching the Earth's surface and atmosphere. Others provide views of the universe that could never be seen from the Earth's surface.

Trash 'n' treasure in orbit

Some of the satellites orbiting the Earth are active and use radio signals to send streams of data down to the surface. Others have stopped working but continue to circle the globe. Some satellites in lower orbits will gradually slow down as a result of the thin atmosphere. They will spiral in towards the Earth in a fiery finish as they burn up on re-entry. The fate of others far beyond the atmosphere is an eternity of circling the Earth. They have joined the pile of 'space junk' gradually accumulating in near-Earth orbit.

The Very Large Array in New Mexico consists of 27 dishes, each with a diameter of 25 metres, arranged in a Y shape. This is the equivalent of a single radio telescope with a diameter of 35 kilometres.

All satellites orbiting the Earth are held there by the Earth's gravitational pull directed to the planet's centre. This means that the centre of every orbit coincides with the centre of the Earth. Some orbits skim as close as a few hundred kilometres above the surface. Others take a more distant view. The time taken for one complete revolution (the period of orbit) of a satellite depends on its height above the Earth. Greater heights result in greater periods.

Looking in, looking out

Artificial satellites can be used to look at the Earth or to look into space. An inward-looking satellite can sweep the surface of the Earth every day, using cameras and remote sensors to observe and measure events on the surface hundreds or thousands of kilometres below. An outward-looking satellite can see directly into space, its view unobstructed by the atmosphere, pollution or dust. Light pollution, an increasing problem for Earth-bound observers as our cities grow, is not an issue for an observer in space.

Inward-looking satellites are used for:

- collecting weather and climate data, providing early warning of events (such as volcanic activity and changing ocean currents) and showing long-term trends
- collecting data used for mineral exploration, crop analysis, mapping, and identifying long-term erosion or degradation
- strategic defence ('spy-in-the-sky') systems
- communications for telephones, television, radio and computer data.
- Outward-looking satellites are used for:
- observing the other planets and bodies circling the sun
- observing stars, galaxies and other remote objects in space
- watching for comets and asteroids that may hit the Earth
- listening for signs of extraterrestrial life.

The International Space Station

The primary purpose of the International Space Station (ISS), completed in 2011 with the support of a number of different space agencies, was to provide laboratories in space for research into microgravity and fields such as medicine, geology and technology. The ISS also provides the opportunity to investigate the effect of a space environment on humans and prepare for the exploration of Mars and beyond by humans. Crew members, which can include astronauts, scientists of all kinds and engineers, are generally on board for several months before returning to Earth.

The International Space Station in orbit

However, the ISS, with the recent addition of sophisticated sensors and other new technology, is now also being used as an inward-looking satellite. It is able to provide images similar to those from other inward-looking satellites, but has the advantage of having a crew who can respond to unfolding events immediately, rather than waiting for further 'instructions' from the ground. This is especially helpful when natural disasters such as volcanic eruptions, earthquakes and tsunamis occur. In addition, the orbit of the ISS is different from most other satellites and is able to collect images at different and often more suitable times.

Hubble Space Telescope

The Hubble Space Telescope is an example of an outward-looking satellite. It was carried into orbit about 600 kilometres above the Earth's surface by the space shuttle *Discovery* in 1990. The Hubble Space Telescope collects images by collecting and analysing data in the form of visible light, **ultraviolet radiation** and **infra-red radiation** from deep space. It produces spectacularly clear images that are relayed back to Earth by radio waves.

The Hubble Space Telescope is the only telescope designed to be serviced and improved in space by astronauts.

The Hubble Space Telescope was the first space telescope that could be serviced while in orbit, and its useful life has been dependent on transporting astronauts to and from Earth aboard space shuttles. Now that NASA's space shuttle program has ceased, servicing is no longer possible. When the orbiting telescope stops functioning it will be 'deorbited' by an unmanned space mission so that it plunges harmlessly into the ocean.

The Hubble Space Telescope will eventually be replaced by the James Webb Space Telescope, which will collect infra-red radiation from the most distant parts of the universe. At the time this book was published, the launch is scheduled for some time in 2018. The uncertainty of the launch date is not surprising, because the James Webb project is a collaboration between three space agencies: NASA, the European

Space Agency (ESA) and the Canadian Space Agency (CSA). Each of these agencies is dependent on government funding, which is often uncertain.

There are several other space telescopes in orbit around the Earth. They all collect radiation from parts of the electromagnetic spectrum and send images and other data back to Earth using radio waves. They include the Chandra X-ray Observatory, carried into orbit by the space shuttle *Columbia* in 1999. Most **X-rays** from space approaching the Earth are absorbed by the atmosphere. By collecting high-energy X-rays coming from neutron stars and black holes, Chandra is able to gather data that could never be collected by X-ray telescopes on Earth's surface.

Data overload?

The unprecedented amount of data coming from telescopes of all types on the ground and in orbit requires processing by **supercomputers** with capabilities well beyond those of personal computers and even large computers used in most industries. IT specialists play a crucial role in developing new and faster computer systems to ensure that exploration of the universe is not limited by data overload.

learnon RESOURCES — ONLINE ONLY

Watch this eLesson: Hubble and the expansion of the universe (eles-1766)

6.6.4 An accelerating expansion

In 1997, a group of astrophysicists, including Australian National University's Professor Brian Schmidt presented evidence that the expansion of the universe was speeding up. Their research took several years to complete and required the analysis of images of the most distant supernovas that could be observed. The images were from light and infra-red telescopes in several locations, including Australia, and from the Hubble Space Telescope. The successful outcome of the research was only possible with the use of advanced digital imaging sensors and powerful state-of-the-art computers. Brian Schmidt and two other members of the team of astrophysicists were awarded the Nobel Prize for Physics in 2011 for their research.

Professor Brian Schmidt

6.6.5 Telescope of the future

The Square Kilometre Array (SKA) is a new radio telescope currently being designed with the cooperation of ten countries including Australia, New Zealand, the United Kingdom, South Africa and China. It is likely that more countries will join the project as it progresses.

The SKA will consist of thousands of dish-shaped antennas in South Africa and Australia, linked together by optical fibre. It will be about 50 times as

An artist's impression of the dishes that will be installed in the Karoo desert in South Africa

sensitive as the best of the current generation of radio telescopes and will be able to scan the sky up to 10 000 times faster. The SKA is scheduled to be in full operation in 2024. It will allow astronomers to investigate events that took place within the first second after the big bang and fill gaps in our knowledge of the period when the universe became dark and the first galaxies formed.

The SKA is sensitive enough to detect weak extraterrestrial signals (if they exist) and planets in other galaxies capable of supporting life. The SKA should also provide answers to questions about distant galaxies, **dark energy**, gravity and magnetism.

The amount of data the SKA will collect in a single day would take almost two million years to play back on an iPod. Its processing power will be the equivalent of about 100 million personal computers.

Of course, the completion of the SKA is dependent on funding from governments of the participating nations. Whether it is completed on time, or at all, depends on the continuation of current funding.

6.6 Exercises: Understanding and inquiring

To answer questions online and to receive **immediate feedback** and **sample responses** for every question, go to your learnON title at www.jacplus.com.au. *Note:* Question numbers may vary slightly.

Remember

1. Describe at least two advantages of Earth-based radio telescopes over light telescopes.
2. Images produced by single radio telescopes are not very sharp. Explain how this problem is solved.
3. List the information that can be revealed by images from radio telescopes.
4. Outline the advantages of telescopes in orbit over telescopes on the Earth's surface.
5. Which part or parts of the electromagnetic spectrum have been collected by the Hubble Space Telescope?
6. State two reasons why the International Space Station is useful as an inward-looking satellite.
7. Explain why fast computers are important to exploration of the universe.

Think

8. Explain why orbiting space telescopes have a limited lifetime.
9. Outline at least four reasons why there can be no certainty about the launch dates of future space missions and projects like the Square Kilometre Array.

Investigate

10. Find out about the research conducted by the Centre for Astrophysics and Supercomputing in Melbourne. Report on:
    (a) the areas of research being conducted
    (b) the range of courses available for students with an interest in the universe
    (c) career opportunities in astrophysics and supercomputing.
11. Australia has always played a crucial role in space missions and space exploration. Use the internet to research and report on the role of each of the following Australian facilities in space exploration and the way in which they are funded.
    (a) Australia Telescope Compact Array
    (b) Canberra Deep Space Communication Complex
    (c) Parkes radio telescope
12. Identify one scientist, engineer or IT specialist who works at one of the facilities listed in question 11 and write a brief report about his or her role in investigating the universe.

6.7 Priority grids and matrixes

6.7.1 Priority grids and matrixes

how to ...?

1. Draw two continuums that cross through each other at right angles.
2. Divide each line into six equal parts.
3. Put a label such as 'Difficult' on the left end of the horizontal line and 'Easy' on the right.
4. Put a label such as 'High reward' at the top of the vertical lines and 'Low reward'at the bottom.
5. Think of an activity and assess it using these two line, placing a mark where you think it fits best. Repeat this for other activities or ideas.
6. Compare and discuss your marked positions with those of others in your class.Share your ideas, values, views and judgements, and listen to those of others.
7. After your discussions and reflections, write your final positions directly onto the grid.

why use?

Helps you make decisions and see how your views and judgements compare with others

question

Which is the best option to follow and why?

also called

Priorities grid; decision grid

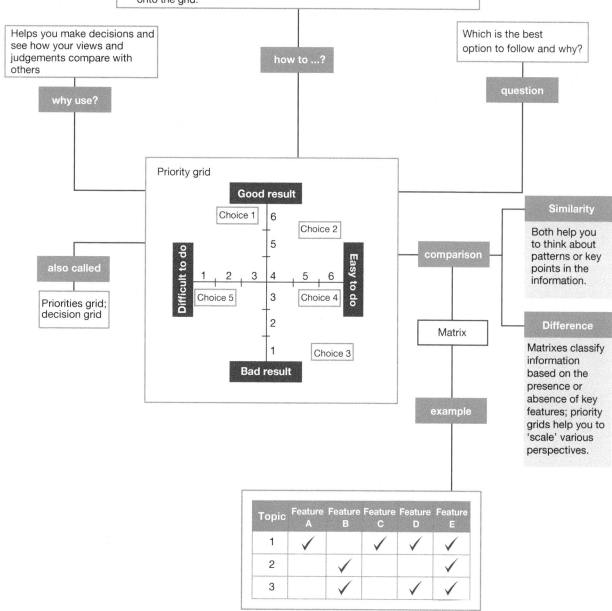

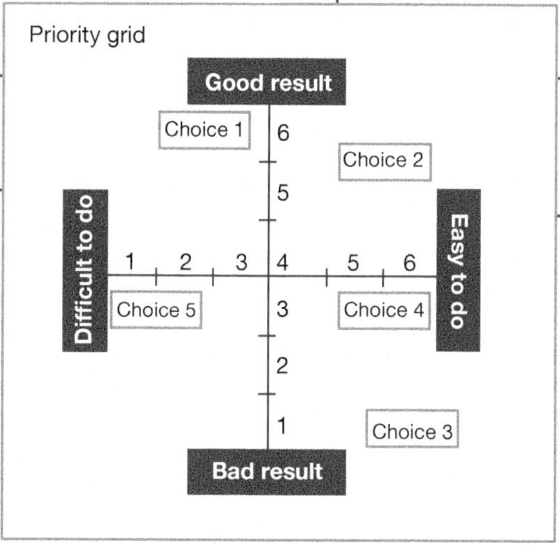

comparison

Similarity

Both help you to think about patterns or key points in the information.

Difference

Matrixes classify information based on the presence or absence of key features; priority grids help you to 'scale' various perspectives.

Matrix

example

Topic	Feature A	Feature B	Feature C	Feature D	Feature E
1	✓		✓	✓	✓
2		✓			✓
3		✓		✓	✓

6.7 Exercises: Understanding and inquiring

To answer questions online and to receive **immediate feedback** and **sample responses** for every question, go to your learnON title at www.jacplus.com.au. *Note:* Question numbers may vary slightly.

Think and create

1. Use a priority grid to evaluate each of the following current and future challenges in space exploration.
   (a) Extending the life of the International Space Station
   (b) Continuing funding for the Square Kilometre Array
   (c) Researching supernovas to find out what happened after the big bang
   (d) Building and operating a permanent base on Mars
   (e) Sending a space probe to Proxima Centauri
   (f) Searching for extraterrestrial life forms
2. A matrix can be used to compare the twentieth century's two competing theories about the universe. Copy and complete the matrix below, using ticks to show which statements apply to one, both or neither of the theories.

A permanent base on Mars is a real possibility. But how important is it? Are the benefits worthwhile? A priority grid can be helpful in answering questions like this.

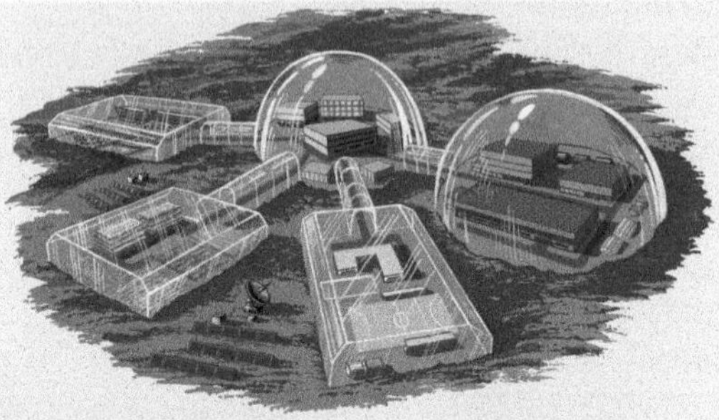

Statement	Big bang theory	Steady state theory
The universe has no beginning.		
The universe began with a single point called singularity.		
The universe is expanding.		
The universe always looks the same.		
The red shift in the spectrum of visible light coming from stars and galaxies provides evidence for the theory.		
New stars and galaxies are created to replace those that move away due to expansion of the universe.		
This theory explains the amount of helium in the universe.		
This theory is supported by the measurement of the current temperature of the universe (about –270 °C).		
This theory was first supported by a Catholic priest.		
The theory will never be proven incorrect.		
An end was put to this theory in 1965.		

3. Use matrixes to compare:
   (a) radio telescopes and light telescopes
   (b) inward-looking satellites to outward-looking satellites
   (c) red giants and white dwarfs
   (d) three theories about how the universe might end
   (e) living in space and living on Earth.

New stars are forming right now in nebulae like this throughout the universe. According to one theory, this has been happening forever; according to another theory it has been happening for only about 14 billion years.

6.8 Review

6.8.1 Study checklist

Stars

- describe and distinguish between planets, stars, constellations, galaxies and nebulae
- describe and explain the motion of stars and planets of the solar system as seen from Earth
- identify the sun as a star
- explain how stars are able to emit energy
- describe the lifetime of stars of different sizes and appreciate the timescale over which changes in stars take place
- interpret the Hertzsprung–Russell diagram in terms of the absolute magnitude, temperature and classification of stars
- distinguish between absolute and apparent magnitude

The changing universe

- identify evidence supporting the big bang theory, such as Edwin Hubble's observations and the detection of background microwave radiation
- compare the big bang theory with the steady state theory
- describe how the universe has changed since the big bang and how it might continue to change in the future

Science as a human endeavour

- describe how radio telescopes and arrays of radio telescopes are used by astronomers and astrophysicists to observe distant parts of the universe
- explain how orbiting space telescopes are used to gather data from deep space and how they compare with Earth-based telescopes
- recognise the role of Australian astronomers and astrophysicists and facilities such as telescopes, arrays and observatories in the exploration and study of the universe
- recognise the importance of IT specialists and the development of fast computers in processing the data obtained by Earth-based and orbiting telescopes
- appreciate that the study of the universe and the exploration of space involves teams of specialists from different branches of science, engineering and technology
- recognise that financial backing from governments or other organisations is required for major scientific investigations and that this can determine if and when research takes place

Individual pathwayws

ACTIVITY 6.1	ACTIVITY 6.2	ACTIVITY 6.3
The mysterious universe doc-8476	Investigating the universe doc-8477	Investigating the universe further doc-8478

learnon ONLINE ONLY

6.8 Review 1: Looking back

To answer questions online and to receive **immediate feedback** and **sample responses** for every question, go to your learnON title at www.jacplus.com.au. *Note:* Question numbers may vary slightly.

1. Solve the crossword puzzle shown.

ACROSS

2. Square Kilometre Array (abbreviation)
5. The constellation the Saucepan is a part of (also known as the Hunter)
6. The name given to the range of colours of visible light
8. The distance travelled by light in a year
12. The closest star to the sun: _______ Centauri
13. The galaxy of which the solar system is a part (two words)
14. Most of the interstellar matter between the stars consists of this element.

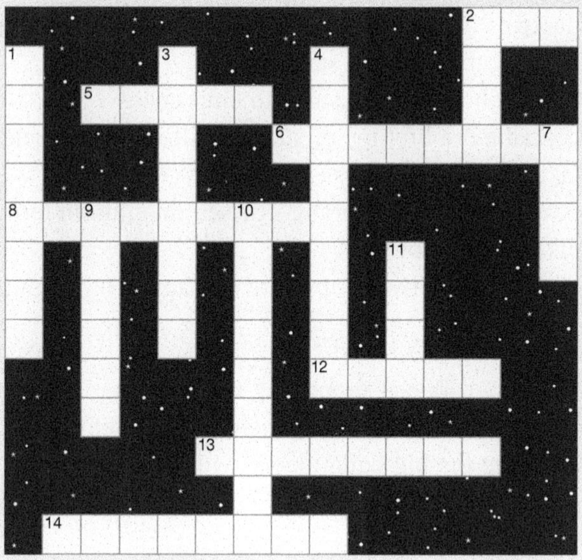

DOWN

1. An effect that shows that some stars are closer to us than others
2. The sun is an example.
3. The famous equation $E = mc^2$ is attributed to this man.
4. The violent fate of some very massive stars
7. Most stars fall into this sequence on the Hertzprung–Russell diagram.
9. A group of stars. The solar system is a tiny speck in one such group.
10. The universe seems to be doing this.
11. A probe launched in 2001 to map cosmic microwave background radition.

2. Why are the constellations we see now so different from the way they were many centuries ago?
3. During which process is energy emitted by stars? Describe the process.
4. Explain the difference between the apparent magnitude of a star and its absolute magnitude.
5. Use the data in the table in subtopic 6.3 to answer the following questions.
   Which of the stars Alpha Centauri, Betelguese and Rigel:
   (a) is brightest when viewed from Earth on a clear night
   (b) has the greatest actual brightness
   (c) is faintest when viewed from the Earth on a clear night?
6. How have scientists gained their knowledge of the life and death of stars if the processes involved take millions of years to occur?
7. What is the difference between a neutron star and a black hole?
8. What makes stars group together to form galaxies?
9. The Doppler effect is most commonly associated with the changing pitch of a sound as its source moves past you. For example, the pitch of the noise made by a speeding train increases as it approaches you and decreases as it moves away from you. Explain how the Doppler effect is relevant to the study of the universe.
10. State Hubble's Law.

11. Two different theories about the beginning of the universe emerged during the twentieth century.
    (a) Name the two theories.
    (b) Which theory proposed that there was no beginning?
    (c) Which theory lost favour in 1965? Why did it lose favour?
12. In your own words, write an account (about 200 words) of the first second after the big bang.
13. Which of the three theories about the end of the universe described in subtopic 6.5 do you think is the most likely to be correct? Give reasons for your answer.
14. For what do each of the following abbreviations stand?
    (a) COBE
    (b) WMAP
    (c) ISS
15. What is cosmic microwave background radiation and why does it exist?
16. At what speed do radio waves travel through space?
17. Outline two major advantages of using radio telescopes instead of light telescopes to view events in deep space from the Earth's surface.
18. Describe the original purpose of the International Space Station.
19. For what contribution to our knowledge of the expansion of the universe was Brian Schmidt awarded the Nobel Prize?

learnON RESOURCES — ONLINE ONLY

Complete this digital doc: Worksheet 6.6: The mysterious universe: Summary (doc-19470)

TOPIC 7
Global systems

7.1 Overview

7.1.1 Why learn this?

We are living in the anthropocene era — an age in which humans are dominating and disrupting many of our planet's natural systems. Is it time for us to recognise our effect and take responsibility for our actions? How much further can we push our global life-support systems? Within the next century, will our species be a mere footprint on what is left of Earth?

7.1.2 Think about global systems

assesson

- Which organism is being blamed for causing the sixth mass extinction?
- What has both a 'layer' and a 'zone' in it?
- When is the 'laughing gas' nitrous oxide nothing to laugh about?
- If global cooling did increase the size of the human brain, what effects might global warming have?
- Are humans still evolving?
- Are you a climate-change sceptic?

LEARNING SEQUENCE

Numerous **videos** and **interactivities** are embedded just where you need them, at the point of learning, in your learnON title at www.jacplus.com.au. They will help you to learn the content and concepts covered in this topic.

7.1.3 Your quest

Are you involved in causing the sixth mass extinction?

It has been suggested that humans have unleashed the sixth mass extinction in Earth's history. Human activities such as destroying habitats, overhunting, overfishing, introducing species, spreading diseases and burning fossil fuels are thought to be the key triggers of this mass destruction.

There have been five other mass extinctions recorded over the past 540 million years. Fossil evidence suggests that in each of these other mass extinctions at least 75 per cent of all animal species were destroyed. These extinctions are thought to have been caused by climate changes.

Scientists suggest that, prior to human expansion about 500 years ago, mammal extinctions were very rare. On average, only two species died out every million years. In the last 500 years, however, at least 80 of 5570 mammal species have become extinct. This is alarming in terms of biodiversity.

Of concern is the increasing list of critically endangered or currently threatened species. If these species become extinct and biodiversity loss continues, scientists suggest that the sixth mass extinction could arrive within 3 to 22 centuries. While this may seem like a long timeframe, compared with all but one of the other five mass extinctions it is considered by palaeobiologists to be fast.

The most abrupt mass extinction, in which an estimated 76 per cent of species (including dinosaurs) were wiped out, occurred around 65 million years ago (at the end of the Cretaceous period). It is generally accepted that this was caused by a comet or asteroid crashing into our planet, resulting in firestorms and dust clouds, which in turn led to global cooling. The four earlier mass extinctions are estimated to have taken hundreds of thousands to millions of years as they were due mainly to naturally caused global cooling or warming.

Investigate, think and discuss

1. (a) List examples of human activities that are suggested to be key triggers for the sixth mass extinction.
   (b) Do you agree or disagree with the suggestion that humans are causing a mass extinction? Justify your response.
2. (a) Compare the rate of mammal extinction prior to and after human expansion.
   (b) Suggest what effect this extinction rate has on biodiversity.
   (c) Suggest why scientists are concerned about loss of biodiversity.
3. (a) Research and construct summary reports on the five recorded mass extinctions.
   (b) Select one of the mass extinctions and write a story that could be acted out by characters living during the time of the extinction. Be sure to include examples of biodiversity prior to the mass extinction and then the biodiversity loss during or at the end of the extinction.
   (c) Communicate your story to others using multimedia (e.g. animation, slowmation or documentary), cartoons or songs.

7.2 Revisiting cycles and spheres

7.2.1 Life's Earth support zone

All habitats on Earth are located in what could be considered a life-support zone. This thin layer of our planet includes the atmosphere, the ocean depths, and the upper part of Earth's crust and its sediments.

7.2.2 The biosphere

The **biosphere** is the life-support system of our planet. It consists of the **atmosphere**, **lithosphere**, **hydrosphere** and **biota** (living things), the interactions between them, and the radiant energy of the sun. The biosphere includes all of the ecosystems on Earth. Interactions within the biosphere include the cyclical movement of essential elements such as carbon, nitrogen and phosphorus.

The biosphere can be considered Earth's life-support system.

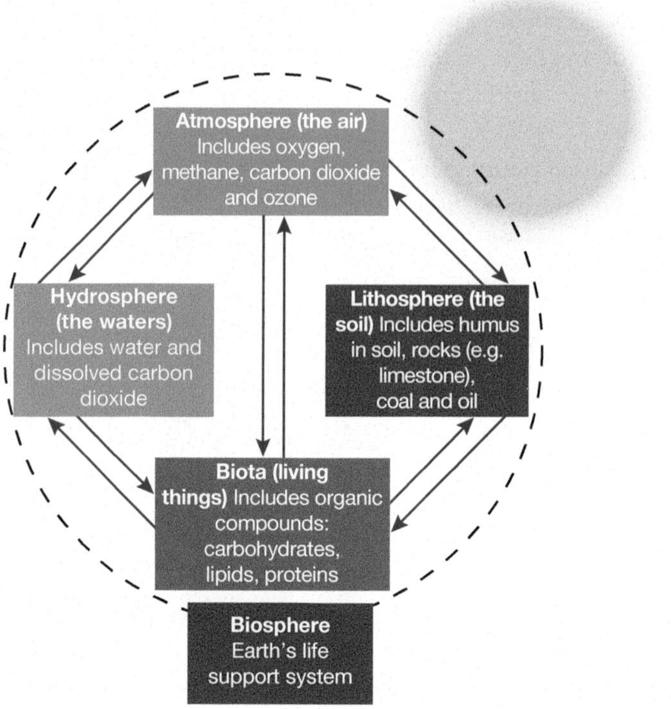

There is pattern, order and organisation within their environments.

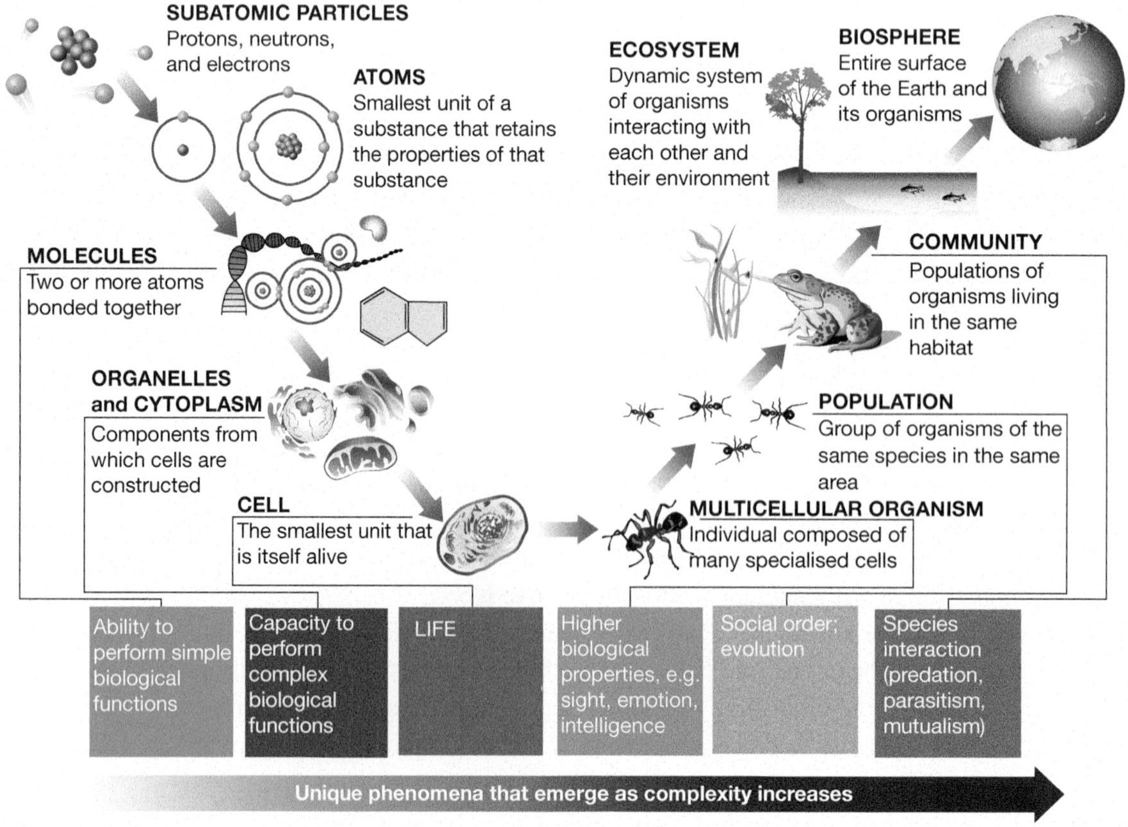

7.2.3 The atmosphere

The Earth's atmosphere is divided into the **troposphere** (lower atmosphere) and the **stratosphere** (upper atmosphere). The troposphere is around 6–17 kilometres high depending on your latitude (how close you are to Earth's equator or the poles). The stratosphere is about 50 kilometres thick and contains an area known as the **ozone layer**. While this layer allows visible and infra-red radiation from the sun through, it absorbs ultraviolet (UV) radiation. This reduces the amount of damaging UV radiation reaching Earth's surface.

Human activity and the atmosphere

Chlorofluorocarbons (CFCs) have been used as cooling agents in refrigerators and air conditioners, as propellants in aerosols, and as industrial solvents. Their use has increased the amount of these compounds being released into the atmosphere. Once in the stratosphere, they are broken down into chlorine atoms, which destroy ozone molecules. This has depleted areas of the ozone layer, increasing the amount of damaging UV rays that get through and cause damage to living organisms.

Layers in the Earth's atmosphere

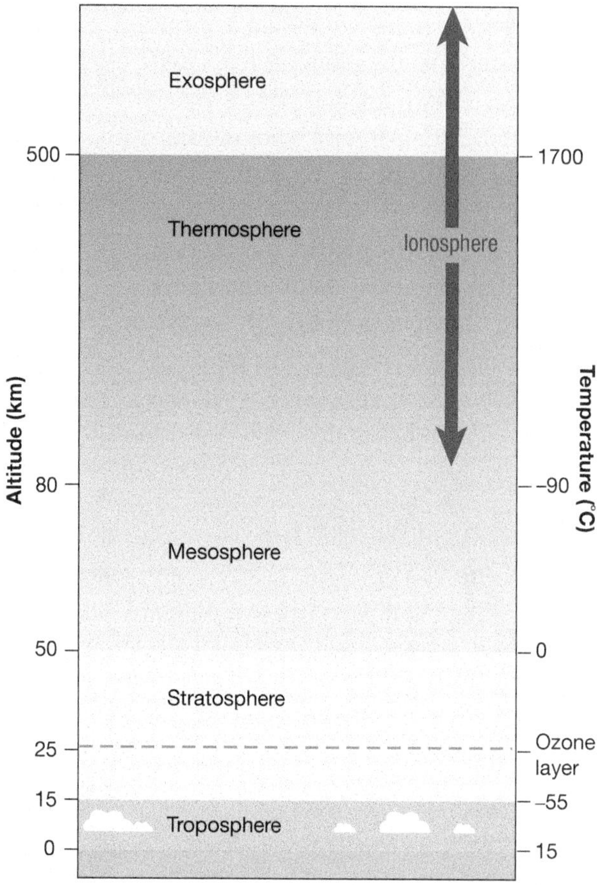

7.2.4 The hydrosphere

The waters of our planet make up the hydrosphere. The simplified figure of the water cycle shown at right describes how water moves through the biosphere.

A simplified view of the water cycle

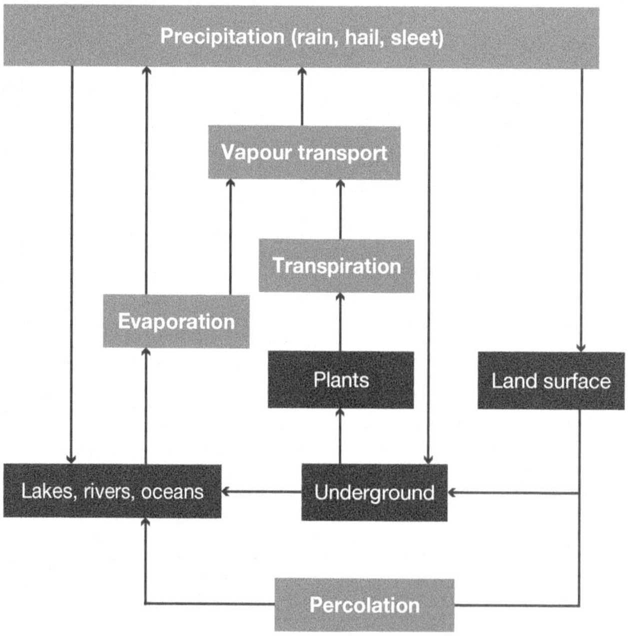

Human activity and the hydrosphere

Toxic or industrial wastes and untreated sewage have made their way into rivers, bays and the ocean, which has had a direct impact on the hydrosphere. Toxins can move along food chains, in some cases being biologically magnified — getting more concentrated — as they move up the chain. While some of these wastes are purposefully dumped, in other cases they enter the water system in run-off from the land or are washed out of the atmosphere in rain.

7.2.5 The lithosphere

Earth's rocky crust and soil make up the lithosphere. It is within this sphere that **igneous**, **sedimentary** and **metamorphic rocks** are formed, broken down and changed from one type to another.

The land surface of our planet is divided into regions called **biomes**. The criterion used to divide regions into biomes is the dominant vegetation type. Environmental factors (such as latitude, temperature and rainfall) influence the type of vegetation that can survive in a particular area and so can be used to predict the type of biome that may exist there. The figure at right shows examples of Earth's biomes and the relationship between the distribution of vegetation types and temperature and rainfall.

The type of dominant vegetation within biomes is influenced by environmental factors.

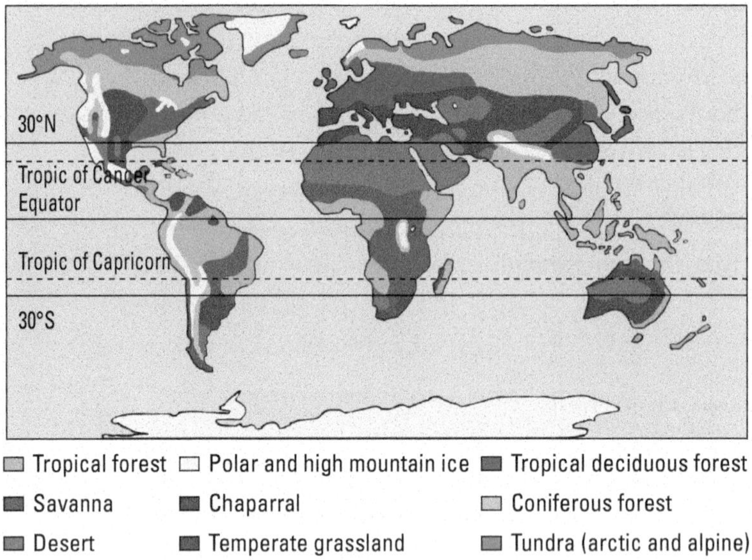

Human activity and the lithosphere

Overstocking, soil exhaustion, salinity, pesticides, unstable landfill, salinisation, toxic seepage, excessive clearing, chemical emissions, deforestation and soil erosion can all be very destructive to the lithosphere. Overgrazing and deforestation may also result in desertification. They can have detrimental effects on habitats and resources and hence the survival of organisms within the ecosystem.

Excessive clearing and deforestation affect the lithosphere.

WHAT DOES IT MEAN?

The term *lithosphere* comes from the Greek words *lithos*, meaning 'stone', and *sphaira*, meaning 'globe' or 'ball'.

7.2.6 The carbon cycle

Carbon is present in various organic and inorganic compounds within the biosphere. It can be found in the hydrosphere as dissolved carbon dioxide, and in the lithosphere as coal or oil deposits and rocks such as limestone. Within the atmosphere it may be present as methane or carbon dioxide, and within living things it may occur as proteins, carbohydrates or lipids.

The carbon cycle models how carbon moves through the biosphere. Carbon travels from the non-living atmosphere to living things when carbon dioxide is absorbed by photosynthetic organisms (such as plants). A simplified version of the carbon cycle is shown at the top of the following page. Can you see the areas within the cycle where the non-living parts of the biosphere (atmosphere, lithosphere and hydrosphere) and the living parts (biota) interact?

Human activity and the carbon cycle

Increased human populations and industrialisation have resulted in an increase in the burning of fossil fuels. Human activity has also led to changed patterns in land use and deforestation. All of these have contributed to an increase in the carbon dioxide that has been added to the atmosphere. Increased levels of this greenhouse gas have added to the enhanced greenhouse effect and global warming. Increased global temperatures may result in melting icecaps, rising sea levels, coastal flooding and unusual weather patterns. These events may threaten the survival of organisms in many ecosystems.

A simplified view of how carbon is cycled within an ecosystem

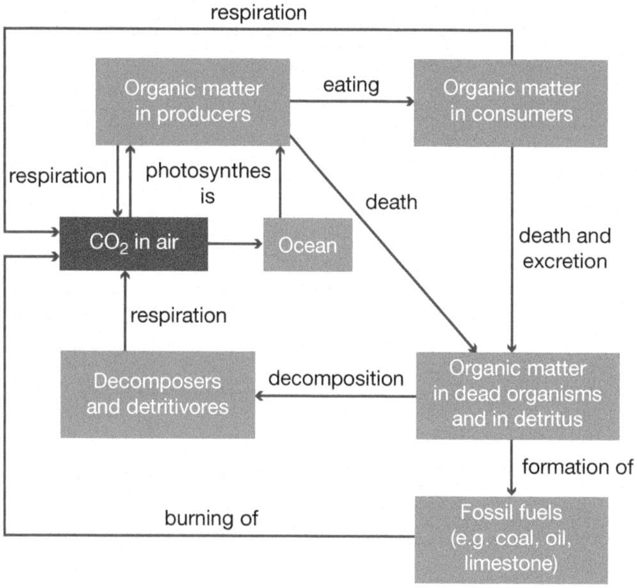

7.2.7 The nitrogen and phosphorus cycles

The nitrogen cycle models how nitrogen moves through the biosphere. A simplified version of this cycle is shown in the figure below. Can you see the areas in which the non-living parts of the biosphere and the living parts interact with each other?

A simplified view of how nitrogen is cycled within an ecosystem

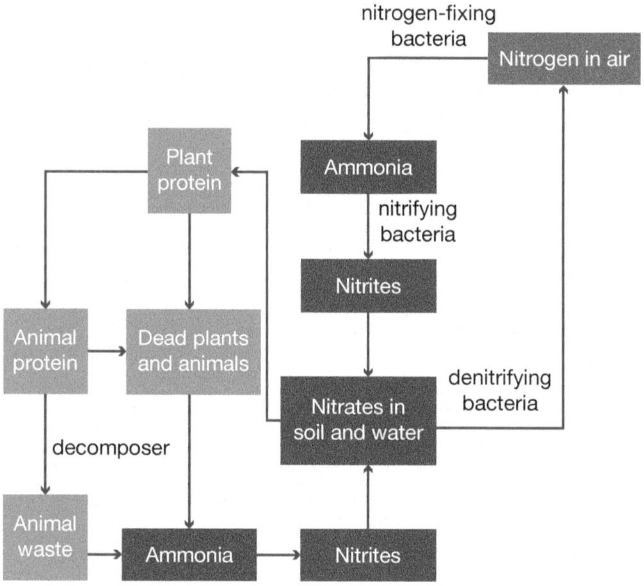

A simplified view of how phosphorus is cycled within an ecosystem

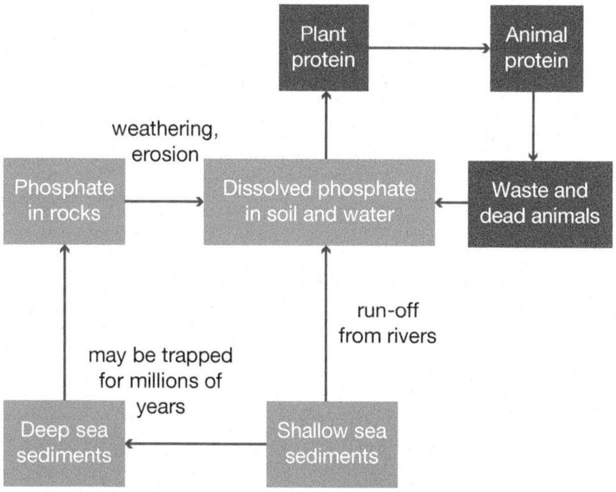

The phosphorus cycle models how phosphorus moves from the lithosphere to the hydrosphere and then through food chains and back.

Human activity and the nitrogen and phosphorus cycles

Large amounts of chemical fertilisers rich in nitrogen and phosphorus have been used on agricultural crops to enhance their growth. The large scale use of these fertilisers has led to considerable quantities of nitrogen and phosphorus moving into lakes, bays and other water systems. In some instances this has led to eutrophication and death of organisms within those ecosystems.

Industrial wastes that contain nitrogen oxides have also been released into the atmosphere. Nitrogen oxide can react with water vapour to form nitric acid and then leave the atmosphere via the water cycle as acid rain. This can change the acidity of water systems, resulting in death of organisms.

The environment of this turtle has been affected by excessive algal growth.

7.2 Exercises: Understanding and inquiring

To answer questions online and to receive **immediate feedback** and **sample responses** for every question, go to your learnON title at www.jacplus.com.au. *Note:* Question numbers may vary slightly.

Remember

1. Identify the term used for the life-support system of our planet.
2. State the four components that make up the biosphere.
3. Suggest how the water, carbon, nitrogen and phosphorus cycles are linked to the biosphere.
4. Suggest what is meant by the term *biota*.
5. Construct a diagram to show the relationship between the atmosphere, lithosphere, hydrosphere and biota.
6. Is the ozone layer in the troposphere or the stratosphere?
7. Outline the importance of the ozone layer to life on Earth.
8. State examples of four gases that you would find in the atmosphere.
9. Suggest why an increase in CFCs in the atmosphere is of concern.
10. Identify the cycle most relevant to the hydrosphere.
11. State examples of precipitation.
12. Provide examples of parts of the Earth that make up the lithosphere.
13. Identify the criterion used to divide regions into biomes.
14. Into which sphere would you place biomes?
15. Provide examples of two environmental factors that contribute to the type of biome that exists in a particular area.
16. Suggest how photosynthesis, cellular respiration and burning of fossil fuels link into the carbon cycle.
17. Distinguish between nitrogen-fixing, nitrifying and denitrifying bacteria.
18. Construct a figure or model to summarise the:
    (a) carbon cycle
    (b) nitrogen cycle
    (c) phosphorus cycle
    (d) water cycle.
19. Suggest a link between the following cycles and the biosphere.
    (a) Carbon cycle
    (b) Nitrogen cycle
    (c) Phosphorus cycle
    (d) Water cycle
20. Outline effects of human activity on the:
    (a) atmosphere
    (b) lithosphere
    (c) hydrosphere
    (d) carbon cycle
    (e) nitrogen and phosphorus cycles.

Think and discuss

21. Suggest a link between your DNA and the phosphorus cycle.

22. The figure below shows a more detailed view of how processes such as photosynthesis (green arrows) and cellular respiration (blue arrows) are involved in interactions between the atmosphere (exchange of gases) and living things. Copy and complete the figure below, inserting the following words: atmosphere, light energy, glucose, oxygen, water, carbon dioxide.

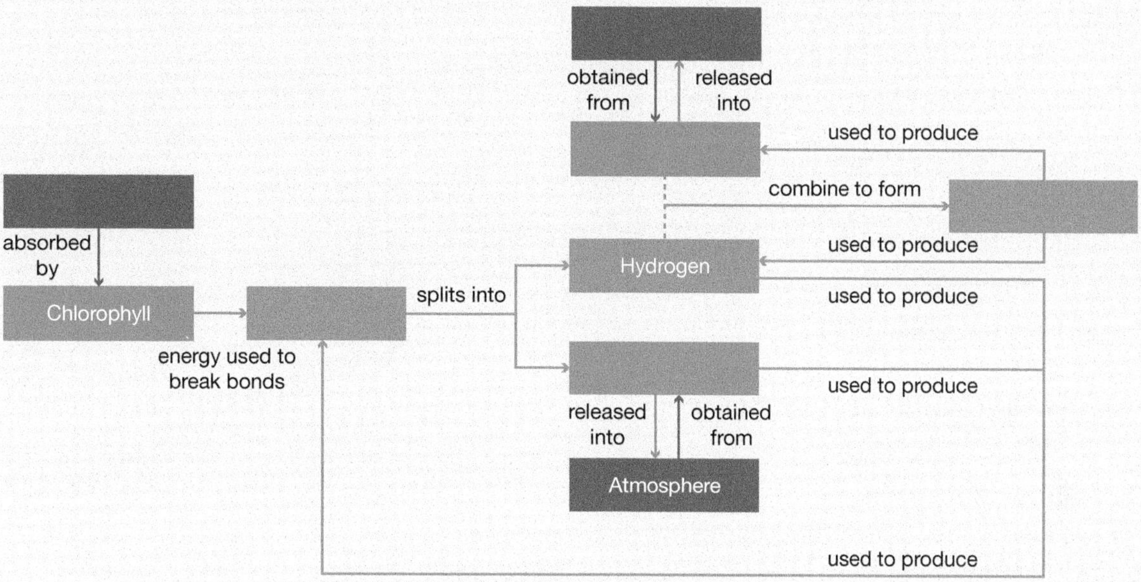

learnon RESOURCES — ONLINE ONLY

Complete this digital doc: Worksheet 7.1: Nature's time machine (doc-19472)

Complete this digital doc: Worksheet 7.2: Cycles in nature (doc-19473)

7.3 Climate patterns

7.3.1 Climate patterns

The Earth's climate is always changing. It always has and always will. So why has climate change become the single most important issue for so many people in the twenty-first century?

The variation of climate over the Earth's surface is largely the result of four major influences.

1. The amount of energy from the sun reaching the surface

Because the Earth is almost spherical in shape, the energy from the sun that reaches the Earth's surface is spread over a larger area in the polar regions than near the equator. That is, the amount of energy reaching each square metre of the Earth's surface in the polar regions is less than near the equator. It is the difference in surface temperature

The spherical shape of Earth results in less of the sun's energy reaching each square metre of the Earth's surface in the polar regions than near the equator.

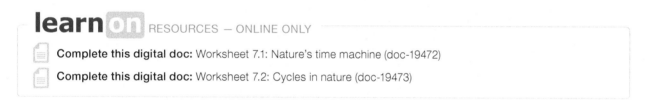

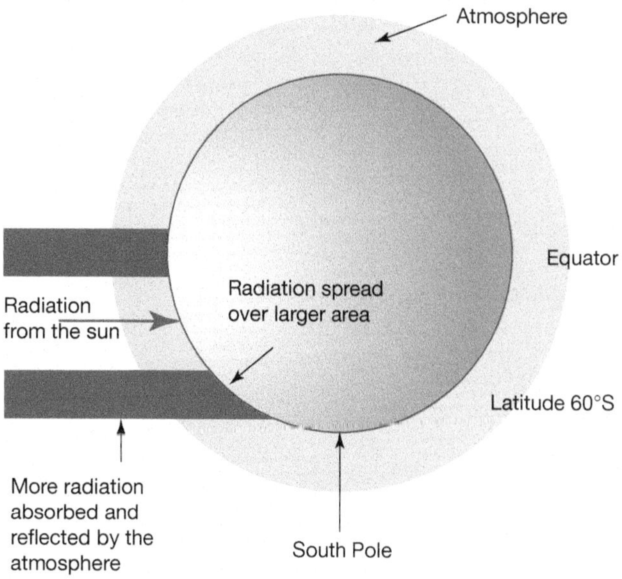

between the poles and the tropics that causes the movement of air that we know as wind.

2. The differing abilities of land and water to absorb and emit radiant heat

During daylight hours the land absorbs **radiant heat** from the sun more quickly than water. At night heat is radiated from the land more quickly than from the water. As a result, the ocean temperature changes less on a daily basis than air and land temperatures, and coastal climates are protected from the high and low temperature extremes of inland climates.

3. The tilt of the Earth's axis

The tilt of the Earth's axis results in the polar regions receiving little or no solar radiation for six months of each year.

4. The features of the land

The temperature of the part of the atmosphere that contains all of the Earth's land masses decreases with increased height above sea level. In addition, mountain ranges have a dramatic effect on the climate of nearby regions. They can block the path of wind blowing towards them, forcing the air to move quickly upwards to form almost permanent clouds, as water vapour in the air condenses quickly. Sandy soils reflect more energy from the sun than dark, fertile soils. Fresh snow reflects up to 90 per cent of the sun's energy that reaches it. Heavily vegetated areas absorb much more of the sun's radiation than bare land because plants use it to photosynthesise.

Weather stations contain devices such as a thermometer to measure temperature, a barometer to measure atmospheric pressure, a hygrometer to measure humidity, an anemometer (pictured; this one has cup-shaped turbines) to measure wind speed and a wind vane to measure wind direction.

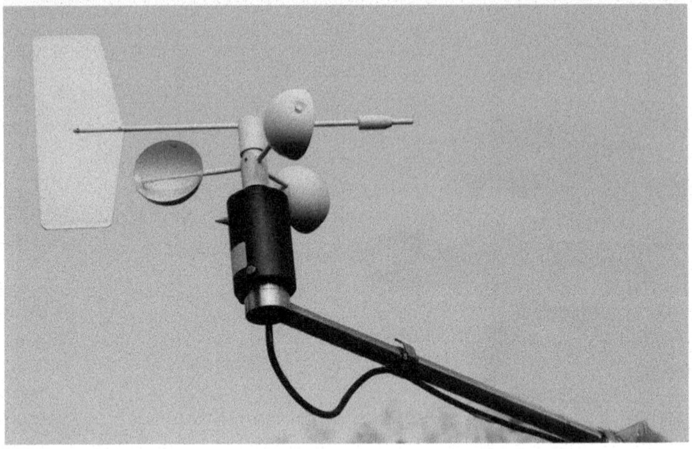

The Earth's ocean currents have a major influence on coastal climates.

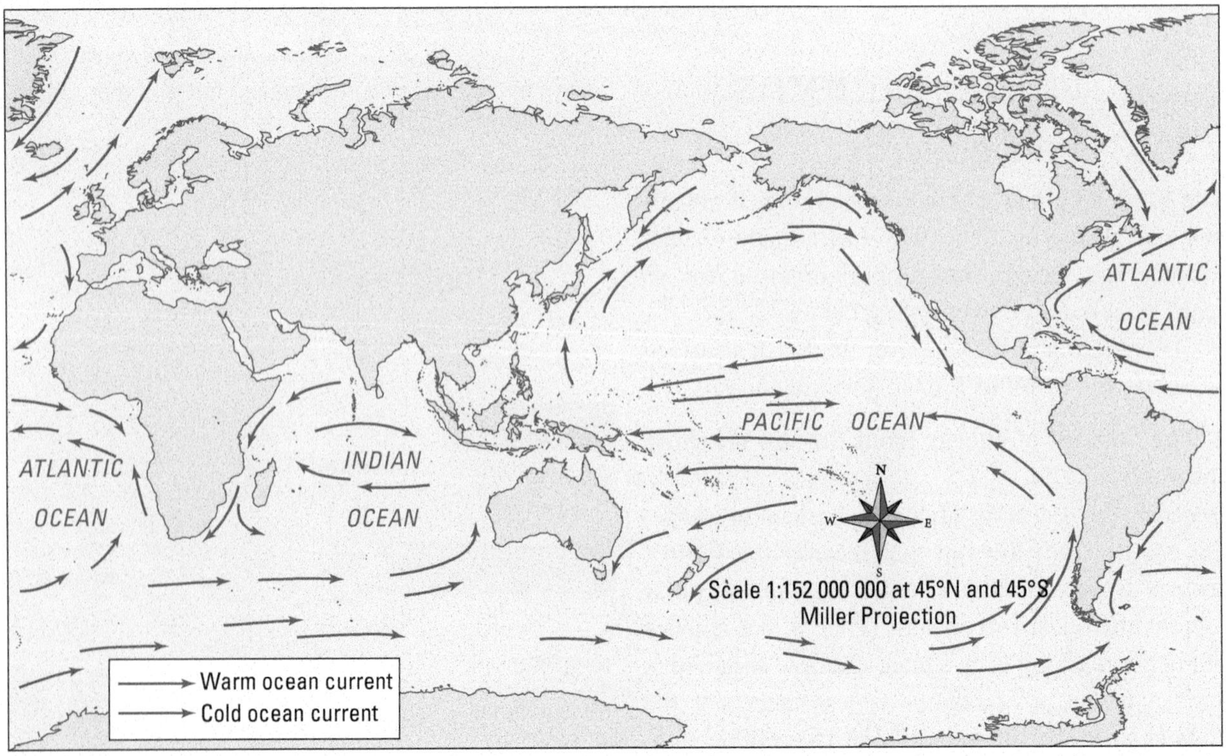

7.3.2 Ocean currents

The water in the Earth's oceans is constantly moving in currents. Ocean currents are the result of the temperature difference between the tropics and poles, and the Earth's rotation. Warm surface water near the equator sinks and cools as it moves towards the poles, while the cold water in polar regions rises and warms as it moves towards the equator. Warm and cold ocean currents move huge volumes of water past coastal regions and have a major influence on their climate. The Gulf Stream (at top left in the map on the previous page), for example, carries warm water from the equator into the North Atlantic Ocean, keeping Great Britain, Norway and Iceland warmer than other regions at similar latitudes. Cold water currents cool coastal regions that would otherwise be hot.

Convection currents carry warm air towards the poles and cool air towards the equator. Wind patterns are complicated by the rotation of the Earth.

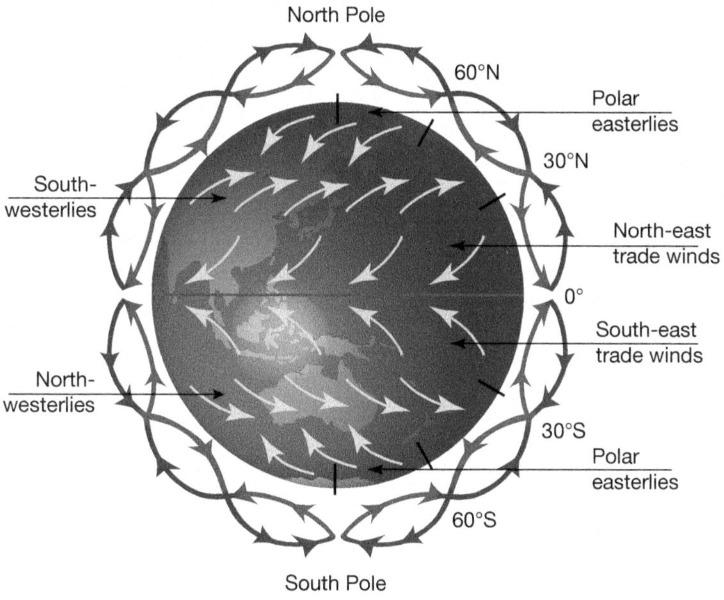

7.3.3 The influence of wind

The differences in surface temperature between the poles and the tropics cause the large-scale convection currents that create wind. Cold air near the poles sinks and moves towards the equator, and hot air near the equator rises and moves towards the poles.

The globe diagram top right shows the effects of these convection currents during March and September, when the sun is directly over the equator. The winds shown are called **prevailing winds** and are generally those most frequently observed in each region. The direction of prevailing winds is complicated by latitude, the rotation of the Earth about its own axis, the tilt of the Earth's axis and the Earth's orbit around the sun. The actual wind direction at any time depends on numerous other factors including the amount of friction caused by the land surface, ocean currents, local variations in air pressure and temperature, variations in water and land temperature, and altitude.

The wind direction in turn influences air temperatures. For example, during the Australian summer, regions along most of the south coast experience high temperatures when the northerly winds bring in hot and dry air from above the land to the north. The same regions can experience cold southerly winds, which bring in cool and damp air from above the oceans to the south.

7.3 Exercises: Understanding and inquiring

To answer questions online and to receive **immediate feedback** and **sample responses** for every question, go to your learnON title at www.jacplus.com.au. *Note:* Question numbers may vary slightly.

Remember

1. List four major factors that influence the variation of climate over the Earth's surface.
2. What causes the large-scale convection currents in the air that create prevailing winds?
3. List five factors that determine the wind direction at any given time or place.

4. Outline the causes of warm and cold ocean currents.
5. Explain why Great Britain, Norway and Iceland experience warmer climates than other regions at similar latitudes.

Think

6. Why do sandy soils reflect more of the sun's radiation than dark, fertile soils?
7. Explain why the average temperature of the Earth's atmosphere was constantly changing for millions of years before humans existed.
8. Outline the likely effect on land-based living organisms caused by:
   (a) rising sea levels
   (b) an increase in average temperatures.

Investigate, discuss and reflect

9. Research, discuss and reflect on each of the following statements about climate change and state your own opinion.
   (a) Australia has vast resources of coal, much of which is exported. The Australian coal industry provides employment and other benefits for the economy. If targets for the reduction of global emissions are high enough to damage the Australian coal industry, the government should not agree to them.
   (b) Developing countries that have little or no industry have not contributed to global warming. These countries should be allowed to increase their carbon dioxide emissions so that they can develop industries and improve their living standards.
10. (a) Carefully examine the table below and suggest what types of vegetation may be found in an environment with a:
    (i) mean annual temperature between 0 °C and 15 °C and a mean annual rainfall around 50–100 cm
    (ii) mean annual temperature between 20 °C and 28 °C and a mean annual rainfall around 250–400 cm
    (iii) mean annual temperature between 20 °C and 28 °C and a mean annual rainfall around 20–30 cm.
    (b) Find out the mean annual temperature and mean annual rainfall of your local environment. What type of vegetation would you expect to find there? Is this the case? If it is not, suggest possible reasons for the difference.
    (c) Find out what climate change is predicted to occur in your local area due to global warming. Which vegetation would be best suited to this type of environment?

Vegetation type	Mean annual temp. (°C)	Mean annual precipitation (cm)
Tundra	−15–−5	0–100
Northern coniferous forests	−5–0	50–150
Mediterranean	−4–17	0–60
Grassland	3–18	50–100
Temperate deciduous forest	3–19	50–300
Desert	−5–30	0–50
Savanna	17–30	50–200
Tropical forests	18–30	100–450

7.4 Global warming

7.4.1 Revisiting the greenhouse effect

Earth's atmosphere acts like a giant invisible blanket that keeps temperatures on our planet's surface within a range that supports life. Within the atmosphere, greenhouse gases trap some of the energy leaving the Earth's surface to help maintain these warm temperatures. The maintenance of Earth's temperatures by these atmospheric gases is called the **greenhouse effect**.

7.4.2 Revisiting global warming

What's the problem?

It's a hot topic. Global temperatures have been increasing and are expected to continue to increase at an accelerated rate. The rising temperature of Earth is known as **global warming**. This may result in melting icecaps, rising sea levels, increased coastal flooding, unusual weather patterns and ocean currents, and consequent threats to the survival of some living things.

What's the cause?

Scientists assert that our increased and growing dependence on fossil fuels since the Industrial Revolution of the nineteenth century is a major cause of global warming. They argue that burning fossil fuels such as coal and oil has resulted in increased levels of **greenhouse gases** (such as nitrous oxide and carbon dioxide) in our atmosphere that are trapping heat, causing the atmosphere to heat up. This is referred to as the **enhanced greenhouse effect**. Some sources of these human-produced greenhouse gases are shown on the following page.

Grazing animals such as cattle and sheep produce large amounts of methane as a waste product. Methane is another powerful greenhouse gas and is also produced by the action of bacteria that live in **landfills** and soils used for crop production.

Much of the nitrous oxide in the atmosphere is produced by the action of bacteria on fertilised soil and the urine of grazing animals.

Greenhouse gases and the enhanced greenhouse effect

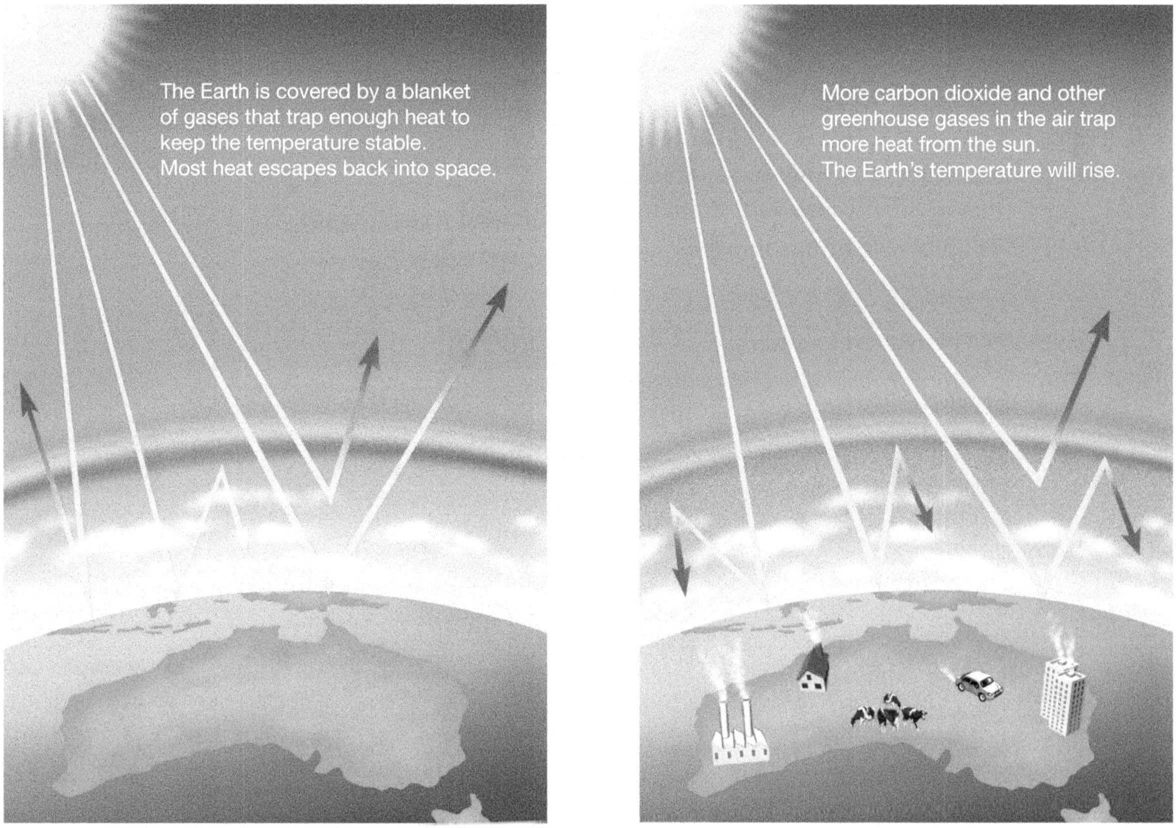

The Earth is covered by a blanket of gases that trap enough heat to keep the temperature stable. Most heat escapes back into space.
More carbon dioxide and other greenhouse gases in the air trap more heat from the sun. The Earth's temperature will rise.

Some sources of greenhouse gases

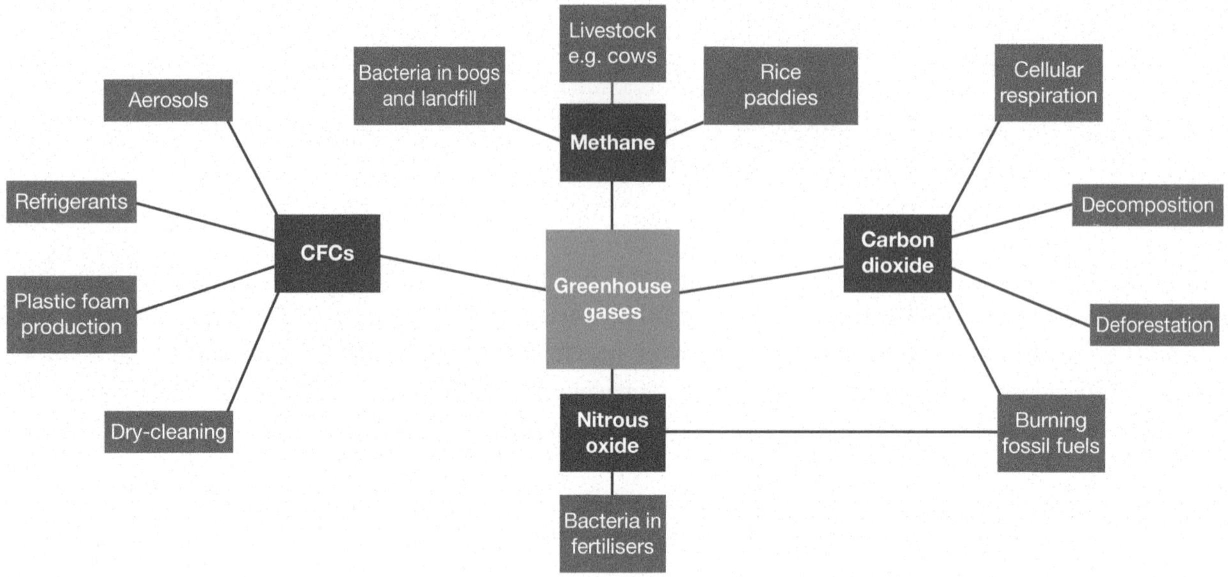

7.4.3 Connecting the carbon cycle to global warming

Photosynthesis and cellular respiration

Light energy, carbon dioxide and water are used by phototrophic organisms such as plants to make glucose and oxygen. This process is called photosynthesis.

$$6CO_2 + 12H_2O \xrightarrow[\text{chlorophyll}]{\text{visible light energy}} C_6H_{12}O_6 + 6O_2 + 6H_2O$$

All living things use **cellular respiration**. During this process glucose is converted into a form of energy that the cells can use. Carbon dioxide is one of the products of this reaction.

$$C_6H_{12}O_6 + 6O_2 \longrightarrow 6CO_2 + 6H_2O + \text{energy}$$

So, in terms of the carbon cycle, carbon dioxide is taken from the atmosphere during photosynthesis and released back during cellular respiration. This suggests that if producers are reduced in number or removed from the atmosphere, there will be less carbon dioxide removed from the atmosphere, resulting in an overall increase in this gas. This explains why cutting down trees and replacing them with buildings or crops with lower photosynthetic rates can contribute to the enhanced greenhouse effect.

Sources of carbon dioxide within the carbon cycle are coloured red.

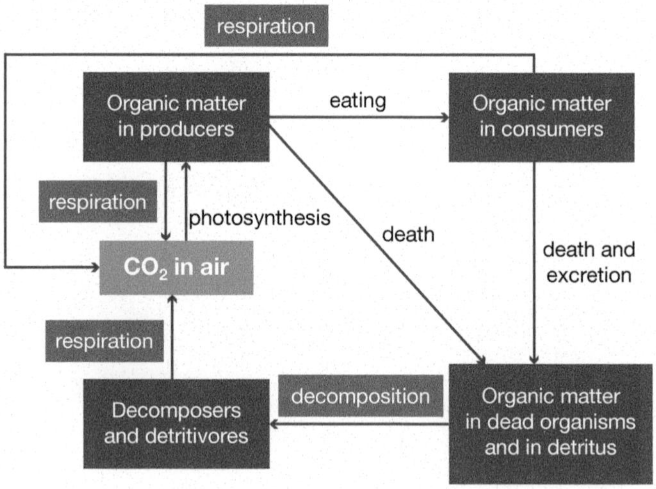

Carbon dioxide is obtained from a variety of sources (coloured red) within an ecosystem.

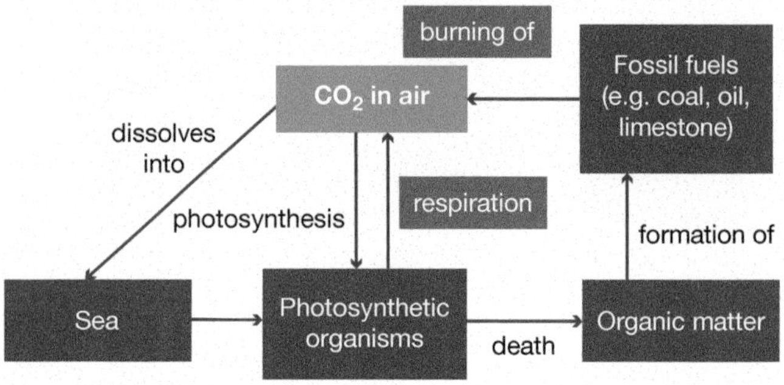

Decomposition and fossil fuels

Carbon dioxide is also released from dead and non-living parts of ecosystems. Some of the carbon dioxide from the atmosphere dissolves into the sea and is absorbed by sea plants and other photosynthetic organisms. These organisms and those that eat them eventually die. Some of their carbon may be used in the formation of fossil fuels. When these fossil fuels are burned, carbon dioxide is released back into the atmosphere.

7.4.4 The ozone factor

Ozone (O_3) in the lower atmosphere is also a significant contributor to the enhanced greenhouse effect. Although ozone occurs naturally, it is also produced by a photochemical reaction that takes place when sunlight falls on emissions from motor vehicles, power stations and bushfires.

Ozone is produced by photochemical reactions involving emissions from motor vehicles and industry.

7.4.5 Secrets in the ice

For thousands of years, snow has fallen in Antarctica. The snow turns to ice, which builds up over time. Dust, gases and other substances from the air become trapped in the ice. The trapped substances provide information about what was in the air at the time the snow fell.

Scientists have used **ice cores** to track the air temperature and concentration of carbon dioxide near the Earth's surface in the past. The graphs below show how these have changed over the 420 000 years leading up to the year 2000.

This ice core was drilled from more than 3.7 km below the surface. Parts of it are more than 150 000 years old.

The carbon dioxide concentration is shown in parts per million (ppm) by volume. The temperature difference shown is the deviation from the average temperature now (represented by 0 on the vertical scale). The pattern of changing temperatures resembles the pattern of the change in carbon dioxide concentrations.

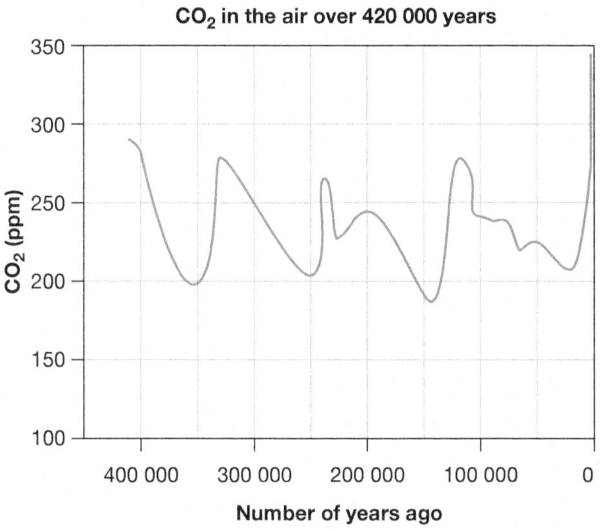

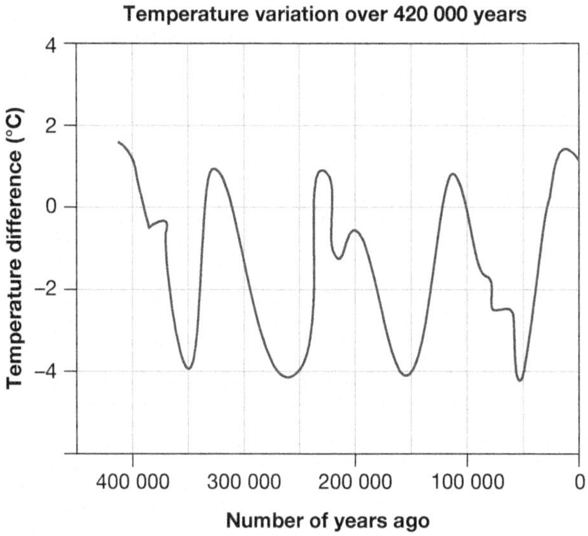

It is clear that there has been a dramatic increase in the amount of carbon dioxide in the atmosphere in recent history. During the current decade the concentration of carbon dioxide has risen to approximately 400 parts per million. There appears to be no significant change in global temperature cycles. However, the graph at right shows that since the Industrial Revolution there has been a dramatic change in the trend of carbon dioxide in the atmosphere.

This graph shows the dramatic increase in atmospheric carbon dioxide since the Industrial Revolution.

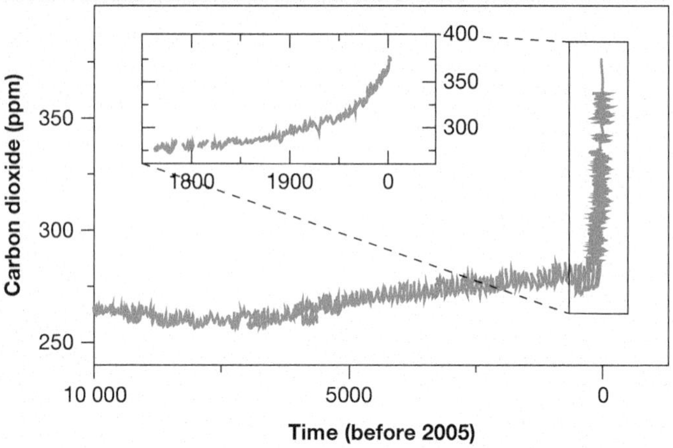

7.4.6 Climate models

Meteorologists and other scientists use computer modelling to make predictions about climate change and the possible consequences. The computer programs used to model climate change simulate the circulation of air in the atmosphere and water in the oceans. An immense amount of data collected from the atmosphere, ocean and land surface is used, together with mathematical equations that describe the circulation. The laws of physics and chemistry, including the laws of conservation of energy and Newton's Laws of Motion, are an important part of the modelling process.

7.4.7 Global temperature

Although the exact future increase in average global temperature is not certain, it is generally agreed that during the next 100 years it could increase by between 1 °C and 4 °C. Although that doesn't sound like much, the consequences are very serious. Computer modelling suggests that the global temperature will not increase evenly across the continents. According to CSIRO, in Australia temperatures could increase by up to 2 °C by 2030 and up to 6 °C by 2070. As a consequence there will be more hot days and fewer cold days, an increase in rainfall in the north-east and a decrease in the south, more bushfires, and more destructive tropical cyclones.

7.4.8 Rising sea levels

According to tide-gauge records, the average global sea level has increased by between 10 cm and 20 cm during the past 100 years. Sea levels are expected to rise further due to:

- the warming ocean water and its resulting thermal expansion
- the melting of glaciers, the polar icecaps and the ice sheets of Greenland and Antarctica. According to NASA, sea ice in the Arctic is melting at the rate of 9 per cent every ten years. Of the world's 88 glaciers, 84 are receding due to melting ice.

Rising sea levels are likely to cause the flooding of low-lying islands and coastal regions.

The low-lying Pacific nation of Kiribati is planning to relocate its population because of the threat of rising sea levels.

7.4.9 Frozen soil

Much of the soil on or below the surface of very high mountains in the polar regions is permanently frozen. Known as **permafrost**, this soil is likely to gradually thaw out as global air temperatures increase. There is a massive amount of carbon stored in permafrost and scientists fear that as it thaws, large quantities of carbon dioxide and methane will be released into the atmosphere. This in turn would increase the rate of climate change.

Another problem associated with the thawing of permafrost is the risk of the collapse of buildings, bridges, roads, pipelines and other structures in populated areas of the northern polar regions. The foundations or bases of many of these structures are embedded in permafrost. As it thaws, any ice present melts, making the soil damp and unstable.

7.4 Exercises: Understanding and inquiring

To answer questions online and to receive **immediate feedback** and **sample responses** for every question, go to your learnON title at www.jacplus.com.au. *Note:* Question numbers may vary slightly.

Remember

1. Suggest why Earth's atmosphere has been described as a giant invisible blanket.
2. What is:
   (a) the greenhouse effect
   (b) the enhanced greenhouse effect
   (c) global warming?
3. Suggest four consequences of global warming.
4. Give examples of three types of greenhouse gases and at least two sources for each.
5. Identify the links between photosynthesis, cellular respiration, decomposition, fossil fuels and global warming.
6. Explain why ozone in the Earth's stratosphere is important to humans and all other life on Earth.
7. Explain how scientists are able to determine the air temperature and the amount of carbon dioxide in the atmosphere hundreds of thousands of years ago when such measurements were never recorded.
8. Explain how the thawing of permafrost could increase the rate of global warming.

Think and discuss

9. Outline the actions that individuals can take to slow the rate of global warming.
10. (a) In your own words, describe what is meant by the term *enhanced greenhouse effect*.
    (b) Suggest a model or simulation that could communicate this concept to others.
11. Suggest how whales that live on plankton could be affected by global warming.
12. (a) Which of the following actions would you be prepared to take so that you can contribute to the fight against global warming?
      - Walk, cycle or use public transport rather than relying on someone to drive you to school, work or leisure activities.
      - Change your diet so that you eat less meat and more fruit and vegetables.
      - Recycle paper, aluminium and steel cans, glass and plastics.
      - Stop using electric clothes dryers and use outdoor clothes lines in dry weather and indoor folding clothes-airers in wet weather to dry clothes.
    (b) Select one of the actions in part (a) that the government could enforce by passing new laws and explain how it could be done.
13. Explain why the average temperature of the Earth's atmosphere was constantly changing for millions of years before humans existed.
14. Outline the likely effect on land-based living things caused by:
    (a) rising sea levels
    (b) an increase in average temperatures
    (c) significantly increased rainfall
    (d) significantly decreased rainfall.
15. Explain why it is necessary for the Australian government to create legislation to address the problem of global warming.

Investigate and report

16. In which industrial processes were CFCs used before they were phased out?
17. Use the internet or other sources to find out how carbon capture can be used to reduce the amount of carbon dioxide in the Earth's atmosphere.
18. There are many people who do not believe that climate change and global warming are taking place. There are others who acknowledge that they are taking place but do not believe they are serious problems. Use the internet and other sources to list the arguments that these two groups of people use to support their beliefs.
19. There are new technologies being developed to reduce the amount of carbon dioxide produced per tonne of coal. Research the integrated gasification combined cycle (IGCC).
20. Tetrachloroethene is a solvent commonly used in the dry-cleaning industry in Australia. Not only is this chemical harmful to our health, it can also contribute to photochemical smog. Find out more about this chemical and new technologies, including 'green dry-cleaning', that are being developed, researched or used as alternatives.
21. Research and report on the contribution of two of the following to climate change research.
    - National Climate Change Adaptation Research Facility
    - Terrestrial Ecosystem Research Network
    - Department of Environment and Primary Industries
    - CSIRO
    - Greenhouse Gas Online
    - Climate Change Research Centre
    - Fisheries Research and Development Corporation
    - Climate Change Research Strategy for Primary Industries.
22. Find out more about what you can do in your home to reduce the amount of carbon dioxide you produce. Create a brochure to teach people how they can help slow global warming.

learnon RESOURCES — ONLINE ONLY

Complete this digital doc: Worksheet 7.3: Ozone layer (doc-19474)

7.5 Heating up for Thermageddon?

7.5.1 Biological implications

Will some parts of Earth get too hot for humans? Computer models are predicting that this could happen in some parts of the tropics in the future. Some scientists have suggested that under these hot and humid conditions, even someone standing in the shade in front of a fan could die of heat stress.

Changes in the Earth's climate due to global warming will probably affect the survival of living organisms. The survival of every living thing on Earth is dependent on the characteristics of its habitat, including some that will be affected by climate change. Some living things will be affected more than others.

7.5.2 Will climate change shape human evolution?

Could Earth get too hot for humans? Is there enough variation within our species so that if things do get too hot to handle at least some of us will survive and our species will continue?

Heat stress threshold

To function normally we need to maintain a core body temperature around 37 °C. If this core temperature rises above 42 °C, we die. Some researchers have used climate computer models to predict the impact of different levels of global warming on populations. Their data suggest that an increase of around 7 °C in the environment may result in heat and humidity making some places on Earth intolerable, and they predict migrations out of these hot and humid countries will occur. They suggest that at increased temperatures of 12 °C about half of the land inhabited today (including Australia) would be too hot to live in.

People living in the affected areas would need to wear 'cooling suits', live underground or stay in constantly air-conditioned environments. Organisms such as livestock or people who cannot afford these buffers may perish.

Relative humidity not only makes a hot day more unbearable, it can also make it more dangerous.

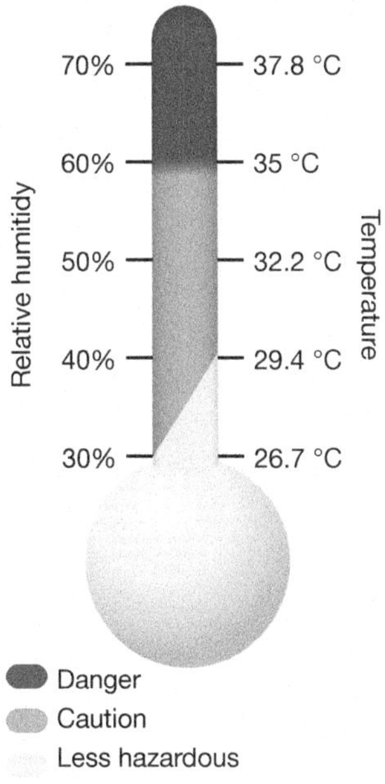

Hot bods?

If Earth keeps warming up, over the long term will we see genetic shifts to select those variations with increased chances of survival? What will a human in a hot future world look like? Some evolutionary biologists have suggested slimmer and taller body shapes that radiate heat better, while at the same time carrying enough fat to be reproductively successful, would be selected for. Some palaeontologists, however, suggest that heat stress would be likely to drive the evolution of smaller mammals.

Disease

With warmer temperatures and global transport and global populations, it is predicted that humans may be more at risk of disease than at any other time in history. There may be an increased incidence of diseases such as food poisoning, skin cancers, eye cataracts and a new range of tropical diseases.

The presence of genes that may provide quick resistance against the onslaught of future diseases is another factor that will determine who survives and who does not.

Are humans still evolving?

A hypothesis has suggested that global cooling was essential for the large brains of humans to evolve. If this hypothesis is supported, does this mean that global warming may lead to a reduction in the size of the human brain? Other scientists suggest that our modern brains have enabled us to develop culture and that, as long as we have culture and technology, we will have a buffer against hot climates.

Research suggests that the human brain is still evolving. Scientists have identified two genes involved in regulating brain size that have been subject to recent natural selection.

7.5.3 Climate sensitivity

How hot things get will depend on how much more carbon dioxide is pumped into the atmosphere and how much warming it produces. This is known as **climate sensitivity**. The Intergovernmental Panel on Climate Change (IPCC) suggests that temperatures may rise between 1.9 and 4.5 °C (around 3 °C) for every doubling of carbon dioxide concentration in the atmosphere. However, the IPCC's computer model is based only on fast feedback processes and excludes slower processes such as the release of methane from thawing permafrost.

With a climate sensitivity of around 1.9 °C, it may take centuries for our planet to warm by 7 °C. With a climate sensitivity of around 4.5 °C, however, the increase could reach 7 °C within a century if we continue with our current levels of carbon dioxide production.

Have you read about any of these indicators or already observed some of them?

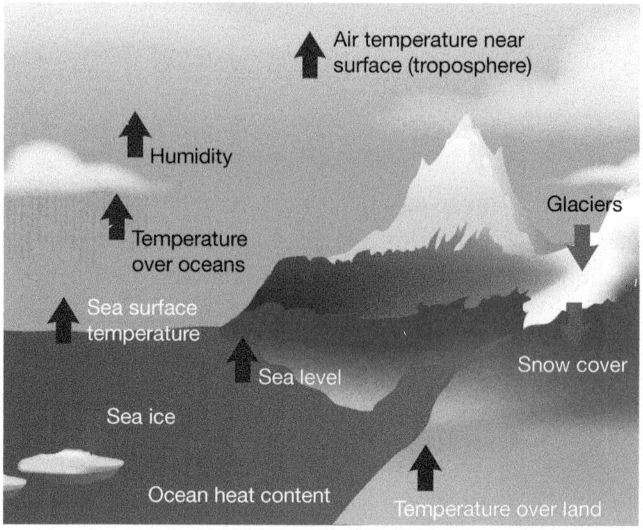

An increase in heat and humidity due to climate change could render half the world uninhabitable. In regions where the 'wet-bulb' temperature (the temperature to which objects can be cooled by evaporation) exceeds 35 °C (the human heat-stress limit), it would be impossible for people to survive without some kind of cooling system.

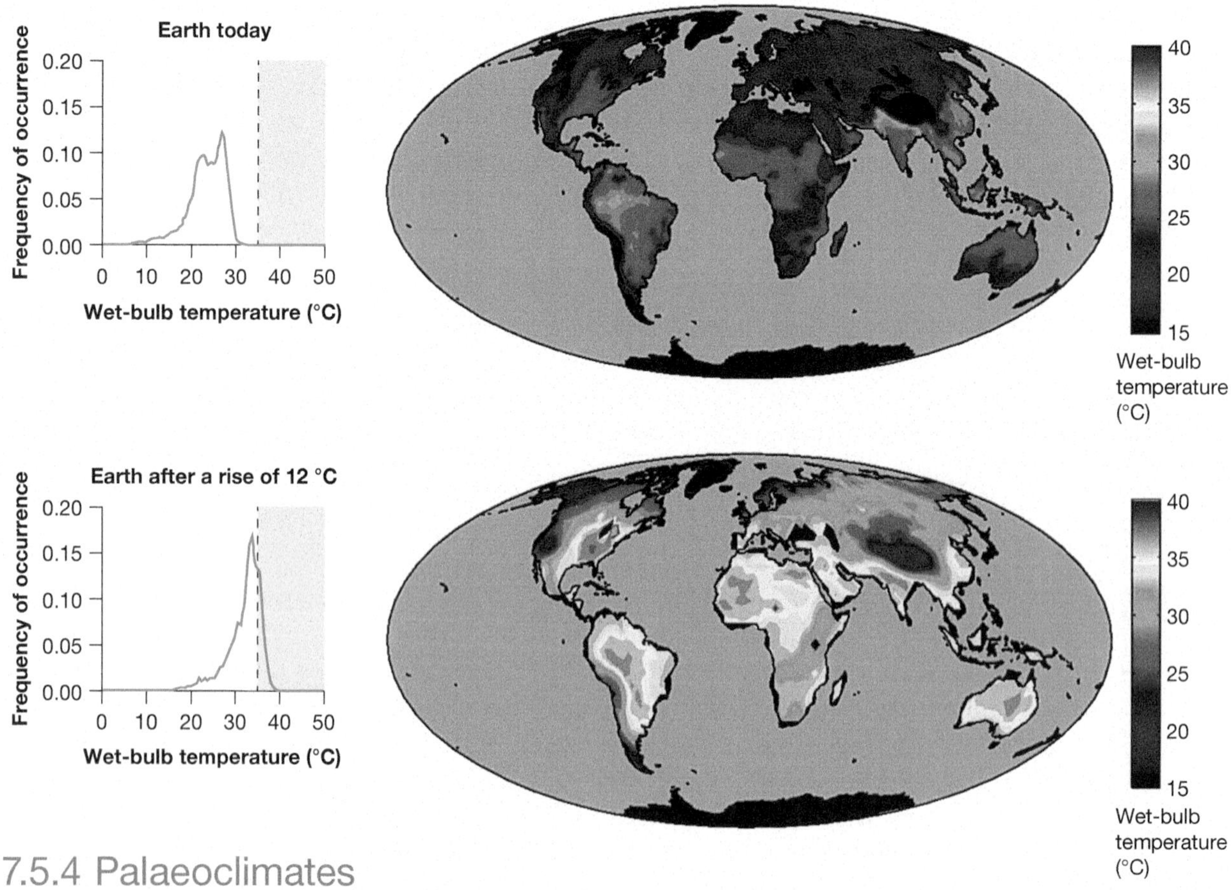

7.5.4 Palaeoclimates

Palaeoclimates offer a unique perspective in that they can show the wide range of climates over various time scales, and transitions between them. This information can be used to develop climate models for future climate studies. The figure on the next page shows examples of various palaeoclimates throughout Earth's history.

Will the study of palaeoclimates throughout history help us develop climate models to predict climates of the future? (Ka = thousand years ago; Ma = million years ago)

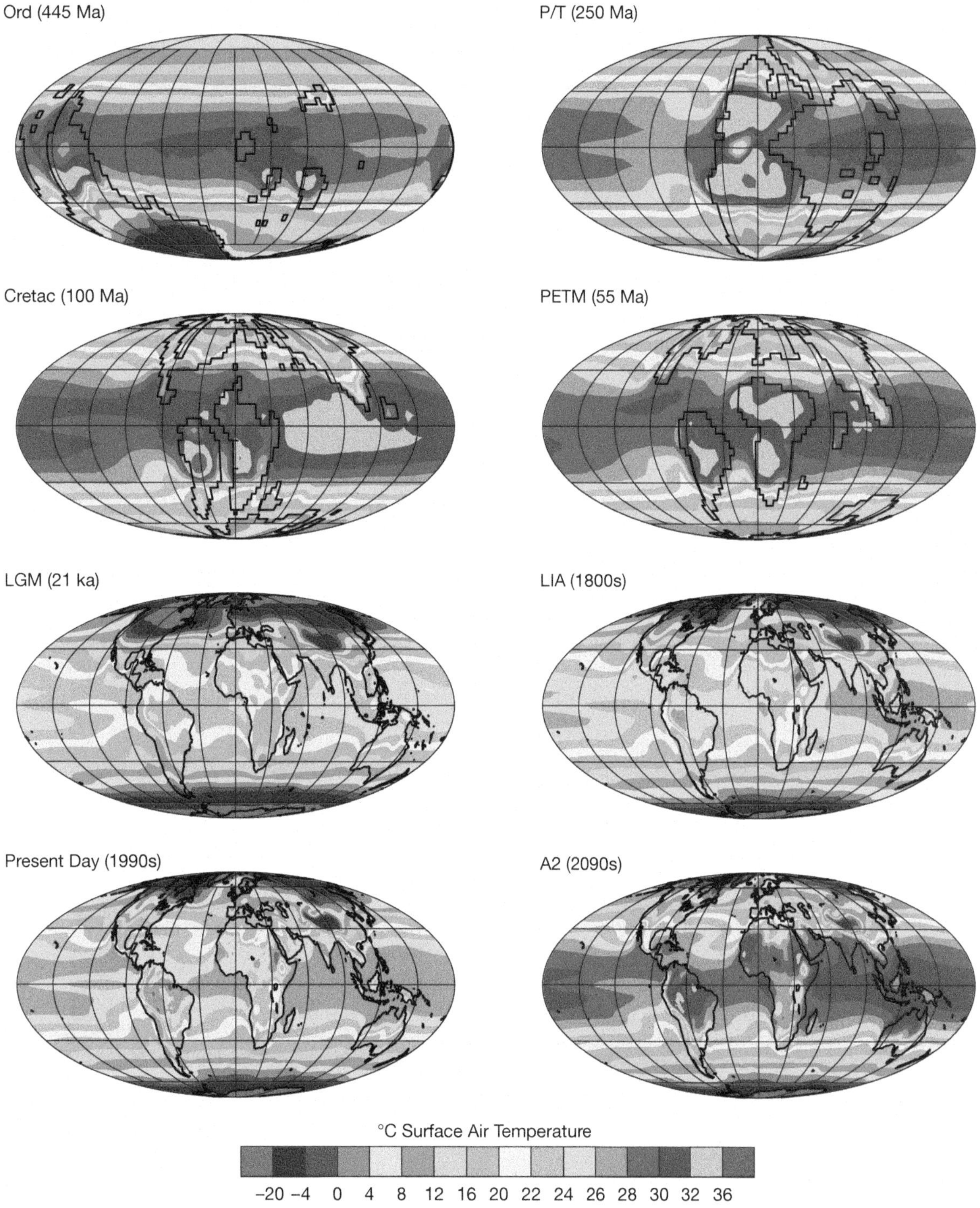

7.5.5 Ocean life

Some marine life will suffer and could even become extinct because of changes in water temperature. Changing temperatures and ocean currents could separate some marine species from their food source. Some marine animals depend on microscopic plankton that float along with the currents. Others depend on species from warmer or colder layers of water than the layer in which they live. It is also possible that some species will suffer from the reduction of oxygen dissolved in ocean water because of increases in temperature. The habitats of some species could be destroyed by rising sea levels.

7.5.6 Biodiversity

Habitats in mangrove swamps, coastal wetlands, coral reefs and other coastal areas may be reduced or lost because of rising sea levels and changed weather patterns. Plants, animals and other organisms adapted to low temperatures and high or low rainfall will have to migrate to other regions. In some cases, where migration is not possible or fails, species could become extinct.

Extinctions due to climate change are likely to add significantly to the loss of biodiversity already caused by loss of habitats due to deforestation and other human activities.

7.5 Exercises: Understanding and inquiring

To answer questions online and to receive **immediate feedback** and **sample responses** for every question, go to your learnON title at www.jacplus.com.au. *Note:* Question numbers may vary slightly.

Remember

1. State what every living thing is dependent on.
2. State the core body temperature that humans need to maintain.
3. Suggest what happens if the core body temperature of a human rises above 42 °C.
4. Suggest strategies that people living in areas affected by extreme heat and humidity use to survive.
5. What is meant by the term *climate sensitivity*?
6. Outline some possible effects of extreme heat and humidity on:
   (a) humans
   (b) life in the ocean
   (c) biodiversity.

Think, investigate and discuss

7. Find out more about palaeoclimates and related types of research that scientists are currently involved in.
8. While the yields of some types of crops, such as wheat and rice, may increase in conditions where there are higher carbon dioxide concentrations, increases in temperatures may be detrimental to other types of crops. Research and report on the effects of global warming on at least three different types of crops.
9. Research suggests that the human brain is still evolving. Scientists have identified two genes involved in regulating brain size that have been subject to recent natural selection. Research and report on recent relevant studies.
10. A warm period of time from Earth's past was the Palaeocene-Eocene Thermal Maximum (PETM) 55 million years ago. Investigate the PETM and report on the types of life forms living at that time and how they coped with warm temperatures.
11. Some palaeontologists suggest that mammals get smaller as the climate gets warmer. Investigate this hypothesis and record your evidence for or against it with current examples.
12. The advice of some scientists is that, as evolution is a slow process, it is unlikely that any adaptation would save us from global warming in time to escape it. They suggest that the answer to surviving climate change is in our skulls. Research and report on the following.
    (a) Did global cooling allow humans to evolve their big brains?
    (b) Can we use an Earth-systems computer to investigate the hypothesis in part (a)?

7.6 Some cool solutions

7.6.1 Finding solutions

Okay, so there might be a climate change problem. What can we do to fix it?

No-one can be certain about the actual consequences of global warming. There are so many variables that influence climate that computer modelling cannot provide completely accurate predictions. However, there is plenty of evidence to indicate that the levels of the greenhouse gases carbon dioxide, methane and nitrous oxide have been increasing over the past 100 years and will continue to increase.

Wind energy is one of several alternative energy sources that do not produce greenhouse gases.

It is clear that global warming must be slowed by reducing the emission of greenhouse gases. This is no easy task and requires:

- a significant reduction in our use of fossil fuels. Not only does this require a reduction in our use of electricity, natural gas and motor fuels, it also requires an increase in our use of alternative energy sources such as wind, solar and wave energy.
- It also requires the development of more energy-efficient devices to ensure that less energy is wasted. a change in our consumption of food to reduce our dependence on livestock that release methane and nitrous oxide into the atmosphere. We may have to eat less meat and more locally grown fruit and vegetables.
- the recycling of products such as glass, paper, metals and plastic that require the burning of fossil fuels for their production and distribution.

7.6.2 Geosequestration

Geosequestration is a process that involves separating carbon dioxide from other flue gases in fossil fuel power stations, compressing it and piping it to a suitable site. There are at least 65 suitable sites (e.g. depleted oil and gas wells) that have been identified in Australia that are capable of taking up to 115 million tonnes of carbon dioxide each year.

Research on this process dates back to the 1970s. Although there are considerable problems with the technology, there is renewed interest in further developing it. It is hoped that it may be used to remove carbon dioxide from the atmosphere and hence reduce global warming.

WHAT DOES IT MEAN?

The word *geosequestration* comes from the Greek term *geo*, meaning 'of the Earth', and the Latin term *sequestrare*, meaning 'to separate'. *Sequestrare* comes from an earlier Latin word meaning 'depositary'.

7.6.3 To chop or not to chop?

We live in a consumer society. The things that we want and need often require large amounts of energy to manufacture and consequently result in the emission of carbon dioxide into the atmosphere. Scientists in the forestry and related industries have suggested that one way to reduce carbon dioxide emissions is to produce and use wood products that have been grown under sustainable forest management strategies. Nick Roberts, Forests NSW chief executive, is passionate about the role that sustainably harvested native forests can play in combating climate change. The view that wood products produced under this sustainable management have the potential to maintain or increase forest carbon stocks is also supported by the IPCC.

In 2009, Fabiano Ximenes, a forest research scientist, and his colleagues from the NSW Department of Primary Industries (DPI) analysed the carbon content of paper and wood products in landfill and found that at least 82 per cent of the carbon originally in the sawn timber remained stored in the wood. This research suggested that wood products could act as a carbon 'sink', not only during use, but even after disposal.

Fabiano Ximenes, Research Officer
— Life Cycle Assessment, DPI

7.6.4 Earth's nine lives

Is it time to think about our relationship with our environment in a new way? Researchers at the Stockholm Environment Institute in Sweden have identified nine planetary life-support systems that provide planetary boundaries that they argue should be adhered to in order to live sustainably. These are:

- rate of biodiversity loss
- climate change
- nitrogen and phosphorus cycles
- stratospheric ozone depletion
- atmospheric aerosol loading
- chemical pollution
- ocean acidification
- fresh-water use
- change in land use.

7.6.5 Metagenomics

Australian agriculture accounts for about 16 per cent of our national greenhouse emissions. Sixty-seven per cent of this is methane emissions from livestock. CSIRO Livestock Industries (CLI) is excited about its research that aims to characterise the microbiome (assortment of microbes in the foregut) of Australian marsupials such as the Tammar wallaby (*Macropus eugenii*). One project involves **metagenomics**, a technology that combines DNA sequencing with molecular and computational biology. This technology is being used by the scientists to study methanogens — bacteria that are involved in breaking down plant fibre in the wallaby's gut. While these bacteria

Tammar wallaby

produce methane, the levels are a lot lower than those produced by cows and sheep. CSIRO's research may lead to discoveries about why marsupials produce far fewer greenhouse emissions that cows and sheep, and contribute to new biotechnologies that may help us to reduce agricultural greenhouse emissions.

7.6.6 The Kyoto Protocol

In 1997, at a meeting in the city of Kyoto, Japan, most of the world leaders signed a document known as the **Kyoto Protocol**. The document was a historic agreement to reduce the amount of greenhouse gases produced by industrialised nations. It set targets for reduction of greenhouse gas production up to the year 2012. The targets varied from nation to nation according to a number of factors, including the nation's stage of industrial development. For example, the target for the United States was a reduction of 7 per cent from 1990 levels. For Japan and Canada it was a reduction of 6 per cent. For the Russian Federation and New Zealand it was 0 per cent.

A sustainable plantation forest of eucalypt trees

However, a signature on the Kyoto Protocol was only an agreement in principle and was not legally binding. The agreement could not come into force until countries producing more than 55 per cent of the world's greenhouse gases confirmed their commitment by ratifying the agreement, thus formally agreeing to the targets set. This took until February 2005. Australia did not ratify the Kyoto Protocol until 2007. The United States refused to ratify it.

The signing of the Kyoto Protocol marked the beginning of ongoing cooperation between most of the world's nations to reduce emissions of carbon dioxide and other greenhouse gases and slow down global warming. Regular conferences are held with the support of the United Nations to monitor progress and review targets.

HOW ABOUT THAT!

Do you use a computer often? Have you ever wondered where all the data you can access through the internet is actually kept? The answer is: on a computer server. Many schools have their own server and most students are allocated a certain amount of storage space on it. The problem is that all these servers need to be kept cool to operate correctly. Servers produce heat and keeping them cool requires a lot of electricity. Much of the electricity needed is produced using fossil fuels, so this contributes to global warming. It has been estimated that, worldwide, computer servers contribute as much as the aviation industry to global warming. One solution is to use less energy in data storage and make efficient use of energy in the IT industry.

7.6 Exercises: Understanding and inquiring

To answer questions online and to receive **immediate feedback** and **sample responses** for every question, go to your learnON title at www.jacplus.com.au. *Note:* Question numbers may vary slightly.

Remember

1. Suggest why no-one can be certain about the actual consequences of global warming.
2. If we can't be certain about the consequences of global warming, why bother about it?
3. Suggest three things that can be done to reduce the emission of greenhouse gases.
4. What is geosequestration and why is it important?
5. Suggest how manufacturing and using wooden products that have been produced using sustainable management may help fight global warming.
6. List the nine planetary boundaries that promote sustainable lifestyles that have been suggested by the Stockholm Environment Institute in Sweden.

7. What is metagenomics?
8. Explain why CSIRO scientists are studying Tammar wallabies in their research related to global warming.
9. What is the Kyoto Protocol and why is it important?

Investigate, think and discuss

10. Explain why it is necessary for the Australian government to create legislation to address the problem of global warming.
11. Find out more about Johan Rockstrom's contributions to science, including the concept of planetary boundaries.

7.7 Global warming — believe it or not?

7.7.1 Global warming is a hot topic

As the physicist Niels Bohr reportedly said, 'Prediction is very difficult, especially of the future.'

While most scientists agree that an increase in the amount of carbon dioxide in the atmosphere is the main cause of global warming, they argue about the details of the cause and about the effects of global warming. The key arguments that scientists are involved in investigating and discussing can be divided into three categories:

1. Are humans responsible for global warming?
2. What will the effects of global warming be?
3. What can be done to stop global warming?

7.7.2 Climate science

Climate scientists are trying to find evidence against the hypothesis that global warming is caused mainly by humans dumping greenhouse gases into the atmosphere. That is, they are considering that the hypothesis may be wrong and are assessing other ways in which this warming may be occurring. Over the last 40 years, however, no evidence against the hypothesis has been found.

A difficulty for climate scientists is not just about predicting how the climate will change, but also in estimating the level of uncertainty within the prediction.

7.7.3 Climate science and policy

Global warming is a thorny problem. There are also clashes over climate science and policy. While some refer to this as the climate debate, to those deeply immersed in it, it may feel more like an ugly war. It has included frontline battles between science and opinion, politics, media and human psychology. There has been scepticism, outright denial, disrespect and even name-calling!

An Australian newspaper reported that, in one country, scientists trying to present evidence for human involvement in climate change were accused of holding elitist, arrogant views. The media has also reported that even in our own country some leading scientists have felt ignored and excluded from contributing to the development of key climate policies and discussions.

7.7.4 Alternative theories

Alternative theories about climate change have been developed. Climate change sceptics, for example, believe that humans are not to blame for rising global temperatures and that what is being experienced is merely part of a natural cycle.

7.7 Exercises: Understanding and inquiring

To answer questions online and to receive **immediate feedback** and **sample responses** for every question, go to your learnON title at www.jacplus.com.au. *Note:* Question numbers may vary slightly.

Investigate, think and discuss

1. In 2010, the IPCC concluded that the increase in the Earth's surface temperature during the second half of the twentieth century needed to be simulated by models that included anthropogenic forcing as well as natural factors. Find out more about anthropogenic forcing and why the IPCC argues that it should be considered in the climate models. Do you agree with the IPCC? Justify your response.
2. In 2011, the IPCC estimated that if we continue as we currently are then average global temperatures will rise by 1.8–4.0 °C by 2100 and sea levels will rise an estimated 23–47 cm.
   (a) Research predicted rises in temperature and sea levels. Do you consider the IPCC's estimates to be conservative, exaggerated or in the middle of the two? Justify your response.
   (b) Do you Think the IPCC is a credible authority on climate change? Provide reasons for your opinion.

Projected consequences of climate change

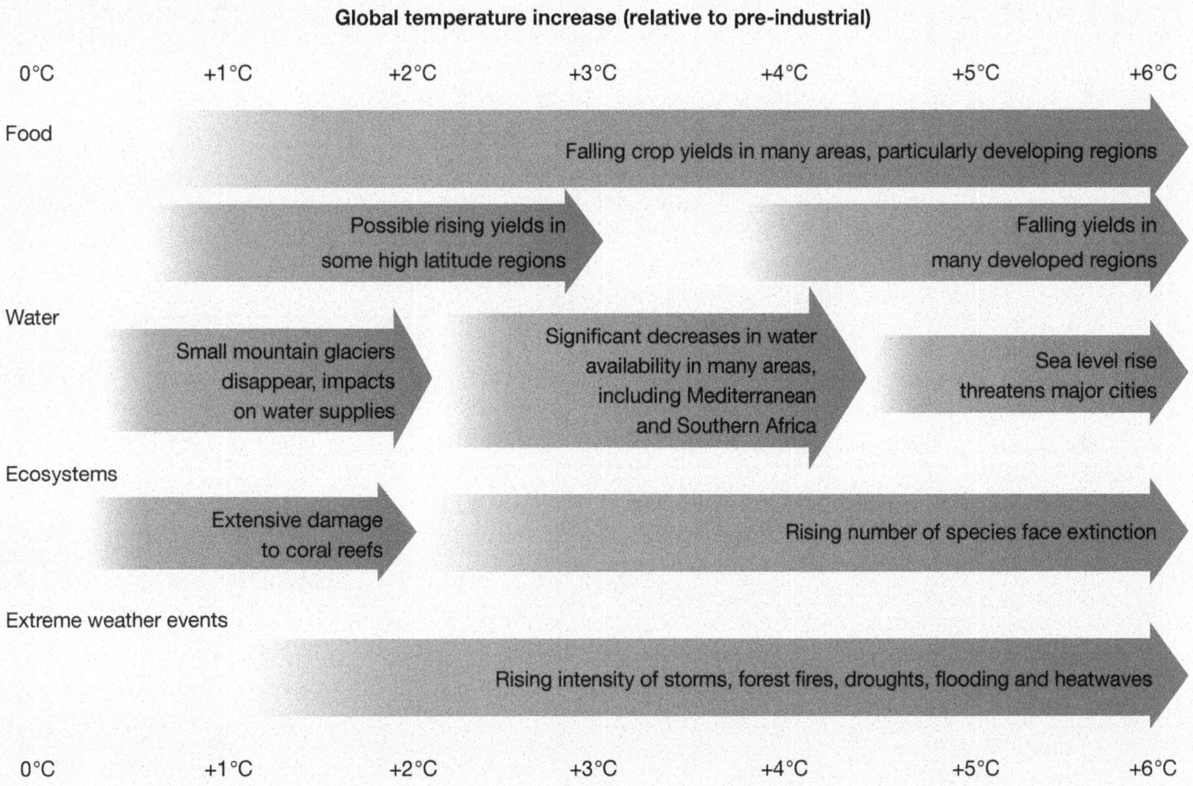

3. It is generally agreed that global warming will lead to worldwide changes in weather patterns, gradual melting of icecaps and rising sea levels. Do you agree with this statement? What is the evidence?
4. One of the difficulties of using models to predict future events such as carbon dioxide emissions is that they need to make assumptions about a series of possible future states based on known facts, rather than on accurate measurements of events from the past. This provides the opportunity for bias in selection. Find out more about the computer models used to predict these events and whether there may be any bias. Share and discuss your findings with others.
5. There have been suggestions that the funders of climate research are only supporting studies that set out to prove that global warming is caused by humans. Find out more about the types of climate research being performed and who is funding them. On the basis of your findings, do you agree or disagree with the suggestion? Justify your response.

6. Find out what *peer review of research findings* is and discuss your findings with others. Construct a PMI chart to evaluate the usefulness of peer review.
7. Find out more about these court cases for and against a greener world.
   - Kivalina vs ExxonMobil
   - Comer vs Murphy Oil
   - Texas vs Environmental Protection Agency (EPA)
   - Connecticut vs American Electric Power (AEP)

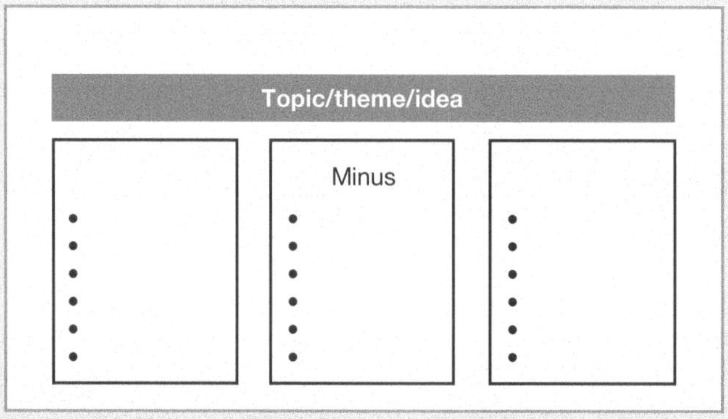

8. Distinguish between *environmentalist* and *environmental scientist*. Make a list of the types of comments that each may have about global warming or climate change.
9. There have been suggestions that belief is frequently obscuring fact in regard to the climate change issues.
   (a) Discuss with others the difference between *belief* and *fact*.
   (b) Suggest criteria that could be used for each of these terms that would enable them to be identified in articles written about climate change.
   (c) Using your criteria for these terms and internet research, find examples of beliefs and facts in climate change articles.
   (d) Share your examples with others in the class.
   (e) As a class, decide on a specific statement or issue that could be used in a class debate.
   (f) Write a presentation that could be used in a debate on climate change. Include a variety of beliefs and facts in your arguments.
   (g) Conduct a class debate on the topic decided on in part (e). Each member of the class is to have a green and a red card. During the debate, when a belief statement or argument is made students are to hold up a red card, and when a fact statement or argument is made they are to hold up a green card.
   (h) Reflect on your experiences regarding the debate and share your reflection with others.
10. Climate change is a natural event and not caused by human activity.
   (a) Research information related to this statement.
   (b) Using a table like the one shown below, and criteria that you have discussed with others and agreed on, evaluate each reference you use for:
      - authority/reputable source
      - bias
      - validity/accuracy.
   (c) Organise your material into a PMI chart or SWOT analysis.
   (d) Organise a class debate on the statement.

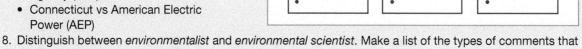

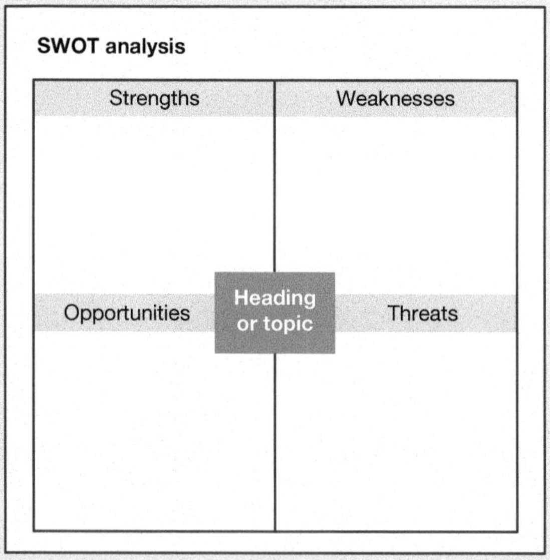

Reference title, author, date	Plus	Minus	Interesting	Other comments	Reputable? (0 = not reputable, 3 = very reputable)	Bias? (0 = very biased, 3 = no bias)	Accuracy/validity? (0 = not accurate or valid, 3 = very accurate and valid)
					0 1 2 3	0 1 2 3	0 1 2 3

11. Professor Michael Raupach is an atmospheric scientist who is co-chairman of the Australian Academy of Science's climate change working group. In 2011, he made the comment: 'There is an enormous difference between a scientific proposition, for which truth is decided on the basis of empirical evidence, and a political proposition, which is adopted or fails depending on the strength of people's convictions. Both of these forms of truth are important in our society, but we're in a lot of trouble if we mix them up — unlike human law, the laws of nature can be read, but not redrafted.'
   (a) Find out what each of the following terms mean and give an example that could be used to demonstrate it: scientific proposition, political proposition, empirical evidence, conviction (not in the criminal sense), truth, human law, law of nature, redrafted.
   (b) In a group, re-read Raupach's statement and discuss its meaning and how it could be rephrased into the language of a Year 10 student.
   (c) Share your rephrased statement with others.
   (d) Do you agree with Raupach's statement? Justify your response.

learnon RESOURCES — ONLINE ONLY

Complete this digital doc: Worksheet 7.4: Global warming (doc-19475)

7.8 Ozone alert!

7.8.1 What's the problem?

What's the problem with a hole in the sky?

About 90 per cent of the ozone in the atmosphere lies in the stratosphere, which extends from about 10 kilometres to 50 kilometres above the Earth's surface, where it blocks out more than 95 per cent of the ultraviolet (UV) rays entering the atmosphere.

During the 1980s it was discovered that the amount of ozone (O_3) in the upper atmosphere was decreasing rapidly. Any decrease in the amount of ozone in the ozone layer is damaging to all living things as they are adapted to being protected from ultraviolet radiation by ozone. For humans, the damage is in the form of sunburn and skin cancer.

This image shows how large the hole in the ozone layer can be.

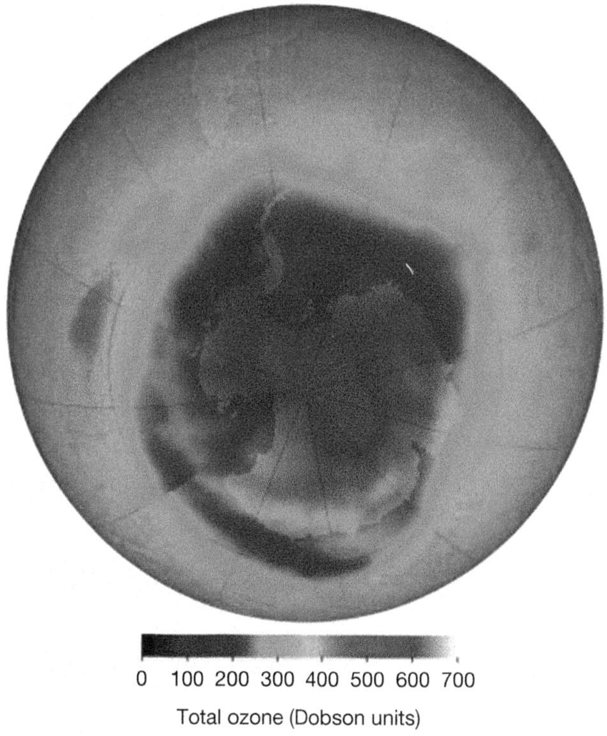

7.8.2 What's the cause?

The main cause of the rapid depletion of ozone in the stratosphere is the emission of chlorine and bromine compounds, particularly chlorofluorocarbons (CFCs), which were once used widely in aerosol spray cans, refrigerators and air conditioners.

In the stratosphere, bonds in CFC molecules are broken and free chlorine atoms are released. These chlorine atoms are involved in reactions that destroy ozone. They are then released back into the atmosphere where they continue to be involved in ozone destruction.

Not long after the discovery of the decrease in ozone, measurements taken by instruments in weather balloons and satellite images showed that the problem was far more serious than initially thought. As a result of international cooperation and recognition that the problem was urgent, the **Montreal Protocol** came into force in 1989.

Chlorine atoms are involved in reactions that lead to the destruction of ozone.

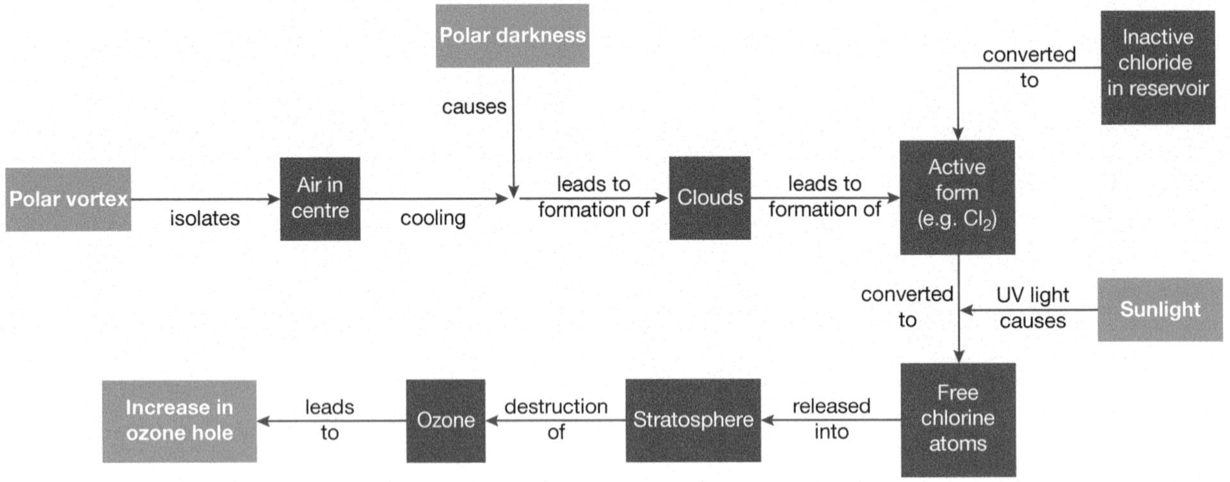

Total ozone levels measured on 10 April 2011. Based on satellite observations, the Total Ozone Mapping Spectrometer (TOMS) provides information on global and regional trends in ozone and other tropospheric aerosols. On the basis of the information shown in this figure, how does Australia rate in terms of its total ozone measurement? Suggest implications of your interpretation of these data.

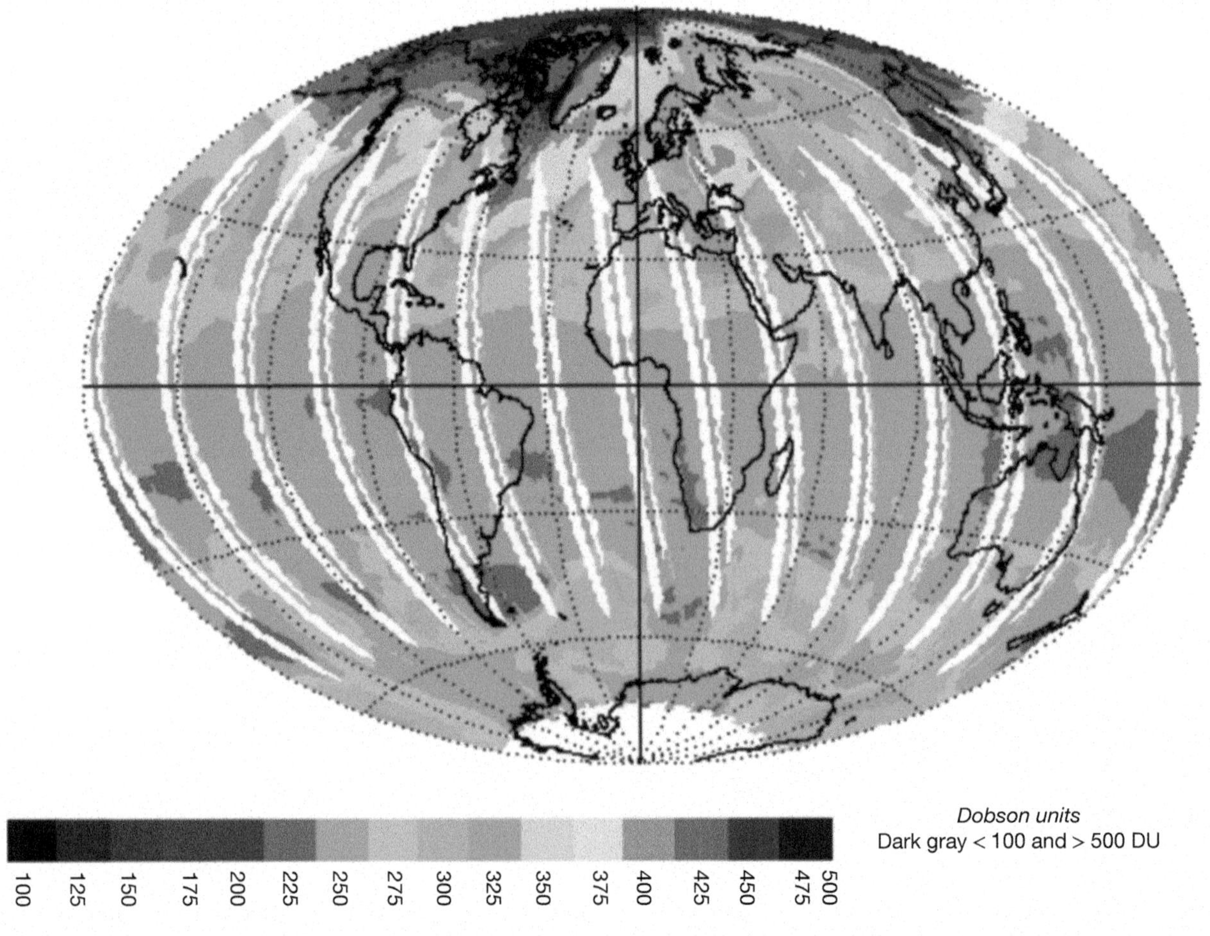

7.8.3 Ozone friendly

Throughout most of the world CFCs have been phased out and replaced in many cases with hydrochlorofluorocarbons (HCFCs), which deplete ozone to a lesser extent than CFCs but which are also greenhouse gases. These in turn are now being replaced by less harmful chemicals and new technology. The depletion of the ozone layer has already slowed, and if governments throughout the world continue to honour their agreements to phase out the use of chemicals that threaten the ozone layer, life on Earth will continue to be adequately protected from ultraviolet radiation.

The figure on the previous page shows an image from the Total Ozone Mapping Spectrometer (TOMS). These data are based on satellite observations that monitor global and regional trends in ozone and other tropospheric aerosols. The Dobson unit (DU) is a measure of total ozone. In the figure the darker reddish colours indicate a higher ozone concentration than the blue and purple colours.

The ozone layer has been significantly depleted since the 1970s.

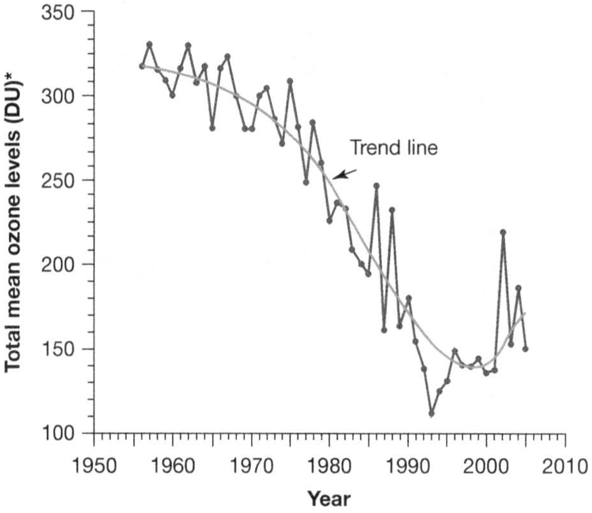

*Dobson units over Halley Bay, Antarctica October.

Colour-coded image of the sea surface temperature as revealed by an AVHRR (Advanced Very High Resolution Radiometer) carried on a satellite. Red represents the hottest and purple the coolest sea surface temperature.

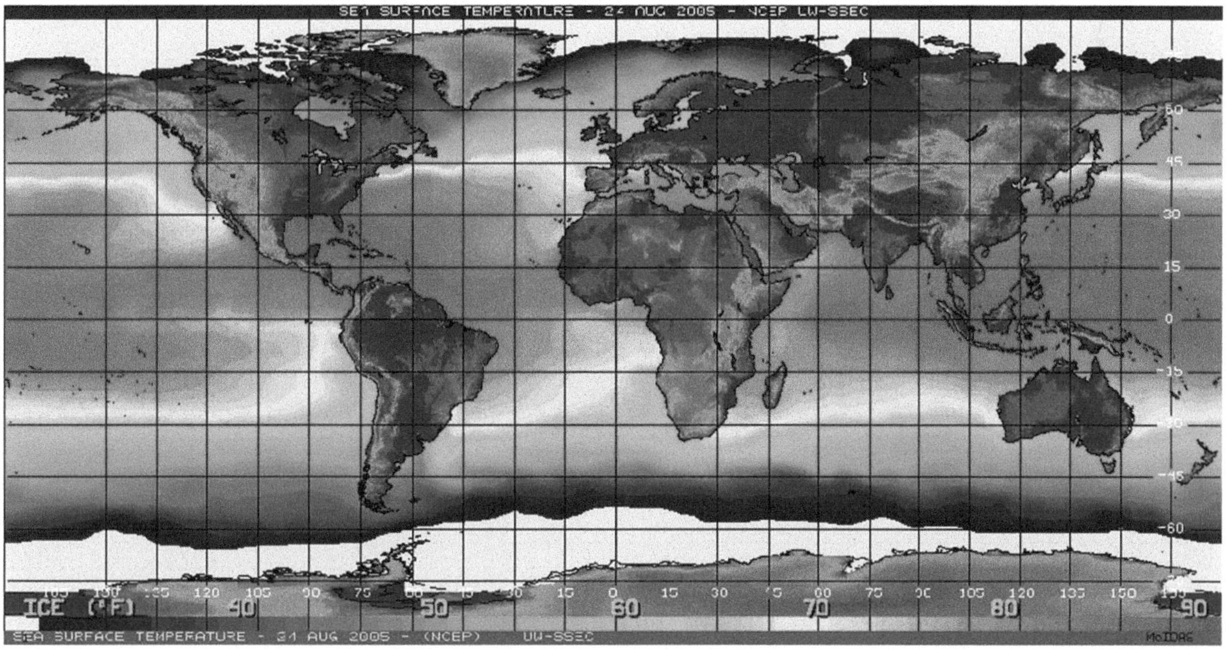

learnon RESOURCES — ONLINE ONLY

Watch this eLesson: Global warming in Australia (eles-0057)

7.8.4 Eyes in space

There are a number of other satellites that are gathering data on Earth's biosphere from a distance. This type of data collection is called **remote sensing**. The satellite *Terra*, for example, has a number of different instruments that gather different types of data on how Earth is changing in response to both natural changes and those caused by humans. Scientists from different fields are also working together on collaborative projects that use data from remote-sensing observations to improve forecasting systems such as those that warn of future floods.

Terra, the flagship satellite of the Earth Observing System. Specialised instruments carried by *Terra* collect data on the land, oceans and atmosphere of our planet that will provide a record of changes over time.

7.8 Exercises: Understanding and inquiring

To answer questions online and to receive **immediate feedback** and **sample responses** for every question, go to your learnON title at www.jacplus.com.au. *Note:* Question numbers may vary slightly.

Remember

1. In which part of the biosphere would you find the most ozone?
2. Outline why the ozone layer is important to life on Earth.
3. (a) Which types of chemicals are likely to cause a depletion in the ozone layer?
   (b) Construct a flowchart to show how these chemicals are involved in ozone destruction.
4. Suggest why the depletion of the ozone layer has slowed.

Analyse, think and discuss

5. (a) What does TOMS stand for?
   (b) How does TOMS get its data?
   (c) What is a Dobson unit?
   (d) Carefully observe the NASA TOMS figure and:
      (i) describe patterns of ozone coverage
      (ii) interpret the patterns of ozone coverage
      (iii) state the Dobson unit range for Australia
      (iv) interpret Australia's ozone pattern in terms of how effectively we may be protected against harmful UV rays.
6. The figure at right shows variations in the annual record of the hole in the ozone layer since 1979. In a group, carefully observe any patterns and discuss possible interpretations.
7. Use the graphs on the next page to answer the following.
   (a) Describe the patterns observed in the graphs.
   (b) Interpret the information in the graphs.
   (c) In which part of the biosphere is the ozone layer located?

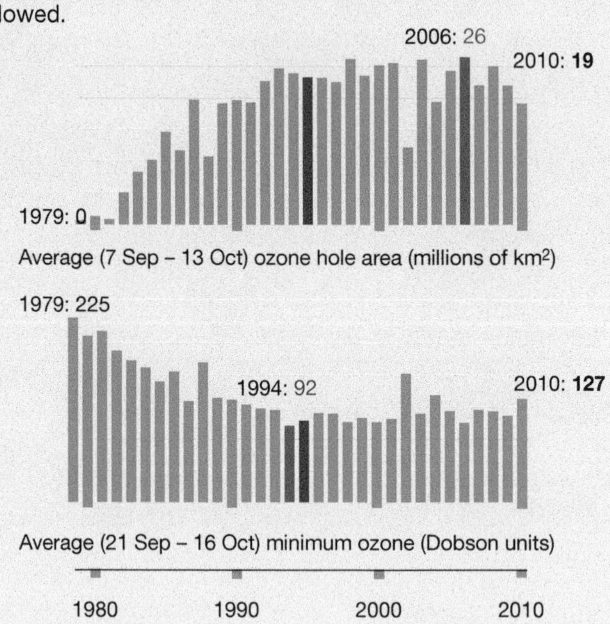

Note: No data were acquired during the 1995 season.

8. (a) Explain why there is concern about the thinning of the ozone layer.
   (b) List examples of three sources of CFCs.
   (c) Outline how CFCs contribute to the development of the ozone hole.
   (d) Explain why temperature and the amount of sunlight influences the depth and size of the ozone hole.

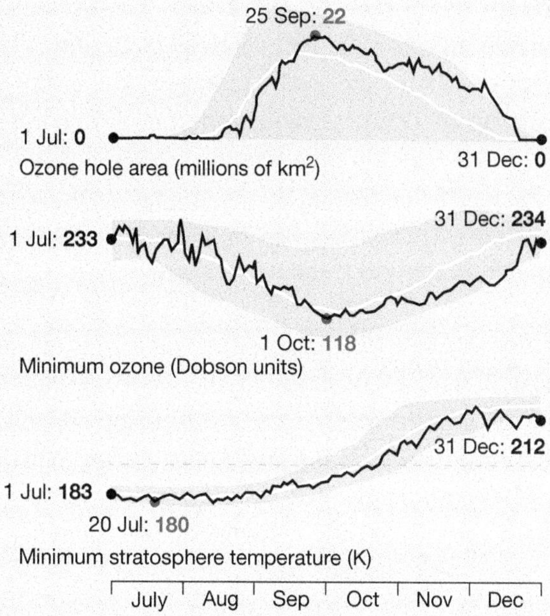

Investigate, think, discuss and report

9. Go online to find the 2010 Sustainable Cities Index developed by the Australian Conservation Foundation (ACF). This index is based on a range of environmental, social and economic issues. It provides a snapshot of the performance of 20 of our largest cities and ranks them on their sustainability.
   (a) Select the city closest to where you live. How did it rate in this index? Do you agree with the ACF's findings? Justify your response. Suggest ways in which your city's score could be improved.
   (b) Select one of the criteria used and find out more about the method used to collect the data.
   (c) Which of the 20 Australian cities scored as our most sustainable city? For which criterion did it score the highest? Suggest reasons for its high score.
   (d) Which city scored the lowest? Suggest reasons for its low score and what it could do to increase its score in the future.
10. Various satellites and data collection instruments are used to measure changes in our environment. Research and report on at least two from each group.
   (a) OMI, TOMS, GOME, NOAA SBUV/2, MLS, Balloon Sondes
   (b) MODIS, MISR, MOPITT, CERES, ASTES

7.9 Biodiversity and climate change

7.9.1 Natural climate change

When the first traces of life appeared on Earth about 3500 million years ago, the climate was hostile. Lightning bolts blasted through a warm atmosphere of hydrogen, methane, ammonia, water vapour and carbon dioxide. There was no oxygen until the first living organisms produced it through photosynthesis. Since then, the composition of gases in the Earth's atmosphere and its temperature have been constantly changing.

7.9.2 Biodiversity

The evolution of life forms on Earth has occurred because some organisms are better suited to a particular environment than others. For some to be better suited than others, there needs to be variation or diversity.

In a global sense, **biodiversity** refers to the total variety of living things on Earth, their genes and the ecosystems in which they live. Biodiversity (or biological diversity) exists at the gene, species and ecosystem level.

7.9.3 Genetic diversity

Genetic diversity can be considered in terms of variation within the genes (alleles), which are made up of DNA. Genetic variation is important for the long-term survival of a species as it increases the chance that at least one of the variations will enable some of the population to survive to reproduce the next generation.

Earth was a hostile place 3500 million years ago. Fossils provide evidence of structures called stromatolites. They existed in warm sea water and consisted of cyanobacteria, one of the earliest forms of life.

7.9.4 Diversity in DNA

Each individual contains their own combination of genetic material in the form of DNA. This information is organised into coding and non-coding regions. The coding regions, called genes, contain genetic information for the synthesis of proteins that contribute to the expression of particular features or traits.

7.9.5 Diversity in alleles

Individuals within a species share the same genes that code (with an environmental influence) for a particular feature or characteristic. However, there can be alternative forms of these genes within the individuals. Alternative forms of genes are called alleles. For example, an individual within a species may have a gene for beak shape. The alleles for beak shape may code for hooked or straight shape. So, some individuals may contain the alleles for hook-shaped beaks, some the alleles for straight-shaped beaks and others the alleles for each type.

The particular combination of alleles for a particular trait (or phenotype) within an individual is called the genotype. For example, if the allele for the hooked beaks is given the symbol *H* and the allele for the straight beaks is given the symbol *h*, then an individual could have a genotype of *HH* or *Hh* or *hh*.

7.9.6 Species diversity

Species diversity can be considered in terms of diversity in populations. While the combination of alleles for a trait within an individual is called a genotype, the combination of all the alleles within a group of individuals of the same species living in a particular place at a particular time (population) is called a **gene pool**.

All environments change over time. It is the diversity (or variation) of the alleles within the gene pool that contributes to the number of possible combinations that could be used to produce the next generation. Increased variety in the expression of these alleles as phenotypes (traits) of the offspring means an increased chance that some of these offspring will be able to survive in the environment in which they are born and will live — even if that environment changes.

If there is little variation in the gene pool, there is less chance of the offspring being able to survive possible changes in their environment such as climate and the availability of habitat, food, mates or other resources. The consequences of this limited diversity within the population may lead to the **extinction** of the species.

7.9.7 Ecological diversity

Ecological diversity can be considered in terms of the diversity in ecosystems. The extinction of a particular species within an ecosystem may affect the survival of other species within that ecosystem. The extinct species' disappearance will have consequences for the food supplies of others within its food web. Unless there are other species that can take its place without having a negative effect on others, the survival of other species may be threatened.

Increased biodiversity within ecosystems can reduce the consequences of losing a species to which the survival of others is linked. Likewise, reduced biodiversity in these ecosystems can lead to the extinction of other species.

7.9.8 Australia's biodiversity

Biodiversity within Australian ecosystems is influenced by both biotic factors and abiotic factors. Abiotic factors, including those that contribute to climate, such as temperature and annual rainfall, can affect the abundance, distribution and types of species within a particular ecosystem. Organisms have particular tolerance ranges for abiotic factors, outside of which they cannot survive.

If global warming results in the development of climatic conditions that are outside a species' tolerance range, and if they are unable to migrate or adapt to the new conditions, then there is a threat that the species may become extinct. Species that are most at risk are those that have low genetic variability, long life cycles and low fertility, a narrow range of physiological tolerance and geographic range, and specialist resource requirements.

7.9.9 Global warming and Australia's biodiversity

Changes in Australia's biodiversity that may be due to climate change include changes in species' ranges and migration patterns, shifts in genetic composition of some species that have short life cycles, and changes in lifestyle and reproduction rates.

Many plants and their pollinators have **coevolved**. Studies have suggested that climate change has upset the life cycles of pollinators (such as bees). Other studies suggest that climate change is causing the flowering times of some plants to be out of synchronisation with their pollinators. With fewer plants being pollinated, fewer are bearing fruit containing seeds essential to produce the next generation of plants.

CLIMATE CHANGE HITS SE AUSTRALIAN FISH SPECIES

Significant changes in distribution of about 30 per cent of coastal fish species in south-east Australia are being blamed on climate change ... Scientists from both the CSIRO Climate Adaptation Flagship and the Wealth from Oceans Flagship have identified shifts in 43 species.

Ecos, October–November 2010

7.9.10 Preparing to adapt to unavoidable climate change

The National Climate Change Adaptation Research Facility (NCCARF) has identified eight priority areas for adaptation research. These are terrestrial biodiversity; primary industries; water resources and freshwater biodiversity; marine biodiversity and resources; human health; cities and infrastructure; emergency management; and social and economic issues.

7.9.11 Mass extinctions

Many scientists believe that we are currently experiencing the sixth mass extinction. Five other mass extinctions have occurred as a result of global climate change. Some argue that humans are responsible for the current mass extinction. The International Union for Conservation of Nature has reported that species are dying out 1000 to 10 000 times faster than they would without human intervention.

Those with the view that humans are to blame divide this sixth extinction into two phases. The first phase began about 100 000 years ago when the first modern humans began to spread throughout the world. The second phase began when humans started to use agriculture around 10 000 years ago.

There have been five mass extinctions in the past — are we currently experiencing a sixth and, if so, is it caused by humans?

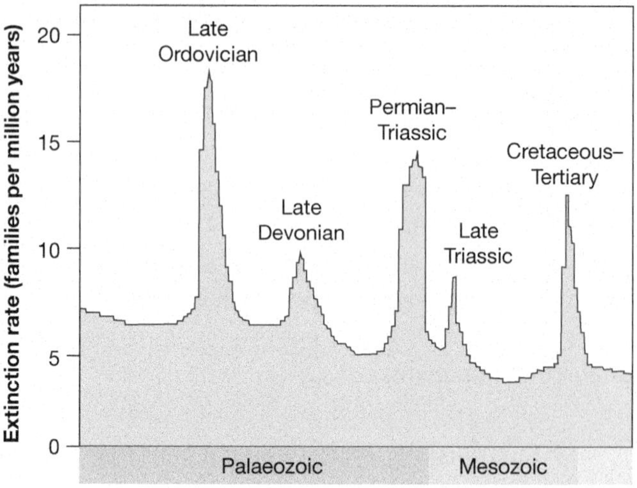

ULTRAVIOLET LIGHT EXPOSURE DAMAGES TADPOLES

Depletion of the ozone layer has been revived as an explanation for the extinction of amphibians after the discovery that increased ultraviolet-B radiation makes striped marsh frog tadpoles more vulnerable to predators.

Since 1980 more than 150 species of amphibians have become extinct. This compares poorly with background extinctions of 1 every 250 years. 'With amphibians being the most threatened of all vertebrates, and also important indicators of environmental health, understanding the causes of their declines is critical for their conservation, and possibly the conservation of other species,' says Lesley Alton, a PhD Student at the University of Queensland's School of Biological Sciences.

Australasian Science, April 2011

7.9 Exercises: Understanding and inquiring

To answer questions online and to receive **immediate feedback** and **sample responses** for every question, go to your learnON title at www.jacplus.com.au. *Note:* Question numbers may vary slightly.

Remember

1. There was no oxygen in Earth's early atmosphere, but there is now. Where did it come from?
2. Suggest a connection between the concepts of *diversity* and *better suited*.
3. Define the following terms.
   (a) Biodiversity
   (b) Genetic diversity
   (c) Species diversity
   (d) Ecosystem diversity
4. State the three levels at which biodiversity can exist.
5. Outline the importance of genetic variation to the survival of a species.
6. State the form in which genetic material exists in all species.

7. Describe the function of genes.
8. Describe the relationship between DNA, genes, proteins and traits using a flow diagram.
9. Distinguish between the following terms.
   (a) Genes and alleles
   (b) Genotype and phenotype
   (c) Genotype and gene pool
   (d) Survival and extinction
10. Compare the survival chances of a species showing low diversity and a species showing high diversity.
11. Suggest the consequences of limited diversity in a population.
12. Suggest how diversity within an ecosystem may increase the survival chances of species within it.
13. State examples of abiotic factors that can affect the survival of an organism.
14. Define the term *tolerance range* and suggest an example.
15. Suggest a connection between global warming and changed abiotic factors within ecosystems.
16. State the features of species that would be most at risk of extinction in changing climatic conditions such as global warming.
17. Suggest changes in Australia's biodiversity that may be due to climate change.
18. Suggest a connection between reduced pollination of some types of plants and climate change.
19. List the eight priority areas identified by NCCARF for adaptation research.
20. Outline the two phases of human contribution to the sixth mass extinction.

Investigate, think and discuss

21. Are you concerned about the arrival of the Earth's sixth mass extinction? One survey asked people to respond to this question by choosing 'Yes', 'No' or 'Sort of but I won't see the effects in my lifetime'. How would you have responded? Justify your response.
22. Do living organisms always have a negative effect on their environment? Justify your response and include a supporting example.
23. Suggest ways in which organisms could be better suited to survive in a particular enviroment than others.
24. Research and report on coevolution and the possible effect that global warming may have on organisms that are linked by this type of evolution.
25. Research and report on examples of life forms that are able to survive in an oxygen-free environment, both throughout Earth's history and today.
26. Identify sources of variation for (a) asexually reproducing and (b) sexually reproducing organisms.
27. Select and research the topic of one of the article extracts.
28. (a) Find out more about:
       (i) coevolution
       (ii) pollination
       (iii) flowering plant life cycles
       (iv) bee life cycles
       (v) extinction
       (vi) climate change
       (vii) pollinator decline.
    (b) Link the terms in part (a) using a mind map or fishbone diagram.
    (c) Research possible implications of pollinator decline for:
       (i) farming and food supplies
       (ii) plant biodiversity on Earth
       (iii) humans.
29. Research and report on the role that museums play in the identification and preservation of species and how this contributes to Australia's biodiversity.
30. The biggest problem connected to the effects of climate in Kakadu's coastal floodplain is the rise in sea level. Salt water has already intruded in various parts of the park and has affected the local populations of *Melaleuca* (paperbark) trees and magpie geese. Research and report on the current and possible effects of rising sea levels in Kakadu.

learnon RESOURCES — ONLINE ONLY

Try out this interactivity: Threats to Earth (int-0218)

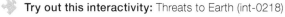

7.10 Biosphere 2

Science as a human endeavour

7.10.1 Biospherics

Humans living in biospheric systems such as small spacecraft and submarines use physical and chemical techniques to recycle clean air and fresh water and remove accumulating wastes. As biospheric systems increase in size, however, the basic concepts of cycling of elements and the importance of biodiversity have direct implications on a number of different issues. These include global warming, the protection of endangered species, sufficient food supplies, effective waste removal and clean water requirements.

Biospherics is an exciting and essential new science. It was first envisioned by Vladimir Vernadsky in Russia in the 1920s. The biosphere project was inspired by John Allen, an American football player turned Beat poet (Johnny Dolphin), who had worked on a number of projects related to the synthesis of ecology and technology. In the early 1980s, along with several colleagues, he formed Space Biospheres Ventures. John Allen and his team designed and built an artificial world — Biosphere 2 — to develop a closed ecological system for research and education. Perhaps eventually this information will be used to sustain human life on other planets, such as Mars.

Plan of Biosphere 2. The glass and structure components acted as a filter for incoming solar radiation so that almost all UV radiation was absorbed.

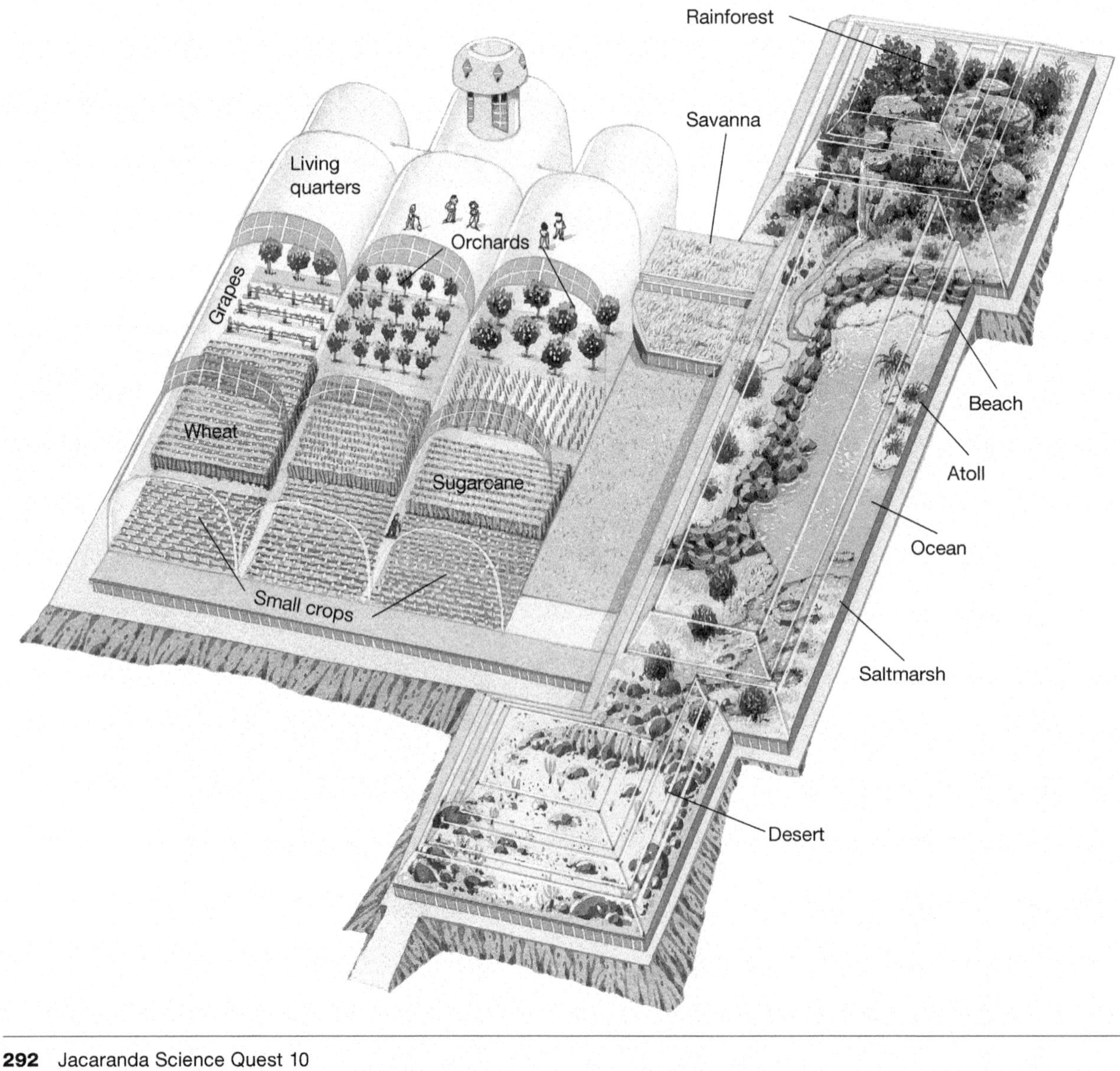

7.10.2 What does it look like?

Biosphere 2 covers 13 000 square metres and contains living quarters and greenhouses containing food crops. Five different artificial environments are enclosed within the structure: a desert, a salt marsh, a tropical savanna, an ocean and a rainforest.

Biosphere 2, southern Arizona

7.10.3 What is it for?

Earth is a natural biosphere. The Earth's biosphere (Biosphere 1) has existed for at least 3.8 billion years. Some have called Biosphere 2 a type of cyber-Earth. Biosphere 2 is an artificially made structural biosphere located at an elevation of 1200 metres above sea level in a temperate desert region in southern Arizona, United States of America. Biosphere 2 was designed as an eco-technological model for space exploration and colonisation. This bioengineered facility was intended to grow food, cleanse the air, and recirculate and purify water for its inhabitants. This was to be achieved without exchange of materials (including atmospheric gases) with the outside world. The purpose of this cyber-Earth was for scientists to gather information to assist in the development of strategies to solve some of Earth's environmental problems and the hurdles of developing human colonies in space.

7.10.4 Closed systems

Biosphere 2 and Earth are similar because they are both closed systems. The space frame of Biosphere 2 has the same job as the Earth's atmosphere, which acts as a giant hollow globe that keeps the Earth a closed system. No event in a closed system (such as Earth's atmosphere or Biosphere 2's special frame) is isolated. If 40 people were to enter the desert biome of Biosphere 2, the sensors would quickly record a decrease in the oxygen levels and an increase in carbon dioxide levels throughout all of the biomes in Biosphere 2. This is because the people would breathe faster than the plants could take up the excess carbon dioxide. Could a similar thing happen outside Biosphere 2?

Abigail Alling stopped her graduate work at Yale University on blue whales to enter Biosphere 2 as the manager of oceans and marshes. She created and operated the world's largest artificial ecological marine system, a mangrove marsh and ocean coral reef, for the Biosphere 2 project. She was one of the original eight people to live inside Biosphere 2 — the artificial cyber-Earth system.

7.10.5 What happened?

Shortly after sunrise on 26 September 1991, eight people and 3800 species of plants and animals were locked inside this artificial world for two years. Worldwide, millions of television viewers watched. The crew had been prepared by years of training and working on developing systems for Biosphere 2. They had also had nine preliminary one-week semiclosed experiments over the previous five months.

7.10.6 Gasping for oxygen

By the end of the first year of their mission, the Biospherians reported deteriorating air and water quality. Oxygen concentrations in the air had fallen from 21 per cent to 14 per cent. This oxygen level was barely enough to keep them alive and functioning. At the same time, carbon dioxide concentrations were undergoing large daily and seasonal variations and nitrous oxide in the air had reached mind-numbing levels. In January 1993, fresh air was pumped in to replenish the dome's atmosphere and rescue the inhabitants. Investigations indicated that the missing oxygen was being consumed by microbes in the excessively rich food crop soil.

It was very fortunate that the fresh concrete used in the structure's construction absorbed carbon dioxide released by microbial metabolism. If this carbon dioxide sink hadn't been available, the air would have become unbreatheable much earlier.

Air flow and carbon dioxide movement through Biosphere 2

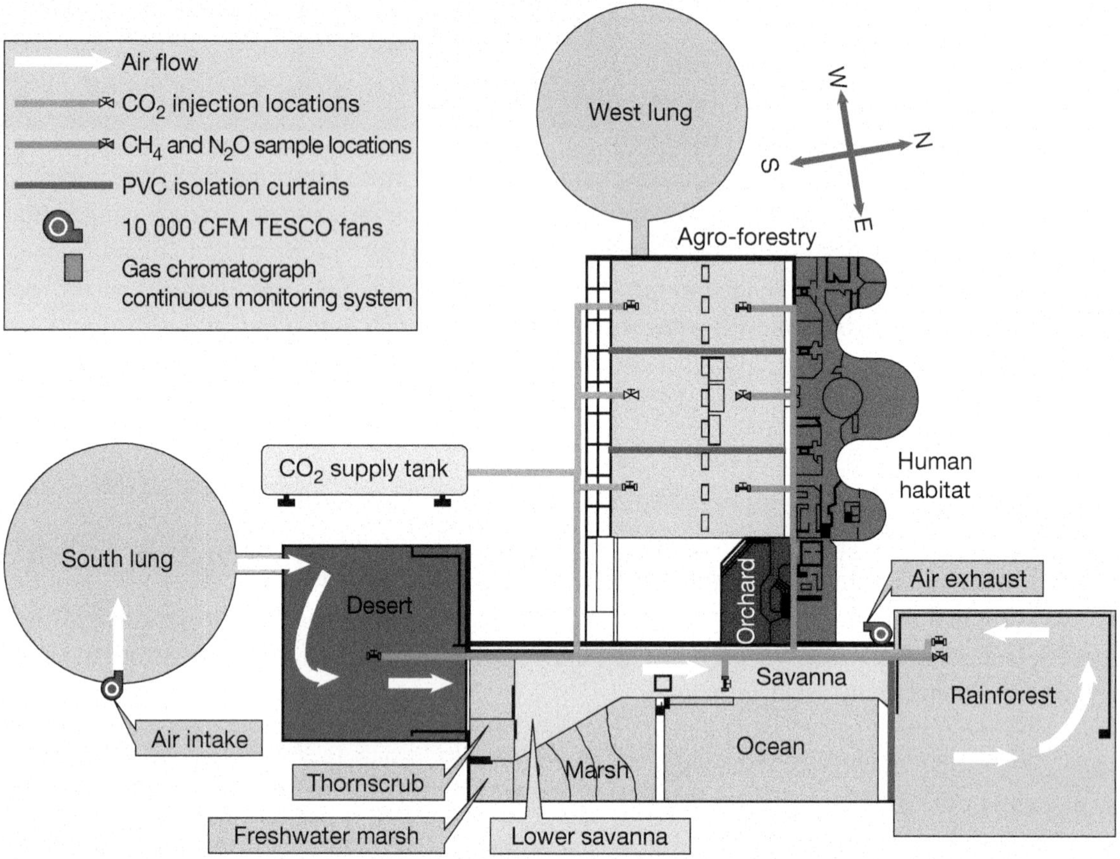

7.10.7 More carbon cycling

Due to a forceful El Niño current, one of Arizona's cloudiest seasons on record was experienced between October 1991 and February 1992. The carbon dioxide concentration inside Biosphere 2 rose to about 3400 ppm (parts per million). The combination of this effect and an unusually dark cloudy period in the last week of December greatly reduced photosynthesis. During this period, the rise in carbon dioxide was kept below 4000 ppm by the operation of a recycler, which captured carbon dioxide and precipitated it into calcium carbonate (limestone). The calcium carbonate could later be released into the air by heating the limestone. This experience provided an insight into how to maximise photosynthesis and minimise soil respiration. Hence, Biosphere 2's goal to maintain its atmosphere was achieved despite the low light conditions.

7.10.8 Getting hungry

Ideally, the chemical-free agriculture system inside Biosphere 2 recycled all human and domestic animal waste products. It also initially included dozens of crop varieties to provide nutritional balance and allow for crop rotation. Biosphere 2, however, encountered considerable food production problems.

One article written about the Biosphere 2 project stated: 'Seal a group of scientists inside Biosphere 2, the futuristic glass-and-dome experiment, for two years and what do you get? Fights over food.' Comments from the Biosphere 2 botanist suggested that personality differences and crop failures made life difficult and that 'food distribution became a very tense issue … I think that made us all a little cranky, always being hungry'.

Due to unexpected crop failures, far less food was produced than had been projected. Only 60 per cent of the sunlight made it through the glass pane of Biosphere 2's space frame. Cloudy days also reduced the light available to plants for photosynthesis. A combination of unprecedented cloudy weather for the second straight year (20 per cent below the low rate of sunshine of 1992) and increased insect pest problems contributed to reduced food production.

An interview with one of the Biospherians in February 1992 described their surprise at their initial weight loss and desire for more food than they were supposed to have. They dipped into their stored food, believing that a better summer harvest would allow them to replenish it later. Unfortunately, the harvest did not improve. A lock was placed on the refrigerator to keep them from sneaking food. When the mission ended, the average weight loss per person was around 13 kg.

7.10.9 Survivors

Eighteen of the 25 introduced vertebrate species became extinct. All of the insect pollinators died, which prevented most plants from producing seeds. This led to food supplies falling to dangerous levels. Weedy vines flourished in the carbon-rich atmosphere and threatened to choke out more desirable plants. Although the majority of insects disappeared, ants and cockroaches thrived and overran everything, including workers.

Water was conserved inside the Biosphere 2 wilderness environments. Condensation, artificial rain or irrigation (by sprinkler systems), evapotranspiration and sub-soil drainage were the major internal water cycling components. Water systems, however, became polluted with excess nutrients. This led to degraded aquatic habitats and contaminated drinking water supplies.

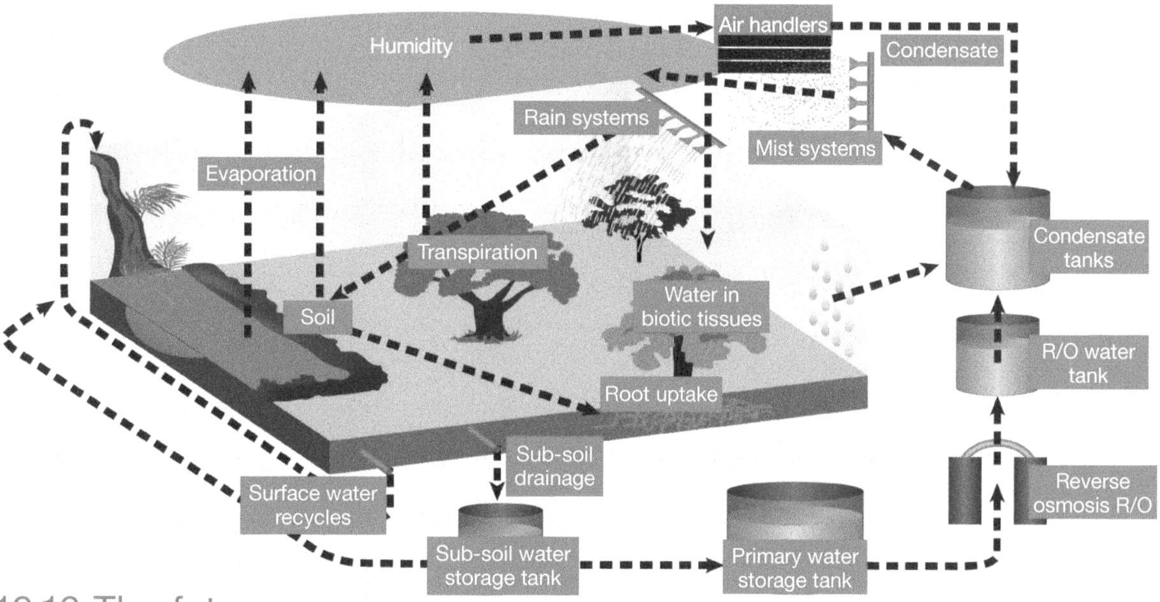

7.10.10 The future

More recent plans for Biosphere 2 include flushing it with carbon dioxide and using it to predict the Earth's future.

As carbon dioxide is a fundamental requirement for photosynthesis, scientists have long suspected that higher carbon dioxide levels will fuel extra plant growth. Some of them have even suggested that rising carbon dioxide levels may boost global harvests. Other scientists have suggested that trees and shrubs around the world will help alleviate the problems of global warming by soaking up some of the additional carbon dioxide. This brings some thought-provoking questions to mind.

- If extra plant growth does appear, will all crop plants be affected in the same way; if not, what are the implications?
- If extra carbon is taken up by the natural biosphere, how long will it stay there?

- What would happen if the carbon dioxide quickly went into the soil and was then returned to the atmosphere?
- Will the carbon dioxide be safely locked up in the forests?
- Are there carbon dioxide levels that may kill off trees and shrubs, resulting in release of their accumulated carbon in one catastrophic burst?
- How long (and with what effects) can a group of people live in an artificial closed system?
- Does the experience of Biosphere 2 bring us any closer to living on Mars?

Will extra CO_2 cause faster growth for crops such as this sugarcane?

7.10 Exercises: Understanding and inquiring

To answer questions online and to receive **immediate feedback** and **sample responses** for every question, go to your learnON title at www.jacplus.com.au. *Note:* Question numbers may vary slightly.

Remember

1. Why was Biosphere 2 not called Biosphere 1?
2. List the five different artificial environments enclosed within Biosphere 2.
3. What was the purpose of Biosphere 2?
4. State one way in which Biosphere 2 is similar to Earth.
5. Why would you expect an increase in carbon dioxide levels and a decrease in oxygen levels if a large number of people entered Biosphere 2?
6. Why was fresh air pumped into Biosphere 2 in 1993?
7. How did microbes affect the carbon dioxide levels?
8. Why was it fortunate that the fresh concrete in the structure absorbed carbon dioxide?
9. How did clouds affect food production?
10. How was water cycled through Biosphere 2?
11. Suggest ways in which the experience and findings of the Biosphere 2 project can be useful.

Think and discuss

12. Due to the moist, artificially generated climate, shrubs and grasses, rather than desert plants, overran the desert area.
    (a) Suggest why this occurred.
    (b) If you were the scientist assigned to solve this problem, suggest how you could increase the number of desert plants.
13. The rainforest in the Biosphere 2 prospered, doubling in size. Job's tears, a grass that normally grows about 60 centimetres tall in the tropics, became a giant of around 4 metres.
    (a) Suggest how this outcome could be advantageous to Biosphere 2.
    (b) Suggest how this outcome could be disadvantageous.

Imagine and create

14. Make a biosphere using a plastic soft drink container.
15. Imagine that you were one of the Biospherians. Write a diary about your two years in Biosphere 2.
16. Imagine you are one member of the first colony to live on Mars in 2050. Write a letter back to your family or friends about your new life.
17. Imagine that a meteor will hit Earth in two years' time and that all of human life needs to be moved off the planet by this time.
    (a) Make a list of all of the things that would be required to support human life.
    (b) Design a spacecraft that can meet these needs and keep you and your fellow travellers alive until you find (or can modify) a planet or environment that is habitable.
18. Imagine that the combination of the greenhouse effect and the hole in the ozone layer have made Earth uninhabitable. You need to design an artificial world that will meet your needs. What would it look like and how would it work?

learnon RESOURCES — ONLINE ONLY

Try out this interactivity: The survival game (int-0217)

Complete this digital doc: Worksheet 7.5: Slowing global warming — alternatives (doc-19476)

Complete this digital doc: Worksheet 7.6: A dome away from home (doc-19477)

7.11 SWOT analyses and fishbone diagrams

7.11.1 SWOT analyses and fishbone diagrams

1. Draw up a square and divide it into four quarters. In the centre of the diagram write down the topic or issue that you are going to analyse.
2. Think about or brainstorm the positive features and behaviours and record them in the Strengths section.
3. Think about or brainstorm the negative features and behaviours and record them in the Weaknesses section.
4. Think about or brainstorm possible opportunities and record them in the Opportunities section.
5. Think about or brainstorm possible threats and record them in the Threats section.

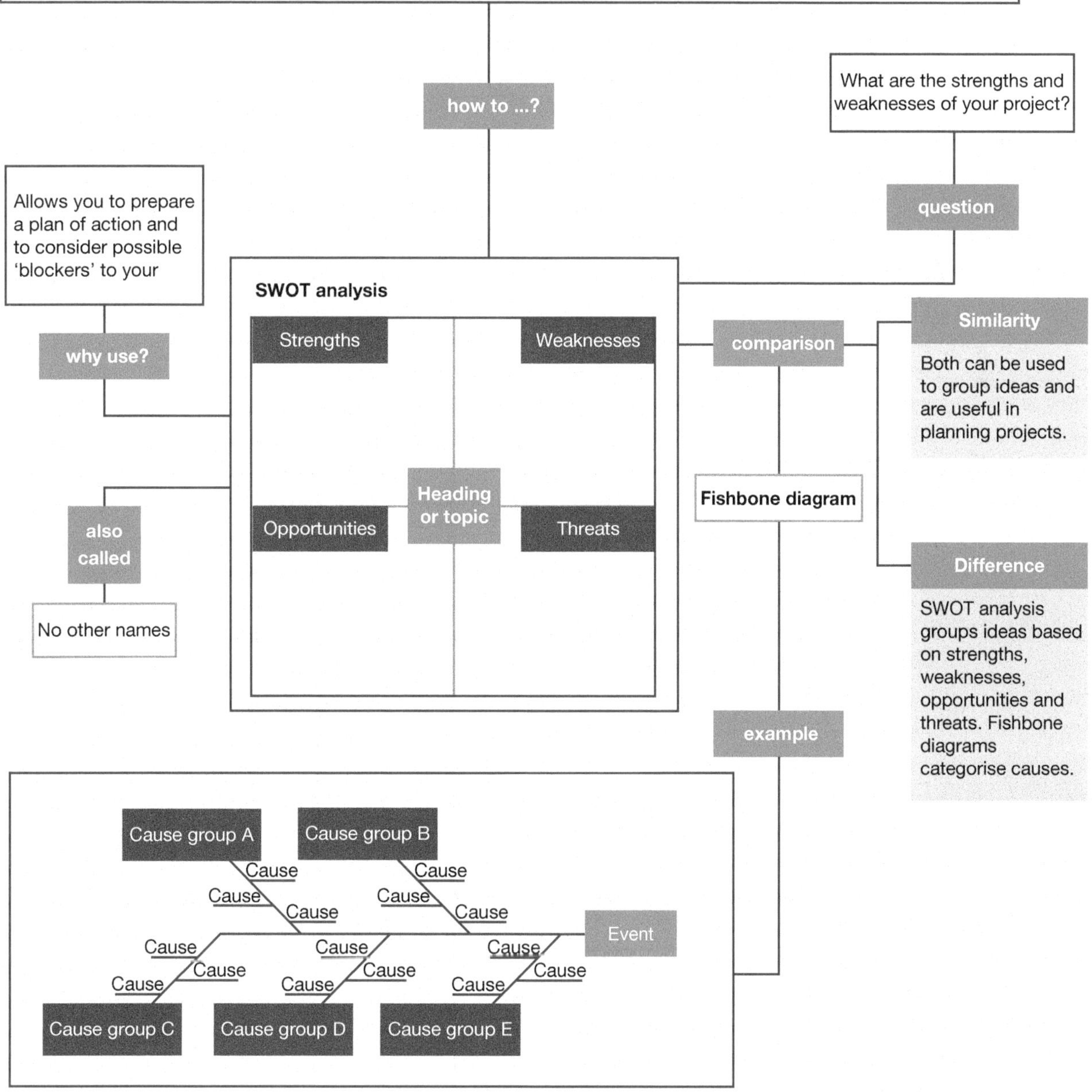

7.11 Exercises: Understanding and inquiring

To answer questions online and to receive **immediate feedback** and **sample responses** for every question, go to your learnON title at www.jacplus.com.au. Note: Question numbers may vary slightly.

Read, think, discuss and create

1. Read the article *Wash clothes with thin air* and use a 321 tool to summarise three interesting points, two important points and one personal point.

WASH CLOTHES WITH THIN AIR

It could be a godsend for drought-stricken communities — a washing machine that needs no water or powder yet cleans clothes in a jiffy.

Scientists in Singapore have invented a revolutionary appliance called the Airwash and it has already caught the eye of one major manufacturer.

The machine works by blasting dirty clothes with jets of air primed with negative ions, which have the effect of clumping dust together, deactivating bacteria and neutralising odours.

The result, the inventors claim, is clean, fresh-smelling clothes that come out of the machine completely dry — meaning an end to clothes lines and perhaps even a death knell for the tumble dryer.

And since no water is involved, fabrics unsuitable for conventional machines — such as leather and suede — can be washed at home instead of having to be dry cleaned.

Negative ions are molecules that have gained an electric charge. Odourless, tasteless and invisible, they are created when molecules in the atmosphere break apart due to fast-moving air and sunlight. In nature, they are found in invigorating environments such as pine forests and where breaking waves pound the seashore.

The Airwash is inspired by the way clothes used to be beaten against river rocks near waterfalls, which are another of nature's negative-ion generators. A prototype has been built by Gabriel Tan and Wendy Chua of the National University of Singapore and Electrolux is watching closely.

The average householder spends nine months in a lifetime doing the washing and the Airwash designers believe any machine that makes the chore easier will be welcomed.

Mr Tan said: 'But as well as being boring, laundry uses up scarce water supplies and pollutes with chemical detergents.'

2. In a team of four, use the 'learning placemat' on the right to:
   (a) write down key points of each individual's summary from question 1
   (b) verbally share your key points with other team members
   (c) agree on a team summary and place it in the middle of your team placemat
   (d) share and discuss your team mat with another team.

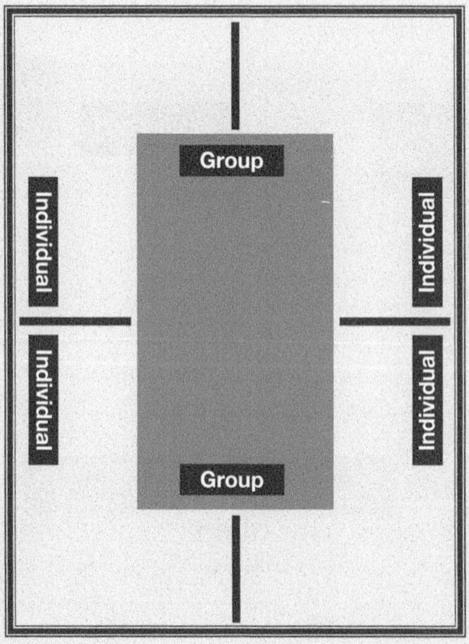

3. (a) Construct your own individual SWOT analysis on the Airwash machine.
   (b) Discuss and compare your SWOT analysis with members from two other groups.
   (c) Report back to your team on what you have found out.
   (d) Construct a team SWOT analysis.
4. In your team, brainstorm other inventions that may result in reduced household water usage.
5. In many states in Australia there are water restrictions to try to conserve the very limited water supply available to us.
   (a) Brainstorm some possible reasons that we have such a limited supply of water.
   (b) In pairs or teams of four, use an affinity diagram to organise your list of reasons into groups.
   (c) Construct a fishbone diagram by putting your title in the head of the fish, labels of the groups of reasons on each of your main side fishbones and the reasons on smaller fishbones off the main bones.

learn on

7.12 Project: The fifty years after...

7.12.1 Scenario

- **260 million years BCE:** A massive volcano in what is presently China erupts, causing atmospheric and oceanic changes leading to the extinction of 95 per cent of life in the oceans and 70 per cent of land-based life.
- **95 million years BCE:** Undersea volcanic activity triggers a mass extinction of marine life and buries a thick mat of organic matter on the sea floor.
- **72 000 BCE:** The Lake Toba volcano in Indonesia ejects nearly 3000 cubic kilometres of material into the atmosphere, cutting off much of the sun's light to the Earth's surface for so long that 50 per cent of humanity dies out.
- **2000 CE:** The UK science program *Horizon* uses the term supervolcano to describe volcanoes capable of massive eruptions covering huge areas with lava and ash and causing long-term weather effects and mass extinctions.
- **2030 CE:** The supervolcano under Yellowstone National Park erupts cataclysmically, destroying half of the US and changing the Earth's atmosphere and surface conditions for centuries to come.
- **The year is now 2080.** Fifty years after the eruption, the gases and ash that the eruption produced, as well as the destruction of large sections of land, have affected the critical environmental cycles of the Earth's environments; human civilisation has had to change its ways in order to survive. Some things remain the same though — we still have radio and television of a sort. Not surprisingly, with the fiftieth anniversary of the Yellowstone eruption (or 'Y-day', as it is known) coming up, lots of TV programs will be focusing on the critical event that changed our world forever.

Grand Prismatic Spring in Yellowstone National Park

7.12.2 Your task

As part of a small documentary film company, you will produce a 5-minute segment for a special edition of a TV science show that will be aired on the fiftieth anniversary of Y-day. In this segment, a science journalist will interview a variety of experts in a retrospective of what happened on Y-day, how the environment has changed over the 50 years since the eruption, and what humanity can expect to happen in the next 50 years.

7.13 Review

7.13.1 Study checklist

Global systems

- provide examples of ways in which human activity has affected global systems
- describe the phosphorus and nitrogen cycles
- outline the processes involved in the carbon cycle
- show the interactions of the carbon, water, phosphorus and nitrogen cycles within the biosphere
- explain the causes and effects of the greenhouse effect
- distinguish between the greenhouse effect and the enhanced greenhouse effect

Biodiversity

- define the term 'biodiversity'
- distinguish between genetic diversity, species diversity and ecological diversity
- outline some sources or causes of genetic diversity
- suggest why species diversity is important to the survival of the species
- suggest why biodiversity is important to the survival of a species
- suggest a link between biodiversity and evolution
- consider the long-term effects of loss of biodiversity
- explain the factors that drive the ocean currents, their role in regulating global climate and their effects on marine life
- outline the effect of climate change on sea levels and biodiversity
- comment on changes to permafrost and sea ice and the impacts of these changes
- suggest how genetic characteristics may have an impact on survival and reproduction
- describe the process of natural selection using examples
- explain the importance of variations in evolution

Global systems and human impacts

- explain the causes and effects of the enhanced greenhouse effect
- suggest a link between the enhanced greenhouse effect and global warming
- outline some human activities that are contributing to global warming
- outline some key issues of the climate change debate
- describe examples of ways in which human activity has affected biodiversity

Science as a human endeavour

- evaluate some strategies for addressing global warming
- comment on the role of science in identifying and explaining the causes of climate change

Individual pathways

ACTIVITY 7.1	ACTIVITY 7.2	ACTIVITY 7.3
Revising global systems doc-8479	Investigating global systems doc-8480	Investigating global systems further doc-8481

learnon ONLINE ONLY

7.13 Review 1: Looking back

To answer questions online and to receive **immediate feedback** and **sample responses** for every question, go to your learnON title at www.jacplus.com.au. *Note:* Question numbers may vary slightly.

1. Global warming is a current issue that is not going away.
   (a) Outline the most accepted view within the scientific community of the cause of global warming.
   (b) Describe examples of effects or consequences of global warming that have been suggested by scientists.
   (c) State your opinion about the possible (i) cause, (ii) effects and (iii) solutions for global warming.
   (d) View the top ten arguments about global warming that are used by sceptics. Rank these statements in order of most like your opinion to least like your opinion. Justify your ranking.
   (e) State the difference between an opinion, a theory and a fact.
   (f) Can scientists have opinions? If you agree, when, how and why should these be shared? If you do not agree, why not?
   (g) Should science play a part in the making of climate policy? Justify your response.
   (h) Suggest possible reasons for the climate debate.
2. Demonstrate your understanding of the following groups of terms by using a visual Thinking tool to show the links between them.
   (a) Species, biodiversity, biodiversity loss, threatened, endangered, extinct, mass extinction
   (b) Biosphere, lithosphere, hydrosphere, biota, atmosphere, troposphere, stratosphere
   (c) Atoms, molecules, organelles, cells, multicellular organisms, species, population, ecosystem, biosphere
   (d) Stratosphere, climate change, greenhouse gas, fossil fuels, global warming, carbon dioxide, methane, nitrous oxide, biodiversity loss, enhanced greenhouse effect, cellular respiration, lithosphere
   (e) Carbon cycle, photosynthesis, cellular respiration, carbon dioxide
   (f) Water cycle, precipitation, transpiration, evaporation, hydrosphere
   (g) Ozone layer, ozone hole, CFCs, stratosphere
   (h) Abiotic factor, biotic factor, temperature, rainfall, climate, multicellular organism, ecosystem, biome
   (i) Greenhouse effect, enhanced greenhouse effect, global warming
3. Constantly changing physical, chemical and biological cycles have contributed to the survival of various forms of life on Earth. Our life-support systems are not in good shape.
   Using knowledge that you have gained from this chapter, comment on the statements above.
4. What is meant by biodiversity and why is loss of biodiversity a concern?
5. (a) The mountain pygmy possum is restricted to an area of 6 km^2 in the Australian Alps. Suggest how such a restricted habitat may influence its chances of survival.
   (b) Suggest abiotic and biotic factors that may affect this possum.
   (c) Suggest how warmer temperatures and reduced snow may affect its lifestyle. Be specific in your response by including examples of different scenarios.
   (d) What is meant by the term *extinction*?
   (e) If this species was to become extinct, suggest implications for other organisms within its ecosystem.
6. Copy the figure at right into your workbook and then use the following terms to complete the links: nitrifying bacteria, uptake by roots, denitrification, decomposition, feeding, nitrogen-fixing bacteria.

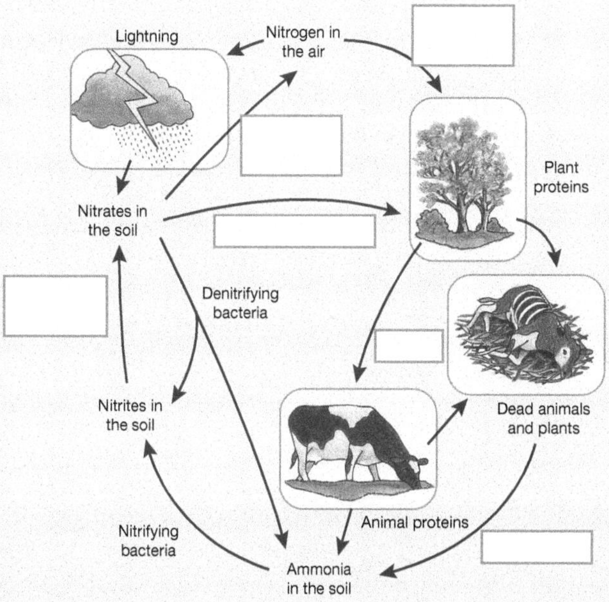

7. Rising sea levels and saltwater intrusion associated with climate change are threats that Kakadu National Park is experiencing.
   (a) Suggest why these threats are associated with climate change.
   (b) Suggest effects that these new threats may have on the (i) biotic and (ii) abiotic parts of this ecosystem.
   (c) Suggest actions that could be taken to reduce the loss of biodiversity within Kakadu National Park.

8. Complete the crossword below.

ACROSS

5. Abbreviation of chlorofluorocarbon
8. Dynamic system of organisms interacting with each other and their environment
9. Planting these may help reduce the effect of global warming.
12. The ozone layer is located in this part of the Earth's atmosphere.
15. These bacteria convert nitrates in soil and water into nitrogen in the air.
16. Plants use this process to make glucose and oxygen.
17. An example of a greenhouse gas
18. Photosynthesis, respiration, death and decomposition are all processes within this cycle.
20. This term relates to the total variety of living things on Earth.

DOWN

1. A group of organisms of the same species in the same area
2. Includes water and dissolved carbon dioxide
3. A layer of this gas helps block out more than 95 per cent of ultraviolet rays entering the atmosphere.
4. The life-support system of our planet
6. Organisms are composed of these.
7. Global warming will lead to a rise in this factor.
10. The loss of a species from Earth
11. Includes rocks, coal and oil deposits, and humus in soil
13. This human activity can result in increased carbon dioxide levels in the atmosphere.
14. Abbreviation of deoxyribonucleic acid
19. Abbreviation of total ozone mapping spectrometer

9. Agriculture has had (and continues to have) a devastating effect on a number of marine ecosystems. Hypoxia in coastal zones from nitrogen and phosphorus outputs of agricultural and livestock industries is one such example.
   (a) Using your knowledge of the nitrogen and phosphorus cycles, explain how these outputs may damage marine ecosystems.
   (b) Suggest strategies that may reduce the negative impact that agriculture has on our ecosystems.

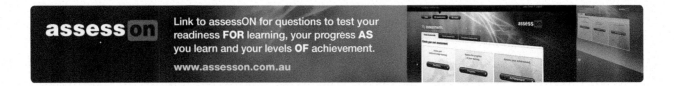

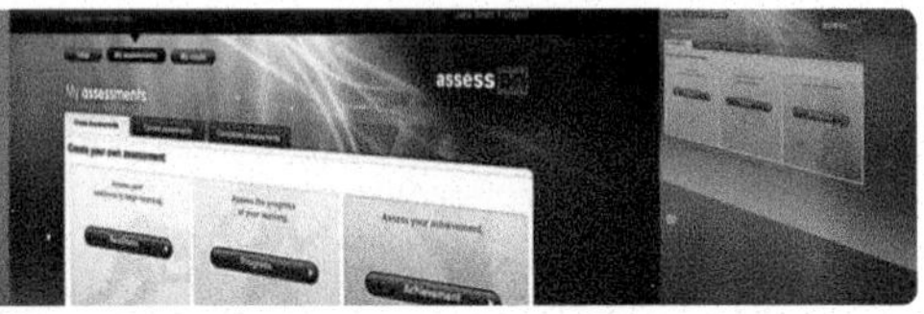

TOPIC 8
Forces, energy and motion

8.1 Overview

8.1.1 Why learn this?

The thrill of a rollercoaster ride allows you to experience sudden changes in motion. When the car suddenly falls, you seem to get left behind just for a while. When you reach the bottom of the track and the car rises, your stomach seems to sink. And when you round a bend, your body seems to be flung sideways. Such a ride raises many questions about the way in which forces affect motion and energy.

8.1.2 Think about forces, energy and motion

assesson

- Could a kangaroo win the Melbourne Cup?
- How do radar guns measure the speed of cars?
- Why do you feel pushed to the left when the bus you are in turns right?
- Why do space rockets seem to take forever to get off the ground?
- Why does it hurt when you catch a fast-moving ball with your bare hands?
- What does doing work really mean?
- Why can't a tennis ball bounce higher than the height from which it is dropped?
- Why are cars deliberately designed to crumple in a crash?

LEARNING SEQUENCE

Numerous **videos** and **interactivities** are embedded just where you need them, at the point of learning, in your learnON title at www.jacplus.com.au. They will help you to learn the content and concepts covered in this topic.

8.1.2 Your quest

A world of forces, energy and motion

Think

Find out what you already know about forces, energy and motion by answering the following questions.

1. Describe your movement when you are a passenger in a car that:
   (a) accelerates suddenly
   (b) stops suddenly
   (c) takes a sharp left turn.
2. Explain your movement in each of the three situations described in question 1.
3. Copy and complete the table below to list as many as possible of the forces acting on each of the people shown enjoying their leisure time in the photos on the right. The number of forces acting on each of them is provided in brackets.

Person	Forces acting on the person
Parachutist (3)	
Bungee jumper (3)	
Skier (2)	
Cyclist (5)	
Reader (2)	
Skater (5)	
Swimmer (4)	

4. Which of the forces listed in your completed table could be described as a non-contact force?
5. Some of the people in the photographs are accelerating (speeding up); others may be travelling at a steady speed or slowing down.
   (a) Which three of the people are the most likely to be accelerating? How do you know?
   (b) Which three people are most likely to be moving at a non-zero constant speed? How do you know?
6. What outside object or substance provides the forward push on the:
   (a) swimmer
   (b) cyclist
   (c) skater?

7. Which three people are clearly losing gravitational potential energy?
8. According to the **Law of Conservation of Energy**, energy cannot be created or destroyed. It can only be transformed into another form of energy or transferred to another object. What happens to the lost gravitational potential energy of each of the three people referred to in question 7?

8.2 Ready, set, go

8.2.1 Calculating speed

Could a kangaroo win the Melbourne Cup? Who would win a race between a sea turtle, a dolphin and an Olympic swimmer? You can answer these questions only if you know the **average speed** of each competitor during the race.

Speed is a measure of the **rate** at which an object moves over a distance. In other words, it tells you how quickly distance is covered. Average speed can be calculated by dividing the distance travelled by the time taken. That is:

$$\text{average speed} = \frac{\text{distance travelled}}{\text{time taken}}.$$

In symbols, this formula is usually expressed as:

$$v = \frac{d}{t} \text{ where } v = \text{speed or velocity.}$$

8.2.2 Which unit?

The speed of vehicles is usually expressed in kilometres per hour (km/h). However, sometimes it is more convenient to express speed in units of metres per second (m/s). The speed at which grass grows could sensibly be expressed in units of millimetres per week. Speed must, however, always be expressed as a unit of distance divided by a unit of time.

Some examples

(a) The average speed of an aeroplane that travels from Perth to Melbourne, a distance of 2730 km by air, in 3 hours is:

$$\begin{aligned} v &= \frac{d}{t} \\ &= \frac{2730\text{ km}}{3\text{h}} \\ &= 910\text{ km/h.} \end{aligned}$$

The formula can also be used to express the speed in m/s.

$$\begin{aligned} v &= \frac{d}{t} \\ &= \frac{2730000\text{ m}}{3 \times 3600\text{ s}} \text{(converting kilometres to metres and hours to seconds)} \\ &= 25\text{ m/s.} \end{aligned}$$

(b) The average speed of a snail that takes 10 minutes to cross an 80 cm concrete paving stone in a straight line is:

$$\begin{aligned} v &= \frac{d}{t} \\ &= \frac{80\text{ cm}}{10\text{ min}} \\ &= 8\text{ cm/min.} \end{aligned}$$

8.2.3 Calculating distance and time

The formula used to calculate average speed can also be used to work out the distance travelled or the time taken.

$$\text{Since } v = \frac{d}{t},$$

$$d = vt \text{ and } t = \frac{d}{v}.$$

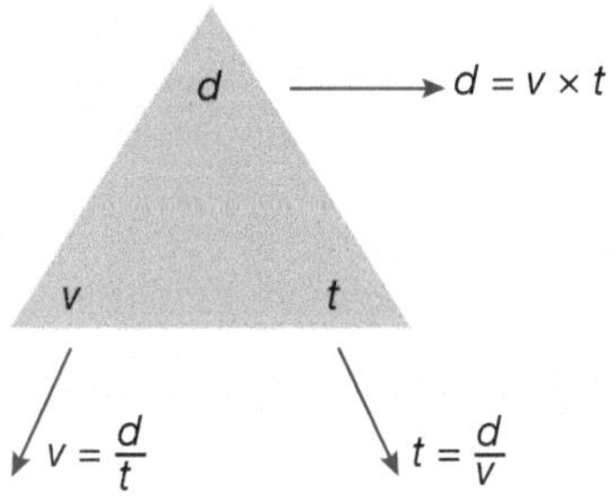

More examples

(a) The distance covered in $2\frac{1}{2}$ hours by a train travelling at an average speed of 70 km/h is:

$$\begin{aligned} d &= vt \\ &= 70 \text{ km/h} \times 2.5 \text{ h} \\ &= 175 \text{ km}. \end{aligned}$$

(b) The time taken for a giant tortoise to cross a 6-metre-wide deserted highway at an average speed of 5.5 cm/s is:

$$\begin{aligned} t &= \frac{d}{v} \\ &= \frac{6.0\,\text{m}}{0.055\,\text{m/s}} \quad \text{(converting 5.5 cm/s to 0.055 m/s)} \\ &= 109\,\text{s (to the nearest second)} \\ &= 1\text{ min } 49\,\text{s}. \end{aligned}$$

8.2.4 When the direction matters

The term **velocity** is often used instead of speed when talking about how fast things move. However, velocity and speed are different quantities. Velocity is a measure of the rate of change in position, whereas speed is a measure of the rate at which distance is covered. To describe a change in position, the direction must be stated. Velocity has a direction as well as a **magnitude** (size). When determining speed, the direction of movement does not matter.

At a snail's pace

Imagine a race between two snails, Bo and Jo, between the points P and R shown in the diagram on the right. Bo, being slower but smarter, takes the direct route. Jo, faster but not as clever, takes an indirect route via Q and travels a greater distance. The race is a dead heat — both snails finish in 1 minute.

The table below describes the motion of the two snails and shows the difference between their speed and velocity.

Notice that when there is no change in direction, the magnitude of the velocity is the same as the speed. If direction changes, velocity and speed will be different.

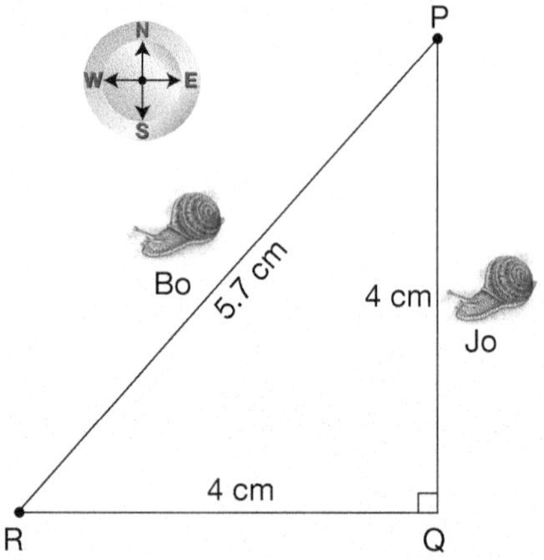

The race between Bo and Jo

	Bo	Jo
Average speed	$\frac{\text{Distance travelled}}{\text{time taken}}$ $= \frac{5.7\,\text{cm}}{1\text{ min}}$ $= 5.7\,\text{cm/ min}$	$\frac{\text{Distance travelled}}{\text{time taken}}$ $= \frac{8.0\,\text{cm}}{1\text{ min}}$ $= 8.0\,\text{cm/ min}$
Average velocity	$\frac{\text{Change in position}}{\text{time taken}}$ $= \frac{5.7\,\text{cm NE}}{1\text{ min}}$ $= 5.7\,\text{cm/ min NE}$	$\frac{\text{Change in position}}{\text{time taken}}$ $= \frac{5.7\,\text{cm NE}}{1\text{ min}}$ $= 5.7\,\text{cm/ min NE}$

8.2 Exercises: Understanding and inquiring

To answer questions online and to receive **immediate feedback** and **sample responses** for every question, go to your learnON title at www.jacplus.com.au. *Note:* Question numbers may vary slightly.

Remember

1. Write the formula used to calculate average speed in symbols and state which quantity each symbol represents.
2. Explain the difference between speed and velocity. Use an example to support your explanation.

Using data

3. Determine the average speed of each of the following.
   (a) A racehorse that wins the 3200 m Melbourne Cup in a time of 3 min 20 s (in m/s)
   (b) A kangaroo fleeing from a dingo, which bounds a distance of 2.5 km in 3 min (in m/s)
   (c) A dolphin that just manages to keep up with a speeding boat for a distance of 2 km for a period of 3 min (in km/h)
   (d) A sea turtle that is able to maintain its maximum speed for 0.5 h. In that time it can swim a distance of 16 km (in km/h).
   (e) An Olympic swimmer who completes a 1500 m training swim in 16 min (in km/h)
   (f) A mosquito that flies a distance of 2 m in 4 s (in cm/s)
4. Use your answers to question 3 to suggest answers to the two questions posed in the introduction to this subtopic.
5. How long would it take you to walk from Melbourne to Sydney, a distance of 900 km, if you walked at an average speed of:
   (a) 5 km/h without stopping
   (b) 5 km/h for 10 h each day
   (c) 1.5 m/s without stopping?
6. How far can a snail crawl if it moves at an average speed of 8.0 cm/min for:
   (a) 3 minutes
   (b) 3 hours?
7. In a heat of a swimming trial, a swimmer swims the 100 m breaststroke event in 68 s. The event is completed in a pool that is 50 m long. She finishes the event at the same end of the pool from which she started. If she begins the event by swimming due north, and takes 35 s to swim the first 50 m, calculate her:
   (a) average speed for the whole swim
   (b) average velocity for the first 50 m
   (c) average velocity for the whole swim.
8. A swimmer completes a 1500 m race in 870 s. Calculate the swimmer's average speed in:
   (a) m/s
   (b) km/h.

Create

9. Design a chart with pictures that compares the speeds of a range of animals.

8.3 Measuring speed

8.3.1 Keeping track of the speed

When German Formula One racing driver Sebastian Vettel broke the Australian Grand Prix qualifying lap record in 2011, he completed a 5.303 km lap in 83.529 seconds.

His average speed was:

$$\begin{aligned} v &= \frac{d}{t} \\ &= \frac{5303\ \text{m}}{83.529\ \text{s}} \\ &= 63.49\ \text{m/s (about 229 km/h).} \end{aligned}$$

However, he was able to speed down the straight at speeds of up to 320 km/h.

Clearly, the average speed does not provide much information about the speed at any particular instant during the race.

The full story of each lap of Sebastian Vettel's race could be more accurately told if his average speed was measured over many short intervals throughout the event. For example, if stopwatches were placed at every 100-metre point along the track, his average speed for each 100-metre section of the circuit could then be calculated. On the other hand, if stopwatches were placed every metre along the track, his average speed for each 1-metre section could be calculated. By using more stopwatches and placing them closer together, a more accurate estimate of his **instantaneous speed** can be obtained. The instantaneous speed is the speed at any particular instant of time.

8.3.2 When time ticks away

A ticker timer provides a simple way of recording motion in a laboratory. When the ticker timer is connected to an AC power supply, its vibrating arm strikes its base 50 times every second. Paper ticker tape attached to a moving object is pulled through the timer. A disc of carbon paper between the paper tape and the vibrating arm ensures that a black dot is left on the paper 50 times every second; that is, a black dot is made every fiftieth of a second.

The average speed between each pair of dots is determined by dividing the distance between them by the time interval. To make calculating the speed easier, every fifth dot can be marked, as shown in the diagram below. Each marked interval on the tape represents five-fiftieths of a second — that is, 0.1 seconds. The average speed during the first interval on the tape shown is:

$$v = \frac{4.3\ \text{cm}}{0.1\ \text{s}}$$
$$= 43\ \text{cm/s}.$$

Motion can be recorded with a ticker timer.

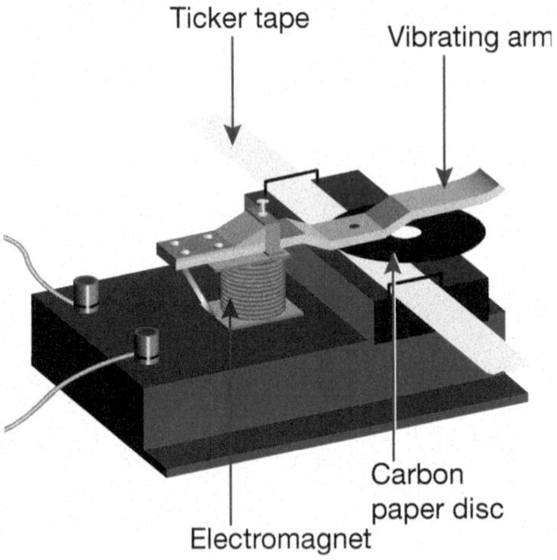

Each marked interval represents a time of 0.1 s.

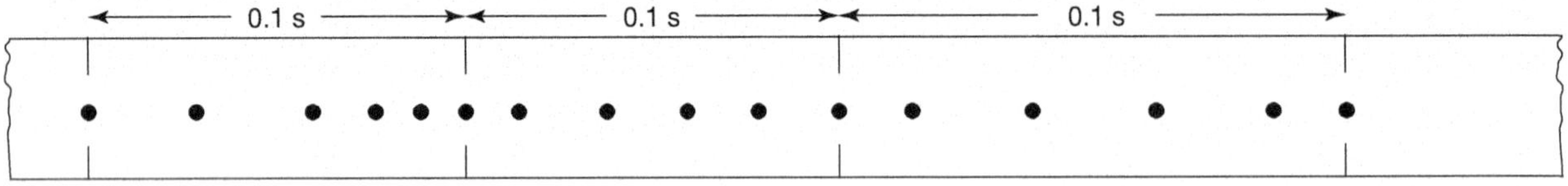

8.3.3 Motion detectors

In many classrooms, ticker timers have been replaced with **sonic motion detectors**. These devices send out pulses of ultrasound at a frequency of about 40 kHz and detect the reflected pulses from the moving object. The device uses the time taken for the pulses to return to calculate the distance between itself and the object. A small computer in the motion detector calculates the speed of the object.

Sonic motion detectors are used on the bumpers of cars to help the driver detect the distance between the car and another object.

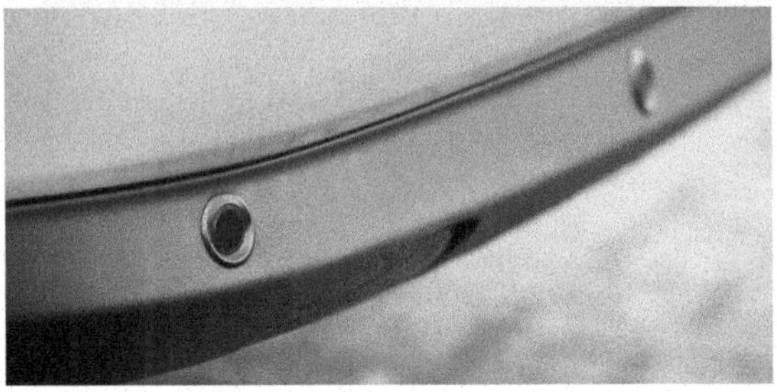

8.3.4 There must be better ways!

Speedometers

The speedometer inside a vehicle has a pointer that rotates further to the right as the wheels of the vehicle turn faster. It provides a measure of the instantaneous speed.

Older speedometers use a rotating magnet that rotates at the same rate as the car's wheels. The rotating magnet creates an electric current in a device connected to the base of the pointer. As the car's speed increases, the magnet rotates faster, the electric current increases and the pointer rotates further to the right.

Newer electronic speedometers use a rotating toothed wheel that interrupts a stationary **magnetic field**. An electronic sensor detects the interruptions and sends a series of pulses to a computer, which calculates the speed using the **frequency** of the pulses.

Car speedometers are not 100 per cent accurate. In Australia, an error of up to 10 per cent is common. Speedometers are manufactured according to the diameter of the tyres on the vehicle. Any change in that diameter will make the reading on the speedometer inaccurate.

Car speedometers provide a measure of instantaneous speed.

8.3.5 Speed and road safety

One of the major causes of road accidents and subsequent fatalities and injuries is excessive speed or driving at speeds that are unsafe for the road or weather conditions. Speed limits and speed advisories are set in an effort to minimise such accidents. The police use three different methods to monitor driving speeds as accurately as possible to ensure that speeding drivers are penalised.

- **Radar guns** and mobile radar units in police cars send out radio waves. The radio waves are reflected from the moving vehicle. However, the frequency of the waves is changed owing to the movement of the vehicle (see subtopic 6.4). The change in the frequency, called the Doppler effect, depends on the speed of the moving vehicle. The altered waves are detected by the radar gun or mobile unit, which calculates the instantaneous speed. One type of radar unit is linked to cameras that automatically photograph any vehicle that the radar detects as travelling above the speed limit.
- Police also use **laser guns** that send out pulses of light that are reflected by the target moving vehicle. The time taken for each pulse to return is recorded and compared with that of previous pulses. This allows the average speed over a very small time interval to be calculated. Laser guns are useful when traffic is heavy because they can target single vehicles with the narrow light beams. Radio waves spread out and, in heavy traffic, it is difficult to tell which car reflected the waves.
- **Digitectors** consist of two cables laid across the road at a measured distance from each other. Each cable contains a small microphone that detects the sound of a moving vehicle as it crosses the cable. The measured time interval between the sounds is used to calculate the average speed of the vehicle between the cables. Although digitectors were phased out after the 1980s, they are regaining popularity as an alternative to radar and laser guns.

8.3.6 The global positioning system

The **global positioning system (GPS)** uses radio signals from at least four of up to thirty-two satellites orbiting the Earth to accurately map your position, whether you are in a vehicle or on foot. Like radar guns, GPS navigation devices use the Doppler effect to calculate instantaneous speed, usually about once every second.

HOW ABOUT THAT!

AFL coaches and sports scientists use GPS locators to track the movement of players around the field. The locators are strapped to the player's upper back. A computer is used to analyse the data to provide information about distance covered, speed, time spent moving at different speeds, maximum speed and acceleration. A built-in sensor also monitors heart rate.

INVESTIGATION 8.1

Ticker timer tapes

AIM: To record and analyse motion using a ticker timer

Materials:

ticker timer
power supply
scissors
G-clamp
ticker tape (in 60 cm lengths)

Method and results

- Clamp the ticker timer firmly to the edge of a table or bench so that you will be able to pull 60 cm of ticker tape through it. Connect the ticker timer to the AC terminals of the power supply and set the voltage as instructed by your teacher.
- Thread one end of the ticker tape through the ticker timer so that it goes under the carbon paper disc.
- Turn on the power supply and check that the ticker timer leaves a black mark on the ticker tape.
- Hold the end of the ticker tape and walk away from the ticker timer so that the ticker tape moves through at a steady speed.
- Remove the ticker tape and mark off the first clear dot made and every fifth dot after the first. (There should be four dots between each of the marked-off dots on the ticker tape.)

1. Measure the distance travelled during each 0.1 s interval and write it on your tape. Label the intervals as interval 1, interval 2, interval 3, etc.
   - Cut your ticker tape into 0.1 s intervals and glue the strips in order onto a sheet of paper. Each strip shows the distance travelled during a 0.1 s time interval. The graph therefore shows how the speed changed with time.

Using ticker tape to plot a graph

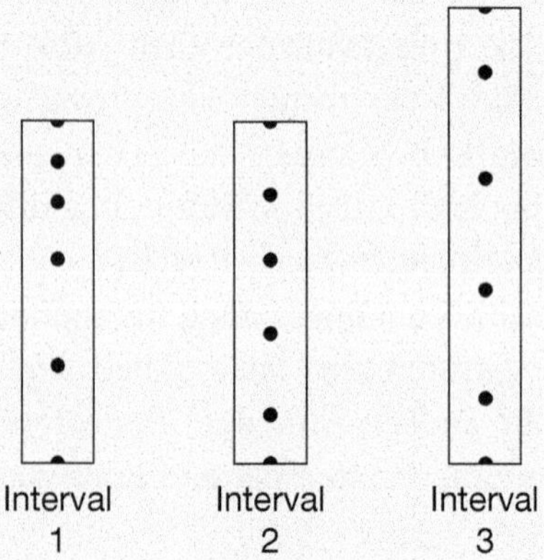

Discuss and explain

2. How much time elapsed between the printing of the first clear dot and the last dot marked off?
3. Calculate the average speed for the motion that took place between the printing of the first clear dot and the last marked dot.
4. Calculate the average speed during each 0.1 s interval.
5. Did you succeed in keeping your speed steady?

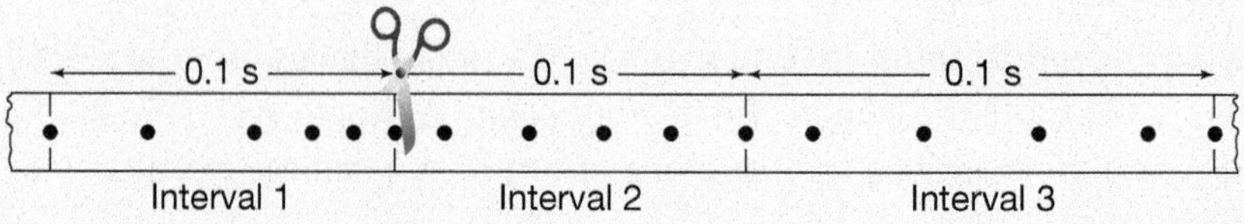

8.3 Exercises: Understanding and inquiring

To answer questions online and to receive **immediate feedback** and **sample responses** for every question, go to your learnON title at www.jacplus.com.au. *Note:* Question numbers may vary slightly.

Remember

1. Explain the difference between instantaneous speed and average speed. Use an example to support your explanation.
2. Which methods are used by the police to measure:
   (a) average speed
   (b) instantaneous speed?

Think

3. Make a list of reasons why a speedometer reading might not be accurate. Include anything that could change the diameter of the vehicle's tyres.
4. After being phased out, digitectors are making a comeback in police detection of speeding drivers. Suggest advantages they might have over radar and laser guns.

Process and analyse

5. Calculate the average speed during the second and third 0.1 s intervals of the ticker tape shown in Investigation 8.1.

Investigate

6. Use data-logging equipment with a motion detector or light gates to record the motion of a toy car or cart down a slope. Use the software to produce a graph of distance versus time and a graph of speed versus time. Comment on the shape of your graphs.

learnon RESOURCES — ONLINE ONLY

Complete this digital doc: Worksheet 8.1: Speed and velocity (doc-19480)

Complete this digital doc: Worksheet 8.2: Ticker tapes (doc-19481)

8.4 Speeding up

8.4.1 Acceleration

The accelerator of a car is given that name because pushing down on it usually makes the car accelerate. When an object moves in a straight line, its **acceleration** is a measure of the rate at which it changes speed.

Acceleration tells you how quickly speed changes. The average acceleration can be calculated by dividing the change in speed by the time taken for the change. That is:

$$\text{average acceleration} = \frac{\text{change in speed}}{\text{time taken}}.$$

We can write this in the form of an equation:

$$a = \frac{\Delta v}{t}.$$

The triangle-shaped symbol is used to represent the change in a value; this symbol (taken from the Greek alphabet) is called delta.

For example, a car travelling at 60 km/h that increases its speed to 100 km/h in 5.0 seconds has an average acceleration of:

$$\text{average acceleration} = \frac{\text{change in speed}}{\text{time taken}}$$

$$a = \frac{\Delta v}{t}$$

$$= \frac{40\ \text{km/h}}{5.0\,\text{s}}$$

$$= 8.0\ \text{km/h per second.}$$

That is, on average, the car increases its speed by 8.0 km/h each second.

If the change in speed is an increase, the acceleration is positive. If the change in speed is a decrease, the acceleration is negative and is called **deceleration**.

8.4.2 Fast starters

The sport of drag racing is a test of acceleration. From a standing start, cars need to cover a distance of 400 metres in the fastest possible time. To do this, they need to reach high speeds very quickly. The fastest drag-racing cars can reach speeds of more than 500 km/h in less than 5.0 seconds.

The average acceleration of a drag-racing car that reaches a speed of 506 km/h in 4.6 seconds is:

$$\text{average acceleration} = \frac{\text{change in speed}}{\text{time taken}}$$

$$= \frac{506\ \text{km/h}}{4.6\,\text{s}} = 110\ \text{km/h/s.}$$

The sport of drag racing is a test of acceleration.

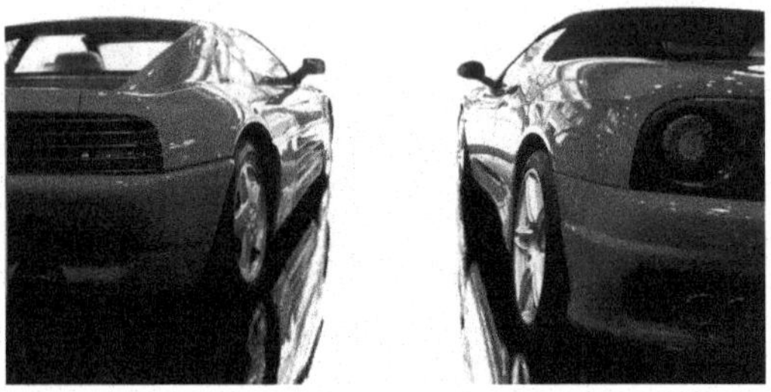

This is read as 110 kilometres per hour per second. It means that, on average, the car increases its speed by 110 kilometres per hour each second. Acceleration can also be expressed in m/s/s (that is, metres per second per second) or m/s^2 (that is, metres per second squared). A change in speed of 506 km/h can be expressed as 141 m/s. The average acceleration of the drag-racing car can therefore be expressed as:

$$\text{average acceleration} = \frac{\text{change in speed}}{\text{time taken}}$$

$$= \frac{141\,\text{m/s}}{4.6\ \text{s}} = 31\,\text{m/s}^2.$$

8.4.3 Slowing down

Once the drag-racing car has completed the required distance of 400 metres, it needs to stop before it reaches the end of the track. The fastest cars release parachutes so that they can stop in time. The acceleration of a car that comes to rest in 5.4 seconds from a speed of 506 km/h is:

$$\text{average acceleration} = \frac{\text{change in speed}}{\text{time taken}}$$

$$= \frac{-506\ \text{km/s}}{5.4\ \text{s}} = -93.7\ \text{km/h/s.}$$

This negative acceleration can be expressed as a deceleration of 93.7 km/h/s.

INVESTIGATION 8.2

Drag strips

AIM: To analyse the motion of an accelerating object using ticker tape

Materials:

ticker timer
G-clamp
power supply
ticker tape (in 60 cm lengths)
toy car (or dynamics trolley)
sticky tape or masking tape
clear length of bench at least 60 cm long

Method and results

- Clamp the ticker timer firmly to the edge of a table or bench so that you will be able to pull 60 cm of ticker tape through it. Connect the ticker timer to the AC terminals of the power supply and set the voltage as instructed by your teacher.
- Thread one end of the ticker tape through the ticker timer so that it goes under the carbon paper disc.
- Attach the end of the ticker tape to the toy car or trolley.
- Turn on the power supply and check that the ticker timer leaves a black mark on the ticker tape.
- Model a drag-racing car by pushing the toy car or trolley forward, starting from rest, so that it reaches a maximum speed near the halfway mark. Make it come to a gradual stop near the end of the 'track'.
- Remove the ticker tape and mark off the first clear dot and every fifth dot after the first. Each interval between the marks represents a time of $\frac{5}{50}$ of a second; that is, 0.1 seconds. Measure the distance travelled during each 0.1 second interval and write it on your tape. Label the intervals as interval 1, interval 2, interval 3 and so on.

1. Construct a table like the one below in which to record your data. Calculate the average speed during each interval and record it in the table.
2. Now cut your ticker tape into 0.1 second intervals and use the strips as described in Investigation 8.1 to construct a graph of speed versus time.

Discuss and explain

3. Describe the motion of the toy car or trolley during the period over which it was recorded. Ensure that the words *speed, accelerated* and *decelerated* are used in your description.
4. Between which intervals was the acceleration:
   (a) positive
   (b) negative?
5. During which interval did the greatest average speed occur?
6. When did the greatest positive acceleration take place?

Drag strip speeds

Interval	Distance travelled (cm)	Average speed (cm/s)
Example	3.6	$\frac{3.6\text{ cm}}{0.1\text{ s}} = 36$
1		
2		
3		

8.4 Exercises: Understanding and inquiring

To answer questions online and to receive **immediate feedback** and **sample responses** for every question, go to your learnON title at www.jacplus.com.au. *Note:* Question numbers may vary slightly.

Remember

1. Why is the accelerator in a car called by this name?
2. Explain the difference between acceleration and deceleration.

Calculate

3. A car that has stopped at a set of traffic lights sets off when the lights turn green. It increases its speed by 5 m/s during each of the first 3 s after it sets off, and by 3 m/s during the following 2 s.
   (a) What is the speed of the car after:
      (i) 1 s
      (ii) 2 s
      (iii) 5 s?
   (b) What is the average acceleration of the car during the first 5 s after it sets off?

Process and analyse

4. Repeat Investigation 8.2 using data-logging equipment and a motion detector or light gates. Use the software to produce a graph of speed versus time. Print a hard copy of your graph.
   (a) Use coloured pencils or a highlighter pen to indicate the part of the graph that shows the car or trolley speeding up. Use a different colour to indicate the part of the graph that shows the car or trolley slowing down.
   (b) What is the maximum speed of the car or trolley?
   (c) At what time was the maximum speed reached?
   (d) How would your graph be different if the trolley sped up more quickly, yet reached the same maximum speed?
   (e) Calculate the average acceleration of the car or trolley used in your own experiment.

learnon RESOURCES — ONLINE ONLY

Complete this digital doc: Worksheet 8.3: Acceleration (doc-19483)

8.5 Let's go for a ride

8.5.1 Forces at work

Imagine sitting in a bus on a school excursion. What happens to you when the bus turns a corner, suddenly stops or accelerates after stopping at a railway crossing?

That's an easy question. But to explain why it happens, you need to look at the forces acting on you and the forces acting on the bus.

The diagram on the top of the next page shows the forces acting on a bus moving along a straight, horizontal road at a constant speed. The upward push of the road must be the same as the **weight**. If that was not the case, the bus would fall through the road or be pushed up into the air. If the **thrust** is greater than the **resistance forces** the bus accelerates. If the resistance forces are greater than the thrust the bus slows down, or decelerates.

Which way is this bus turning — left or right?

There are only two significant forces acting on you:

- your weight; that is, the force applied on you by the Earth's gravitational attraction
- the push of the bus seat, which pushes upwards and forwards. The upward part of this force is just enough to balance your weight. The forwards part of this force is what keeps you moving at the same speed as the bus.

The forces acting on a moving bus. The forces are in balance when the bus is not changing speed or direction.

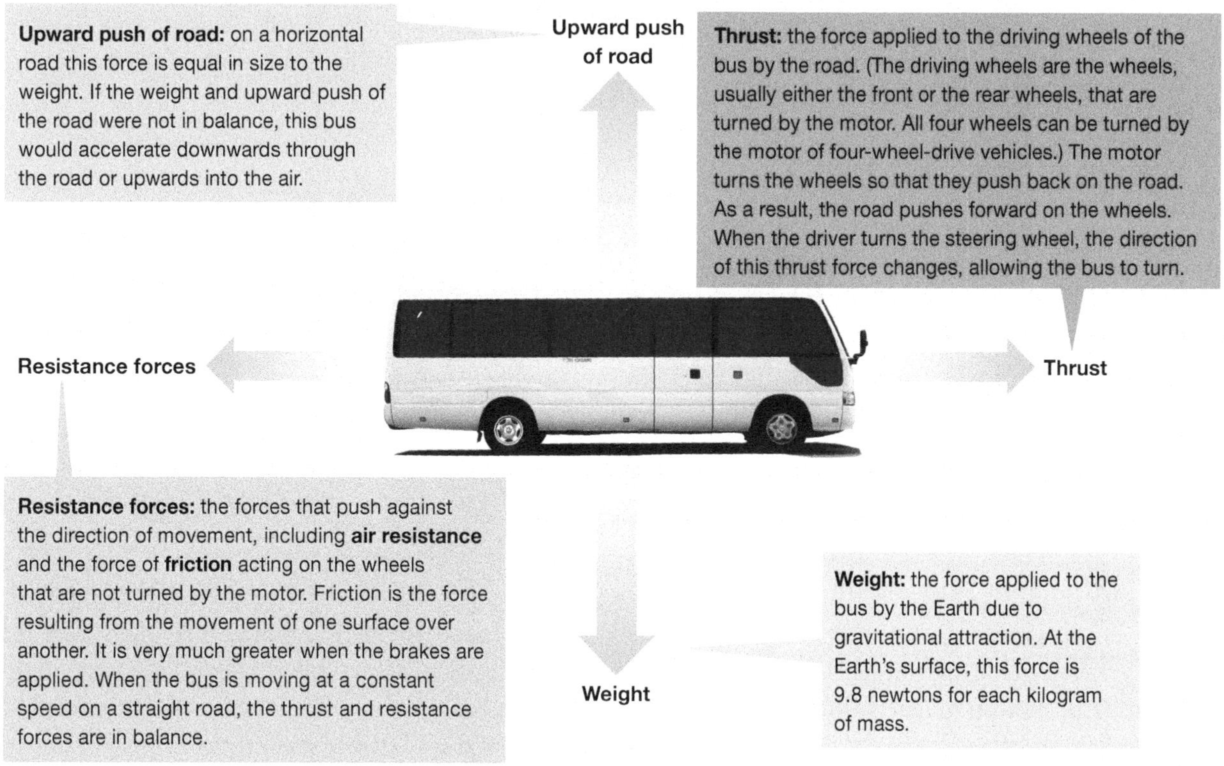
Upward push of road: on a horizontal road this force is equal in size to the weight. If the weight and upward push of the road were not in balance, this bus would accelerate downwards through the road or upwards into the air.
Upward push of road
Thrust: the force applied to the driving wheels of the bus by the road. (The driving wheels are the wheels, usually either the front or the rear wheels, that are turned by the motor. All four wheels can be turned by the motor of four-wheel-drive vehicles.) The motor turns the wheels so that they push back on the road. As a result, the road pushes forward on the wheels. When the driver turns the steering wheel, the direction of this thrust force changes, allowing the bus to turn.
Resistance forces
Thrust
Resistance forces: the forces that push against the direction of movement, including **air resistance** and the force of **friction** acting on the wheels that are not turned by the motor. Friction is the force resulting from the movement of one surface over another. It is very much greater when the brakes are applied. When the bus is moving at a constant speed on a straight road, the thrust and resistance forces are in balance.
Weight
Weight: the force applied to the bus by the Earth due to gravitational attraction. At the Earth's surface, this force is 9.8 newtons for each kilogram of mass.

CALCULATING WEIGHT AT THE EARTH'S SURFACE

The force of gravitational attraction towards a large object like a planet is called weight. The weight (in newtons) of any object can be calculated using the formula:

$$\text{weight} = mg$$

where m = mass (in kilograms)

g = gravitational field strength (in N/kg).

At the Earth's surface, the gravitational field strength is 9.8 N/kg. That is, the gravitational force acting on each kilogram of mass is 9.8 N.

The weight of a 1 kg object is therefore:

$$\text{weight} = mg = 1\,\text{kg} \times 9.8\,\text{N/kg} = 9.8\,\text{N.}$$

The weight of a 2000 kg bus is:

$$\text{weight} = mg = 2000\,\text{kg} \times 9.8\,\text{N/kg} = 19600\,\text{N.}$$

8.5.2 Explaining the rough ride

Sir Isaac Newton's laws of motion were first published in 1687, many years before buses, cars, trains and aeroplanes were invented. However, Sir Isaac would have delighted in explaining the way your body moves in a bus or any other vehicle.

Newton's First Law of Motion states that an object will remain at rest, or will not change its speed or direction, unless it is acted upon by an outside, unbalanced force. In many manoeuvres that you may experience as a passenger on a bus, an unbalanced force is acting on the vehicle to change its speed or direction.

For example, the bus stops suddenly when the brakes are applied. The resistance forces are large and there is no thrust. Your seat is rigidly attached to the bus, so it also stops suddenly. However, the resistance forces are not acting on you. You continue to move forward at the speed that you were travelling at before the brakes were applied until there is a force to stop you. That force could be provided by a seatbelt, the back of any seat in front of you, a passenger in front of you or the windscreen of the bus.

When the bus makes a sudden right turn, the unbalanced force acting on the bus to change its direction is not acting directly on you. You continue to move in the original direction. The inside wall of the bus moves to the right but you don't. So it seems like you've been flung to the left.

8.5.3 Newton's first law at work

The expensive crockery on the table in the illustration on the right is quite safe if the magician is fast enough. Newton's First Law of Motion, also known as the **law of inertia**, provides an explanation. The magician is pulling on the tablecloth — not on the expensive crockery. There is a small unbalanced force acting on the crockery due to friction. However, if the tablecloth is pulled away quickly enough, this force does not act for long enough to make the crockery move. If the tablecloth is pulled too slowly, the force of friction on the crockery will pull it off the table as well.

Don't try this at home!

What is this thing called inertia?

Inertia is the property of objects that makes them resist changes in their motion. The inertia of the crockery keeps it on the table. The inertia of the tablecloth is not large enough to allow it to resist the change in motion. Clearly, the greater the mass of an object, the more inertia it has. For example, it takes a much larger force to change the motion of a heavy train than it does to change the motion of a small car. Try pulling an A4 sheet of paper out from under a 20-cent coin. Then try it with a 5-cent coin.

INVESTIGATION 8.3

Forces on cars

AIM: To investigate the forces acting on a toy car

Materials:

toy car

Method and results

- Rest a toy car on a smooth, level surface.
- Push the car quickly forwards and then let it go.

Discuss and explain

1. What forces are acting on the car while it is at rest?
2. How do you know that there is more than one force acting on the car while it is at rest?
3. Are the forces on the car in balance after you stop pushing? How do you know?
4. Which force or forces cause the car to slow down after you stop pushing it forwards?
5. How would the car's motion be different if you pushed it forwards and let it go on:
   (a) a much smoother surface
   (b) a rough surface?

SEVERAL HURT WHEN UNITED FLIGHT HITS TURBULENCE

WELLINGTON, New Zealand (AP) — A United Airlines flight from Sydney to San Francisco detoured to Auckland late Wednesday local time after several people on board were injured when the plane hit severe turbulence over the Pacific Ocean, an airline spokesman said.

A female cabin attendant broke a leg and a male cabin crew member had back and shoulder injuries from being thrown around in the turbulence.

Three passengers were taken to hospital with 'bruising and muscular discomfort'. Two other passengers with minor injuries were treated by ambulance staff at the airport.

Passenger Julie Greenwood told *The New Zealand Herald* newspaper the turbulence lasted about 30 seconds. 'It was like an earthquake in the air — I was lifted out of my chair twice,' she said.

Airline spokesman Jonathan Tudor in Auckland told *The Associated Press* that United Airlines flight 862, carrying 269 passengers and 21 crew, had taken off from Sydney at 3:35 pm New Zealand time and flew into 'clear air turbulence' after about four hours.

Those who were uninjured would be housed overnight in Auckland hotels, he added. It was not immediately clear when the flight would continue on to San Francisco.

Mary Brander, 77, from Sydney, said the bottom seemed to be falling out of the plane when it struck the turbulence.

'One minute we were in clear blue sky and it hit,' she told *The New Zealand Herald*.

8.5 Exercises: Understanding and inquiring

To answer questions online and to receive **immediate feedback** and **sample responses** for every question, go to your learnON title at www.jacplus.com.au. *Note:* Question numbers may vary slightly.

Remember

1. Which force prevents a bus from falling through the surface of a road?
2. Use your memories of bus trips to help you answer the questions below. Assume that you are comfortably seated, not wearing a seatbelt and that the bus seats have no head restraints.
   (a) What would be your immediate resulting motion (as a passenger on the bus) if the bus performed the following manoeuvres?
      (i) A very quick start from rest
      (ii) A forward motion at constant speed
      (iii) A very sharp right-hand turn
      (iv) A slow left-hand turn
      (v) An emergency stop from a speed of 60 km/h
      (vi) A forward jerk as the bus is struck from behind by another vehicle
   (b) During which type of manoeuvre in part (a) does the bus move with the same speed and direction as the passenger?
   (c) Explain how properly fitted seatbelts would change the resulting motion of the passenger.
   (d) Explain how a head restraint would change the resulting motion of the passenger in a bus that is struck from behind.
3. List two forces that resist the forward motion of a bus.
4. State Newton's First Law of Motion.
5. What is inertia?

Think

6. The gravitational field strength on Mars is only 3.7 N/kg. What would your weight be on Mars?
7. Which is greater, the thrust or the resistance forces, when a bus is moving along a horizontal road with:
   (a) increasing speed
   (b) decreasing speed
   (c) constant speed?
8. Explain in terms of Newton's First Law of Motion why you should never step off a bus, tram or train before it has completely stopped.
9. The car shown on the right is travelling to the left at a constant speed. The four major forces acting on the car are represented by arrows labelled A, B, C and D.

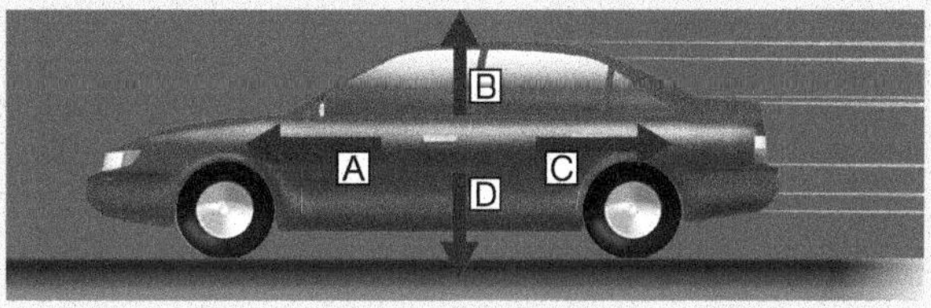

(a) Which two forces combine to provide the force represented by arrow C?
(b) How does the size of the force represented by arrow A compare with that of arrow C?
(c) Describe two different changes in the forces acting on the car that could cause it to slow down.

10. Read the newspaper article on pages 316–17 and answer the following questions.
(a) Explain why the passengers and crew were injured during the flight.
(b) What was really being thrown around — the passengers or the aircraft? Explain your answer.
(c) How would a seatbelt protect a passenger in an incident like the one described?

learnon RESOURCES — ONLINE ONLY

Watch this eLesson: Science demonstrations (eles-1076)

Complete this digital doc: Worksheet 8.4: Inertia and motion (doc-19485)

Complete this digital doc: Worksheet 8.5: Force and gravity (doc-19486)

8.6 Newton's Second Law of Motion

8.6.1 Newton's Second Law of Motion

Newton's Second Law of Motion describes how the mass of an object affects the way that it moves when acted upon by one or more forces.

For example, when two men pull in opposite directions on the box shown in the diagram at right, the net force is 40N to the right. The vertical forces can be ignored because the weight of the box and the upward push of the floor are equal and opposite in direction. If the floor is smooth, friction can be ignored.

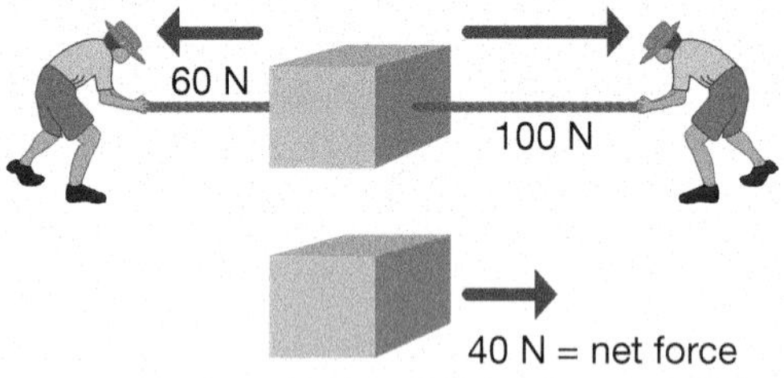

In symbols, Newton's second law can be expressed as:

$$a = \frac{F}{m}$$

where a = acceleration
F = the net force on the object
m = the mass of the object.

The **net force** is the total force acting on the object. If the net force is measured in newtons (N) and the mass is measured in kilograms (kg), the acceleration can be determined in metres per second squared (m/s^2).

This formula describes the observation that larger masses accelerate less rapidly than smaller masses acted on by the same total force. It also describes how a particular object accelerates more rapidly when a larger total force is applied. When all of the forces on an object are balanced, the total force is zero.

Newton's second law is often expressed as $F = ma$.

8.6.2 Newton's second law in action

The launching of any rocket at Cape Canaveral in Florida is a spectacular sight. At launch, a rocket like the one pictured at the top right of the next page has a mass of about 580000 kilograms. Most of this mass is fuel, which is burned during the launch.

There are two forces acting on such a spacecraft, including the main rocket and four or five smaller boosters, as it blasts off:

- the downward pull of gravity (weight). The weight of the spacecraft at launch is about 5.8 million newtons.
- the upward thrust provided by the burning of fuel, which is typically about 12 million newtons.

The thrust (upwards) is about 6.2 million newtons greater than the weight (downwards). That is, the net force on the spacecraft is about 6.2 million newtons upwards.

Newton's second law can be used to estimate the acceleration of the space shuttle at blast-off:

$$a = \frac{F}{m}$$

$$= \frac{6200000\,\text{N upwards}}{580000\,\text{kg}}$$

$$= 10.7\ \text{m/s}^2.$$

In other words, the spacecraft gains speed at the rate of only 10.7 m/s (or 39 km/h) each second.

Newton's second law also explains why the small acceleration at blast-off is not a problem. As the fuel is rapidly burned, the mass of the space shuttle gets smaller. As this happens, the acceleration gradually increases, and the space shuttle gains speed more quickly.

A spacecraft is launched by powerful rockets — yet it seems to take a long time to get off the ground. Newton's second law provides an explanation.

INVESTIGATION 8.4

Force, mass and acceleration

AIM: To investigate Newton's Second Law of Motion

Materials:

dynamics trolley	*pulley*
string	*four 500 g masses*
one 2 kg mass	*stopwatch*
masking tape	*metre ruler*

Method and results

1. Copy the table below.

	Time(s)			
Mass on the trolley (g)	**Trial 1**	**Trial 2**	**Trial 3**	**Average**
0				
500				
1000				
1500				
2000				

- On your benchtop, use the masking tape and the metre ruler to mark out the starting line and finishing line for a 1-metre course for your dynamics trolley.
- Tie one end of the string to your trolley. Place the trolley at the starting line and bring the string forward so that it passes over the finish line and then hangs over a pulley at the end of the bench. Tie the 2 kg mass to this end of the string. Hold the trolley in place while you do this.

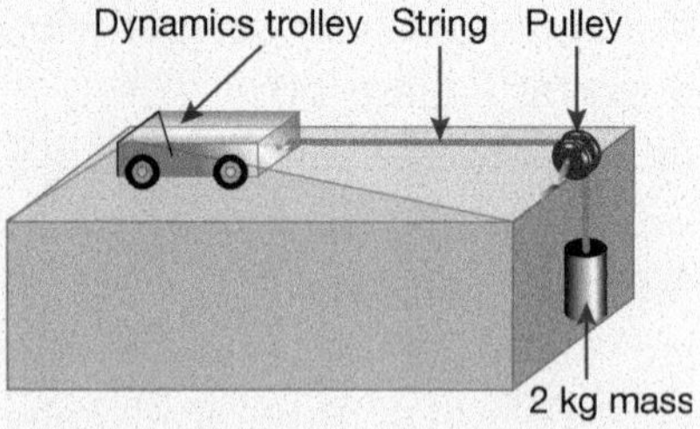

- Start the stopwatch at the same moment the 2 kg mass is dropped and time how long the unladen trolley takes to cross the finish line. Enter the time in the data table. Repeat this step twice more and then determine the average race time.
- Place a 500 g mass on the trolley and repeat the previous step.
- Repeat the experiment for increasing masses of 1000 g and 1500 g.

Discuss and explain

2. Which of the trolleys had the fastest race time? How can you tell if it had the greatest acceleration?
3. The equation $d = ut + \frac{1}{2}at^2$ allows you to calculate the size of the acceleration that was acting on the trolley each time, where d = distance, u = starting speed, and t is the average race time to cover the distance.

   As $d = 1$ m and $u = 0$, this equation simplifies so that we get $a = \frac{2}{t^2}$.

   Use this equation and the average race times in the table to determine the average acceleration of each trolley.
4. The weight of the 2 kg mass provided the force to move the trolley. Calculate the size of this force.
5. Was the weight of the 2 kg mass the only force acting on the trolley? What other forces can you identify that would have affected the trolley's acceleration?
6. Were the forces acting on the trolley each time balanced or unbalanced? How can you tell?
7. Give a general statement about the relationship between the mass of the trolley and its acceleration when a constant force is applied.

8.6 Exercises: Understanding and inquiring

To answer questions online and to receive **immediate feedback** and **sample responses** for every question, go to your learnON title at www.jacplus.com.au. *Note:* Question numbers may vary slightly.

Remember

1. Express Newton's second law in symbols.
2. Why does the acceleration of a spacecraft being launched increase as it rises?

Think

3. What total force would cause a 1.5 kg glass salad bowl to accelerate across a table at 0.30 m/s^2?
4. A 10 kg sled is pulled across the snow so that the total force acting on it is 12 N. What is the average acceleration of the sled?

Using data

5. Two identical toy carts, A and B, each with a mass of 1.0 kg, are pushed across a smooth, level tabletop with the same force. One of them contains a heavy brick. Cart A accelerates more rapidly than Cart B.
   (a) Which toy cart contains the brick? How do you know?
   (b) If the acceleration of cart A is 2.0 m/s^2, what is the total force acting on each cart?
   (c) If the acceleration of cart B is 0.5 m/s^2, what is the mass of the brick?

learnon RESOURCES — ONLINE ONLY

Complete this digital doc: Worksheet 8.6: Newton's second law (doc-19488)

8.7 What's your reaction?

8.7.1 Newton's Third Law of Motion

Newton's Third Law of Motion states that for every **action** there is an equal and opposite **reaction**. That is, when an object applies a force to a second object, the second object applies an equal and opposite force to the first object.

In fact, forces always occur in pairs. Sometimes it is painfully obvious. For example, when you catch a fast-moving softball or cricket ball with your bare hands, your hands apply a force to the ball. The ball applies an equal and opposite force to your hands — causing the pain.

In order to move forward quickly, the athletes need to push back. Why?

On the move with Newton's Third Law of Motion

Whether you are getting around on the ground, in the water, in the air or even in outer space, action and reaction forces are needed.

- When an athlete pushes back and down on the starting block, the starting block pushes the athlete forwards and upwards.
- The forward push of the road on the driving wheels of a car occurs only because the wheels push back on the road.
- When you swim through water, you push back on the water with your arms and legs so that the water pushes you forward.

8.7.2 Up and away

The force that pushes a jet aeroplane forward is provided by the exhaust gases that stream from its engines. As the jet engines push the exhaust gases backwards (an action), the gases push forwards with an equal and opposite force. This forward push is called the thrust. In order to equal or exceed the resistance to the jet's motion, the thrust needs to be very large. The huge blades inside a jet engine compress the air flowing into the engine and push it into the combustion chamber behind the blades. In the combustion chamber, fuel is added and burns rapidly in the compressed air. The exhaust gases are forced out of the engine at very high speed.

Sometimes the fact that forces occur in pairs is painfully obvious.

Blast off

The rockets used to launch spacecraft use an action and reaction pair of forces to propel themselves upwards. Like jet engines, they push exhaust gases rapidly out behind them (an action). As the rocket pushes the gases out, the gases push in the opposite direction on the rocket (a reaction).

Unlike jet engines, rockets used to launch spacecraft do not use air to burn fuel. They carry their own supply of oxygen so that the fuel can burn quickly enough to lift huge loads into space. The oxygen is usually carried as a liquid or a solid. Once spacecraft are in orbit, smaller rockets can be used to make the craft speed up, slow down or change direction.

learnon RESOURCES — ONLINE ONLY

Watch this eLesson: Newton's laws (eles-0036)

HOW ABOUT THAT!

Rockets, believed to have been invented by the Chinese, have been used as weapons since the thirteenth century.

INVESTIGATION 8.5

Just a lot of hot air

AIM: To observe Newton's Third Law of Motion in action

Materials:

a balloon

Method and results

- Inflate a balloon and hold the opening closed.
- Release the balloon and observe its motion through the air.

1. What happens to the air inside the balloon when you release the balloon?
2. Which way does the balloon move as the air is pushed out?

Discuss and explain

3. What provides the force that pushes the balloon through the air?
4. What is similar about the way in which the balloon is propelled and the way a jet engine works?
5. How is the motion of the balloon different from the motion of a jet engine?

INVESTIGATION 8.6

Balloon rocket

AIM: To model an application of Newton's Third Law of Motion

Materials:

drinking straw (plastic)
masking tape
balloon (sausage-shaped if possible)
scissors
fishing line (about 20 m)

Method and results

- Cut two short pieces from the drinking straw and thread a length of fishing line through them. Attach the ends of the fishing line to two fixed points so that the line is taut.
- Inflate the balloon and hold it closed while your partner attaches it to the pieces of straw with masking tape as shown in the diagram at right.
- Release the balloon and observe its motion along the string. Record the total distance travelled by your balloon rocket.

A balloon rocket

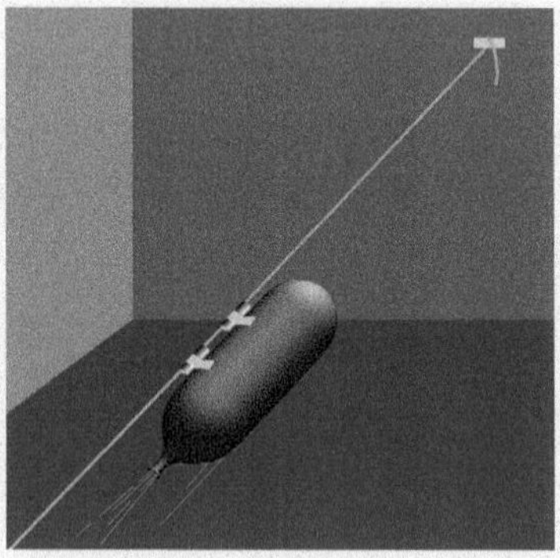

Discuss and explain

1. Compare your balloon rocket with those of others in your class. What features of the balloon rocket seemed to determine its range?
2. Suggest how your balloon rocket could be improved.

8.7 Exercises: Understanding and inquiring

To answer questions online and to receive **immediate feedback** and **sample responses** for every question, go to your learnON title at www.jacplus.com.au. *Note:* Question numbers may vary slightly.

Remember

1. State Newton's Third Law of Motion.
2. List three pairs of action and reaction forces. Be sure to state what each force is acting on.

Think

3. When you walk forwards, what provides the forward push?

4. How are rockets similar to jet engines? How are they different?
5. A yacht uses the push of the air on its sails to propel it forwards. If the push of the air on the sails is an action, what is the corresponding reaction?
6. What reaction force propels a Murray River paddleboat forwards? State clearly the action force that makes up the other part of the action–reaction pair.

Investigate

7. Find out how a hovercraft works. What action–reaction pairs are involved in its operation? Perhaps you could make a working model of one. Start by making a list of the materials that you will need.
8. Test your ability to identify Newton's laws in action by completing the **Newton's laws** interactivity.

learnon RESOURCES — ONLINE ONLY

Try out this interactivity: Newton's laws (int-0055)

Complete this digital doc: Worksheet 8.7: Newton's third law (doc-19491)

8.8 Getting down to work

8.8.1 Who is doing more work?

Whenever you apply a force to an object and the object moves in the direction of the force, **work** is done. A boy does work when he lifts weights — he applies an upward force and the weights move up. However, when he holds the weights still, no work is being done on the weights. He is applying an upward force equal to the downward pull of gravity on each weight, but there is no movement in the direction of that force.

Out of contact

In the examples below, the object on which the work is done is in contact with the object doing the work, and moves in the direction of the force applied to it. However, contact is not always necessary for work to be done. The Earth does work on you when you fall because of the gravitational attraction between it and you. When you pick up a small piece of paper with a charged plastic pen, work is done on the paper because of the attraction between the electric charge in the pen and the paper.

8.8.2 Working out work

The amount of work done on an object by a constant force is the product of the size of the force and the distance moved by the object in the direction of the force. That is:

work done = force × distance travelled in the direction of the force.

If the force is measured in newtons (N) and the distance is measured in metres, work done is measured in units of newton metres (N m).

Work and energy

Energy can be transferred to an object by doing work on it. Doing work on an object can also convert the energy an object possesses from one form to another. In fact, work is a measure of change in energy. The amount of energy transferred or converted when 1 newton of force moves an object 1 metre is 1 joule.

If you lift a 5-kilogram bowling ball with a force of 49 newtons (just enough to overcome its weight) through a height of 40 centimetres, the amount of work done is given by:

$$\begin{aligned} \text{work done} &= \text{force} \times \text{distance} \\ &= 49\,\text{N} \times 0.40\,\text{m} \\ &= 19.6\,\text{N m} \\ &= 19.6\ \text{J}. \end{aligned}$$

By doing work on the bowling ball, you have transferred 19.6 joules of energy to it. The additional energy is stored in the ball as gravitational potential energy. This stored energy has the potential to be converted into other forms of energy or transferred to other objects. For example, if you drop the ball, the force of gravity can do work on the ball, increasing its **kinetic energy**. Kinetic energy is the energy associated with movement. If your toe happens to be in the way when the bowling ball reaches floor level, the kinetic energy is transferred to your toe — ouch!

8.8.3 Storing it up

All stored energy is called **potential energy**. Energy can be stored in several different ways.

- **Elastic potential energy** (also called strain energy) is present in objects when they are stretched or compressed. Stretched rubber bands and springs have elastic potential energy. So do compressed springs like the one shown on the right. When the hand is opened, the elastic potential energy in the compressed spring is converted into kinetic energy.
- **Gravitational potential energy** is present in objects that are in a position from which they could fall as a result of the force of gravity. The water in a hydro-electric dam has gravitational potential energy. When the water is released, the force of gravity pulls down on it, doing work and converting the gravitational potential energy into kinetic energy.
- **Electrical potential energy** is present in objects or groups of objects in which positively and negatively charged particles are separated. It is also present when like electric charges are brought close together. The most obvious evidence of electrical potential energy is in clouds during thunderstorms. When enough electrical potential energy builds up, electrons move as lightning between clouds or to the ground.
- **Chemical potential energy** is present in all substances as a result of the electrical forces that hold atoms together. When chemical reactions take place, the stored energy can be converted to other forms of energy or it can be transferred to other atoms. Chemical potential energy is a form of electrical potential energy.
- **Nuclear energy** is the potential energy stored within the nucleus of all atoms. In radioactive substances, nuclear energy is naturally converted to other forms of energy. In nuclear reactions, such as those in nuclear power stations, in nuclear weapons and in the sun and other stars, nuclei are split or combined. As a result, some of the energy stored in the reacting nuclei is converted into other forms of energy.

8.8 Exercises: Understanding and inquiring

To answer questions online and to receive **immediate feedback** and **sample responses** for every question, go to your learnON title at www.jacplus.com.au. *Note:* Question numbers may vary slightly.

Remember

1. When you 'go to work' you are not always working. When are you really doing work?
2. List two examples of work being done on an object when there is no direct contact with another object.
3. How can the amount of work done on an object be calculated?

Think

4. Which type of force, other than electrical and gravitational forces, can do work on an object without being in contact with it?
5. How much work is done by:
   (a) a gardener as he pushes on a lawnmower 5 m across a level lawn with a horizontal force of 300 N?
   (b) three people as they unsuccessfully try to push a bogged car out of the mud? The car does not move. Each of the three people applies a forward force of 400 N to the car.
6. An Olympic diver with a weight of 540 N dives into a pool from a height of 10 m.
   (a) How much work is done on her by the force of gravity?
   (b) How much of her gravitational potential energy would you expect to be converted into kinetic energy during her dive?
7. Name an example of a child's toy that converts:
   (a) gravitational potential energy into kinetic energy
   (b) elastic potential energy into kinetic energy.

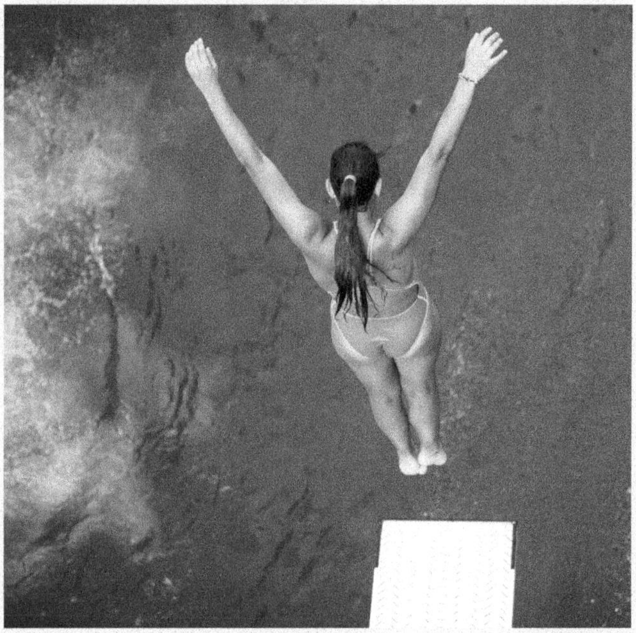

Investigate

8. Research and report on the life of James Joule. What did Joule achieve to deserve the honour of having the unit of energy named after him?
9. Investigate and write a report on the energy changes that take place when you cut the lawn with a lawnmower. Identify the forces that do work while you are mowing the lawn.

8.9 Systems: Energy ups and downs

8.9.1 Ups and downs

The energy changes that take place while bouncing on a trampoline occur when work is done in turn by the force of gravity, the trampoline or the person using it.

When the girl in the illustrations on the top of the next page is in the air, work is being done by the force of gravity to increase either her gravitational potential energy (going up) or her kinetic energy (going down). When she lands on the trampoline, she does work on the trampoline to convert her kinetic energy into elastic potential energy in the trampoline. As she is pushed back up into the air, the trampoline is doing work on her.

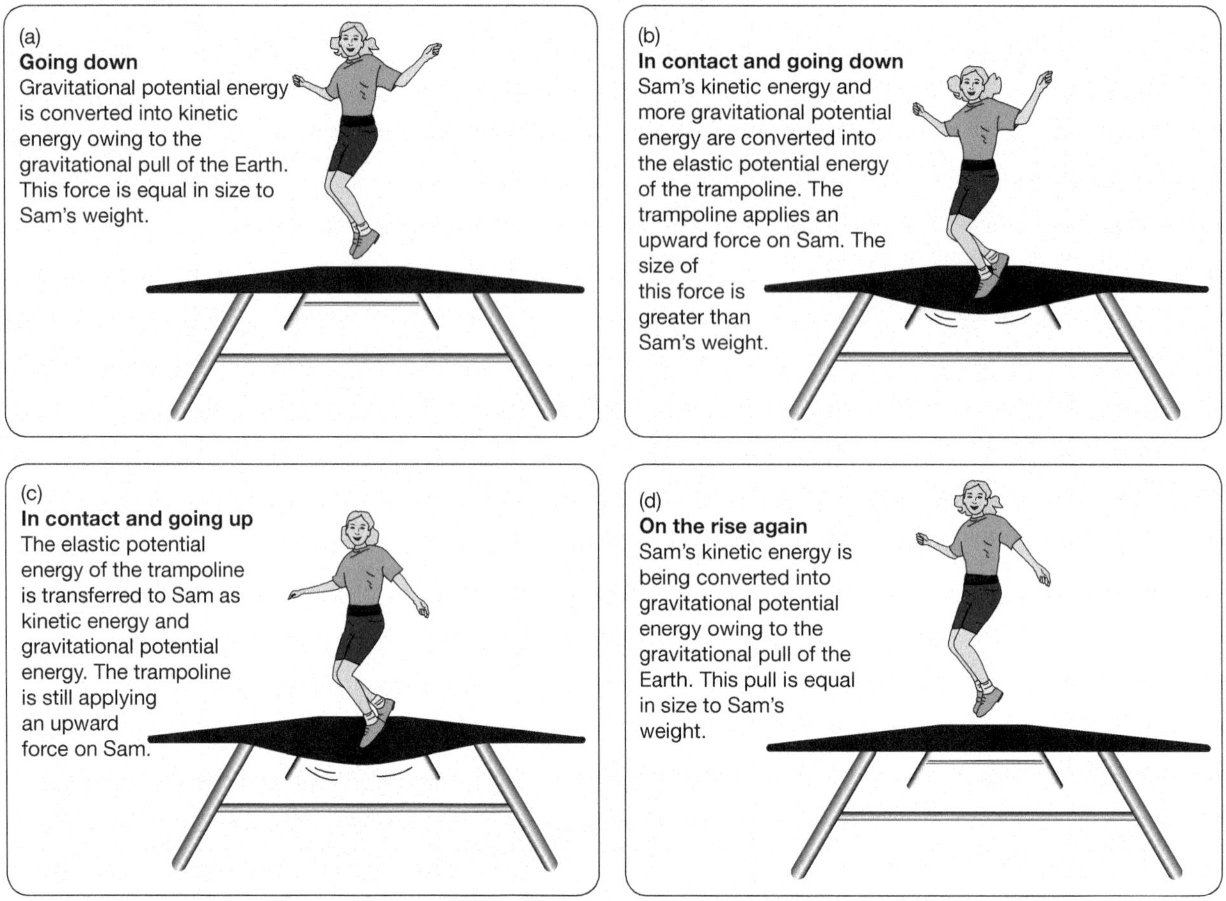
(a)
Going down
Gravitational potential energy is converted into kinetic energy owing to the gravitational pull of the Earth. This force is equal in size to Sam's weight.
(b)
In contact and going down
Sam's kinetic energy and more gravitational potential energy are converted into the elastic potential energy of the trampoline. The trampoline applies an upward force on Sam. The size of this force is greater than Sam's weight.
(c)
In contact and going up
The elastic potential energy of the trampoline is transferred to Sam as kinetic energy and gravitational potential energy. The trampoline is still applying an upward force on Sam.
(d)
On the rise again
Sam's kinetic energy is being converted into gravitational potential energy owing to the gravitational pull of the Earth. This pull is equal in size to Sam's weight.

8.9.2 Never created, never destroyed

Whenever work is done, energy is transferred to another object or converted into another form of energy. Energy is never created — nor is it ever destroyed. In fact, the total amount of energy in the universe remains constant. This observation is known as the Law of Conservation of Energy. The energy of a bouncing basketball is converted from kinetic energy to stored energy and back again. However, because it 'loses' energy during these conversions, it never bounces back to its original height. Nevertheless, its energy is not really lost. It is transferred to other objects — the air and the ground.

8.9.3 Systems and efficiency

Consider the basketball and the concrete surface to be a system. The **efficiency** of a system as an energy converter is defined as:

$$\text{efficiency} = \frac{\text{useful energy output}}{\text{energy input}} \times 100\%$$

The 'useful' energy within the system is the sum of the kinetic energy, elastic potential energy and gravitational potential energy (GPE) of the basketball. The energy input for the basketball is its initial gravitational potential energy. If this system were 100 per cent efficient, the basketball would bounce back to its original height and keep bouncing forever. The useful energy output would be the same as the energy input.

Calculating efficiency

The efficiency of the first bounce of a basketball can be calculated using:

$$\begin{aligned}\text{efficiency} &= \frac{\text{useful energy}}{\text{input energy}} \times 100\% \\ &= \frac{\text{GPE regained after bounce}}{\text{initial GPE}}\end{aligned}$$

The initial gravitational potential energy is the same as the amount of gravitational potential energy lost by the ball as it falls. That is equal to the amount of work done on the ball by the force pulling it down. The work done by the force of gravity as the ball falls is given by:

$$\begin{aligned}\text{work done} &= \text{force } \times \text{ distance travelled in the direction of the force}\\ &= \text{weight of basketball} \times \text{height through which it falls}\\ &= m \times 9.8 \text{ N/kg} \times h_1\end{aligned}$$

where m is the mass of the ball (in kilograms) and h_1 is the height from which the ball was dropped (in metres).

The amount of gravitational potential energy lost, that is, the initial gravitational potential energy, is $9.8\ m \times h_1$ joules.

The amount of gravitational potential energy regained by the ball as it rises to its maximum height is equal to the work done by the force of gravity against its direction of motion. That is, the amount of gravitational potential energy regained by the ball is:

$$\begin{aligned}\text{work done} &= \text{force } \times \text{ distance travelled in the direction of the force}\\ &= \text{weight of basketball} \times \text{height through which it raises}\\ &= m \times 9.8 \text{ N/kg} \times h_2\end{aligned}$$

where h_2 is the height to which the ball rebounds (in metres).

The amount of gravitational potential energy regained is $9.8\ m \times h_2$ joules.

The efficiency of the first bounce is therefore:

$$\begin{aligned}\text{efficiency} &= \frac{9.8m \times h_2}{9.8m \times h_1} \times 100\%\\ &= \frac{h_2}{h_1} \times 100\%\end{aligned}$$

For example, if a basketball is dropped from a height of 1.2 metres and bounces back to a height of 72 cm, its efficiency is:

$$\begin{aligned}\text{efficiency} &= \frac{h_2}{h_1} \times 100\%\\ &= \frac{72}{120} \times 100\%\\ &= 60\%.\end{aligned}$$

INVESTIGATION 8.7

Follow the bouncing ball

AIM: To investigate the energy transfers and transformations from a bouncing ball

Materials:

tennis ball

metre ruler

Method and results

- Drop the tennis ball from a height of one metre onto a hard surface. Watch the top of the ball closely as it hits the ground and rebounds.

1. Measure the height from the top of the ball to the ground and take care that the ball is dropped from rest and not assisted on its way down.

2. Drop the ball again from the same height and measure the height to which it rebounds after a single bounce.
3. Repeat your measurements at least five times and find the average bounce height.

Discuss and explain

4. Construct a flowchart using text descriptions similar to those used in the illustrations in this subtopic of the girl on the trampoline to show the energy changes that take place as the ball is:
   (a) falling through the air
   (b) slowing down while in contact with the ground
   (c) speeding up just before leaving the ground
   (d) rising through the air.
5. Identify the largest force that is acting on the ball as it:
   (a) falls through the air
   (b) is in contact with the ground
   (c) rises through the air.
6. What percentage of the tennis ball's initial gravitational potential energy is regained by the end of the first bounce?
7. Where has the 'lost' energy gone? Is it really lost?

8.9.4 Swing high, swing low

A pendulum is a suspended object that is free to swing to and fro. The most well-known use of pendulums is in clocks, mostly very old ones. A playground swing is simply a large pendulum and provides an excellent model of a system in which energy is alternately transformed from gravitational potential energy to kinetic energy. Without a push, the swing slows to a stop. The potential and kinetic energy of the system have dropped to zero. But that energy is not lost; it has been gradually transferred to the surroundings.

INVESTIGATION 8.8

Swing high, swing low

AIM: To investigate energy changes of a pendulum (a swing)

Method and results

- Push a friend on a swing.

1. Construct a flowchart, including text descriptions like those shown in the illustrations at the beginning of this subtopic to show how the energy of your friend changes through one complete backwards and forwards swing.

Discuss and explain

2. The initial kinetic energy of the person on the swing is zero. Where does the person's initial increase in kinetic energy come from?
3. In order to keep the person swinging up to the same height over and over again, you must continue to push. If energy is conserved, why do you need to continue to provide additional energy by pushing?

8.9 Exercises: Understanding and inquiring

To answer questions online and to receive **immediate feedback** and **sample responses** for every question, go to your learnON title at www.jacplus.com.au. *Note:* Question numbers may vary slightly.

Think

1. When you speed down a playground slide, the amount of kinetic energy that you gain on the way down is less than the amount of gravitational potential energy that you lose.
   (a) Where 'missing' energy go?
   (b) What can be done to maximise the amount of your initial gravitational potential energy that is converted into your kinetic energy?

2. The girl on the trampoline is able to return to the same height after each bounce. Explain why the system of the girl and the trampoline is not really 100 per cent efficient.
3. Consider the energy that is transferred away from a tennis ball bouncing on a concrete surface.
   (a) To what objects or substances is the gravitational potential energy and kinetic energy of the tennis ball transferred?
   (b) Into what forms of energy is the ball's gravitational potential energy and kinetic energy transformed?

Calculate

4. Evaluate the efficiency of the bounce of a cricket ball dropped vertically onto a concrete floor from a height of 1.40 metres if it rebounds to a height of 35 cm.
5. A 50 kg boy drops from a height of 1.0 m onto a trampoline.
   (a) Calculate the weight of the boy. Assume that $g = 9.8$ N/kg.
   (b) How much work is done on the boy by the force of gravity?
   (c) What amount of gravitational potential energy has been lost by the boy at the instant he makes contact with the trampoline?
   (d) What is the boy's kinetic energy when he makes contact with the trampoline?

Create

6. Construct a poster that shows what happens to your gravitational potential energy during either a bungee jump or a rollercoaster ride.
7. Design and build a device that uses the gravitational potential energy stored in a tennis ball (or another type of ball) to perform a simple task.

8.10 Making cars safe

Science as a human endeavour

8.10.1 Engineering for safety

In 1970, there were about 6 million vehicles being driven on Australia's roads. During that year, 3798 people lost their lives in road accidents. Now there are about 17 million vehicles on Australia's roads. Yet the average number of lives lost each year in road accidents is less than 1400.

One of the key reasons for the reduction in the road toll is that the cars we drive today are safer than ever before. Cars are designed by engineers who use scientific knowledge and experimentation to make cars lighter, stronger and, most importantly, safer.

8.10.2 Crash tests

Safety features such as seatbelts, collapsible steering wheels, padded dashboards, head restraints, airbags and crumple zones have to be tested by engineers before being introduced. The testing continues after introduction as car manufacturers strive to make cars even safer.

Testing of safety features involves deliberately crashing cars with crash test dummies as occupants. The dummies are constructed to resemble the human body and numerous sensors are used to detect and measure the effects of a collision.

Before real crash testing takes place, engineers use computer modelling to simulate crashes with virtual cars.

This inertia reel seatbelt is shown in the locked position. If rapid deceleration occurs, the small pendulum is able to swing forward with the inertia of the vehicle and lock the reel holding the seatbelt, preventing the passenger from moving forward.

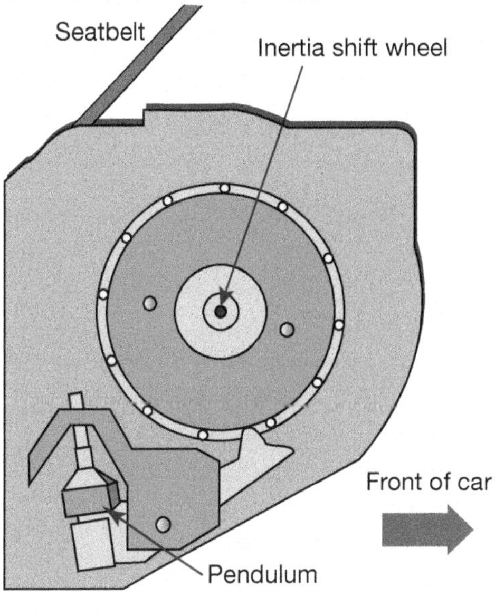

8.10.3 The effect of inertia

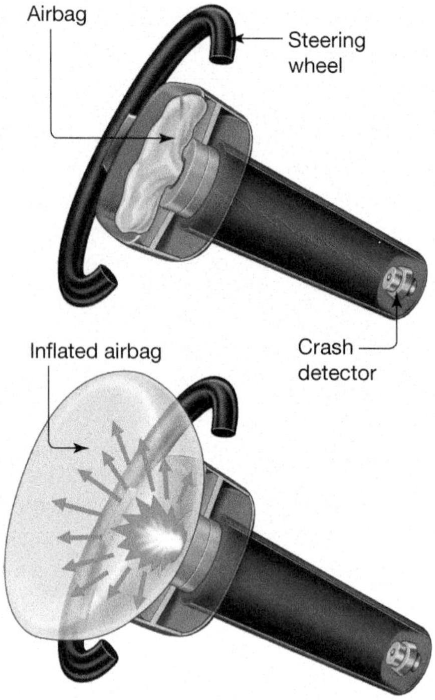

Most deaths and serious injuries in road accidents are caused when the occupants collide with the interior of the vehicle or are thrown from the vehicle. In a head-on collision the vehicle stops suddenly. However, unrestrained occupants continue to move at the pre-collision velocity of the vehicle until they collide with the steering wheel, dashboard or windscreen. Seatbelts provide an immediate force on the occupants so that they don't continue moving forwards. Front airbags reduce injuries caused by collisions between the upper body (which is still moving) and the steering wheel, dashboard or windscreen.

Side airbags are becoming more common. They protect occupants in the event of a side-on collision. More recent airbag technology measures the size of the impact and delays inflation until just the right moment. These improvements are the direct result of computer modelling and real crash testing.

Head restraints on seats reduce neck and spinal injuries, especially in a vehicle that is struck from behind by another vehicle. An impact from the rear pushes a vehicle forwards suddenly. Your body is pushed forwards by your seat. However, without a head restraint, your head remains where it was. The sudden strain on your neck can cause serious spinal injuries. Neck injuries caused this way are often referred to as whiplash injuries. Some new cars have head restraints that automatically move forward and up when a collision occurs.

A stationary car is struck from behind by another vehicle. Without a head restraint, your head remains at rest until pulled forwards by your neck.

8.10.4 The zone defence

The occupants of a car sit in a very strong and rigid protection zone designed to prevent outside objects (including the car's engine, other cars and tree trunks) from entering the passenger 'cell' and causing injuries during a collision. The roof panel is supported by strong pillars to make it less likely to be crushed.

The rigid passenger cell is flanked by **crumple zones** at the front and rear of the vehicle. These zones are deliberately designed to crumple, absorbing and spreading much of the energy transferred to the vehicle during a collision. As a result, less energy is transferred to the protection

The zone defence being tested. The front crumple zone absorbs and spreads energy. It also allows the car to stop more gradually. Crash test dummies are used to model the effect of collisions on the human body.

zone carrying the occupants, reducing the chance of injuries. The crumple zone also allows the vehicle to stop more gradually. Without a crumple zone, the vehicle would stop more suddenly and perhaps even rebound. Occupants would make contact with the interior at a greater speed, and there would be a greater chance of serious injury or death.

Car safety features employed in a frontal collision

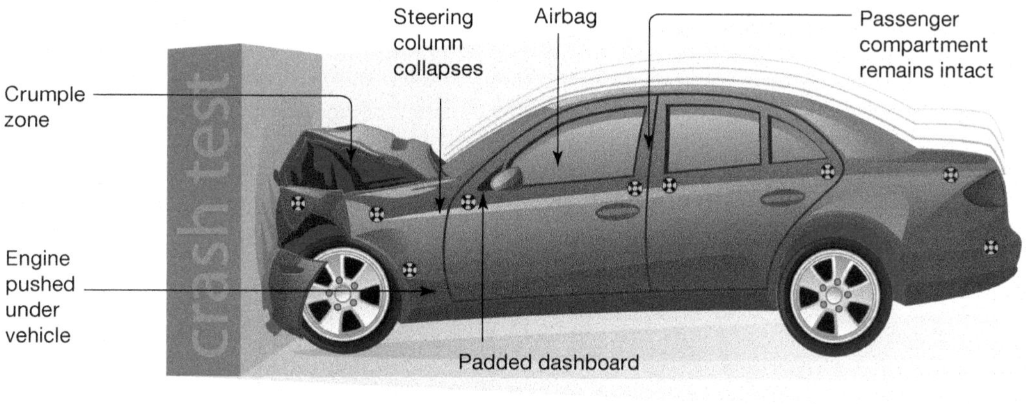

8.10 Exercises: Understanding and inquiring

To answer questions online and to receive **immediate feedback** and **sample responses** for every question, go to your learnON title at www.jacplus.com.au. *Note*: Question numbers may vary slightly.

Remember

1. List six safety features that are designed to make cars safer in the event of a collision.
2. How do engineers test vehicle safety features to make sure that they do what they are designed to do?
3. What happens to the motion of an unrestrained occupant when a car suddenly stops because it has collided head-on with another car or object?
4. How does each of the following features protect occupants during a collision?
   (a) Seatbelts
   (b) Airbags
   (c) Head restraints
5. What are crumple zones, and how do they protect the occupants of a vehicle during a collision?
6. Why is it important that there is a strong and rigid zone between the two crumple zones of a car?

Brainstorm

7. The safety features described in this subtopic are designed to reduce the chances of serious injury or death when a collision takes place. Scientists and engineers have designed many other safety features in cars and other vehicles that reduce the chances of a collision actually taking place. Work in a group to brainstorm these features and complete a table like the one below. Some examples are included to help you get started.

Safety features designed to prevent collisions

Feature	How the feature works
Tyre tread	Increases friction and makes steering and braking more reliable, especially in wet weather. The tread even pushes water out from beneath the tyre when the road is wet.
Windscreen wipers	Keep the windscreen clear to ensure good visibility for the driver.
Speed alarm	The driver selects a maximum speed. If that speed is exceeded an alarm sounds, warning the driver to slow down.

Investigate

8. Use the internet to research and report on:
   (a) how anti-lock brake systems (ABS) make braking in an emergency situation safer
   (b) the benefits of electronic stability control (ESC).

Evaluate

9. Collect two or more advertising brochures for recently manufactured passenger vehicles, then complete the following tasks.
   (a) Outline the features of the cars that relate to the safety of the occupants of the vehicle.
   (b) List the claims made about the safety features of the cars that do not make reference to scientific evidence.
   (c) Comment on missing scientific evidence that you would want to see included to support claims made in the brochures.

8.11 Cycle maps and storyboards

8.11.1 Cycle maps and storyboards

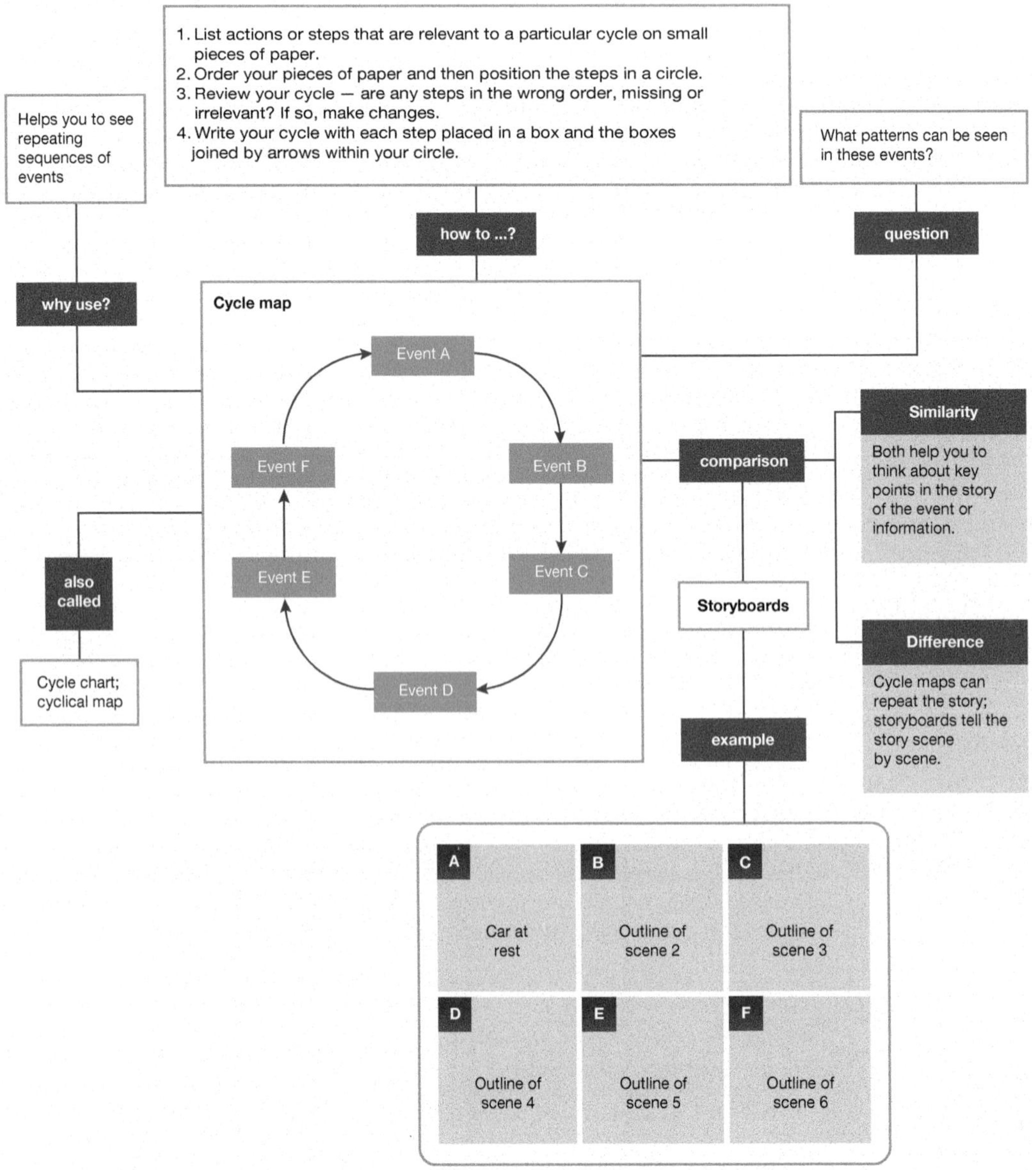

8.11 Exercises: Understanding and inquiring

To answer questions online and to receive **immediate feedback** and **sample responses** for every question, go to your learnON title at www.jacplus.com.au. *Note*: Question numbers may vary slightly.

Think and create

1. The cycle map on the right represents the motion of a girl on a trampoline. The vertical arrows represent the direction of motion of the girl.
   (a) Some of the boxes describing the energy changes that are taking place are empty. Copy the cycle map and complete it by describing the missing energy changes.
   (b) During which stage (or stages) of the cycle is the girl:
      (i) accelerating upwards
      (ii) accelerating downwards
      (iii) decelerating
      (iv) moving with zero velocity?
   (c) During which stage (or stages) of the trampoline cycle is the net force acting on the girl:
      (i) up
      (ii) down?
   (d) During which stages of the trampoline cycle is the force of gravity acting on the girl?
   (e) What is happening to the energy lost by the girl after coming into contact with the trampoline?

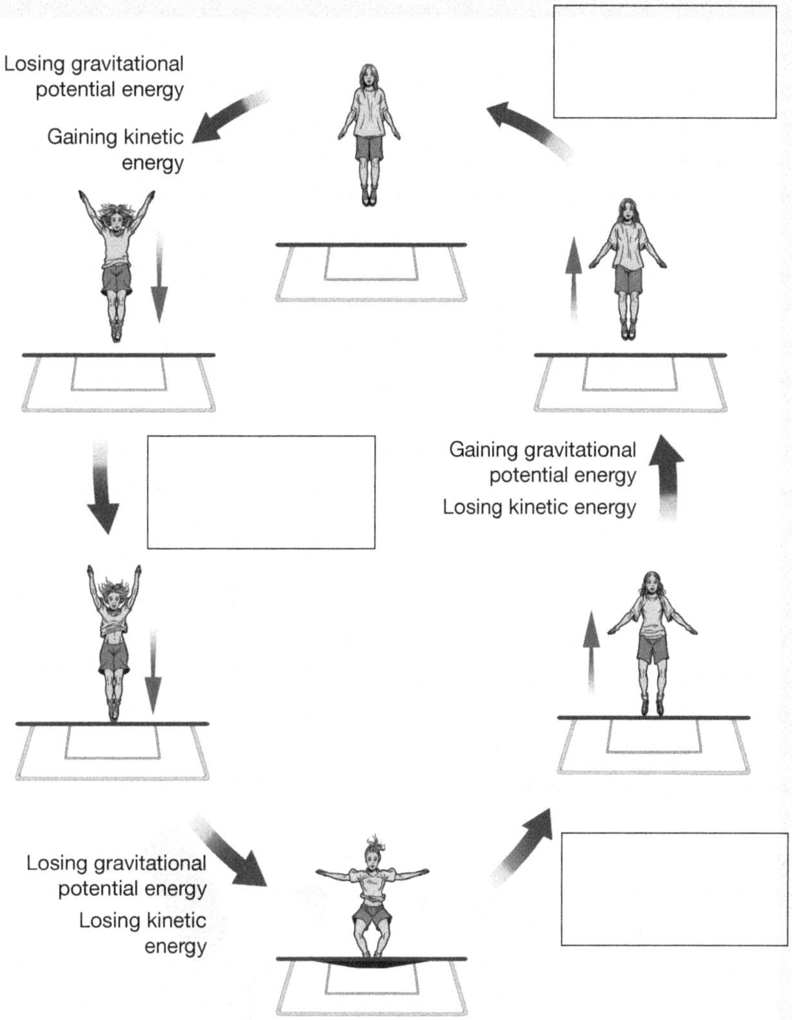

2. Construct a cycle map that describes the energy changes that take place when a tennis ball is dropped from a height and is allowed to bounce several times.
3. Create a storyboard that describes the motion of the passenger in a car as it accelerates from rest, travels in a straight line at constant speed, slows down before turning left, and accelerates again before coming to a sudden stop. Assume that the passenger is wearing a seatbelt. Add a commentary to each of the sketches of your storyboard that describes the forces acting on the passenger.

Use images like these of a passenger in a car to construct your storyboard.

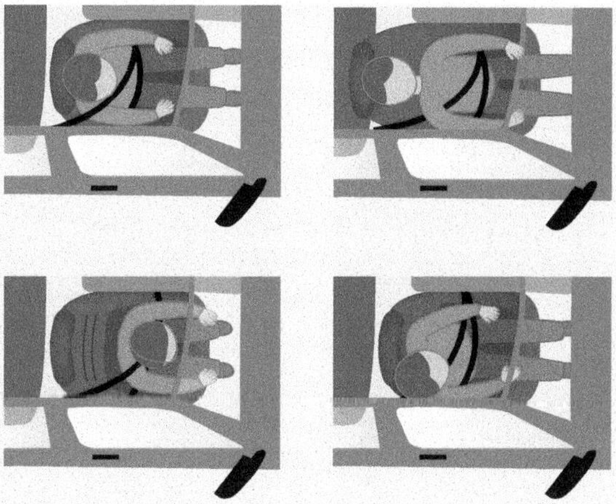

learn on

8.12 Project: Rock'n'rollercoaster

8.12.1 Rock'n'rollercoaster

Scenario

Many psychologists think that the reason rollercoasters are so popular is tied up with the 'rush' that follows stimulation of the fear response. When exposed to the combination of speed, noise, high hills, twists, loops and steep descents of the ride, our brains tell us that there is some element of threat or danger. This triggers our 'fight or flight' instinct, sending adrenaline coursing through our bodies in a way which many people find very stimulating. Of course, some of us just throw up rather than finding it fun!

Thrill-ride engineers say that the aim of a good ride is to provide a simulation of danger without actually putting people at risk. By manipulating the characteristics of gravity, periodic motion and speed, these engineers use physics to trick the body into thinking that it is in a lot more trouble than it really is. But the line between a ride that thrills and a ride that kills is a very narrow one, and the structural and mechanical engineers who design and build these rides must test their designs rigorously by using computer models or even physical models before the first steel rail leaves the factory! So how would you, as a team of rollercoaster engineers working at the new theme park Chunderworld, design a rollercoaster that was high on the thrill but zilch on the kill?

Your task

- Use your knowledge of physics and forces to design and build a model rollercoaster that has a set length and includes a minimum number of loops, hills and turns. Your design will also include a rollercoaster car that will travel along the length of the track. In order for the design to be considered successfully tested, your car must be able to travel the length of the track and then be brought to rest within the last 50 cm without derailing.
- Draw a plan or diagram of your rollercoaster identifying the positions and types of components used — hill, turn, twist or loop — and the points at which the car has the highest and lowest values of kinetic energy and gravitational potential energy.
- Finally, set up a blog that includes:
  (a) a summary of how the kinetic energy and gravitational potential energy values change over the course length
  (b) a log describing the development, building and testing of your rollercoaster and its different sections, including the method used to bring the car to a safe stop at the end
  (c) your drawing/plan of your completed rollercoaster
  (d) video footage of your rollercoaster in action from start to finish.

8.13 Review

8.13.1 Study checklist

Describing motion

- describe straight line motion in terms of distance, change in position, speed, velocity and acceleration using the correct units
- use a ticker timer or other available technology to gather data to determine the speed and acceleration of an object in straight line motion

Forces and Newton's laws of motion

- identify the forces acting on a motor vehicle in straight line motion on a horizontal surface
- describe the involuntary movement of people and objects in moving vehicles in terms of Newton's First Law of Motion
- define inertia as the resistance of objects to changes in their motion
- recall Newton's Second Law of Motion and use it to predict the effect of the net force acting on an object on its motion
- recall and apply Newton's Third Law of Motion to describe the interactions between two objects

Work and energy

- define work as the product of the force acting on an object and the distance travelled by the object in the direction of the force
- equate work to a change in kinetic or potential energy
- distinguish between elastic, gravitational, electrical, chemical and nuclear potential energy
- relate energy transfers and transformations to the Law of Conservation of Energy within a system
- recognise that useful energy is reduced during any energy transfer
- calculate the efficiency of a simple energy transformation
- describe a simple model of energy transformation and transfer within a system

Science as a human endeavour

- recognise the need for accurate methods of measuring the speed of vehicles in order to enforce laws designed to discourage excessive speed
- explain the movement of occupants of a vehicle in a motor accident in terms of inertia
- explain the safety features of cars in terms of energy transfer
- appreciate the role of scientists, engineers and computer modelling in the design of safety features in motor vehicles
- use scientific knowledge to evaluate claims made in the advertising of motor vehicles

Individual pathways

ACTIVITY 8.1
Revising forces, energy and motion
doc-8482

ACTIVITY 8.2
Investigating forces, energy and motion
doc-8483

ACTIVITY 8.3
Analysing forces, energy and motion
doc-8484

learnon ONLINE ONLY

8.13 Review 1: Looking back

To answer questions online and to receive **immediate feedback** and **sample responses** for every question, go to your learnON title at www.jacplus.com.au. *Note:* Question numbers may vary slightly.

1. Complete the equations below.
   (a) Average speed = $\frac{\quad\quad}{\text{time taken}}$
   (b) Average velocity = $\frac{\quad\quad}{\text{time taken}}$
   (c) Average acceleration = $\frac{\quad\quad}{\text{time taken}}$
   (d) Acceleration = $\frac{\quad\quad}{\text{time taken}}$
2. The table below provides information about four laps completed by one of the drivers in an Australian Formula One Grand Prix. The distance covered during one complete circuit of the course is 5.3 km.
   (a) Make a copy of the table and fill in the empty cells.

Lap no.	Time (s)	Average speed (m/s)	Average speed (km/h)
5	90		
15		60	216
25	110		
35	92	57.6	

   (b) Suggest two likely reasons for the lower average speed during lap 25.
3. Complete these statements about Newton's laws of motion in your workbook.
   (a) An object remains at rest, or will not change its speed or direction, unless …
   (b) When an unbalanced force acts on an object, the mass of an object affects …
   (c) For every action, there is …
4. Explain why the following statement is false: *A car travelling along a straight road has no forces acting on it.*
5. The diagram below shows part of the ticker tape record of the motion of a toy car as it is pushed along a table. As the tape moves through the ticker timer, a new black dot is produced every fiftieth of a second (0.02 s). The ticker tape has been divided into four equal time intervals labelled A, B, C and D.
   (a) During which time intervals is the speed of the toy car:
      (i) increasing
      (ii) decreasing?
   (b) During which of the four time intervals is the:
      (i) speed of the toy car constant
      (ii) acceleration of the toy car constant?
   (c) During which of the four intervals was the total force acting on the toy car zero?
   (d) During which of the four intervals was the unbalanced force acting on the toy car in the same direction as the car was moving?
   (e) What period of time is represented by each of the four time intervals? Express your answer in decimal form.
   (f) What is the average speed during the entire time interval represented by the ticker tape?

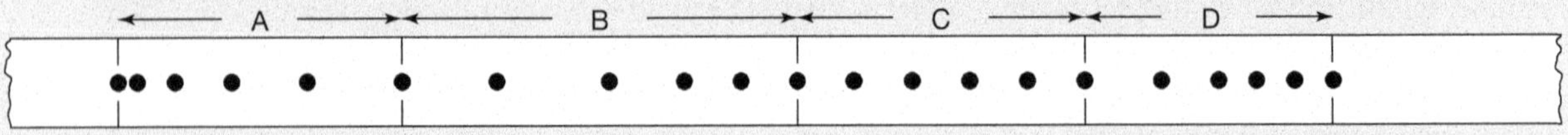

6. Explain in terms of Newton's First Law of Motion why it is dangerous to have loose objects inside a moving car.
7. Newton's Second Law of Motion can be expressed by the formula $F = ma$. What quantities do each of the symbols in the formula represent?
8. Who is doing more work — a body-builder holding an 80 kg barbell above her head or a student writing the answer to this question? Explain your answer.
9. List the energy changes that take place as a tennis ball dropped from a height of 2 m:
   (a) falls towards the ground
   (b) is in contact with the ground
   (c) rebounds upwards through the air.

10. List the forces acting on the tennis ball described in question 9 while it is:
    (a) moving through the air
    (b) in contact with the ground.

11. Many older people who drove cars more than 50 years ago make the comment 'They don't make them like they used to' in discussions about crumple zones. They describe how older cars were stronger and tougher, and therefore safer. Write a letter to a person who has made such a statement explaining why it is better that 'they don't make them like they used to'.
12. Use Newton's second law to calculate answers for the following.
    (a) A 1400 kg car accelerates at 3 m/s^2. What size force is needed to cause this acceleration?
    (b) A force of 160 N causes an object to accelerate at 2 m/s^2. What is the object's mass?
    (c) A force of 210 N acts on a mass of 70 kg. What is the acceleration?
13. Complete the crossword that follows using the clues provided.

ACROSS

1. This type of timer provides an easy way of recording motion on paper tape in the school laboratory.
4. You will find these used on ticker-timer tape.
8. A method used by police to measure the instantaneous speed of vehicles
9. This is done whenever you move an object.
10. Another word for size
12. The Chinese invented these to use as weapons.
14. You must use this whenever you are in a car.
15. It cannot be created or destroyed but can be transferred to another object or changed to another form when work is done.
16. This always causes a reaction.
17. Potential energy (abbreviation)

DOWN

2. The property of an object that makes it resist changes in motion
3. Kinetic energy (abbreviation)
5. The force that pushes a car forward
6. Used to slow very fast drag-racing cars in a short time
7. The type of energy that all moving objects have
9. This quantity is equal to the force of gravitational attraction towards the centre of the Earth.
11. The unit of force and a very famous name
13. This type of balance can be used in the laboratory to measure weight.
14. Always wait until trains do this before stepping off.

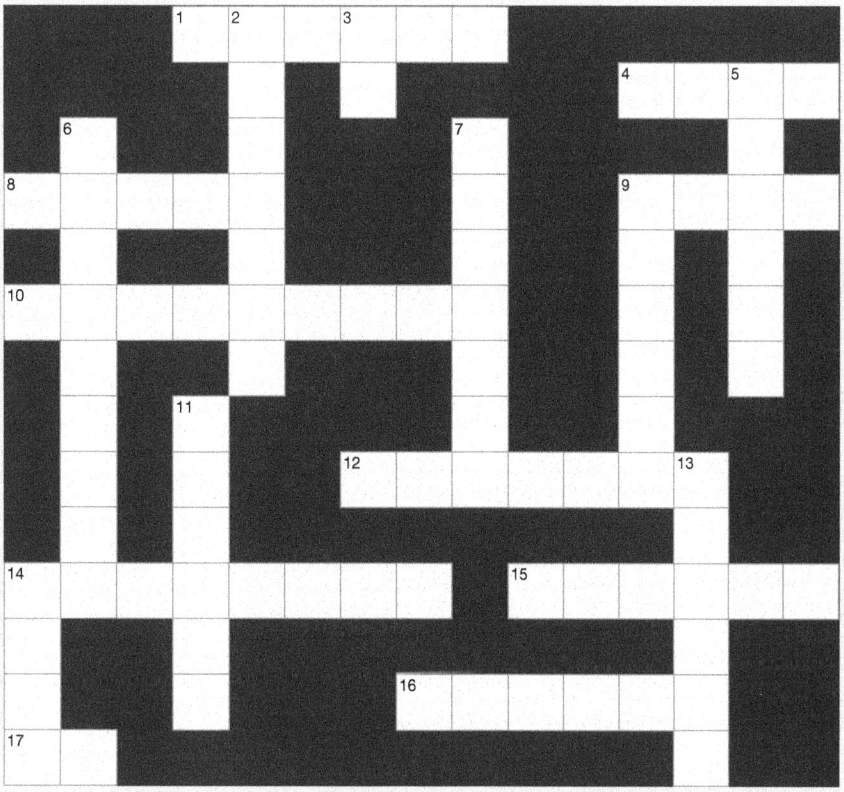

14. By increasing the time over which a collision occurs, the force of the impact on a car and its passengers can be decreased. List and describe at least two safety features of a car that decrease the force of impact in this way.
15. True or false?
    (a) Energy can be created but never destroyed.
    (b) Energy can be transferred from one object to another.
    (c) Energy can be transformed from one type to another.
    (d) Energy cannot be stored.
    (e) Energy is measured in joules.
16. Write the rule for calculating work.
17. How much work is done to lift a 5 kg box onto a shelf 1 m above the ground?

learnON RESOURCES — ONLINE ONLY

Complete this digital doc: Worksheet 8.8: Forces, energy and motion: Summary (doc-19494)

TOPIC 9
Science quests

9.1 Overview

9.1.1 Why learn this?

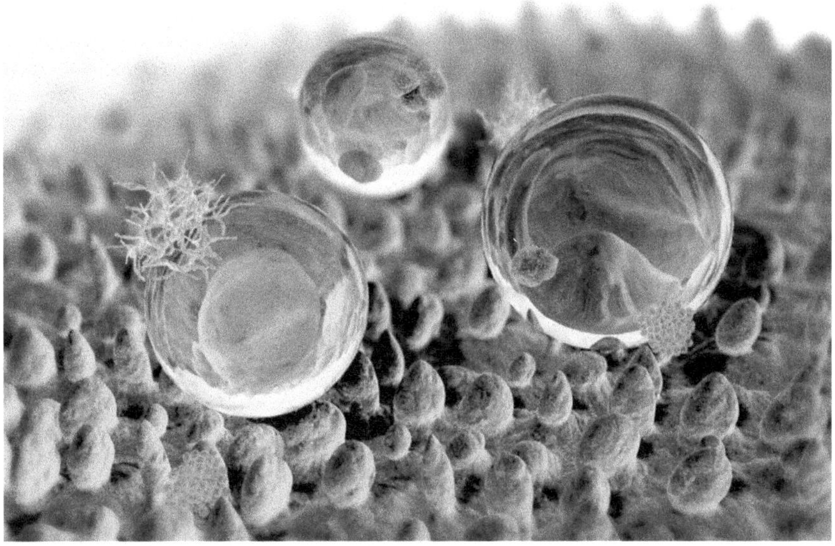

We live in a time of great change. The future will present us with many challenges and discoveries — some so exciting and absurd that they may be beyond our wildest hopes and dreams …

Are these crystal balls? No, but they help us glimpse the future. They are water droplets resting on a synthetic surface that is self-cleaning. The water droplets pick up any dirt and roll off. What new applications can you think of for this surface?

9.1.2 Think about these

assessON

- Why did Superman's parents send him to Earth?
- What is the relationship between the Incredible Hulk, nanotechnology and gamma radiation?
- Could X-Men interbreed with humans?
- What does Spider-Man have in common with sea slugs?
- What sorts of spiders might be crawling inside your body in the future?
- Do you have music in your genes?
- What does the movie *Avatar* have to do with microbes and nanowires?
- How can killer tomatoes protect you from disease?

LEARNING SEQUENCE

Numerous **videos** and **interactivities** are embedded just where you need them, at the point of learning, in your learnON title at www.jacplus.com.au. They will help you to learn the content and concepts covered in this topic.

9.1.3 Your quest

Towards immortality

Is artificial evolution of our species possible? DNA technology, drugs and implants for existing or experimental therapies could make this a reality. We can already insert new genes into various parts of the adult human body. In the future, this may also include gametes and embryos. We have the technology to cut and paste various genetic sequences, not only within the same species, but between species. How might these modifications affect future generations?

We also have the power to replace body parts with natural organs, mechanical organs or tissues derived from stem cells. We already have drugs such as steroids to enhance physical performance, and psychoactive drugs that can alter our powers of cognition such as memory, mood, appetite, libido and attention. Where will our next discoveries and technologies take us?

INVESTIGATION 9.1

Life in 2050

AIM: To research potential future technologies and write a story about their implications

1. Research and make notes on how current scientific research may affect life in the future. You may wish to consider the following questions.
   - What future technologies and applications will change our lives?
   - How far may our life spans be extended?
   - Will we all be disease-free?
   - Will some people be more entitled to medical services than others?
   - How much of us will remain organic and how much integration with computers and other synthetic materials can occur before we are no longer considered human?
2. Drawing on your note from question complete the following tasks
   (a) Creatively weave your findings from question 1 into a science fiction story about life in 2050.
   (b) Design a cover page for your story.
   (c) Put together some promotional material for your story and include this on the back cover of your book.
   (d) Share your story with your class.

9.2 Daring to dream

Science as a human endeavour

9.2.1 Creatively critical

Why not go on a journey that will take you beyond your wildest dreams?

A clever scientist realises that science is not just about critical thinking with clarity, accuracy and detail. It is also about thinking flexibly and creatively, with an open mind. Reading, writing or watching science fiction can help unlock your mind's doors to take a step outside reality. Science fiction can take you to another universe where anything is possible. It provides you with the opportunity to dream and imagine endless possibilities and creations.

Science fiction authors in the past, such as Aldous Huxley (1894–1963) and Arthur C. Clarke (1917–2008), have had a considerable effect by taking others on a journey beyond their wildest dreams. It is only in our time that some of their dreams are becoming a reality.

9.2.2 In our image

Mary Shelley's novel *Frankenstein*, in which scientist Victor Frankenstein creates a monster who eventually turns on and kills Frankenstein and those he loves, has led to a genre of stories in which the creation destroys the creator. The popularity of such a theme has contributed to technophobia, or the fear of technological advances. These advances may not be only in robotics, but also in other types of creation such as new chemical compounds or transgenic organisms produced by genetic manipulation.

Science fiction tells tales of how humans attempt to outdo nature and are often then confronted by menaces of their own making. The tales also describe how the human spirit, determination and imagination are used to solve and conquer these menaces.

Isaac Asimov has been universally acknowledged as the father of robotics. He wrote many stories about the human fear of robots. His story *Evidence* (1946) suggests that a well-programmed robot not only could look human but, if programmed with the Three Laws of Robotics, would be more ethical than many politicians.

> By using human ova and hormone control, one can grow human flesh and skin over a skeleton of porous silicone plastics ... The eyes, the hair, the skin, would be really human, not humanoid ... if you put a positronic brain and such other gadgets ... inside, you have a humanoid robot.
>
> From *Evidence* by Isaac Asimov

In another of Asimov's stories, *The Bicentennial Man* (1976), he deals with the issues of the ethical responsibility that humans have for their creations, the relationship between organic and inorganic matter, and the futuristic line between living and non-living. The novel is about a humanoid robot, Andrew, who, unlike other robots, desperately wants to become human. The story begins with him questioning another robot:

> 'Have you ever thought you would like to be a man?' Andrew asked. The surgeon hesitated a moment, as though the question fitted nowhere in his allotted positronic pathways. 'But I am a robot, sir.'
>
> From *The Bicentennial Man* by Isaac Asimov

In the 1994 novel *The Ship Who Searched* by Anne McCaffrey and Mercedes Lackey, the benefits of becoming more 'machine' were explored. In this story, a paralysing alien virus leaves the heroine unable to live without a mechanical support system and she is transformed into a 'shell person' or 'brain ship' to adventure throughout the universe.

> No amount of simulator training conveyed what it really felt, to have a living, breathing ship wrapped around you ... Never mind that her 'skin' was duralloy metal, her 'legs' were engines, her 'arms' the servos she used to maintain herself inside and out ... That all of her senses were ship's sensors linked through brainstem relays. None of that mattered. She had a body again!
>
> From *The Ship Who Searched* by Anne McCaffrey and Mercedes Lackey

9.2.3 New worlds

It is believed that biotechnology promises the greatest revolution in human history. The commercialisation of molecular biology has occurred with astonishing speed and is considered to be the most stunning ethical event in the history of science.

Since the discovery of DNA, science fiction authors have incorporated the new scientific concepts into stories with futuristic worlds in which humans are forced to cope with their creations.

> Problem-solving was his specialty, and he had been selected for it before birth. Gene analysis had chosen the best DNA chain from his parents' sperm-and-ovum bank. This, and subsequent training, had fitted him perfectly for command.
>
> From *War with the Robots* by Harry Harrison

Aldous Huxley's *Brave New World* (1931), Harry Harrison's *War with the Robots* (1962), Michael Crichton's *Jurassic Park* (1991) and Julian May's *Jack the Bodiless* (1992) have all addressed the topic of futuristic genetics.

> If this insect has any foreign blood cells, we may be able to extract them and obtain paleo-DNA, the DNA of an extinct creature. We won't know for sure, of course, until we get whatever is in there, replicate it, and test it. That is what we have been doing for five years now. It has been a long, slow process — but it has paid off.
>
> From *Jurassic Park* by Michael Crichton

> All four of us children inherit from the Remillard side of the family a dominant polygenic mutant complex: we're smart, we have extremely high metafunctions, and our bodies age up to a certain point and then persistently rejuvenate. The traits have a reduced penetrance and exhibit variable expressivity. You know what that means?
>
> From *Jack the Bodiless* by Julian May

Of all of these authors, Aldous Huxley gave the most thorough description of the impact that genetic manipulation may have on our society. His book *Brave New World* also described a society in which different human castes were created, produced and brainwashed to happily meet different needs and services. The following text is written from the point of view of a 'Beta individual'.

> Alpha children wear grey. They work much harder than we do, because they're so frightfully clever. I'm awfully glad I'm a Beta, because I don't work so hard. And then we are much better than the Gammas and Deltas. And Epsilons are still worse. They're too stupid to be able ...
>
> From *Brave New World* by Aldous Huxley

In 1976, Robert Swanson, a venture capitalist, and Herbert Boyer, a biochemist, formed a commercial company to exploit Boyer's gene-splicing techniques. Their company, Genentech, quickly became the largest and most successful of the genetic engineering start-ups. Since this time, many similar companies have sprung up with the purpose of creating genetically modified organisms that can be utilised for financial gain. What will your future hold? How will the new technologies affect you? And what about your children? That is, of course, if the government allows you to have them!

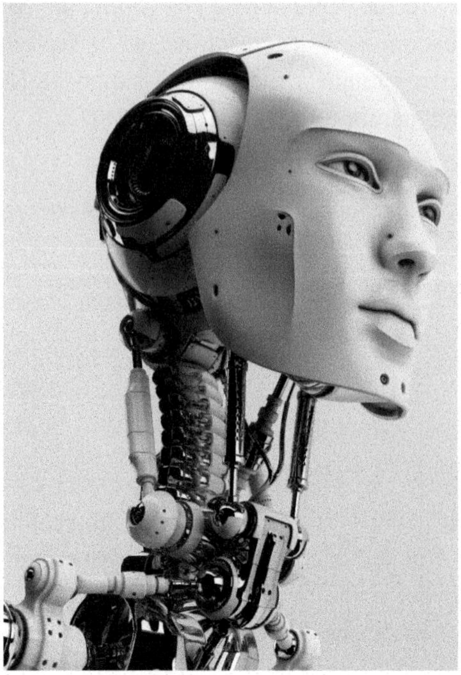

9.2.4 Punk in science fiction

Films such as *Blade Runner* and the *Matrix* trilogy provide examples of cyberpunk. This form of science fiction is a blend of **cybernetics** and punk. The term became widespread in the 1980s, especially to describe novels from authors such as William Gibson and Bruce Sterling. Cyberpunk often explores possible near-futures of Earth that resemble dystopia rather than utopia.

In many cyberpunk stories, there is a conflict between computer hackers, artificial intelligence and big corporations. The main characters in these stories are often alienated loners, living on the edge of their society. Their lives are ravaged by the negative effects of advanced technology, with even their own bodies often having undergone some form of invasive modification. In these possible futures, the advanced technology may be blended with a loss of social order, control and morality.

Over the last few years, offshoots of the cyberpunk genre have resulted in the birth of related genres such as biopunk, steampunk and postcyberpunk. In postcyberpunk, it is heartening that there are characters who act to improve social conditions or at least try to prevent their further decay. Will your creative science fiction writing produce another science fiction genre?

He'd operated on an almost permanent adrenaline high, a byproduct of youth and proficiency, jacked into a custom cyberspace deck that projected his disembodied consciousness into the consensual hallucination that was the matrix.

From *Neuromancer* by William Gibson

9.2.5 Now showing

Many early science fiction stories are being remade as movies, while new science fiction stories continue to inspire and set your synapses firing. Some science fiction movies explore DNA and genetic engineering (for example, *Gattaca, Jurassic Park* and *X-Men*). Other movies take us on journeys through time and space (for example, *Star Trek, The Matrix* and *The Time Machine*). The issues and chaos of disease are expressed in *Outbreak, Contagion* and *The Andromeda Strain* and the potential dangers of robotics are woven into movies such as *I, Robot* and *2001: A Space Odyssey.*

Inside the matrix

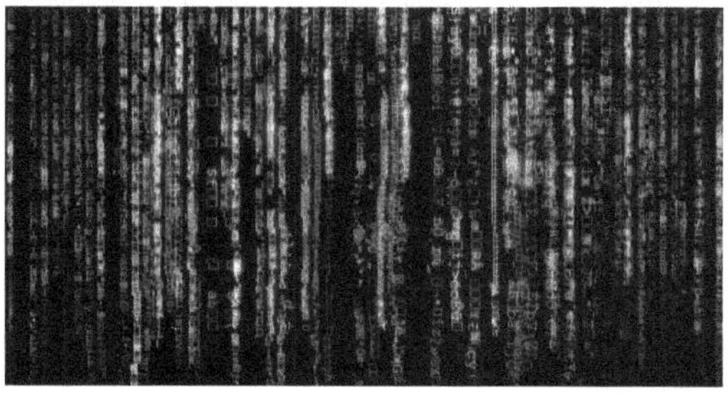

learnON RESOURCES — ONLINE ONLY

Watch this eLesson: Faster computing (eles-1077)

9.2 Exercises: Understanding and inquiring

To answer questions online and to receive **immediate feedback** and **sample responses** for every question, go to your learnON title at www.jacplus.com.au. *Note:* Question numbers may vary slightly.

Think, discuss and create

1. (a) Watch a science fiction movie of your choice. For example: *Gattaca; X-Men; Star Trek; The Time Machine; Outbreak; I, Robot; Frankenstein; 20 000 Leagues Under the Sea.*
   (b) Construct a mind map to summarise what the movie was about.
   (c) Construct a PMI chart in which the 'P' lists the accurate science in the movie and the 'M' lists the inaccurate, false or misleading science. Under 'I', list things that you found interesting or inspirational about the movie.
   (d) Share your PMI chart with others, adding any new ideas that you agree with to your chart.
   (e) Suggest changes that you would include if you were to remake the movie.

2. On your own or in a team, write your own science fiction story.
   (a) To start, think about some science concepts, principles or theories that you think would be good in a story. Research these ideas, so that you can give substance to your story and so that its science content is not superficial.
   (b) Design a cover page for your story that is representative of the plot.
   (c) Publish a class magazine that contains contributions from all students.
   (d) Read a different class member's story each night and construct a PMI chart of it.
   (e) Provide a copy of the story's PMI chart to its author, so that they have some feedback.
3. Write a short story that leads on from the following: *I woke up at 2.30 in the afternoon in a hospital bed. Two things were different. My skin had been peeled off my left arm, exposing electronic circuitry, and my right foot was missing.*

Think and discuss

4. Suggest fears that people may have about future robots, computers, chemicals or genetically engineered organisms.
5. Will computers become more intelligent than human beings? What implications may this have?
6. What is artificial intelligence?
7. *Nature never appeals to intelligence until habit and instinct are useless. There is no intelligence where there is no change and no need of change.* Discuss what you think H. G. Wells meant by this in his novel *The Time Machine*.
8. Two students observed the figure at right. One student suggested that the red sphere indicated a mutation in the DNA sequence of a gene; the other student, however, disagreed. Which student do you think is correct? Justify your response.

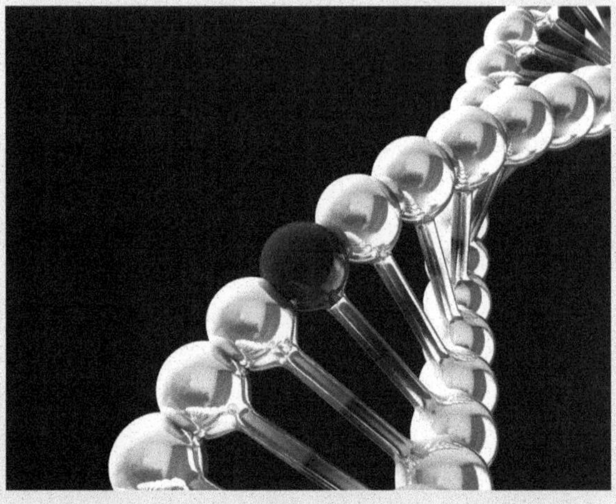

Investigate

9. Read and comment on the social implications of any of the novels described in this subtopic.
10. What are the rules for the modern robot? Security, safety and sex are considered the big concerns. As robots will be entering homes and workplaces, the need for strict guidelines is hugely important. Discuss this with your team and construct a robot rule book.
11. How important do you think science fiction is in addressing issues that may affect humanity in the future? Can it help us deal with the ethical dilemmas we may face with increasing progress in biotechnology, genetic modification and computing? If you believe that science fiction authors should not be the ones addressing these concerns for us, who should?

Investigate and create

12. Imagine that you have created a new gene. Its modified DNA sequence codes for a feature that humans do not currently possess. Use the internet and your own imagination to address the following.
    (a) Describe the new feature that this gene codes for.
    (b) Is the gene going to be inserted into the somatic or the sex cells? Explain why.
    (c) Outline reasons for the creation of this new gene.
    (d) Identify how this gene is going to be inserted.
    (e) Who will have access to this gene technology? Explain why.
    (f) Write a science fiction story that includes your responses to parts (a)–(e).

learnon RESOURCES — ONLINE ONLY

Complete this digital doc: Worksheet 9.1: Inventions and innovations (doc-20051)

9.3 Superheroes to super science

Science as a human endeavour

9.3.1 Super science?

Is there science in the stories of superheroes? Where do the ideas of superpowers come from? Where do scientists get ideas and inspiration for new technologies and products?

9.3.2 Superman

Superman was created in 1938 by Jerry Siegel and Joe Shuster. The story starts with Superman being sent as a baby from his home planet, Krypton, by his parents as his planet is about to explode. It would make sense that he was sent to Earth because the lower gravity and high energy from our sun might give him an increased chance of surviving. The gravity difference between Earth and Krypton contributes to some of his superpowers. Superman's energy is thought to come from the sun. The creators may have taken this idea from the way in which plants use light energy in their process of photosynthesis. If this was the case, did Superman's cells use the light energy to trigger some kind of nuclear reaction (like cold fusion), or did he have some way of storing the massive amount of energy that he would have required for all of his super-activities? When Superman is exposed to kryptonite, some of his symptoms are similar to radiation sickness; does that mean that kryptonite is radioactive? Where would kryptonite fit into our periodic table? Would it fit among the very heavy elements?

9.3.3 The Incredible Hulk

Dr Bruce Banner, a nuclear scientist, was accidently exposed to gamma rays and became the Incredible Hulk, a giant with tremendous strength and green skin. The tales behind how his transformation occurs vary. In a recent movie version, his father had modified DNA in his germline cells, which then passed to Bruce. When Bruce's experiments in nanotechnology led to his exposure to a massive dose of gamma radiation, the radiation acted as a catalyst to express the modified DNA. The incredibly fast and immense cell replication required for Bruce to become the Incredible Hulk and then somehow lose the extra mass to become the meek Bruce again is difficult to explain.

The character of the Incredible Hulk was created by Stan Lee and Jack Kirby in 1962. Science has changed a lot since the 1960s and that has affected how Bruce Banner's transformation is explained. What type of superheroes could be created using pioneering science that is happening now?

9.3.4 X-Men

The theory of evolution on Earth suggests that it took billions of years for life to evolve from single-celled organisms to the life that we see today. In the X-Men, however, it took only a couple of generations for significant mutations in hundreds of individuals to give them unique powers and abilities. This burst of mutations that radically changed some of our species into X-Men matches the idea of punctuated evolution suggested by Niles Eldridge and Stephen Jay Gould, rather than Darwin's evolution by a slow process of gradual change.

The X-Men go by the name *Homo sapiens superior* rather than *Homo sapiens*, suggesting that they are a human subspecies. If so, they would be able to interbreed with humans and produce fertile offspring.

Different types of radiation are involved in the explanation for how the Incredible Hulk and Spider-Man gained their superpowers. Could radiation have been involved in the radical mutations that created the X-Men?

9.3.5 Spider-Man

After being accidentally bitten by a radioactive spider, Peter Parker gains incredible powers. He has incredible speed, amazing strength, a knack for climbing walls, and he can fire sticky, silky thread from his wrists. Although at first it seems far-fetched that the venom of this spider had such an effect, think about it a little more. Did the radiation alter the spider's DNA so that the venom produced (a protein) was able to trigger mutations within the DNA in Peter's cells? If these mutations were then expressed, could they lead to spider proteins being synthesised, resulting in particular characteristics that Peter did not have before?

How does Spider-Man climb up walls? Spiders are covered in tiny hair-like structures called setae. Molecular interactions between these and surfaces such as walls and ceilings enable spiders to climb easily. Peter's webbing is an adhesive polymer that mimics spider silk, which has a tensile strength almost equal to steel.

The melding of Spider-Man's superpowers can inspire scientists to use their imaginations and apply new thinking to fields such as bionics or **biomimicry**, using inspirations in nature to develop new products and technologies. Already a team of Italian scientists are suggesting that their latest nanotechnology discovery may unlock the secret to a wall-scaling Spider-Man suit.

9.3.6 Biomimicry

Biomimicry is the practice of developing new products and technologies that are based on replicating or imitating designs in nature. Millions of years of evolution has often solved problems in ways we can harness to create useful technologies. The most famous example of biomimicry is the invention of Velcro in 1941 by George de Mestral, a Swiss engineer. He was inspired by burrs that stuck to his dog's hair.

Another example is superhydrophobicity or the **lotus effect**. The surfaces of lotus leaves are bumpy, causing water to bead and roll off. Scientists have developed a way to chemically treat the surface of plastics and metals in a similar way. Imagine all of the applications this process could have.

Scientists have marvelled at the intricate patterns of **silica** within the cell walls of tiny single-celled algae called **diatoms**. They have also been impressed with the ability of diatoms to manipulate silicon at nanoscale levels (around one billionth of a metre), and they have genetically engineered some

A close-up of a burr. The casing of the seed allows it to hitch a ride on animals and disperse.

Velcro mimics the action of burrs.

diatoms to manufacture working valves of specific shapes and sizes. These valves are then used in silicon-based nanodevices that can deliver drugs to target cells within human bodies. This is an example of **biosilification**.

9.3.7 How about that? Science mimics life?

Sangbae Kim is an expert in rapid prototyping methods for biologically inspired robotic systems. He uses ideas of mechanisms used by animals, such as how they move, to create mobile robots. One of his robots is the Stickybot, which has foot pads inspired by the feet of the gecko that allow it to climb walls at a speed of about 1 m/s.

The tiny hairs (setae and spatulae) on the pads of a gecko's feet cling to surfaces using a molecular interaction known as the **Van der Waals force**. This helps to support the gecko's weight as it scurries up walls (in the same way that similar structures work for spiders). Kim has covered Stickybot's feet with hairs made of silicone rubber. Later models of this design could be used for repairing underground oil pipelines or cleaning windows in multistorey buildings.

Another of Kim's robots is his cockroach-inspired hexapod, iSprawl, which can run up to 15 body-lengths per second over rough terrain.

Dr Sangbae Kim gets his inspiration for his robots from the animal kingdom. Geckoes have almost half a million setae on each foot, enabling them to climb up even very smooth surfaces.

9.3 Exercises: Understanding and inquiring

To answer questions online and to receive **immediate feedback** and **sample responses** for every question, go to your learnON title at www.jacplus.com.au. *Note:* Question numbers may vary slightly.

Investigate, think and discuss

1. Suggest why going to a planet with a lower gravity might give Superman an increased chance of survival.
2. What is cold fusion? Comment on related research or experiments.
3. Describe the symptoms of radiation sickness and ways to treat it.
4. Find out more about Superman's kryptonite. Based on the information you have found, suggest where it would fit into the periodic table.
5. What are gamma rays and why are they dangerous to living things?
6. If Bruce Banner's father's modified DNA was in somatic cells rather than germline cells, would it still have been passed on to Bruce? Explain.
7. What is nanotechnology? Identify four different types of applications, research or products that involve nanotechnology.
8. Suggest why the Incredible Hulk would have to have incredibly fast cell replication. Suggest where these cells go when he shrinks back to being Bruce.
9. Research theories about gradualism versus punctuated evolution. Which do you think is the best theory? Justify your response.
10. What is the scientific name for X-Men? Does this mean that they are a different species to non-mutant humans? What are the implications of this?
11. Is it possible for foreign DNA to make its way into the human genome? Justify your response.
12. If spider DNA was inserted into Peter Parker's DNA, suggest how that could result in Peter expressing spider characteristics.
13. Find out more about the possibility of a Spider-Man suit.
14. What is meant by the term *biomimicry*? Provide two examples.
15. Find out more about the Van der Waals force and how it helps geckoes to climb walls.
16. Use the internet to find examples of robots that have been inspired by animals.
17. Find at least two products or technologies that are based on mimicking nature.

learnon RESOURCES — ONLINE ONLY

Complete this digital doc: Worksheet 9.2: Science under scrutiny (doc-20052)

Complete this digital doc: Worksheet 9.3: Great ideas and lucky breaks! (doc-20053)

9.4 Nano news

Science as a human endeavour

9.4.1 Tiny, but packing a powerful punch

When we immerse ourselves into the world of nanotechnology, we need to learn to think very, very small. So small that many of nature's laws that work in the big world no longer work in the same way!

When we talk about **nanotechnology**, we need to think in **nanometres** (one billionth of a metre, or $\frac{1}{1000000000}$ m). Although our thinking about nanotechnology needs to be small, the implications and potential applications of nanotechnology are enormous. In fact they are so fantastically enormous, that it is very hard to imagine what they all will be.

Nanotechnology has already enabled us to develop super-smart and super-strong materials and medicines, but what new technologies are yet to come from this technology? Will you be one of the scientists who will contribute to the creation, development and application of technologies that are currently beyond our wildest dreams?

9.4.2 Nanobots to the rescue

An artist's impression of what a medical nanobot may look like

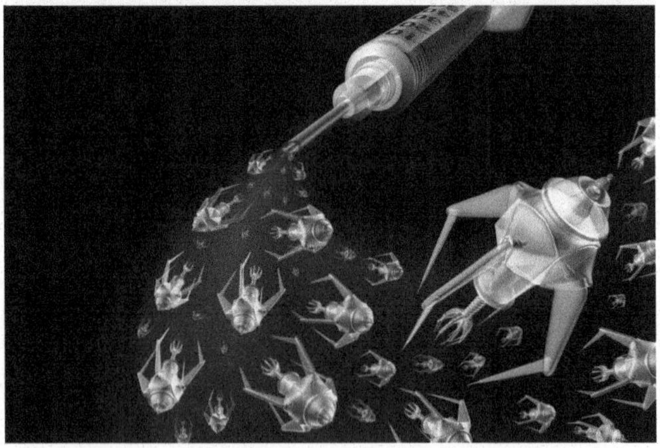

Nanotechnology is infecting many other technologies. There are many amazing applications of this technology when it is combined with others. One exciting area is the melding together of nanotechnology, information technology and biotechnology.

Nanotechnology enables us to create and use materials and devices that work at the level of molecules and atoms. Imagine minuscule machinery that could be injected to perform surgery on your cells — from the inside! These nanomachines, or nanobots, could be programmed to seek out and destroy invaders such as bacteria, protozoans or even your own cancerous cells. Heart attacks or strokes caused by blockages in your arteries might also become a thing of the past. These nanobots may be able to cruise through your bloodstream to clear plaque from your artery walls before it has a chance to build up. Could these nanobots also be programmed to repair our telomeres and prevent ageing of our cells, act as antioxidants destroying dangerous free radicals, repair DNA mutations — even stop us from growing old? Does nanotechnology hold the secrets of our immortality?

9.4.3 Nano-spiders

Let's hope that you don't have a fear of spiders, because one day they might be crawling through your body to deliver a drug directly to specific cells or kill cancerous cells. Scientists have created microscopic robots that look like spiders and are about 100 000 times smaller than the diameter of a human hair. These spider-bots can walk, turn and even create products of their own. Their body is made out of the protein streptavidin; attached to it are three legs of DNA and a fourth leg acting as an anchoring strand.

Remote-controlled power pistons

Scientists have built nanoscopic DNA pyramids that respond to different chemicals by changing their shape. They suggest that these structures could be used as motors for nanoscale robots.

Nanotechnology probes treating red blood cells

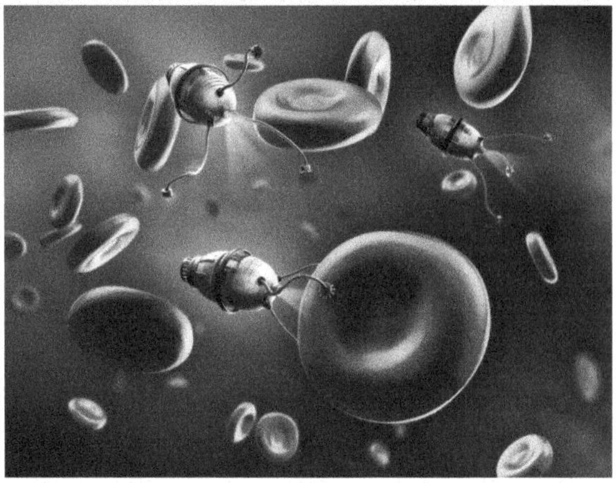

9.4.4 Nanocells

Scientists have already designed artificial cells. One of these is an artificial red blood cell. These tiny machines carry stores of oxygen and carbon dioxide with sensors to detect levels of these gases. When levels of oxygen are low they release oxygen, and when carbon dioxide levels are high they absorb carbon dioxide. These artificial cells are around 200 times more efficient than our current red blood cells; this may allow us to swim underwater or sprint for 15 minutes without needing to take a breath. If they were available and you could purchase them, would you use them?

9.4.5 Death-delivering dendrimers

Dendrimers floating with cells

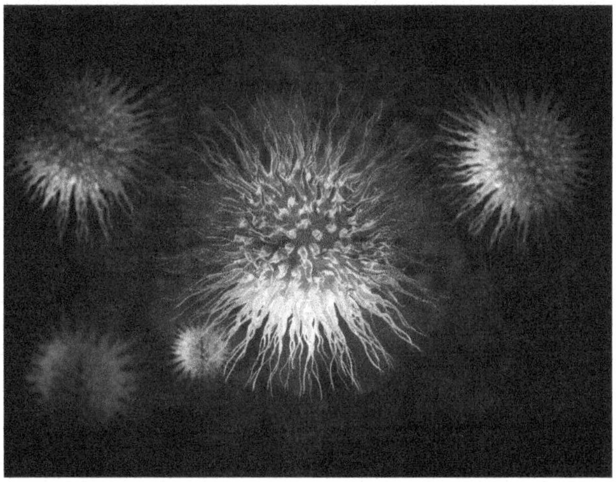

Nanoparticles around the size of 0.1–100 nm — small enough to pass through our cell membranes — are being developed to deliver drugs directly to cancer cells. The basic structure of such nanoparticles is called a **dendrimer**. Scientists have attached folic acid, methotrexate and a fluorescent dye to dendrimers. Folic acid is essential in cell division, and as cancer cells are actively dividing, they have a high demand for it. The folic acid in the nanoparticle acts as bait to attract the cancer cells. Methotrexate is a drug that kills cancer cells. When the cancer cells accept the nanoparticle, the methotrexate poisons the cell, killing it. The fluorescent dye allows the process to be monitored. The size of these dendrimers allows them to be filtered out of the blood by the kidneys and eventually excreted in urine.

9.4.6 Nanomusic

Do you have musical genes? Can you hear your own personal symphony of life? Researchers at Project Evolution have converted the language of DNA and proteins into music. The pattern in the music of Huntington's disease, a triplet repeat disorder, shows up as a repeated musical theme. The tones and rhythms hint at the code behind the code. For example, hydrophilic amino acids have a lower note than hydrophobic amino acids.

9.4.7 Nanoscaffolds

Nanoscaffolds could be implanted into different parts of the body to encourage the regrowth of damaged tissue. An example use could be to encourage nerve tissue to regrow optic nerves. Your optic nerve connects your eye to your brain. It can be severed by a traumatic injury (such as in a car crash) or damaged by glaucoma causing excessive pressure in your eyeball. These traumas can lead to vision loss. By using nanoscaffolds like a garden trellis, the growth of axons of optic nerve cells could be encouraged so that the communication gap can be bridged.

9.4.8 Nanofactories

What will future production factories look like? Imagine millions of tiny robots working together on an invisible, submicroscopic production line. Not only could they assemble almost anything and do it atom by atom, but they could also be programmed to make more of themselves! If you have watched *Star Trek*, maybe this is the technology that they use to generate their clothing and food supplies. Is this yet another example of fiction becoming fact?

9.4.9 Smart stuff

How about wearing clothing that literally reflects your mood? Or your wallpaper or lighting changing colour with mood changes or programming? Intelligent fibres, interactive textiles and smart fabrics that have been created with nanotechnology may change colour in a flash.

What if 'one size fits all' really was the case? Imagine shoes that instantly changed shape to mould perfectly around your feet and clothing that changed texture or shape to match your environmental conditions.

Imagine the effects of paint that has musical nanomachines mixed into it. After putting this paint onto your walls or furniture, you could program it to produce music, tones or vibrations to suit your moods.

9.4.10 Tiny but tough

Imagine a car, train or plane made of diamond and just as strong as current models but 50 times lighter. Nanotechnology may be used to rearrange carbon atoms into an inexpensive workable diamond material. This may lead to the production of a car so light that you could pick it up and carry it!

9.4.11 Nanotubes come on strong

It's not science fiction anymore! Scientists at the University of Texas at Dallas have developed a way to make carbon nanotube 'muscles'. These ribbons of tangled carbon nanotubes can act like artificial fibres in a robot, expanding and contracting to create movement. Not only are they stretchy along their width, they are extremely stiff and strong along their length. Their ability to maintain these properties in temperatures ranging from about 196 °C to more than 1500 °C would enable robots with these nanotube muscles to function in extreme conditions.

Carbon nanotube muscles

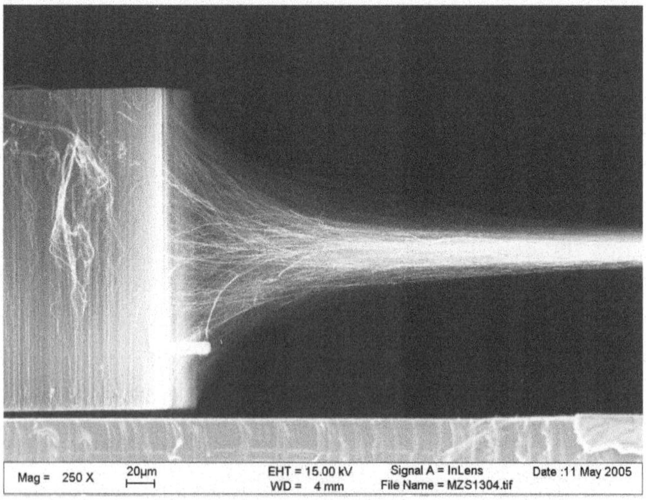

9.4.12 Nanowires

In the movie *Avatar*, the indigenous people of the planet Pandora are connected into a network that links all elements of the biosphere. They are in tune with all other life forms on the planet, from phosphorescent plants to pterodactyl-like birds.

On Earth we have a parallel interconnected ecosystem similar to that in the movie. Some researchers suggest that sulfur-eating bacteria living in the muddy sediments of the sea floor are connected by a network of microbial nanowires. These scientists suggest that these fine

protein filaments are involved in shuttling electrons back and forth, allowing these communities of bacteria to function as one super-organism. Lars Peter Nielsen of Aarhus University in Denmark and his team have discovered evidence that may support this controversial theory that he calls electrical symbiosis.

Could this idea inspire the development of another type of communication technology? Will we be connected to each other and possibly other life forms by implanted nanoparticles or nanobots? Could we also be connected to the **abiotic factors** in our environment, being sensitive to their needs and changes? Would such an interconnected ecosystem where we are all in tune with each other and our biosphere help us look after our planet better? Will it save us from extinction?

Will we be able to continue to control nanotechnology, or will it control us? Who is in whose hands?

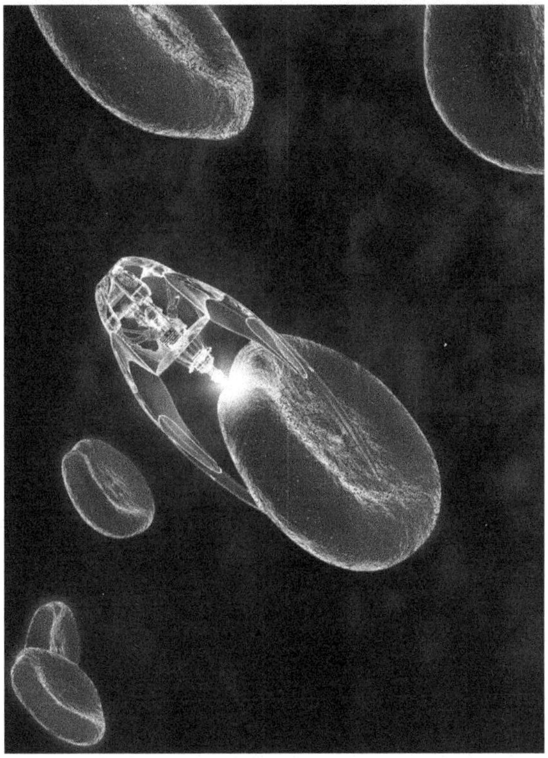

Tiny size, enormous responsibility

Will nanotechnology be our technological saviour or our exterminator? At the level of atoms and molecules, many of the laws that we accept and use do not necessarily apply. Scientists are still trying to figure out what these laws are, and their implications for the types of technologies that could be — and have been — developed. Have we opened a 'Pandora's box' that will lead to a future of unexpected disasters, or one full of great wonders?

Who is regulating the research, development and application of nanotechnology? Who has ownership of the products of technology and responsibility for any unforseen dangerous consequences? Are there ethical implications in the types of nanotechnologies that should be allowed? Who decides? Who are the guinea pigs for many of these new nanotechnology products that are largely untested over the long term? What are the implications on future generations of our use of these nanotechnologies?

9.4 Exercises: Understanding and inquiring

To answer questions online and to receive **immediate feedback** and **sample responses** for every question, go to your learnON title at www.jacplus.com.au. *Note:* Question numbers may vary slightly.

Remember

1. State how many nanometres are in one metre.
2. Suggest why it is difficult to know what all of the potential applications of nanotechnology will be.
3. Identify examples of other technologies with which nanotechnology is being combined.
4. At what level does nanotechnology allow us to work?
5. Describe possible medical applications of nanomachines or nanobots.
6. (a) Compared with the diameter of a human hair, how big are the nano-spiders developed by scientists to destroy cancerous cells?
   (b) Identify the organic components of these nano-spiders.
7. Suggest a use for nanoscopic DNA pyramids that can be triggered to change shape by responding to different chemicals.
8. Provide an example of an artificially designed cell and how it could be used.

9. Describe how dendrimers can be used to kill cancerous cells.
10. Describe a potential application of:
    (a) nanoscaffolding
    (b) nanowires
    (c) nanotubes.
11. Suggest why research, development and potential applications of nanotechnology may need to be regulated.
12. Suggest why the pattern in the 'music' of Huntington's disease shows up as a repeated musical theme.
13. Unscramble the following nano terms.
    (a) thogaynecnolon
    (b) matoneresn
    (c) renimderd
    (d) wonairen

Investigate, think and create

14. Research the technology used to make the spider-bots move. Construct your own robot that can move. Your model may be built out of motorised Lego (or similar) or shown in animation, slowmation or any other multimedia format.
15. Nano-spiders have drawn huge interest because they are able to sense their environment and react to it. Research online to find out how this is achieved. Create your own picture book, animation or documentary of this process.
16. (a) Many science fiction stories and shows include ideas based on nanotechnology. Examples include the Borg in *Star Trek: The Next Generation* and the Replicators in *Stargate: SG-1*. Watch a science fiction TV show and evaluate the plausibility of the science within it.
    (b) Suggest your own story for another episode in one of these shows.
    (c) Use puppets or multimedia to share your story with others.
17. (a) Research a possible application of nanotechnology that interests you, including its implications. Possible topics include:
    - spacesuits
    - clothing
    - nanobots
    - nanomachinery
    - nanofactories
    - nanodevices
    - medicines, e.g. drug delivery, biosensors, bioresorbable materials
    - cosmetics, e.g. anti-ageing, skin care
    - artificial cells or body parts
    - aeroplanes or spacecraft
    - nanomaterials, e.g. quantum dots, dendrimers, nanotubes, fullerenes.

    (b) Share and discuss your findings with other members of your team. Construct a PMI chart to summarise your discussion.
    (c) Create a science fiction story that incorporates your findings. Share your story with the class.

Investigate, think and discuss

18. What sorts of research are being performed into the safety of using nanotechnology?
19. Who is regulating nanotechnology? Who is determining the safety of this technology? What are their criteria?
20. Find out more about the theory of electrical symbiosis. Do you think that such a process really exists? Justify your response.
21. If a form of electrical symbiosis or some other type of interconnectedness between individual life forms and the biosphere could exist on Earth, suggest implications of this for survival.
22. If dendrimers used to treat diseases are excreted in urine, what happens to them after excretion? Suggest ecological consequences that should be considered.
23. (a) Watch the movie *Avatar* and evaluate the accuracy or plausibility of the science within it.
    (b) Discuss ideas that the movie inspires for possible scientific research.

learnon RESOURCES — ONLINE ONLY

Explore more with this weblink: Gene2Music: Listen to the music of genes

Complete this digital doc: Worksheet 9.4: Killer-bot (doc-20054)

9.5 Gardening in the laboratory

Science as a human endeavour

9.5.1 Grow it back

Would you like to become a laboratory gardener? Want to grow some new cells, tissues, organs or organisms? Just follow the instructions, plant the seeds and watch them grow!

Want to grow back some missing body parts or create spares? Researchers are racing to create skin, cartilage, heart valves, breasts, ears and other body tissues in tissue-engineering laboratories. Some burns victims have had uninjured skin shaved off their bodies and grown in laboratories, while others have had pre-grown skin grafted onto their bodies.

One method used to grow new tissues involves the injection of synthetic proteins that induce tissue to grow and change. These proteins give messages, depending on the combination, to make more fat or bone. Using these methods, spare parts could be grown on-site, or grown elsewhere and transferred later. Some would need surgery for shaping, others would grow using scaffolding (such as freeze-dried joints or cartilage) to give shape.

Imagine a future in which injection of growth proteins into areas where a person is missing a body part would enable bone, joints, fat, tissue, nerves and blood vessels to grow. Imagine being able to grow an ear, joint, nose or finger for immediate use or as a spare!

Reconstructed skin tissue

Cardiac researcher Doris Taylor has revived the dead. The process involved rinsing rat hearts with a detergent solution to strip the cells, until all that remained was a protein skeleton of translucent tissue — a 'ghost heart'. She then injected this scaffold with fresh heart cells from newborn rats and waited. Four days later, she saw little areas beginning to beat. After eight days, the whole heart was beating. Could this research lead to new transplant technologies that could be used in humans?

Gabor Forgacs, a tissue engineer at University of Missouri, is making blood vessel networks by using a 3D printer to print them out. He cultures three types of blood vessel cells and then loads them into a fridge-sized bioprinter that has been programmed with the pattern of the vessels to be printed.

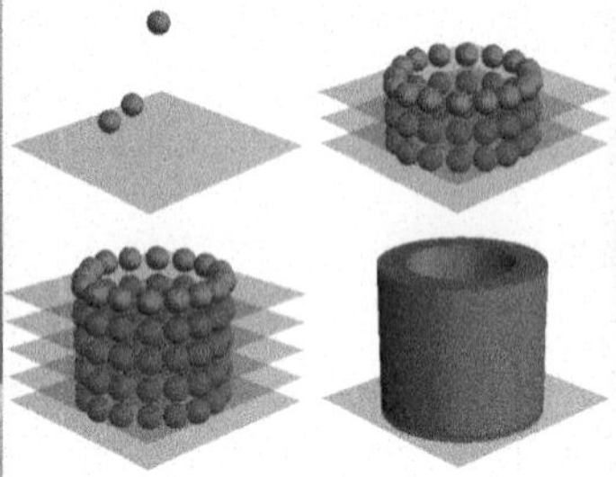

Bearing a resemblance to a jelly baby, a 'living doll' has been created from liver cancer cells. These cells are held in place by 100 000 capsules of collagen. Shoji Takeuchi's team built this figure so that drugs could be tested in conditions closer to those inside the body rather than in a dish.

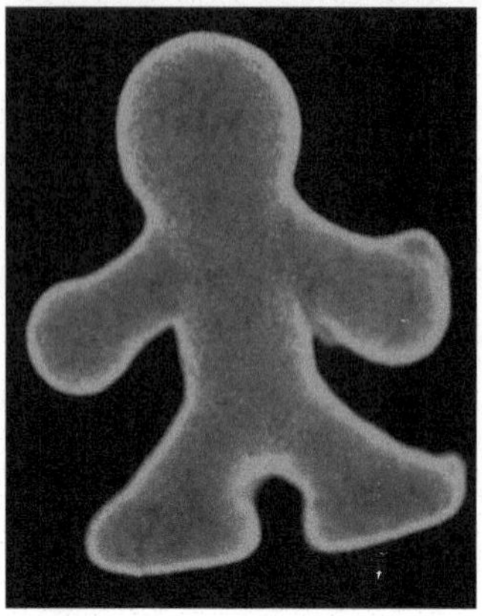

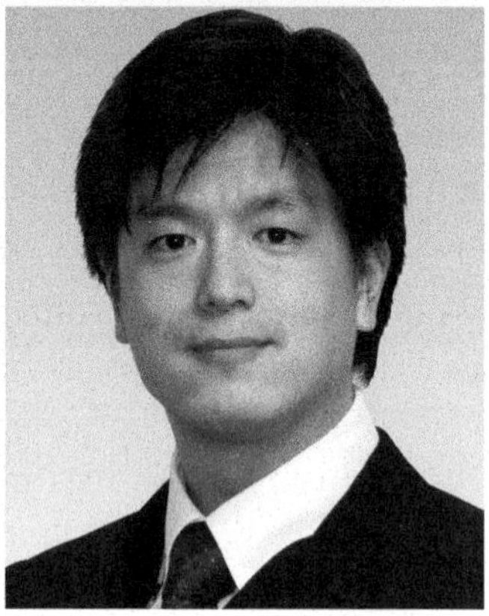

9.5.2 Mix 'n' match

Transgenic organisms result from combining the genetic information of two different organisms. Genetic information from Arctic fish has been added to tomatoes to make them frost resistant, and genetic information from a bacterium has been added to cotton and potatoes to give them resistance against certain insect pests. Do these altered organisms belong to the same species as those that are not altered? What other changes may result from these new DNA additions?

Keith McLean is a CSIRO scientist whose work involves the development of biomedical materials to replace, repair and regenerate damaged body parts. His current research is focused on novel biomedical adhesives, ophthalmic biomaterials, bioactive scaffolds for cell therapy and platforms for the propagation of stem cells. The main image here shows mouse fibroblasts (cells of connective tissue).

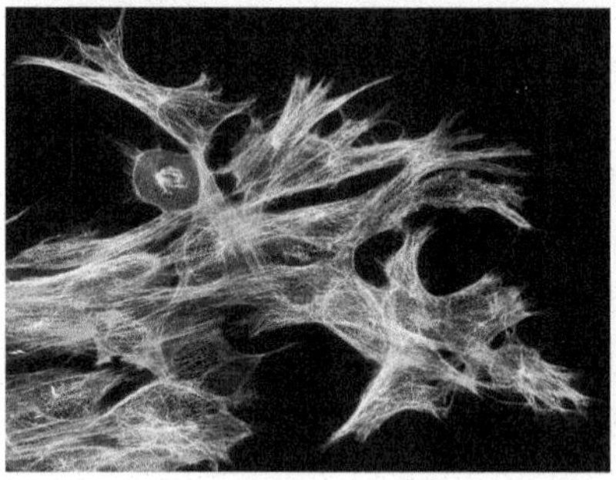

Recently, the Genetic Manipulation Advisory Committee (GMAC) approved the release of more than 500 transgenic tomato plants into Queensland and Victoria. A gene has been added to these tomatoes so that caterpillars that would normally eat the tomatoes are killed if they eat the plants. It is hoped that this addition will reduce the amount of pesticide that needs to be used.

CSIRO scientists have modified potatoes by switching off the gene that causes cut or bruised potatoes to go brown. They have achieved this by copying the gene, reversing it and then replacing it in the potato plant. This technology could also be applied to other fruit and vegetables.

9.5.3 Cloning around

Genetic engineering techniques enable the DNA in organisms to be altered to produce proteins and medicines that humans need. These techniques can also produce organs with minimal possibility of rejection for transplants. Cows have been genetically altered to produce more milk or leaner meat, and bacteria have had human genetic information added to them so that they produce insulin for diabetics and blood clotting factors for haemophiliacs. Cloning can then be used to produce these altered organisms in large numbers.

Killer tomatoes? Scientists are developing genetically modified tomatoes that contain edible vaccines. Diseases that are currently being focused on are Alzheimer's, cholera and hepatitis B. The availability of affordable vaccines for these diseases that can be easily grown and processed in countries where they are most needed will save many lives.

A healthy glow? Ruppy is the world's first transgenic dog. She and four other cloned beagles have a gene that makes them glow under UV light. Dog fibroblast cells were infected with a virus that inserted the fluorescence gene into their nuclei. These nuclei were transferred into egg cells that had been emptied of their original nuclei; they were then implanted into surrogate mothers. In the future, the same technique could be used to clone dogs with genes for human diseases or knock certain genes out.

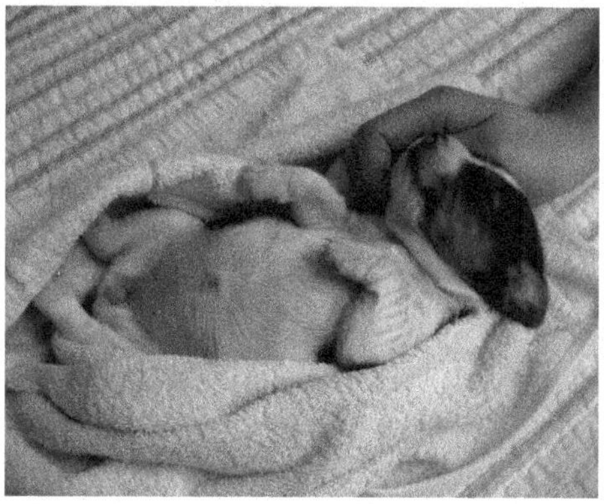

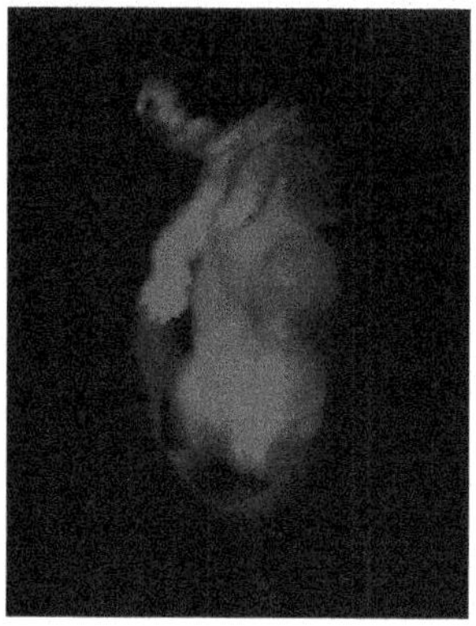

For all of its advantages, cloning may be considered a form of **asexual reproduction** because it does not involve the fusion of two cells. Although it may enable the production of large numbers of identical off-spring, their similarities may lead to a decrease in biodiversity and reduce their genetic survival if they are exposed to unfavourable circumstances. The effect of the environment on the cloned individuals also needs to be considered, as genetic inheritance is not the sole factor in determining the phenotype of an organism.

9.5.4 Stem wars

What's the stem of the trouble? What are stem cells and why are people arguing about them?

Stem cells are important because they are so versatile. They have the ability to differentiate into many different and specialised cell types. They may differentiate into blood cells, bone cells, heart cells, liver cells, nerve cells or skin cells. This ability makes them invaluable in the treatment and possible cure of a variety of diseases where they may replace faulty, diseased or dead cells.

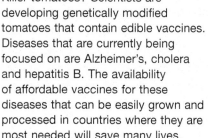

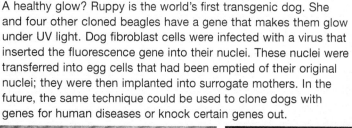

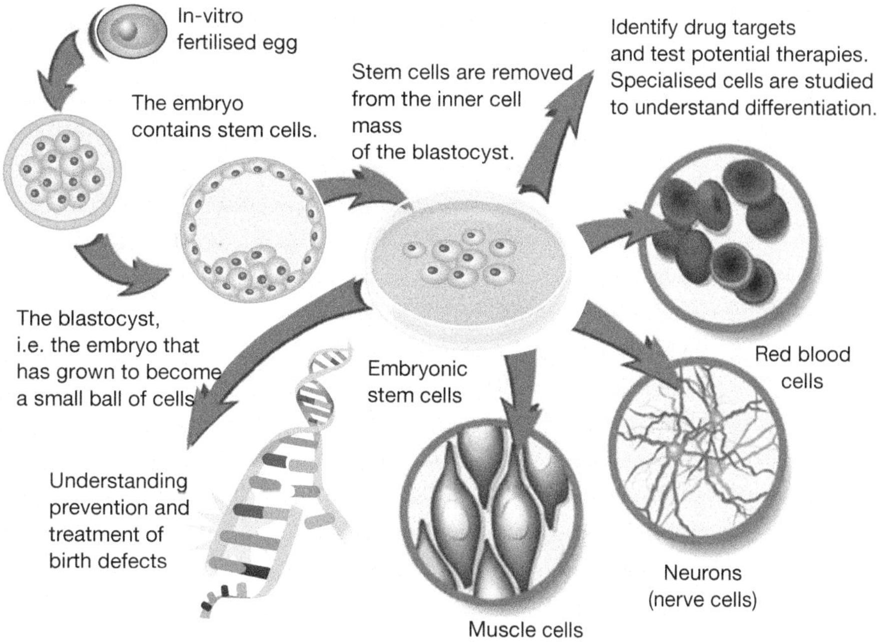

Stem cells can be divided into categories based on their ability to produce different cell types.

- **Totipotent stem cells** are the most powerful as they can give rise to all cell types.
- **Pluripotent stem cells** can give rise to most cell types (e.g. blood cells, skin cells and liver cells).
- **Multipotent stem cells** can give rise to only certain cell types (e.g. various types of blood or skin cells).

HOW ABOUT THAT!

In 1998, it was reported that a researcher from the University of Wisconsin had found a way to isolate cells from the inner mass of an early human embryo and develop the first embryonic stem cell lines. The stem cell issue had entered the public arena.

9.5.5 Stem cell sources

Stem cells can be described as being **embryonic stem cells** or **somatic stem cells**. Embryonic stem cells are pluripotent and can be obtained from the inner cell mass of a blastocyst (the mass of cells formed at an early stage of an embryo's development). Somatic stem cells are multipotent and can be obtained from bone marrow, skin and umbilical cord blood.

9.5.6 What's the problem?

The source of embryonic stem cells raises many ethical issues. Embryonic stem cells can be taken from spare human embryos that are left over from fertility treatments or from embryos that have been cloned in the laboratory. Some say that this artificial creation of an embryo solely for the purpose of obtaining stem cells is unethical. There has also been concern about the fate of the embryo. In the process of obtaining stem cells, the embryo is destroyed.

Some parents have decided to have another child for the sole purpose of being able to provide a diseased or ill child with stem cells. In this case, the blood from the umbilical cord or placenta is used as a source of stem cells. Some suggest that this is not the right reason to have a child and that children should not be considered as a source of spare parts for their siblings.

9.5.7 Endogenous stem cells

Most research so far has been on creating stem cells from embryos or adult tissues in labs and manipulating their development using chemical growth factors and implanting them where needed. But there could be another way: awakening our bodies' own endogenous stem cells to achieve natural regeneration. Imagine being able to regrow entire lost limbs, as some amphibians can!

Research in human stem cells faces a number of challenges. There is much about these cells that we do not fully understand, such as the mechanisms that drive cell specialisation and the interaction between host and transplanted cells; the long-term stability of transplanted cells also needs to be established.

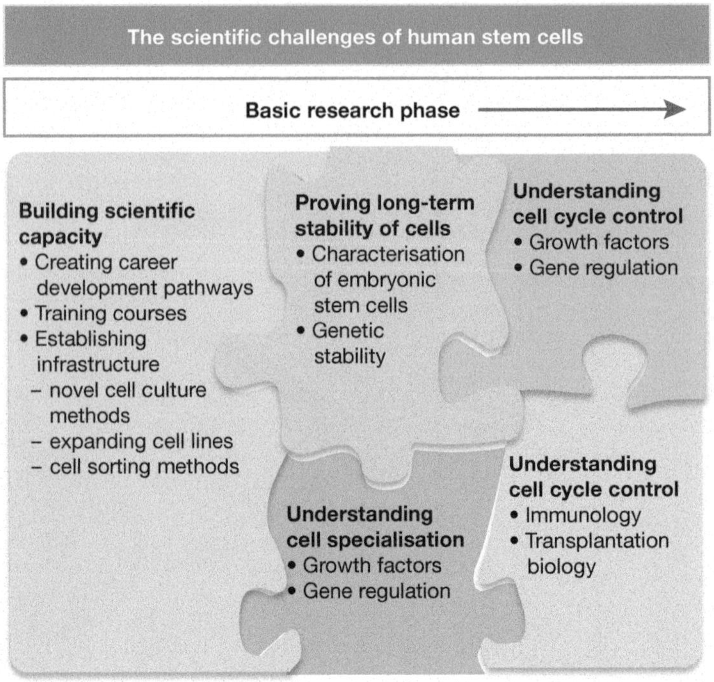

HOW ABOUT THAT!

Scientists at CSIRO are studying Australian frogs and their ability to produce a sticky glue-like substance. These frogs, from the genus, *Notaden*, produce this substance as a protective measure against predators. The CSIRO scientists want to mimic the design of this glue so that they can produce a medical adhesive that could be used to repair damaged cartilage, close up wounds and bond tendons and bone. Could their research also provide us with a substance to help us stick in our replacement spare parts?

9.5 Exercises: Understanding and inquiring

To answer questions online and to receive **immediate feedback** and **sample responses** for every question, go to your learnON title at www.jacplus.com.au. *Note:* Question numbers may vary slightly.

Think

1. If cloning takes over as the main form of human reproduction, sperm and eggs would no longer be essential.
   (a) Suggest advantages and disadvantages of this. Give reasons for your suggestions.
   (b) If 100 clones were made of a single individual, would they all look and act the same? Explain your answer.
   (c) Suggest the evolutionary consequence that this may have on the human race.
2. Should human DNA be inserted into the DNA of other organisms? Give reasons for your opinion.
3. Suggest improvements to the human design. List both advantages and disadvantages of these improvements.
4. Would you eat genetically modified food? Find out published arguments for and against this issue. Once researched, decide on your viewpoint and present your arguments in a class debate on the issue.

Imagine

5. Design a futuristic human. Give reasons for your changes to the original human design and present your information in a poster.
6. Imagine you are living in the future and your partner has asked you to have his or her children. You think that this is wonderful until you realise that he or she means clones. What would you say?
7. Imagine that you have just woken up in a laboratory and have been told that you are a clone. Write a story about your life.

Investigate, share and discuss

8. Find out about the relationship between cloning and biotechnology.
9. For one of the listed transgenic organisms, find out the following information and present your findings in a report and as a class presentation: tomatoes, potatoes, pigs, cows, fish, tobacco, cotton, bacteria.
   (a) How and why they were made
   (b) Any biological, social, economic or ethical issues that may result from the change to the organism
   (c) Your own comments on what you think and feel about these changes
10. Debate issues related to:
    (a) the development of transgenic organisms
    (b) the cloning of human tissues and organs.
11. Find out the requirements for space travel. Suggest how a human could be genetically engineered to be well suited to travelling in space.
12. Investigate some of the following questions.
    (a) Which inherited genetic diseases are potentially treatable with stem cells?
    (b) How many different kinds of adult stem cells exist and in which tissues can they be found?
    (c) Why have the adult stem cells remained undifferentiated?
    (d) What are the factors that stimulate adult stem cells to move to sites of injury or damage?

13. In your team, discuss the following questions to suggest a variety of perspectives.
    (a) Is it morally acceptable to produce and/or use living human embryos to obtain stem cells?
    (b) Each stem cell line comes from a single embryo. A single cell line allows hundreds of researchers to work on stem cells. Suggest and discuss the advantages and disadvantages of this.
    (c) If the use of human multipotent stem cells provides the ability to heal humans without having to kill another, how can this technology be bad?
    (d) Parents of a child with a genetic disease plan a sibling whose cells can be used to help the diseased child. Is it wrong for them to have another child for this reason?
14. Find out how stem cell research is regulated in Australia and one other country. What are the similarities and differences of the regulations? Discuss the implications of this with your team-mates.
15. Research aspects of stem cell research and put together an argument for or against the research and its applications. Find a class member with the opposing view and present your key points to each other. Ask questions to probe any statements that you do not understand or would like to clarify. Construct a PMI chart to summarise your discussion.
16. Snuppy, the world's first cloned dog, was created by Woo Suk Hwang's team at Seoul University in South Korea.
    (a) Find out more about Hwang's scientific achievements.
    (b) Find out more about Hwang's research and a reason for his arrest and jail sentence. Were his actions justified? Justify your response.

9.6 Space trekking

Science as a human endeavour

9.6.1 Space trekking

Mars Rover

Will humans one day venture out into space and colonise other planets? What new technologies will we need, not just to get there, but to be able to survive and thrive?

H. G. Wells (1866–1946) was an extraordinarily powerful and imaginative storyteller. *The Time Machine* (1895), *The Island of Dr Moreau* (1896), *The Invisible Man* (1897), *The War of the Worlds* (1898) and *The First Men in the Moon* (1901) are some of his ingeniously creative stories. H. G. Wells had his own ideas on what life on Mars would look like and he incorporated these into his writings.

No life has yet been discovered on Mars. This does not mean that it does not or has not existed. In 1997, new technologies enabled a data-collecting mission to begin to send large amounts of information back to Earth as to what Mars is really like — and if creatures similar to those from H. G. Wells's imagination really do glide across its surface!

9.6.2 The red planet

Throughout history, sky watchers have observed the bright red dot in the sky. Some believed that it carried war or pestilence or, for some cultures, even the need for human sacrifice. Perhaps as a consequence of this, the Romans named Mars after their god of war. Other observers theorised that Mars held intelligent life, which had created canals to channel water to cities. H. G. Wells's *The War of the Worlds*, in which Martians attack the Earth, and Tim Burton's 1996 comedy *Mars Attacks* are two stories that use imagination and the science of their times.

Almost 30 Earth missions have failed to get to Mars, the next planet beyond Earth. In 1965, *Mariner 4* made the first successful attempt when it flew within 10 000 kilometres of Mars, transmitted back photographs of Martian meteor craters and revealed that the planet did not have a measurable magnetic field. About seven years later, the first soft landing was made by the Soviet probe *Mars 3*. This landing allowed the first television pictures to be sent from the surface of the red planet.

Spectacular 3-D colour images of the Martian surface were taken by the European Space Agency's probe *Mars Express* in 2004. These images showed gullies that appear to have been carved by water. Also in 2004, NASA landed two rovers, *Spirit* and *Opportunity*, on the Martian surface with the task of examining the chemical composition of rocks. In 2014, *Opportunity* was still exploring the frigid wasteland of Mars. Already it has survived almost 30 times longer and driven 50 times further than it was designed for. But the rovers may be in for some stiff competition in the future — space scientists are increasingly considering using balloons to explore planets, especially those with hostile environments like Mars.

9.6.3 Move over, red rover

If we ever make it to Mars ourselves, we will need to come up with new strategies and technologies to survive the Martian environment. We will need to be able to leave our spacecraft and land vehicles. Specially designed spacesuits will need to provide us with a personal environment that supplies our nutrients, recycles our wastes, protects us from the outside and enables us to do what we want or need to do.

Mars exploration

Australian aerospace engineers are involved in the development of futuristic Martian spacesuits such as MarsSkin and mechanical counter-pressure (MCP) suits. MarsSkin suits are tight and elastic, contain an inner layer of lycra microfibres that transports heat and moisture away from the body and include sensors to monitor hydration, heart rate and body temperature. Their tough outer layer provides protection against harsh environments. MCP suits have electronic polymer fibres within tight-fitting elastic material and use electricity to mould the suit to the human body. The extra pressure this suit provides reduces the chance of the wearer losing consciousness in the much lower Martian atmospheric pressure.

9.6.4 Red planet greenhouses

In 2010, the United States President Barack Obama announced his intention to send humans to Mars by the mid 2030s. Scientists have already begun experiments to simulate Martian conditions on Earth. Their experiments are investigating the possible use of micro-organisms to convert Martian rocks into soil, generate oxygen, purify oxygen and recycle waste. These microbial colonies will be our first gardens on Mars.

9.6.5 The future challenge

If we travel to another planet, what other challenges do we need to plan for? Should we be altering our DNA by introducing genes that may give us an increased chance of survival in the different environmental conditions that we are likely to encounter on other planets such as Mars? Should we be genetically engineering specific organisms to take with us? If so, what features should they possess? Maybe we should be cloning ourselves or making replacement parts just in case something goes wrong. Should we be developing new technologies that will help keep us alive in environments that we have not evolved to exist in?

Stocking up on replacements

Maybe we should be cloning ourselves so that we have spare body parts. If the journey takes a number of generations, then we can still be there at the end. Or maybe we should develop a range of bionic replacement parts and just insert them when the old ones wear out.

Our survival may depend on the applications of old and new technologies to supply requirements essential for life. Some of these technologies may involve the conversion of human wastes into essential nutrients.

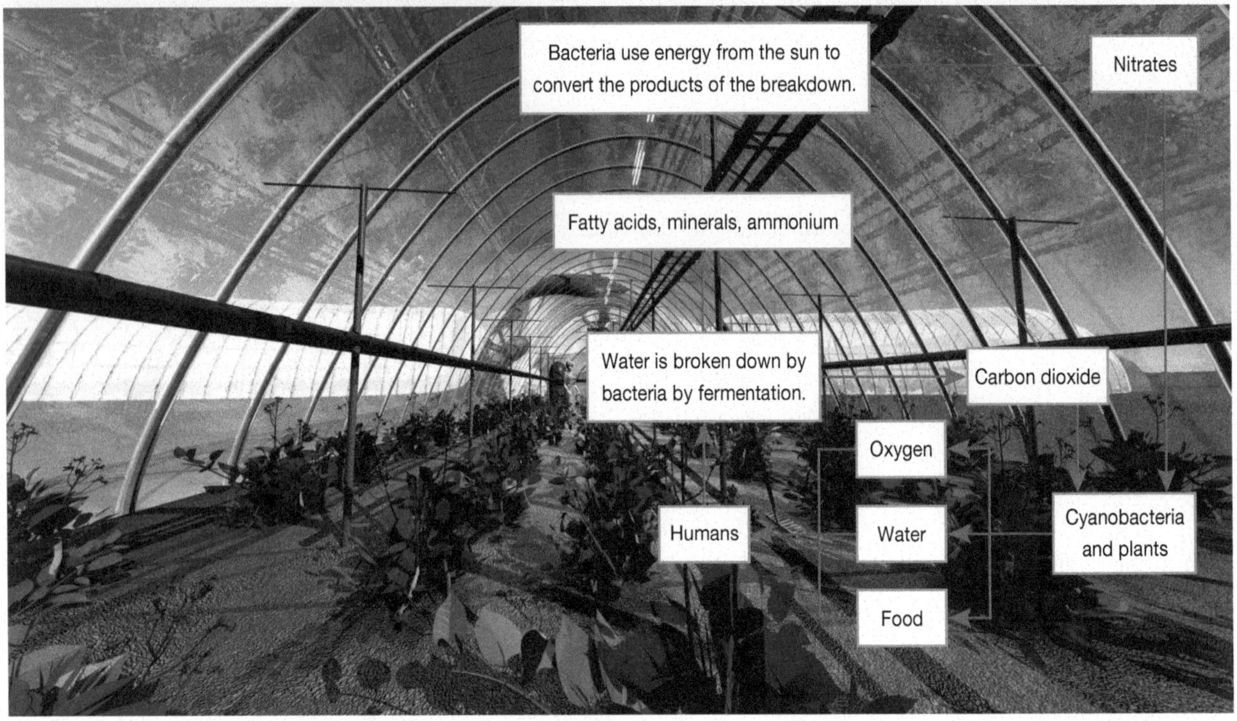

Can we fight alien diseases?

Should we develop our applications of nanotechnology to defend us against alien disease? If we haven't evolved with these alien pathogens, will they be able to infect us? What will their biology be?

If they do invade us, perhaps we could use nanobots or artificial defences. There have already been designs of these types of defence systems put forward; should we develop these ideas further?

Plastic antibodies

Antibodies made entirely from plastic have already been used to save the lives of mice injected with bee venom. These artificial antibodies contain cavities moulded to match the shapes of their target molecules. In the future, these could be used in humans to combat toxins or even to protect against proteins that cause allergic reactions to pollen and some foods.

These plastic antibodies were made by a process called **molecular imprinting** — a process similar to making a plaster cast of your hand but at a nanoscale. It involves taking a plastic cast of the target molecule (such as a toxin) by mixing it with monomers (small molecules) that mould themselves around it. When the original target molecule is dissolved, these casts have the specific shape for trapping the target molecule.

It is anticipated that after these plastic antibodies have been injected into the body and have captured their target molecule (the venom or toxin), they would be engulfed by white blood cells and removed from the bloodstream to be destroyed by the liver. If these artificial antibodies are made of biodegradable materials, then their destruction will be less risky.

Nanobots to the rescue

Will nanotechnology take us to the stars? If we come across invasive alien life forms, just their physical intrusion into our bodies may cause us harm. We will not have evolved strategies to defend ourselves. This is where nanobots may come to our rescue.

Who are you and where are you going?

Do we need to develop a separate set of ethics to live by while we travel in space and when we get to our destination? If so, what should they be? Are you interested in going on a journey that will take you further from Earth than anyone has been before? If so, what do you have to contribute to this new world? Now is the time to begin dreaming and planning who you will become and where you want to go.

The immune system creates antibodies to fight off invading microbes or toxins. Plastic antibodies, developed by UC Irvine researchers, imitate this behaviour — in this case attacking toxins of bee venom in the bodies of mice.

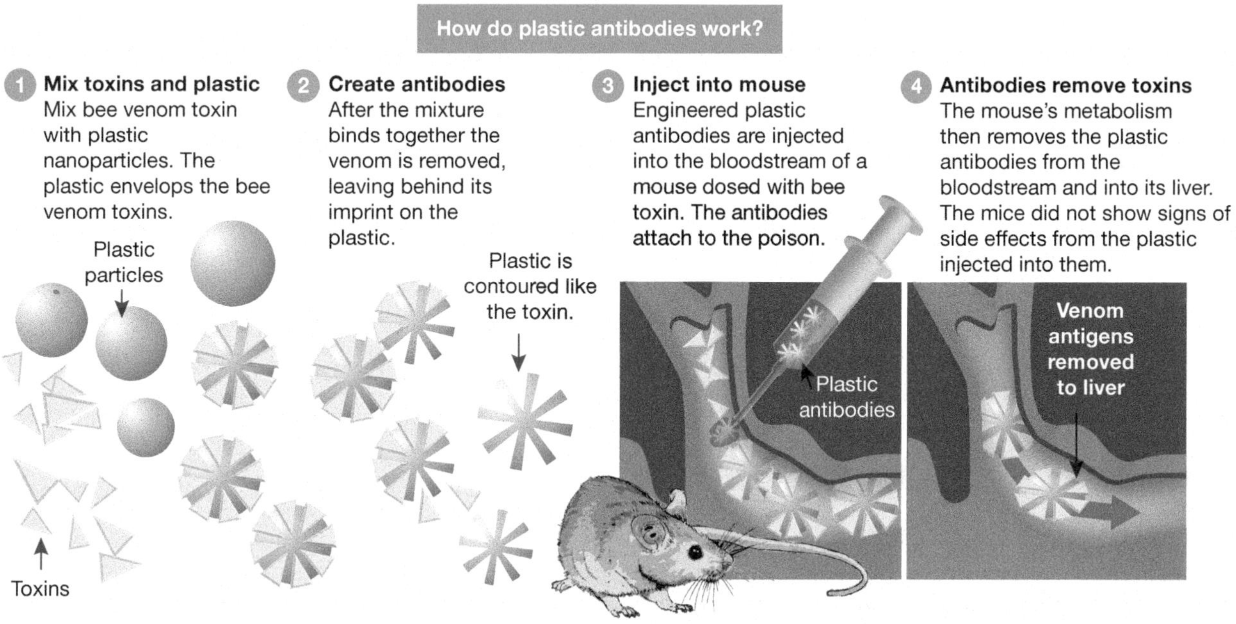

How do plastic antibodies work?
1 Mix toxins and plastic
Mix bee venom toxin with plastic nanoparticles. The plastic envelops the bee venom toxins.
Plastic particles
Toxins
2 Create antibodies
After the mixture binds together the venom is removed, leaving behind its imprint on the plastic.
Plastic is contoured like the toxin.
3 Inject into mouse
Engineered plastic antibodies are injected into the bloodstream of a mouse dosed with bee toxin. The antibodies attach to the poison.
Plastic antibodies
4 Antibodies remove toxins
The mouse's metabolism then removes the plastic antibodies from the bloodstream and into its liver. The mice did not show signs of side effects from the plastic injected into them.
Venom antigens removed to liver

What might the future look like seen through bionic eyes? *Source*: Image courtesy of Bionic Vision Australia.

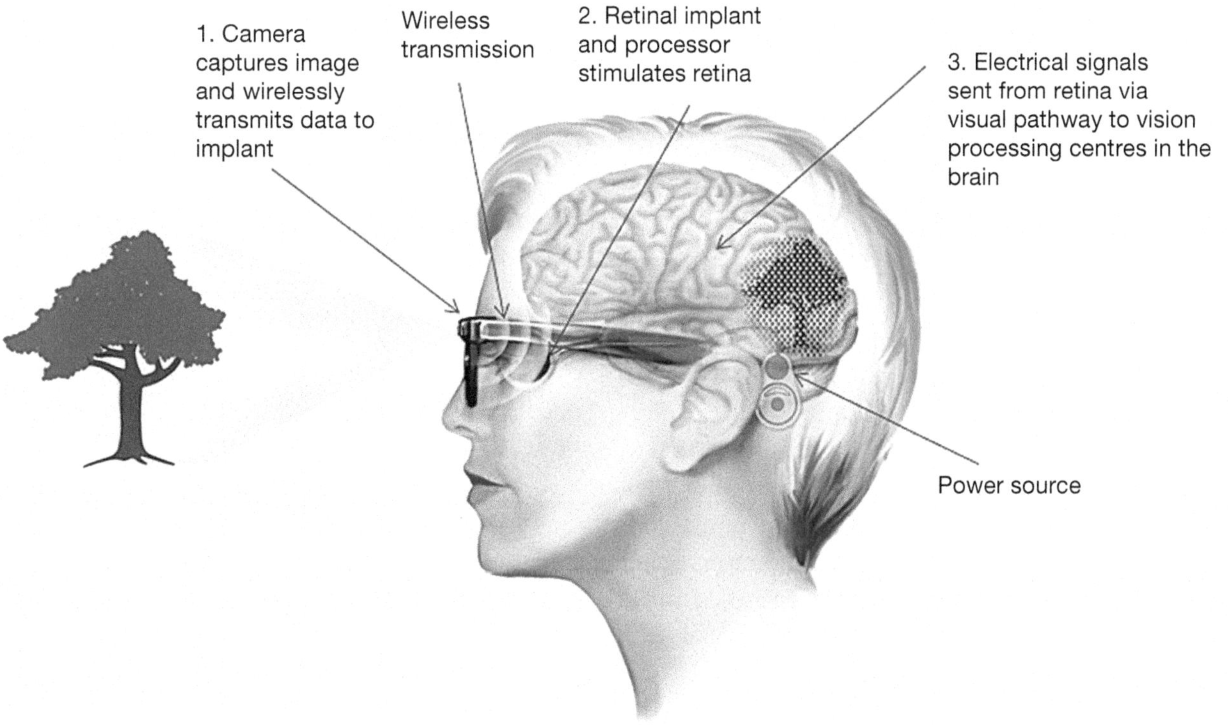

9.6 Exercises: Understanding and inquiring

To answer questions online and to receive **immediate feedback** and **sample responses** for every question, go to your learnON title at www.jacplus.com.au. *Note:* Question numbers may vary slightly.

Investigate, imagine, discuss and create

1. Locate, read and summarise the similarities and differences between science fiction novels about life on Mars or Martians. Try reading Brian Aldiss's *The Forgotten Life*, Ray Bradbury's *Martian Chronicles*, Arthur C. Clarke's *The Snows of Mount Olympus* and *Sands of Mars,* and H. G. Wells's *The War of the Worlds*.
2. What sorts of projects and research are aerospace engineers involved in?
3. (a) Research the differences between current conventional spacesuits and possible future spacesuits.
   (b) Research the possible effects of the Martian atmosphere on the human body.
   (c) Suggest spacesuit modifications on the basis of your findings. What sorts of science technologies need to be developed to allow this?
   (d) Use your findings to design, sketch and label your future Martian spacesuit.
4. Find out about organisations that are focused on research about travel to and colonisation of Mars. Report your findings in a *Mars Mania* field guide brochure, PowerPoint presentation, web page or creative story.
5. Research and creatively report on:
   (a) the NASA Haughton–Mars Project (HMP)
   (b) the MELiSSA loop life-support system
   (c) NASA research on the feasibility of living in space
   (d) research and development of artificial life.
6. (a) If you found a self-replicating organism living within your computer, what would you do? Discuss and justify your response.
   (b) What defines life? Explain.
   (c) If you were to travel to another planet, what protocols would you have with regards to how you were to treat any life you encountered on that planet?
7. Design your own nanobots that will help us to survive venturing out into space.
   (a) Research and report on examples of current research into humans and space travel.
   (b) Use the internet and your research to identify three questions that could be investigated further.
   (c) Write your own science fiction story that is packed with challenges, new technologies and excitement.
8. The first inhabited outpost on Mars will possibly depend on microbes for its essential functions. The MELiSSA loop life-support system may use micro-organisms to convert crew waste into resources that can be recycled. Find out more about this system and use the information to design your own system to support human life on Mars.

9.7 The quest continues

Science as a human endeavour

9.7.1 The quest continues

Why question? Why bother asking questions about the world around you? Why not just accept the way things are? What makes scientists do what they do? Do we really need science?

Science does not occur in a vacuum. Science is about people and their quest to find out about the hows, whys and wheres of the world around them. Science employs as its most important tools imagination, insight and the desire to understand or find out for the individual or for others. Some discoveries have been made accidentally, others after sequential and thorough use of scientific methods and procedures. The societies in which scientists live greatly influence the science in which they become involved.

9.7.2 Science quests throughout history

Year	Scientific events	Other events
1784	Benjamin Franklin invents bifocals.	Life expectancy is about 35.5 years.
1829	Stephenson's *Rocket* launches the start of the railway age.	Jean-Baptiste de Larmack dies and Jules Verne turns 1.
1831	Faraday discovers electromagnetic induction.	
1833	Gauss's electric telegraph key is invented. Anselme Payen isolates the first enzyme — diastase.	
1834	Lindsay achieves continuous electric light.	
1837	Photography is invented.	
1838	Matthias Schleiden suggests that all plants are made up of cells.	
1842	Ether anaesthesia is invented.	
1845	Parson's giant telescope begins a new era of astronomy.	
1846	Guncotton, the first modern explosive, is invented.	
1848		The Year of Revolutions: Second Republic in France
1852	Vulcanite (hard rubber) is created.	Third Empire begins under Louis Napoleon.
1853	The internal combustion gas engine is invented.	
1854	Chemically produced aluminium is created.	
1855	Ruhmkorff's bichromate battery is created.	
1857	Singer's domestic sewing machine is invented. The first electric street lighting is installed in Lyon.	
1858	Rudolf Virchow suggests that cells can only arise from pre-existing cells.	
1859	Darwin's *On the Origin of Species* is published.	Gregor Mendel turns 37 and Charles Darwin turns 50.
1863	Huxley's *Man's Place in Nature* is published and TNT is invented.	Jules Verne's *Five Weeks in a Balloon* is first published.
1864	Nobel introduces nitroglycerine and Pasteur introduces pasteurisation.	Jules Verne's *A Journey to the Centre of the Earth* is first published.
1865	Aerophore, a compressed-air diving apparatus, is invented by Rouquayrol and Denayrouze. Gregor Mendel discovers patterns of inherited characteristics.	Jules Verne's *From the Earth to the Moon* is first published.
1868	Leclanché invents a dry-cell non-rechargeable battery.	
1869	Celluloid is discovered. Dmitri Mendeleev proposes the periodic table.	Marie Curie turns 2 and H.G. Wells turns 3.
1870	Chewing gum and the washout toilet are introduced.	Jules Verne's *Twenty Thousand Leagues Under the Sea* is published.
1873	Remington introduces the mass-produced typewriter.	Jules Verne's *Around the World in 80 Days* is published.

(*continued*)

Year	Scientific events	Other events
1876–1878	Thomas Edison makes his first talking machine — the phonograph.	
1878	The cathode-ray tube (later the basis for television) is invented and Alexander Graham Bell invents the telephone.	
1879	Swan makes the first practical electric light bulbs in London and Edison makes them in the USA.	
1885	Daimler and Benz work on the first motorcar with an internal combustion engine.	
1887	Hertz discovers electromagnetic waves (the basis for radio).	
1889	Data-processing computer using punched cards is invented.	
1892	Ivanovsky discovers the virus. Diesel engine is invented.	
1893	The solar-electric cell is invented. The first open-heart surgery is performed.	
1894	Marconi's wireless telegraphy is invented.	Aldous Huxley is born.
1895	Röntgen discovers X-rays.	H. G. Wells's *The Time Machine* is published.
1896	Cavendish discovers electrons.	
1898	Krypton and neon are discovered. Holland's submarine is invented. Curie discovers the two radioactive elements radium and polonium.	H. G. Wells's *The War of the Worlds* is published. Albert Einstein turns 19, Alexander Fleming turns 17 and Howard Florey is born.
1899	Electric wave wireless telephone and the wire tape-recorder are invented. Guglielmo Marconi invents the 'wireless'.	H. G. Wells's *The First Men in the Moon* is published.The Boer War begins.
1900	Planck's quantum theory is proposed. Von Zeppelin's dirigible airship is invented.	Life expectancy is about 45 years.
1901	The first signals are sent across the Atlantic Ocean and received. The first electric hearing aid is invented.Wilhelm Röntgen's discovery of X-rays wins him one of the first Nobel Prizes.	Australia becomes a federation.
1902	The ionosphere is discovered by Kennelly-Heaviside and hormones are discovered by Bayliss and Starling.	
1903	The Wright brothers fly in their first successful 'heavier than air' machine.	
1905	Einstein's *Special Theory of Relativity* is published and the first artificial joint is used in arthritic patient's hip.	Jules Verne dies.
1911	Ernest Rutherford ('father of nuclear energy') proposes a model for the atom.	Aldous Huxley has his 17th birthday.
1914		World War I begins.
1927	George Lemaître theorises that the universe has been expanding from a 'primal atom'. His theory is later popularised as the 'big bang'.	
1928	Penicillin is accidentally discovered by Alexander Fleming.	
1929	Electroencephalogram is first introduced and Hubble's Law, a strong pillar of the 'big bang' theory, is discovered.	

Year	Scientific events	Other events
1930	Pluto is discovered.	
1931		Aldous Huxley's *Brave New World* novel is first published (written in 1930).
1933	The first electron microscope is built.	
1936	The first artificial heart is invented.	James Watson turns 4, Rosalind Franklin and Isaac Asimov turn 16, Francis Crick turns 20.
1939		World War II starts.
1943	Barbara McClintock suggests the existence of 'jumping genes' in her studies on maize.	
1944	Pfizer is the first to mass-produce penicillin.	Infant deaths steadily decline.
1945	Kidney dialysis machine is first used and the first atomic bomb is detonated during a secret test in Alamagordo, New Mexico.	World War II ends.
1947	The sound barrier is broken. The supersonic age begins.	
1948		George Orwell's *Nineteen Eighty-Four* is written.
1953	Watson and Crick decipher the structure of DNA.	
1957	*Sputnik* is launched.	
1963		Robert A. Heinlein's *Time for the Stars* is first published. John F. Kennedy is assassinated.
1965		Frank Herbert's *Dune* is first published.
1969	'One small step for a man — one giant leap for mankind' — Neil Armstrong is the first man to walk on the Moon.	
1971	Nuclear magnetic resonance imaging is used to diagnose illnesses. Black holes are discovered by sensors of the *Explorer 42* spacecraft in the Cygnus constellation.	
1972	The first global views of Mars are returned by *Mariner 9*.	
1976	Genentech company is formed by the venture capitalist Swanson and the biochemist Boyer to exploit Boyer's gene-splicing techniques. The *Viking 1* makes the first landing on Mars.	Isaac Asimov's *The Bicentennial Man* is published.
1977		George Lucas' *Star Wars* is released.
1980		Life expectancy is about 75 years.
1982	Sally Ride is the first US woman in space.	Robert A. Heinlein's *Friday* is first published. Anne McCaffrey's *The Crystal Singer* is published (written 1974–1975).
1983	Barbara McClintock wins the Nobel Prize for Medicine for her discovery of 'jumping genes'.	
1984	Meteorite ALH-84000 is discovered in Antarctica.	
1988	The National Institutes of Health and the Department of Energy embark upon the International Human Genome Project.	
1989	Physicist Stephen Hawking's *A Brief History of Time* is published.	

(continued)

Year	Scientific events	Other events
1990	Gene therapy is first attempted on a human.	Retirees outnumber teenagers for the first time in history.
1995	The first electronic atlas of the human body is created — the visible man whose frozen cadaver was sliced into one-millimetre increments.	
1996	The first complete genome of a life form, a yeast, is sequenced. US scientists reveal that meteorite ALH-84000 is Martian and contains organic compounds and microfossils.	
1997	The cloning of Dolly the sheep is revealed.	
1998	Evidence of planets orbiting stars in other galaxies is found.	
1999	Simple organic molecules discovered in outer space, leading to new hypothesis about extraterrestrial life.	Human chromosome 22 is sequenced.
2000	Australia's first cloned animal is born — Suzi the calf. Scientists at Monash University are involved in developing a method of growing body parts from embryonic stem cells.	Olympic Games are held in Sydney. Human chromosome 21 is sequenced.
2001		'9/11': Hijacked planes crash into the World Trade Center in New York.
2003	The Human Genome Project completed. More than 20 000 human genes mapped.	Australia's population reaches 20 million.
2004	The remains of an 18 000 year old, one-metre tall hominin skeleton found on the Indonesian island of Flores is formally named *Homo floresiensis* and nicknamed the 'hobbit'.	
2006	A paralysed man has a brain implant that allows him to control a computer using the power of thought. The definition of a planet is changed and Pluto is now considered a dwarf planet. A genetic study suggests that human and chimp ancestors may have interbred long after their lineages had split.	Commonwealth Games are held in Melbourne. Steve Irwin (known as The Crocodile Hunter) dies after being stung by a stingray while filming a marine documentary.
2008	The first direct observations of exoplanets are made. Ice is discovered on Mars.	US president Barack Obama is elected.
2010	The first self-replicating synthetic bacterial cell is created.	

9.7 Exercises: Understanding and inquiring

To answer questions online and to receive **immediate feedback** and **sample responses** for every question, go to your learnON title at www.jacplus.com.au. *Note:* Question numbers may vary slightly.

Investigate

1. Find out more about one of the scientific quests in the timeline in this subtopic and present your findings as a poster to the class.
2. Add other scientific quests, events or Australian Nobel Prize winners to the science quests timeline.
3. (a) Select a year (or time period) and find out as many different scientific discoveries as you can.
   (b) Find out what life was like for people who lived at this time and take note of any other events that were occurring during that time.
   (c) Suggest implications of the scientific discoveries or events on the people of that time and on people in future times.

4. Find out about the winners of scientific Nobel Prizes and their work. Present your information as an autobiography.
5. Find out about the Nobel Prize winners' sperm bank. Discuss the ethics associated with it.

Think and create

6. Use the information in this subtopic to produce a crossword or scientific trivial pursuit game.
7. Create a timeline for the events that you feel were the most important science quests.

Investigate, think and create

8. Find out about a scientific discovery and what life was like during the time of this discovery. Write a story or play about the event and then act it out to the class.
9. Read one of the novels shown in the science quests timeline and suggest future inventions or discoveries that the ideas in the novel may lead to.

9.8 Review

9.8.1 Study checklist

Science fiction

- explore famous works of science fiction
- explore new genres of media including cyberpunk

Superheroes

- examine the science which underlies the powers of superheroes
- examine the links between nature and manufacturing i.e. biomimicry
- explore the concept of biologically inspired robotic systems

Nanotechnology

- explore the concepts of nanostructures including nanobots and nanotubes

Producing new cells

- describe some techniques of growing new cells
- examine how transgenic organisms are made
- distinguish between stem cells and cloning

New technologies

- explore which new technologies will be need for space exploration

Scientific events

- examine some important scientific events from the late eighteenth century to the present

Individual pathways

ACTIVITY 9.1
Revising science quests
doc-8485

ACTIVITY 9.2
Investigating science quests
doc-8486

ACTIVITY 9.3
Analysing science quests
doc-8487

learnON ONLINE ONLY

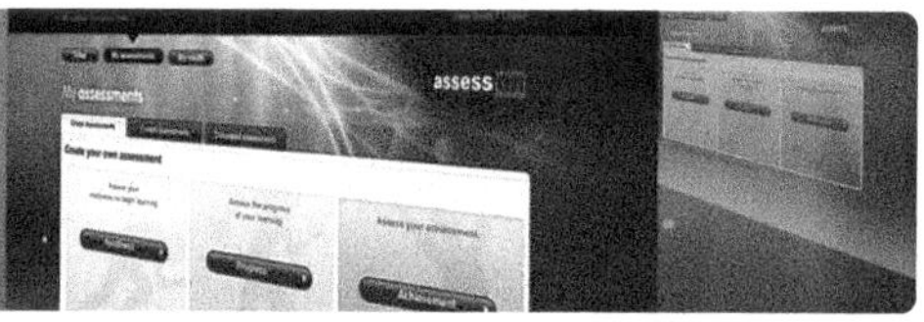

GLOSSARY

$2n^2$ rule: a rule that states that the nth shell of an atom can hold $2n^2$ electrons
abiotic factors: describes the non-living things in an ecosystem
absolute age: number of years since the formation of a rock or fossil
absolute dating: determining the age of a fossil and the rock in which it is found using the remaining amount of unchanged radioactive carbon
absolute magnitude: actual brightness of a star
absolute zero: temperature at which the particles that make up an object or substance have no kinetic energy, approximately –273.15 °C
acceleration: rate of change in speed
action: a force
activity series: classification of metals that places the elements in decreasing order of reactivity
adenine: a purine nucleobase that binds to thymine in DNA
alkali metals: very reactive metals in group 1 of the periodic table
alkaline earth metals: reactive metals in group 2 of the periodic table
alleles: alternative forms of a gene for a particular characteristic. Each allele is characterised by a slightly different nucleotide sequence.
alloy: mixture of several metals or sometimes a metal and a non-metal, such as carbon
amino acid: an organic compound that contains both a carboxyl and an amino chemical group. Amino acids are the building blocks of proteins.
amylase: enzyme in saliva that breaks down starch into sugar
analogous structures: body structures that perform a similar function but may not have a similar basic structure
angiosperms: plants that produce flowers and seeds after fertilisation
anions: atoms or groups of atoms that have gained electrons and are negatively charged
antigen: substance that stimulates the production of antibodies
apparent magnitude: brightness of a star as seen from Earth
aqueous solutions: solutions in which water is the solvent
asexual reproduction: reproduction that does not involve fusion of sex cells (gametes)
atmosphere: the layer of gases around the Earth
atomic number: number of protons in the nucleus of an atom. The atomic number determines which element an atom is.
atoms: very small particles that make up all things. Atoms have the same properties as the objects they make up.
autosomal inheritance: an inherited trait coded for by genes located on autosomes
autosomal recessive: recessive trait with alleles located on an autosome (not a sex chromosome)
autosomes: non-sex chromosomes
average speed: distance travelled divided by time taken
base-pairing rule: the concept that in DNA every adenine (A) binds to a thymine (T), and every cytosine (C) binds to a guanine (G). Also known as Chargaff's rule.
big bang theory: a theory which states that the universe began about 15 billion years ago with the explosive expansion of a singularity
big chill theory: a theory which proposes that the expansion of the universe will continue indefinitely until stars use up their fuel and burn out
big crunch theory: a theory which proposes that the universe will snap back on itself resulting in another singularity

big rip theory: a theory which proposes that the universe will rip itself apart due to accelerating expansion
binomial system of nomenclature: system devised by Carl von Linné giving organisms two names, the genus and another specific name
biodiversity: variation in the many different communities and their environments on Earth
biofuels: fuels manufactured from plant or animal material
biogas: gas made from plant or animal waste
biogeography: geographical distribution of species
bioluminescence: the release of light energy from a living thing
biomass: material produced by living organisms
biomes: regions of the Earth divided according to dominant vegetation type
biomimicry: the practice of developing new products and technologies based on replicating or imitating designs in nature
biosilification: the ability to manipulate silicon at the nanoscale that involves the use of living cells
biosphere: the life-support system of the Earth
biota: living things
black hole: the remains of a star, which forms when the force of gravity of a large neutron star is so great that not even light can escape
blue shift: shift of lines of a spectral pattern towards the blue end when a light source is moving rapidly towards the observer
bonding electrons: shared electrons holding two atoms together
branching evolution: when a population is divided into two or more new populations that are prevented from interbreeding. Also known as divergent evolution.
brittle: breaks easily into many pieces
bronze: an alloy that is a mixture of copper and tin
carbon dating: a radiometric dating technique that uses an isotope of carbon-14 to determine the absolute age of fossils
carrier: in terms of genetics, a person who is heterozygous for a characteristic and therefore does not display the recessive trait
catalase: enzyme in the liver involved in the breakdown of hydrogen peroxide, a toxic waste product from cells in the body
catalyst: chemical that speeds up reactions but is not consumed in the reaction
catastrophism: the theory that the Earth was changed only by sudden catastrophes rather than evolutionary processes
cations: atoms or groups of atoms that have lost electrons and are positively charged
cell division: a process that results in the production of new cells
cellular respiration: the chemical reaction involving oxygen that moves the energy in glucose into the compound ATP. The body is able to use the energy contained in ATP.
char: become coated with black carbon when heated
Chargaff's rule: *see* base-pairing rule
chemical formula: shorthand statement of the elements in a substance showing the relative number of atoms of each kind of element
chemical potential energy: energy present in all substances as a result of the electrical forces that hold atoms together
chemiluminescence: light produced from a chemical reaction
chlorofluorocarbons (CFCs): organic compounds used as coolant agents, propellants in aerosols, and solvents. Their manufacture is being phased out as they also cause damage to the ozone layer.
chlorophyll: the green-coloured chemical in plants that absorbs the light energy used in photosynthesis

chloroplasts: membrane-bound organelles found in plant and some protoctistan cells that contain light-capturing pigments (such as chlorophyll); the site of photosynthesis
chromosomes: tiny thread-like structures inside the nucleus of a cell. Chromosomes contain the DNA that carries genetic information.
climate sensitivity: the measure of temperature change in the climate, dependent on the amount of carbon dioxide released into the atmosphere
clones: genetically identical copies
coal: a sedimentary rock formed from dead plants and animals that were buried before rotting completely
codominance: type of inheritance in which the heterozygote shows the expression of both alleles in its phenotype
codominant inheritance: *see* codominance
codon: sequence of three bases in mRNA that contains information to start or stop protein synthesis or for the addition of a specific amino acid
coevolved: two species whose evolution has been influenced by selective pressures exerted on each other, e.g. bees and the flowers they pollinate
combination reactions: chemical reactions in which two substances, usually elements, combine to form a compound
combustion reactions: chemical reactions in which a substance reacts with oxygen and heat is released
complementary base pairs: in DNA, specific base pairs will form between the nitrogenous bases adenine (A) and thymine (T) and between the bases cytosine (C) and guanine (G).
complete dominance: type of inheritance in which the dominant trait requires only one allele to be present for its expression. It masks the allele for the recessive trait.
conductors: materials that allow electric charge to flow through them
constellations: groups of stars that were given a particular name because of the shape the stars seem to form in the sky when viewed from Earth
convergent evolution: tendency of unrelated organisms to acquire similar structures due to similar environmental pressures
co-polymer: polymer in which two different monomers alternate
corrosion: chemical reactions that wear away a metal. Air, water or chemicals contained in air and water can be corrosive.
cosmology: the study of the beginning and end of the universe
covalent bond: shared pair of electrons holding two atoms together
covalent compounds: compounds in which the atoms are held together by covalent bonds
crossing over: exchange of alleles (alternative forms of genes) between maternal and paternal chromosomes
crosslinks: chemical bonds formed between polymer chains
crumple zones: zones in motor vehicles that are deliberately designed to crumple, absorbing and spreading much of the energy transferred to the vehicle during collision
crystals: geometrically shaped substances made up of atoms and molecules arranged in one of seven different shapes
cybernetics: the science of communications and automatic control systems in both machines and living things
cytogenetics: the study of heredity at a cellular level, focusing on cellular components such as chromosomes
cytokinesis: division of the cytoplasm of a cell
cytosine: a pyrimidine nucleobase that binds to guanine in DNA
dangerous goods: chemicals that could be dangerous to people, property or the environment. They are divided into classes including explosive, toxic, corrosive and flammable.
dark energy: a theoretical form of energy that may exist throughout the universe

deceleration: decrease in speed, resulting in negative acceleration

decomposition reactions: chemical reactions in which one single compound breaks down into two or more simpler chemicals

deforestation: removal of trees from the land

deletion: a type of mutation where one nucleotide is deleted from the DNA sequence

dendrimer: basic structure of a nanoparticle that is being developed to deliver drugs to cells

deoxyribonucleic acid: *see* DNA

deoxyribose: the sugar in the nucleotides that make up DNA

diatoms: tiny, single-celled algae

digitectors: devices that consist of two cables laid across a road at a measured distance from each other and microphones to detect sound. The time interval between sounds is used to calculate the average speed of a vehicle between the cables.

diploid: the paired set of chromosomes within a somatic cell

diploid zygote: product of the fusion of two haploid gametes

displacement reactions: chemical reactions involving the transfer of electrons from the atoms of a more reactive metal to the ions of a less reactive metal

divergent evolution: when a population is divided into two or more new populations that are prevented from interbreeding. Also known as branching evolution.

DNA: the abbreviation of deoxyribonucleic acid, it is the chemical substance found in all living things that encodes the genetic information of an organism. DNA is composed of building blocks called nucleotides, which are linked together in a chain.

DNA fingerprinting: involves isolating and separating DNA fragments into their specific lengths to form a pattern that may be used to identify an individual or to determine the presence of a particular allele

DNA hybridisation: technique that can be used to compare the DNA in different species to determine how closely related they are

DNA ligase: a method of pasting DNA fragments together

DNA replication: process that results in DNA making a precise copy of itself

dominant: refers to a trait (phenotype) that requires only one allele to be present for its expression in a heterozygote

Doppler effect: observed change in frequency of a wave when the wave source and observer are moving in relation to each other

double helix: DNA molecules have the appearance of a spiral ladder or double helix, a sugar–phosphate backbone or frame, and rungs or steps that are made up of nitrogenous bases joined together by hydrogen bonds.

ductile: capable of being drawn into wires or threads; a property of most metals

ecliptic: the path that the sun traces in the sky during the year

efficiency: the fraction of energy supplied to a device as useful energy. It is usually expressed as a percentage.

elastic potential energy: the potential energy stored in a stretched elastic material

electrical potential energy: energy present in objects or groups of objects in which positively and negatively charged particles are separated

electromagnetic radiation: heat, light, X-rays, radio waves and other forms of radiation made up of electromagnetic waves. These waves are produced by the acceleration of an electric charge and have an electric field and a magnetic field at right angles to each other.

electron configuration: an ordered list of the number of electrons in each electron shell, from inner (low energy) to outer (higher energy) shells

electron dot diagrams: diagrams using dots to represent the electrons in the outer shell of atoms and to show the bonds between atoms in molecules

electron shell diagram: diagram showing electrons in their shells around the nucleus of an atom

electron shells: energy levels around the nucleus of an atom in which electrons are found
electron transfer: movement of electrons from one atom, ion or molecule to another
electrophoresis: technique used to separate the fragments of DNA on the basis of their size and charge
electrovalency: the number of positive or negative charges on an ion
embryonic stem cells: stem cells derived from the inner cell mass of a blastocyst. These cells are pluripotent and can give rise to most cell types.
emigration: the act of leaving one country or region to settle in another
endogenisation: incorporation of a foreign genome into the chromosomes of another organism. An example of this is a retrovirus that converts their RNA genome into DNA before implanting it into their host's chromosomes. If this is incorporated into their host's germline, it can become a part of the genome of future generations.
endosymbiosis: a process that describes the evolutionary origin of mitochondria and chloroplasts within eukaryotic cells
endosymbiotic theory: a theory that can be used to describe the evolutionary origin of mitochondria and chloroplasts in eukaryotic cells. The ancestors of mitochondria and chloroplasts may have been prokaryotes (bacteria) that were ingested by a host cell. Over time, these ingested prokaryotes may have evolved with their hosts to such an extent that both were dependent on each other for their survival.
enhanced greenhouse effect: an intensification of the greenhouse effect caused by pollution adding more carbon dioxide and other greenhouse gases to the atmosphere; associated with global warming
environment: the living and non-living things in a particular place at a particular time; that is, the surroundings of a living thing
enzymes: biological catalysts
epigenetics: a new branch of science that involves studying the effect of our environment and experiences on the expression of our genetic information
equations: one-line statements describing a chemical reaction, with the reactants on the left and the products on the right separated by an arrow
eras: divisions of geological time defined by specific events in the Earth's history. Eras are divided into periods.
ethanol: the common drinking alcohol with molecules containing two carbon atoms. Alcohols are a group of carbon compounds with the —OH functional group attached.
eugenics: the theory and practice of improving the human species by means of selective breeding
eukaryotic cells: cells that possess membrane-bound organelles such as a nucleus and mitochondria (e.g. animals, plants, fungi and protoctistans)
extinction: complete loss of a species when the last organism of the species dies. The Tasmanian tiger is believed to be extinct but there are still occasional reports of sightings.
fault: a break in a rock structure causing a sliding movement of the rocks along the break
fertilisation: penetration of the ovum by a sperm
fireflies: insects that release light
fluorescent: describes substances that release light when given energy
fold: a layer of rock bent into a curved shape, which occurs when rocks are under pressure from both sides
fossil fuels: substances such as coal, oil and natural gas that have formed from the remains of ancient organisms. They are used as fuel when burnt in order to produce heat.
fossils: evidence of life in the past
fraction: part of a mixture
fractional distillation: distillation of a mixture of substances with different boiling points. The heavier hydrocarbons have higher melting points and condense to a liquid at the lowest part of the cooling column.
frequency: the number of waves passing a single location in one second
galaxies: very large groups of stars and dust held together by gravity

galvanised: describes a metal coated with a more reactive metal that will corrode first; often zinc covering iron

gametes: reproductive or sex cells such as sperm or ova

gene: segment of a DNA molecule with a coded set of instructions in its base sequence for a specific protein product; when expressed, may determine the characteristics of an organism

gene cloning: the process of making genetically identical copies of a gene. An application of gene cloning is the insertion of a specific gene into bacteria, so that the bacteria will act as microfactories and produce considerable quantities of desired proteins.

gene pool: all the genetic information for a particular species

gene sequencing: involves the identification of the order of nucleotides along a gene

genetic drift: changes due to chance events such as floods and fires

genetic engineering: one type of biotechnology that involves working with DNA

genetic engineers: use special tools to cut, join, copy and separate DNA

genetics: study of inheritance

genome: the complete set of genes present in an organism or somatic cell; the entire genetic make-up

genome maps: maps that describe the order and spacing of genes on each chromosome

genomics: the study of genomes

genotype: genetic instructions (contained in DNA) inherited from parents at a particular gene locus

genotyping: a process that determines the alleles at various locations within the human genome

geology: the science that deals with the study of the Earth and how it evolves

geosequestration: the process that involves separating carbon dioxide from other flue gases, compressing it and piping it to a suitable site

global positioning system (GPS): device that uses radio signals from satellites orbiting the Earth to accurately map the position of a vehicle or individual

global warming: the observed rise in the average near-surface temperature of the Earth

Gondwana: one of the continents formed when Pangaea broke up. Part of Gondwana became Australia.

gradualism: the theory that suggests the Earth's geological features were due to the cumulative product of slow but continuous processes

gravitational potential energy: energy stored due to the height of an object above a base level

greenhouse effect: a natural effect of the Earth's atmosphere trapping heat, which keeps the Earth's temperature stable. The sun's energy passes through the atmosphere and warms the Earth. Heat energy radiated from the Earth cannot pass through the atmosphere and is trapped.

greenhouse gases: gases found in the atmosphere that contribute to the greenhouse effect, trapping the sun's heat (for example, carbon dioxide)

groups: columns of the periodic table containing elements with similar properties

guanine: a purine nucleobase that binds to cytosine in DNA. Also found in guano.

gymnosperms: plants that have unenclosed seeds while in their unfertilised state

haemoglobin: the red pigment in red blood cells that carries oxygen

half-life: time taken for half the radioactive atoms in a sample to decay; that is, change into atoms of a different element

halogens: non-metal elements in group 17 of the periodic table

haploid: the possession of one copy of each chromosome in a cell

haploid gamete: a sex cell containing only one set of chromosomes

hazardous substances: chemicals that have an effect on human health. This effect may be immediate such as poisoning, or long term like cancer.

heterozygote: two different alleles are present in the genotype

heterozygote advantage: the possession of both alleles for a particular trait infers some type of survival advantage. For example, individuals who are heterozygous for sickle-cell anaemia may have less chance of dying from malaria.

heterozygous: having two different alleles for a characteristic
holobiont: sum of genetic information of its host and its microbiota
hologenome theory of evolution: emphasises the role that micro-organisms have within our evolution. The hologenome is made up of the combined genomes of the host and the microbes within it.
homologous: used to describe members of each matching pair of chromosomes
homologous structures: body structures that perform a different function but have a similar basic structure
homology: similar characteristics that result from common ancestry
homozygous: having two identical alleles for a characteristic within the genotype
homozygous dominant: having two alleles for the dominant trait in the genotype
homozygous recessive: having two alleles for the recessive trait in the genotype
human endogenous retroviruses (HERVs): viral genomes remaining from viral invasions throughout human evolutionary history
hybrid: in reference to allele combinations for a particular trait, this would be a heterozygous organism. In reference to crossbreeding, this could be the offspring between two different types of organisms.
hydrocarbons: compounds containing only carbon and hydrogen
hydrosphere: the water on the Earth's surface
ice cores: samples of ice extracted from ice sheets containing a build-up of dust, gases and other substances trapped over time
igneous rocks: rocks that form from the cooling of lava or magma as it is thrown through the air from a volcanic eruption
immigration: the act of passing or coming into a new habitat or place of residence
incomplete dominance: type of inheritance in which the heterozygote shows the expression of the two alleles in its phenotype in a blending of the characteristics
induced mutation: a mutation of DNA that can be explained or identified
inertia: property of objects that makes them resist changes in their motion
infra-red radiation: invisible radiation emitted by all warm objects. You feel infra-red radiation as heat.
inheritance: genetic transmission of characteristics from parents to offspring
insertion: a type of mutation where a nucleotide is inserted into the original nucleotide sequence of DNA
instantaneous speed: speed at any particular instant of time
introduced species: a species that is not native to an ecosystem. It has been brought in from another ecosystem.
inversion: a reversal in the sequence of a number of genes on a chromosome
ionic bond: attractive force between ions with opposite electrical charge
ionic compounds: compounds containing positive and negative ions held together by the electrostatic force
ionosphere: highest layer of the atmosphere where the air is extremely thin. This layer reflects radio waves, enabling communication between many parts of the Earth.
ions: atoms or groups of atoms that have lost or gained electrons
isotopes: atoms of the same element that differ in the number of neutrons in the nucleus
karyotype: the number and general appearance (size, shape and banding) of a set of chromosomes in a somatic cell
kinetic energy: energy due to motion of an object
kleptoplasts: captured plastids (*see* kleptoplasty)
kleptoplasty: a process in which one organism captures the plastids from another organism and integrates them into its cellular structure. The captured plastids are not passed onto the next generation of the organism.
Kyoto Protocol: an international agreement with the goal of reducing the amount of greenhouse gases produced by industrialised nations

landfills: areas set aside for the dumping of rubbish

laser guns: devices that send out pulses of light, which are then reflected by the target moving vehicle. Laser guns can target single vehicles with narrow light beams.

last universal common ancestor (LUCA): cell from which all living things could have descended

Laurasia: one of the continents formed when Pangaea broke up

Law of Conservation of Energy: law that states that energy cannot be created or destroyed. However, energy can be transformed from one type to another or transferred between objects.

law of inertia: *see* Newton's First Law of Motion

linkage analysis: use of markers to scan the genome and map genes on chromosomes

linked: used to describe genes located on the same chromosome

lithosphere: the outermost layer of the Earth; includes the crust and uppermost part of the mantle

locus: position occupied by a gene on a chromosome

lotus effect: describes the repellence of water on the surface of lotus leaves. Also known as superhydrophobicity.

luciferases: enzymes involved in the light-producing chemical reaction in fireflies

lustre: appearance of a mineral caused by the way it reflects light. A mineral can appear glassy, waxy, metallic, dull, pearly, silky or brilliant.

magnetic field: area where a magnetic force is experienced by another magnet. The direction of the magnetic force is shown by drawing field lines; the size of the force is shown by how close together the lines are.

magnitude: size. Also a measure of the brightness of a star.

main sequence: area on the Hertzsprung–Russell where the majority of stars are plotted. Stars on the main sequence produce energy by fusing hydrogen to form helium. Such stars are at times referred to as being in their 'adult' stage, one of stability.

malleable: able to be beaten, bent or flattened into shape

mass number: number of protons and neutrons in the nucleus of an atom

material safety data sheet (MSDS): a document containing important information about hazardous chemicals

maternal chromosomes: chromosomes from the ovum

meiosis: cell division process that results in new cells with half the number of chromosomes of the original cell

messenger RNA (mRNA): single-stranded RNA transcribed from a DNA template that then carries the genetic to a ribosome to be translated into a protein

metagenomics: a technology that combines DNA sequencing with molecular and computational biology

metalloids: elements that have the appearance of metals but not all the other properties of metals

metals: elements that conduct heat and electricity; shiny solids which can be made into thin wires and sheets that bend easily. Mercury is the only liquid metal.

metamorphic rocks: rocks formed from another rock that has been under great heat or pressure (or both)

microsatellites: short, repetitive DNA sequences on a chromosome that are also called short tandem repeats (STR). Patterns of these sequences can be used in DNA fingerprinting.

Milky Way: galaxy that includes our sun as well as 200–400 billion other stars

mineral ores: rocks mined to obtain a metal or other chemical within them

mitochondria: the process of cellular respiration occurs in these membrane-bound organelles, which are found in all eukaryotic cells. Singular: mitochondrion.

mitosis: cell division process that results in new genetically identical cells with the same number of chromosomes as the original cell

molecular compounds: compounds in which the atoms are held together by covalent bonds

molecular formula: shorthand statement of the elements in a molecule showing the relative number of atoms of each kind of element

molecular genetics: study of genetics at a molecular level

molecular imprinting: a process by which molecules that match a target molecule can be made. This process may be used to make plastic antibodies.

molecular ions: groups of atoms that have an overall charge and are treated as an entity, e.g. OH^-, SO_4^{2-}, NH_4^+

molecular markers: markers that enable identification of particular molecules

molecules: particles with two or more atoms joined (bonded) together

monohybrid ratio: the 3:1 ratio of a particular characteristic for offspring produced by heterozygous parents, controlled by autosomal complete dominant inheritance

monomers: small repeating units that make up a polymer. A monomer is a molecule, usually containing carbon and hydrogen, and sometimes other elements.

monosomy: a condition in which there is only one copy of a particular chromosome (rather than two) in a cell, e.g. Turner's syndrome, which results in only one sex chromosome (XO)

Montreal Protocol: an international agreement with the goal of reducing and eliminating the use of substances that contribute to the depletion of the ozone layer

multipotent stem cells: stem cells that can give rise to only certain cell types, e.g. various types of blood or skin cells

mutagen: agent or factor that can induce or increase the rate of mutations

mutagenic agent: *see* mutagen

mutations: changes to DNA sequence, at the gene or chromosomal level

nanometres: one billionth of a metre

nanotechnology: a rapidly developing field that includes studying and investigating our environment at a molecular level and then applying what we learn to a variety of technologies. Some of these involve the use of nanotechnology in medicine.

natural selection: process by which a species gives rise to new species that has characteristics that make them better adapted for survival in a particular environment. This is also called 'survival of the fittest'.

nebulae: clouds of dust and gas that may be pulled together by gravity and heat up to form a star

net force: the resultant force of two or more forces acting on an object

neutral: having equal amounts of negative and positive electric charge and, therefore, no overall electric charge. Atoms are neutral whereas ions have either a positive or negative electric charge.

neutralisation: reaction between an acid and a base. A salt and water are the products of this type of reaction.

neutron star: extremely dense remnants of a supernova in which protons and electrons in atoms are fused to form neutrons

Newton's First Law of Motion: law that states that an object remains at rest or continues to move with the same speed in the same direction unless acted on by an outside, unbalanced force

Newton's Second Law of Motion: law that states that the acceleration of an object equals the total force on the object divided by its mass

Newton's Third Law of Motion: law that states that for every action there is an equal and opposite reaction

nitrogenous base: adenine, thymine, cytosine, guanine and uracil are examples of nitrogenous bases that may be found in nucleotides.

noble gases: elements in the last column of the periodic table. They are extremely inert.

non-homologous: used to describe chromosomes that do not match

non-metals: elements that do not conduct electricity or heat; they melt and turn into gases easily, and are brittle and often coloured

nuclear energy: the energy stored at the centre of atoms, the tiny particles that make up all substances.

nuclear fusion: joining together of the nuclei of lighter elements to form another element, with the release of energy

nuclear reactions: reactions involving the breaking of bonds between the particles (protons and neutrons) inside the nuclei of atoms

nucleic acids: molecules composed of building blocks called nucleotides, which are linked together in a chain

nucleotides: compounds (DNA building blocks) containing a sugar part (deoxyribose or ribose), a phosphate part and a nitrogen-containing base that varies

nucleus: central part of an atom, made up of protons and neutrons

nucleus (in eukaryotic cells): a membrane-bound organelle containing the genetic material DNA

nylon: one of a group of polymers called polyamides and used to make clothing, rope, guitar strings and machine parts

OIL RIG: mnemonic used for remembering electron transfer for redox reactions: Oxidation Is Loss, Reduction Is Gain

organelles: small structures which a specific function, located inside a cell

ova: female gametes or sex cells. Singular: ovum.

oxidation: chemical reaction involving the loss of electrons by a substance

ozone: a gas in the atmosphere made up of particles with three oxygen atoms

ozone layer: a layer in the stratosphere, about 25 km above Earth, that has high concentrations of ozone gas. The ozone layer absorbs over 90 per cent of the sun's ultraviolet light.

palaeontologists: scientists who study fossils

palaeontology: study of organisms of the geological past as represented by their fossil remains

Pangaea: a giant continent that existed 200 million years ago and broke into two parts called Laurasia and Gondwana

parallax: apparent movement of close stars against the background of distant stars when viewed from different positions around the Earth's orbit

partial dominance: heterozygous offspring show a phenotype that is different from the phenotype of an individual with either homozygous genotype

paternal chromosomes: chromosomes carried in the sperm

pedigree chart: diagram showing the family tree and a particular inherited characteristic for family members

perforin: a protein responsible for forming pores in diseased cells

periodic law: statement made by Mendeleev that elements with similar properties occur at regular intervals when all elements are listed in order of atomic mass

periodic table: table listing all known elements. The elements are grouped according to their properties and in order of the number of protons in their nucleus.

periods: subdivisions of geological time. Periods are divided into epochs.

periods (periodic table): rows of elements on the periodic table

permafrost: soil on or below the surface of very high mountains in the polar regions that is permanently frozen

phenotype: characteristics that result from the expression of an organism's genotype. Phenotype depends on both the genotype and the environment.

photochromic: describes lenses made from glass that darkens in bright light

photosynthesis: food-making process in plants that takes place in chloroplasts. The process uses carbon dioxide, water and energy from the sun.

phyletic evolution: when a population of a species progressively changes over time to become a new species

planet: large object that orbits a star. Planets do not produce their own light.

planetary nebula: ring of expanding gas caused by the outer layers of a star less than eight times the mass of our sun being thrown off into space

plastics: synthetic substances capable of being moulded

plate tectonics: theory describing the movement of parts of the Earth's crust, called plates, and explaining the events at the boundaries between the plates

pluripotent stem cells: stem cells that can give rise to most cell types, e.g. blood cells, skin cells and liver cells
point mutation: a mutation at one particular point in the DNA sequence that may cause a substitution, deletion, or insertion mutation
Polyethylene: a material formed from the monomer ethylene. Used to make plastic bags, soft-drink bottles and other household products
polymerase chain reaction (PCR): enables amplification of small amounts of DNA, increasing the amount of DNA available
polymerisation: chemical reaction joining monomers in long chains to form a polymer
polymers: plastics made up of monomers joined together in long chains
polyvinyl chloride: polymer formed from the monomer vinyl chloride; also known as PVC
positrons: positively charged electrons
potassium–argon dating: a radiometric dating technique based on the measure ratio of potassium-40 to radiogenic argon-40
potential energy: energy that is stored due to the position or state of an object
precipitate: solid product of a chemical reaction that is insoluble in water
precipitation reactions: chemical reactions in which there is a water-insoluble product
prevailing winds: winds most frequently observed in each region of the Earth
prokaryotic cells: cells that do not have a nucleus or other membrane-bound organelles, e.g. bacteria
protostar: the final stage of development of a star in which the temperature is not quite high enough for nuclear fusion to occur
pulsar: a spinning neutron star. Pulsars can be detected using radio telescopes.
pulsating star: star that periodically expands due to an increase in core temperature and then cools and contracts under gravity. As the star cools, its colour becomes redder.
Punnett square: a diagram that is used to predict the outcome of a genetic cross. It shows possible combinations of the alleles for a particular trait that are present in the gametes of each parent.
pure breeding: an organism for which two alleles for a particular gene are the same
quasars: one of many extremely distant, very massive sources of high-energy radio-frequency electromagnetic radiation, of unknown structure
radar guns: devices used commonly in police cards that send out radio waves that are reflected from moving vehicles. The frequency of the waves is changed depending on the movement of the vehicle.
radiant heat: heat transferred by radiation, as from the sun to the Earth
radio telescope: a telescope that can detect radio waves from distant objects
radiometric dating: a technique in which radioactive substances are used to calculate the age of rocks or dead plants and animals
rate: how fast an event happens, e.g. the speed of a reaction
reaction: force acting in response to another force
recessive: refers to a trait (phenotype) that will only be expressed in the absence of the allele for the dominant trait
recombinant DNA technology: technology that can form DNA that does not exist naturally, by combining DNA sequences that would not normally occur together
red giant: star in the late stage of its life. Red giants are cooler than main sequence stars and in their core helium is fused to form carbon and other heavier elements.
red shift: shift of lines of a spectral pattern towards the red end when the source is moving away from the observer
redox reactions: chemical reactions involving oxidation and reduction; that is, electron transfer
reduction: chemical reaction involving the gain of electrons by a substance
relative age: the age of a rock compared with the age of another rock

relative atomic mass: a number that compares the mass of atoms to an agreed mass; such as 112 of the mass of a carbon-12 isotope
relative dating: method of dating that determines the age of a rock layer by relating it to another layer using superposition and the fossils contained
remote sensing: data collection about Earth's biosphere completed from space by devices such as satellites
resistance: the ability of an organism to survive the actions or effects of a certain activity (e.g. disease or chemicals such as antibiotics or pesticides)
resistance force: force that an effort force must overcome in order to do work on an object
restriction enzymes: enzymes that cut DNA at specific base sequences (recognition sites)
restriction fragment length polymorphisms (RFLPs): variations in the lengths of DNA fragments in individuals with different alleles of a gene. Also known as RFLP.
retroposons: segments of DNA that can break off a chromosome and paste themselves elsewhere in the genome
ribose: a sugar contained in the nucleotides that make up RNA
ribosomes: mRNA is 'read' or translated into proteins in these organelles
risk assessment: a procedure that identifies the potential hazards of an experiment and gives protective measures to minimise the risk
salt: one of the products of the reaction between an acid and a base. The salt contains the positive metal ion from the base and the negative non-metal ion from the acid.
scientific notation: a way of writing very large or small numbers, using power notation; e.g. 1.5×10^8 km.
sedimentary rocks: rocks formed from sediments deposited by water, wind or ice. The sediments are cemented together in layers, under pressure.
selective agents: the different living (biotic) and non-living (abiotic) agents that influence the survival of organisms
selective pressures: factors that contribute to selecting which variations will provide the individual with an increase chance of surviving over others
semi-conservative model: a model describing the process of DNA replication in which the new DNA comprises of one old and one new DNA strand
sex chromosomes: chromosomes that differ in males and females in a species. In humans, for example, females contain two X chromosomes whereas males contain an X and a Y chromosome.
sex-linked inheritance: an inherited trait coded for by genes located on sex chromosomes
sexual reproduction: involves the joining together of male and female reproductive cells (gametes)
shells: energy levels surrounding the nucleus of an atom into which electrons are arranged
silica: silicon dioxide; also known for its hardness
single nucleotide polymorphisms (SNPs): genetic differences between individuals that can result from single base changes in their DNA sequences
singularity: a single point of immense energy present at the time of the big bang
smelting: production of a metal in its molten state
somatic cell nuclear transfer: a type of therapeutic cloning used to clone embryos
somatic cells: cells of the body that are not sex cells
somatic stem cells: stem cells derived from bone marrow, skin and umbilical cord blood. These cells are multipotent and can give rise to only certain types of cells.
sonic motion detectors: devices that send out pulses of ultrasound at a frequency of about 40 kHz and then detect the reflected pulses from the movıng object
speciation: formation of new species
species diversity: the number of different species within an ecosystem
spectrum: light from a source separated out into the sequence of colours, showing the different frequencies

sperm: male reproductive cell. It consists of a head, a middle section and a tail (flagellum) used to swim towards the ovum (egg).
spontaneous mutation: a mutation of DNA that cannot be explained or identified
steady state theory: a theory which states that there was no beginning to the universe and that the universe does not change in appearance
stratosphere: the second layer of the atmosphere up to about 55 km above the Earth's surface, between the troposphere and the mesosphere
stromatolites: layered rock structures formed in shallow water over long periods of time
structural formula: diagram showing the arrangement of atoms in a substance with covalent bonds drawn as dashes
substitution: a type of mutation where one nucleotide is substituted for a different nucleotide within a DNA sequence
supercomputers: computers with the fastest processing capacity available
supergiants: very large stars that are expanding while running out of fuel, and will eventually explode
supernova: huge explosion that happens at the end of the life cycle of supergiant stars. A neutron star or black hole remains.
symbionts: organism within a symbiotic relationship
symbiosis: very close relationship between two organisms of different species. It may benefit or harm one of the partners.
symbols: one- or two-letter code(s) used for the elements, often an abbreviation of their name
synthetic: manufactured by people
taxonomy: a formal classification system of living things
telomerase: an enzyme involved in maintaining and repairing a telomere
telomere: a cap of DNA on the tip of a chromosome that enables DNA to be replicated safely without losing valuable information
tetraploid: each cell contains four sets of chromosomes
theory of evolution: a theory about how change occurs in the inherited characteristics of a group of organisms
thermoplastic: describes plastics that soften when heated
thermosetting: describes plastics that char when heated instead of softening
thrust: forward push
thymine: a pyrimidine nucleobase that binds to adenine in DNA
totipotent stem cells: the most powerful stem cells that can give rise to all cell types
transcription: the process by which the genetic message in DNA is copied into a mRNA molecule
transfer RNA (tRNA): specific tRNA molecules located in the cytosol transport specific amino acids to complementary mRNA codons
transgenic: organism with genetic information from another species inserted into its genome
transition metal block: block of metallic elements in the middle of the periodic table
translated: the genetic message in mRNA is decoded into amino acids, which results in the synthesis of a protein
Transposition: the ability of a gene to change position on the chromosome
transposons: a section of chromosome that moves about the chromosome within a cell through the method of transposition. Also known as jumping genes.
triplet: a sequence of three nucleotides in DNA that can code for an amino acid. For example, the triplet base sequence CAA codes for the amino acid, valine.
trisomy: three copies of a chromosome instead of the normal pair of two, e.g. the addition of a number 21 chromosome that results in Down syndrome within each cell
troposphere: the layer of the atmosphere closest to the Earth's surface. The particles of the air are packed most closely in this layer and they spread out further away from the surface.

ultraviolet radiation: invisible radiation similar to light but with a slightly higher frequency and more energy

uniformitarianism theory: the antithesis of catastrophism; based on the concept that the Earth is changed by natural forces that occur gradually over time

uracil: a nitrogenous base that may be found in the nucleotides of RNA

UVB radiation: one of the wavelengths of radiation emitted by the sun

valency: equal to the number of electrons that each atom needs to gain, lose or share to fill its outer shell

Van der Waals force: very weak molecular interactions

variations: different forms of something

vector: in terms of genetic engineering, refers to an agent (e.g. plasmid or virus) that carries donor DNA into a cell

velocity: a measure of rate of change in position. Unlike speed, it has a direction as well as a magnitude.

wavelength: the distance between two adjacent waves

weight: a measure of the size of the gravity force pulling an object towards the centre of a massive body, such as the Earth. The weight of an object depends on the object's mass.

weighted mean: average mass of an element that is calculated from the percentage of each isotope in nature

white dwarf: the core remaining after a red giant has shed layers of gases. A white dwarf has no nuclear reactions and its only energy source is gravity that pulls it into a core of very dense matter, a jumble of tightly packed electrons, protons and neutrons.

work: transfer in energy that occurs when a force moves an object in the direction of the force. The object doing work loses energy and the object on which work is done gains energy.

X-linked recessive: a trait located on the X chromosome and inherited recessively

X-linked trait: genetic information for the trait is located on the X chromosome

X-rays: high energy electromagnetic waves that can be transmitted through solids and provide information about their structure

zeolites: crystalline substances consisting of aluminium, silicon and oxygen that are used as catalysts to break up large molecules in crude oil

zygote: formed by the fusion of male and female reproductive cells

INDEX

W

X

Z